STUDENT STUDY GUIDE
&
SELECTED SOLUTIONS MANUAL

FRANK L. H. WOLFS

UNIVERSITY OF ROCHESTER

FOURTH EDITION

VOLUMES II & III

PHYSICS

for

SCIENTISTS & ENGINEERS

with Modern Physics

GIANCOLI

PEARSON

Prentice
Hall

Upper Saddle River, NJ 07458

Sponsoring Editor: Christian Botting
Operations Specialist: Amanda Smith
Supplement Cover Manager: Paul Gourhan
Supplement Cover Designer: Paul Gourhan
Cover photographs top left clockwise: Daly & Newton/Getty Images, Mahaux
Photography/Getty Images, Inc. - Image Bank, The Microwave Sky:
NASA/WMAP Science Team, Giuseppe Molesini, Instituto Nazionale di Ottica
Florence

© 2009 Pearson Education, Inc.
Pearson Prentice Hall
Pearson Education, Inc.
Upper Saddle River, NJ 07458

Printed in the United States of America

10 9 8 7 6 5 4 3 2 1

ISBN-13: 978-0-13-227325-1
ISBN-10: 0-13-227325-X

Pearson Education Ltd., *London*
Pearson Education Australia Pty. Ltd., *Sydney*
Pearson Education Singapore, Pte. Ltd.
Pearson Education North Asia Ltd., *Hong Kong*
Pearson Education Canada, Inc., *Toronto*
Pearson Educación de Mexico, S.A. de C.V.
Pearson Education—Japan, *Tokyo*
Pearson Education Malaysia, Pte. Ltd.

Contents

Preface

This *Student Study Guide & Selected Solutions Manual* has been prepared to accompany Volumes 2 and 3 of the fourth edition of the textbook *Physics for Scientists & Engineers with Modern Physics* by Douglas C. Giancoli.

The study guide includes the following components for each chapter:

- **Chapter overview and objectives**. The list of objectives is comprehensive and your course may omit some of these objectives.
- **Summary of equations**. This list includes both the fundamental laws of physics in equation form and equations that result from application of the laws or principles to specific situations. It is important to know what each equation is stating and to what situations it can be applied.
- **Chapter summary**. The chapter summary provides a brief review of the principles covered in the chapter. Examples, illustrating these principles, are discussed.
- **Practice quiz**. A practice quiz is included at the end of each chapter. The quiz consists of several conceptual questions and several quantitative problems. The level of difficulty of both the conceptual questions and quantitative problems vary from relatively easy to difficult. The answers to the quiz questions are in the back of the study guide.
- **Question responses**. The responses to select odd end-of-chapter questions from the textbook are included in the study guide. These solutions were prepared by Katherine Whatley and Judith Beck.
- **Problem solutions**. The solutions to select odd end-of-chapter problems from the textbook are included in the study guide. These solutions were prepared by Bob Davis, J. Erik Hendrickson, and Michael Ottinger. To make the best use of these solutions, make an honest attempt to solve the problems directly from the text before looking at the solutions. Use the solutions to verify that you have correctly solved the problems or to find the step in the reasoning used to solve the problem that is giving you problems.

I have tried to assure that everything in this study guide is correct. I am solely responsible for any errors or omissions that remain in this study guide. I am happy to receive any feedback that readers send, both positive and negative, that may help me improve any future editions of this study guide.

I would like to thank everyone who assisted me directly or indirectly while preparing this study guide. I like to thank Christian Botting at Pearson for his assistance during this project. I very much appreciated being able to include select solutions to the end-of-chapter questions and problems, prepared by Katherine Whatley, Judith Beck, Bob Davis, J. Erik Hendrickson, and Michael Ottinger. And finally, I like to thank my wife Jean and my children Frank, Madeline, and Jeanie for giving me the time to work on this project and keeping me on track by continuing to ask "What chapter are you on?" Thanks Jean, Frank, Madeline, and Jeanie!!!!

Finally, I like to acknowledge three people who have had a significant influence on the career path I have chosen to follow. I like to thank my first physics teacher, Dr. Henk Klein Haneveld, who got me excited about physics. Henk, bedankt! I also like to thank my parents, who always have been very interested in what I am doing and taught me to follow my interests. Pama, bedankt!

Frank L. H. Wolfs
Department of Physics and Astronomy
University of Rochester
wolfs@pas.rochester.edu

Chapter 21: Electric Charge and Electric Field

Chapter Overview and Objectives

In this chapter, the concepts of electric charge, Coulomb's law, and electric fields are introduced. Methods to calculate the electrostatic forces associated with discrete and continuous charge distributions are discussed.

After completing study of this chapter you should:
- Know that there are two signs of electric charge.
- Know that like charges repel and unlike charges attract.
- Know that electric charge is conserved in physical processes.
- Know the difference between conductors, insulators, and semiconductors.
- Know what induced charge is.
- Know Coulomb's law and how to calculate the force on charged particles caused by other charged particles.
- Know the definition of the electric field.
- Know how to calculate the force on a charged particle using the electric field.
- Know how to calculate electric fields for distributions of point charges and continuous charge distributions.
- Know what electric field lines are and what they represent.
- Know that, in static conditions, the electric field inside a conductor is zero and that the electric field just outside a conductor is perpendicular to its surface.
- Know what an electric dipole is and how to determine its electric dipole moment.

Summary of Equations

Scalar form of Coulomb's law:
$$F_{12} = \frac{1}{4\pi\varepsilon_0} \frac{Q_1 Q_2}{r^2}$$
(Section 21-5)

Vector form of Coulomb's law:
$$\vec{\mathbf{F}}_{12} = \frac{1}{4\pi\varepsilon_0} \frac{Q_1 Q_2}{r_{21}^2} \hat{\mathbf{r}}_{21}$$
(Section 21-5)

Definition of electric field:
$$\vec{\mathbf{E}} = \frac{\vec{\mathbf{F}}}{q}$$
(Section 21-6)

Electric field of a point charge:
$$\vec{\mathbf{E}} = \frac{1}{4\pi\varepsilon_0} \frac{Q}{r^2} \hat{\mathbf{r}}$$
(Section 21-6)

Electric field of a continuous distribution of charge:
$$\vec{\mathbf{E}} = \int d\vec{\mathbf{E}} = \int \frac{1}{4\pi\varepsilon_0} \frac{\hat{\mathbf{r}}}{r^2} dq$$
(Section 21-7)

Force on a charged particle in an electric field:
$$\vec{\mathbf{F}} = q\vec{\mathbf{E}}$$
(Section 21-10)

Electric dipole moment:
$$\vec{\mathbf{p}} = Q\vec{\mathbf{d}}$$
(Section 21-11)

Torque on an electric dipole in an electric field:
$$\vec{\tau} = \vec{\mathbf{p}} \times \vec{\mathbf{E}}$$
(Section 21-11)

Potential energy of a dipole in an electric field:
$$U = -\vec{\mathbf{p}} \cdot \vec{\mathbf{E}}$$
(Section 21-11)

Chapter Summary

Section 21-1. Static Electricity; Electric Charge and Its Conservation

A property of matter is **electric charge**. There are two types of electric charges: positive and negative charges. The force between positive charges and the force between negative charges is repulsive. The force between positive and negative charges is attractive. This is usually stated as **like charges repel and unlike charges attract**.

In all observed processes, the total electric charge remains unchanged. This observation is called **the law of conservation of electric charge**. If a particle with charge $+Q$ is created in a process, another particle with charge $-Q$ must be created at the same time.

Section 21-2. Electric Charge in the Atom

The current accepted model of ordinary matter is that matter consists of atoms. Atoms have a massive nucleus that accounts for most of the mass of the atom and is positively charged. The nucleus consists of positively charged protons and uncharged neutrons. The diameter of the nucleus is about one hundred thousandth of the diameter of the atom. Negatively charged particles, called electrons, orbit the nucleus. The charge of an electron has exactly the same magnitude, but the opposite sign, as the charge of the proton. Neutral atoms have the same number of protons and electrons and, as a consequence, atoms are neutral. Atoms can lose or gain electrons and obtain a net electric charge. We call atoms that have a net electric charge **ions**.

Molecules may have no net charge, but will still interact with other charges. This can happen if the charge in the molecule is distributed asymmetrically. If the charges in the molecule are distributed asymmetrically, we say the molecule is a **polar** molecule.

Section 21-3. Insulators and Conductors

Materials through which electric charge is easily transported are called **conductors**. Materials through which it is very difficult to transport electric charge are called **insulators**. An intermediate class of materials are called **semiconductors**.

Section 21-4. Induced Charge; the Electroscope

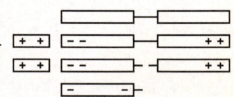

Charge can be transferred from one electrically charged object to another by direct contact. Conductors can also be charged without direct contact. Consider sequence of pictures in the Figure on the right. A pair of conductive bodies, attached to each other by a thin conductive wire, has no net electric charge. In the second picture, a positively charged body is brought near one end of the pair of conductive bodies. The positive charge attracts negative charges to the end it is near and repels positive charges toward the far end. In the third picture, the thin conductive wire between the two bodies is cut, leaving a net negative charge on the body on the left and a net positive charge on the body on the right. In the fourth picture, when the other bodies are removed, the charge redistributes itself because the negative charges repel each other. This process of charging bodies is called **charging by induction**.

Electroscopes and **electrometers** are instruments that can be used to measure electric charge.

Section 21-5. Coulomb's Law

Coulomb's law is a quantitative relationship of the size of the force between charges to the amount of charge and the distance separating the charges. The electric force is proportional to the product of the two interacting charges divided by the square of the distance between them. We write this as

$$F = k\frac{Q_1 Q_2}{r^2}$$

where F is the magnitude of the force between the two charges, Q_1 and Q_2 are the magnitudes of the two interacting charges, and r is the distance between these charges. The charge is measured in units called coulombs (C). The charge of an electron is 1.602×10^{-19} C. The constant k is called **Coulomb's constant** and has a value of

$$k = 8.988 \times 10^9 \, \text{N} \cdot \text{m}^2/\text{C}^2$$

Frequently, k is expressed in terms of another constant:

$$k = \frac{1}{4\pi\varepsilon_0}$$

where ε_0 is called the **permittivity of free space** and has a value

$$\varepsilon_0 = 8.85 \times 10^{-12} \ \text{C}^2/\text{N} \cdot \text{m}^2$$

We can rewrite Coulomb's law in vector notation to include information about the direction of the force:

$$\vec{\mathbf{F}}_{12} = \frac{1}{4\pi\varepsilon_0} \frac{Q_1 Q_2}{r_{21}^2} \hat{\mathbf{r}}_{21}$$

where $\vec{\mathbf{F}}_{12}$ is the force of charge Q_2 on charge Q_1, r_{21} is the distance from charge Q_2 to charge Q_1, and $\hat{\mathbf{r}}_{21}$ is the unit vector that points in the direction from charge Q_2 to charge Q_1. Note that Q_1 and Q_2 carry the sign of the charge in contrast to the non-vector form of Coulomb's law for which the signs are dropped.

Example 21-5-A. Applying the principle of superposition. A charge $Q = -3.6 \times 10^{-4}$ C is located a distance $d = 0.45$ m from a charge $q = -8.3 \times 10^{-4}$ C. At what position along the line between the two charges is the total force on any third charge equal to zero?

Approach: The net force on any charge between the two charges is the vector sum of the forces due to each charge. In the region between the charges q and Q, the two forces are pointing in opposite directions. In this problem we will determine at what location these two forces cancel.

Solution: The geometry of the charge configuration is shown in the Figure on the right. Consider the force on a charge q', located between charge q and Q. At the position where the net force is 0, the magnitudes of the two forces are the same:

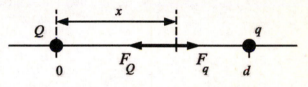

$$\frac{1}{4\pi\varepsilon_0} \frac{q'Q}{x^2} = \frac{1}{4\pi\varepsilon_0} \frac{q'q}{(d-x)^2}$$

This equation can be rewritten in the following way:

$$Q(d-x)^2 = qx^2 \quad \Rightarrow \quad Q\left(d^2 - 2dx + x^2\right) - qx^2 = 0 \quad \Rightarrow \quad (Q-q)x^2 - 2dQx + Qd^2 = 0$$

The distance x can now be determined:

$$x = \frac{2dQ \pm \sqrt{(-2dQ)^2 - 4(Q-q)(Qd^2)}}{2(Q-q)} = \frac{2dQ \pm \sqrt{4Q^2d^2 - 4Q^2d^2 + 4qQd^2}}{2(Q-q)} =$$

$$= \frac{dQ \pm \sqrt{qQ}d}{(Q-q)} = \frac{1 \pm \sqrt{\dfrac{q}{Q}}}{\left(1 - \dfrac{q}{Q}\right)}d = \frac{1 \pm \sqrt{\dfrac{q}{Q}}}{\left(1 + \sqrt{\dfrac{q}{Q}}\right)\left(1 - \sqrt{\dfrac{q}{Q}}\right)}d = \frac{1}{\left(1 - \sqrt{\dfrac{q}{Q}}\right)}d \quad \text{or} \quad \frac{1}{\left(1 + \sqrt{\dfrac{q}{Q}}\right)}d$$

The two solutions for x are:

$$x_1 = \frac{1}{\left(1 - \sqrt{\dfrac{q}{Q}}\right)}d = \frac{1}{1 - \sqrt{\dfrac{-8.3 \times 10^{-4}}{-3.6 \times 10^{-4}}}}(0.45) = -0.87 \text{ m}$$

$$= \frac{1}{\left(1 + \sqrt{\dfrac{q}{Q}}\right)}d = \frac{1}{1 + \sqrt{\dfrac{-8.3 \times 10^{-4}}{-3.6 \times 10^{-4}}}}(0.45) = 0.18 \text{ m}$$

Because we know that the position at which the net force is zero must be located between the charges q and Q, the 0.†8 m solution is the answer to the question. The other solution corresponds to a position at which the magnitudes of the two forces are equal, but the directions of the forces are the same and the vector sum will not be equal to zero.

Example 21-5-B. Calculating the net force generated by a pair of point charges. A $q_1 = +3.4$ μC charge is located at position (0.0, 0.0, 0.0). A $q_2 = -4.5$ μC charge is located at position (0.0, 2.0 m, 3.0 m). Determine the electric force exerted by these two charges on a point charge $q_3 = -1.3$ μC, located at (1.0 m, 1.0 m, 0.0 m).

Approach: The total force exerted on charge q_3 is the vector sum of the forces exerted on charge q_3 by charges q_1 and q_2. Care must be taken to determine the proper direction of each of the forces.

Solution: The position vector that points from charge q_1 to charge q_3 is

$$\vec{r}_{13} = \vec{r}_3 - \vec{r}_1 = \left(1.0\,\hat{i} + 1.0\,\hat{j} + 0.0\,\hat{k}\right) - \left(0.0\,\hat{i} + 0.0\,\hat{j} + 0.0\,\hat{k}\right) = \left(1.0\,\hat{i} + 1.0\,\hat{j}\right) \text{ m}$$

The electric force exerted on charge q_3 by charge q_1 can be determined using Coulomb's law and is equal to

$$\vec{F}_{13} = \frac{1}{4\pi\varepsilon_0}\frac{q_1 q_3}{r_{13}^2}\hat{r}_{13} = \frac{1}{4\pi\varepsilon_0}\frac{q_1 q_3}{r_{13}^3}\vec{r}_{13} = \left(9.0\times 10^9\right)\frac{\left(3.4\times 10^{-6}\right)\left(-1.3\times 10^{-6}\right)}{\left(\sqrt{1.0^2 + 1.0^2 + 0.0^2}\right)^3}\left(1.0\,\hat{i} + 1.0\,\hat{j}\right) =$$

$$= \left(-1.41\times 10^{-2}\,\hat{i} - 1.41\times 10^{-2}\,\hat{j}\right)\text{N}$$

The position vector that points from charge q_2 to charge q_3 is

$$\vec{r}_{23} = \vec{r}_3 - \vec{r}_2 = \left(1.0\,\hat{i} + 1.0\,\hat{j} + 0.0\,\hat{k}\right) - \left(0.0\,\hat{i} + 2.0\,\hat{j} + 3.0\,\hat{k}\right) = \left(1.0\,\hat{i} - 1.0\,\hat{j} - 3.0\,\hat{k}\right)\text{m}$$

The electric force exerted on charge q_3 by charge q_2 can be determined using Coulomb's law and is equal to

$$\vec{F}_{23} = \frac{1}{4\pi\varepsilon_0}\frac{q_2 q_3}{r_{23}^2}\hat{r}_{23} = \frac{1}{4\pi\varepsilon_0}\frac{q_2 q_3}{r_{23}^3}\vec{r}_{23} = \left(9.0\times 10^9\right)\frac{\left(-4.5\times 10^{-6}\right)\left(-1.3\times 10^{-6}\right)}{\left(\sqrt{1.0^2 + 1.0^2 + 3.0^2}\right)^3}\left(1.0\,\hat{i} - 1.0\,\hat{j} - 3.0\,\hat{k}\right) =$$

$$= \left(1.44\times 10^{-3}\,\hat{i} - 1.44\times 10^{-3}\,\hat{j} - 4.33\times 10^{-3}\,\hat{k}\right)\text{N}$$

The net force on the -1.3 μC charge is the vector sum of the two forces:

$$\vec{F} = \vec{F}_{13} + \vec{F}_{23} = \left(-1.41\times 10^{-2}\,\hat{i} - 1.41\times 10^{-2}\,\hat{j}\right) + \left(1.44\times 10^{-3}\,\hat{i} - 1.44\times 10^{-3}\,\hat{j} - 4.33\times 10^{-3}\,\hat{k}\right) =$$

$$= \left(-1.27\times 10^{-2}\,\hat{i} - 1.55\times 10^{-2}\,\hat{j} - 4.33\times 10^{-3}\,\hat{k}\right)\text{N}$$

Section 21-6. The Electric Field

The **electric field** $\vec{E}$ at a given point in space is defined as the electric force $\vec{F}$ on a small test charge q divided by q:

$$\vec{E} = \frac{\vec{F}}{q}$$

The test charge must have a small enough charge in order not to change the distribution of charge(s) that is responsible for the force on the test charge. The electric field created by a point charge Q is

$$\vec{E} = \frac{\vec{F}}{q} = \frac{1}{4\pi\varepsilon_0}\frac{Qq}{qr^2}\hat{r} = \frac{1}{4\pi\varepsilon_0}\frac{Q}{r^2}\hat{r}$$

where r is the distance from the charge Q to the point at which the electric field is evaluated and $\hat{r}$ is the unit vector that points from the charge Q to the point at which the electric field is evaluated.

The total electric field due to several electric charges is the vector sum due to the electric field associated with each individual charge:

$$\vec{E} = \vec{E}_1 + \vec{E}_2 + \vec{E}_3 + \cdots$$

where $\vec{E}$ is the total electric field and $\vec{E}_1$, $\vec{E}_2$, $\vec{E}_3$, ... are the electric fields due to the individual point charges. This is called **the principle of superposition for electric fields**.

Example 21-6-A. Calculating the electric field. A charge $q_x = 2.17$ μC is located on the x-axis a distance $x = 0.345$ m from the origin. A second charge $q_y = -4.21$ μC charge is located on the y-axis a distance $y = 0.576$ m from the origin. What is the electric field at the origin?

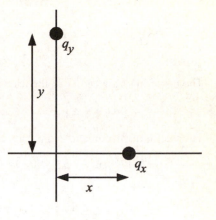

Approach: The electric field at the origin is the vector sum of the electric field due to charge q_x and the electric field due to charge q_y. To determine the direction of the electric field at the origin due to each point charge, determine the direction in which a positive charge placed at the origin would move.

Solution: The electric field at the origin due to q_x is directed in the $-\hat{\mathbf{i}}$ direction. The electric field at the origin due to q_y is directed in the $+\hat{\mathbf{j}}$ direction. The net electric field at the origin is

$$\vec{E} = \frac{1}{4\pi\varepsilon_0}\frac{q_x}{r_x^2}\left(-\hat{\mathbf{i}}\right)_x + \frac{1}{4\pi\varepsilon_0}\frac{q_y}{r_y^2}\left(-\hat{\mathbf{j}}\right) = \left(9.0\times10^9\right)\frac{\left(2.17\times10^{-6}\right)}{\left(0.345\right)^2}\left(-\hat{\mathbf{i}}\right) + \left(9.0\times10^9\right)\frac{\left(-4.21\times10^{-6}\right)}{\left(0.576\right)^2}\left(-\hat{\mathbf{j}}\right) =$$

$$= \left(-1.64\times10^5\,\hat{\mathbf{i}} + 1.14\times10^5\,\hat{\mathbf{j}}\right)\text{N/C}$$

Example 21-6-B. Principle of superposition. An equilateral triangle with sides of length a has charges at each of its vertices. One vertex has a charge $-q$ and the two other vertices have a charge $+q$. What is the electric field at the center of the equilateral triangle?

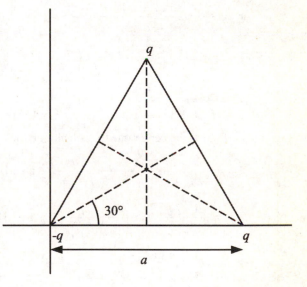

Approach: In this problem we need to determine the electric field due to each charge separately, and calculate the vector sum of these fields. Since the magnitudes of each charge and their distance from the center of the triangle are the same, the magnitude of the electric field due to each charge will be the same. By determining the components of the field along the chosen coordinate axes, we can quickly determine the vector sum of the fields.

Solution: The magnitude of the electric field at the center of the triangle due to a single charge is equal to

$$\left|\vec{E}_q\right| = \left|\vec{E}_{-q}\right| = \frac{1}{4\pi\varepsilon_0}\frac{q}{\left(\dfrac{\frac{1}{2}a}{\cos 30°}\right)^2} = \frac{1}{4\pi\varepsilon_0}\frac{q}{\left(\dfrac{\frac{1}{2}a}{\frac{1}{2}\sqrt{3}}\right)^2} = \frac{3}{4\pi\varepsilon_0}\frac{q}{a^2}$$

The directions of the electric fields due to the three charges are shown in the Figure on the right. As is clear from the Figure, the net field at the center of the triangle will be pointing towards the bottom-left corner, where charge $-q$ is located. The magnitude of the total electric field is equal to

$$\left|\vec{E}_{net}\right| = \left|\vec{E}_{-q}\right| + 2\left|\vec{E}_q\right|\cos 60° = \frac{3}{4\pi\varepsilon_0}\frac{q}{a^2} + 2\left(\frac{3}{4\pi\varepsilon_0}\frac{q}{a^2}\right)\left(\frac{1}{2}\right) = \frac{3}{2\pi\varepsilon_0}\frac{q}{a^2}$$

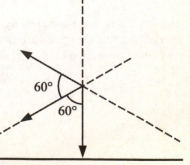

An alternative approach to solve this problem relies in expressing all fields in vector notation and then calculating the vector sum by adding the components along the coordinate axes. We choose our coordinate system such that charge $-q$

is located at the origin, and one of the other charges is located on the x axis. For this system, we can write the location of each of the charges as follows:

$$-q \quad at \quad \vec{r}_1 = \vec{0}$$

$$+q \quad at \quad \vec{r}_2 = a\,\hat{i}$$

$$+q \quad at \quad \vec{r}_3 = a\cos 60°\,\hat{i} + a\sin 60°\,\hat{j} = \tfrac{1}{2}a\,\hat{i} + \tfrac{\sqrt{3}}{2}a\,\hat{j}$$

The position of the center of the triangle is

$$\vec{r}_c = \tfrac{1}{2}a\,\hat{i} + \tfrac{\sqrt{3}}{6}a\,\hat{j}$$

The vector displacements from each charge to the center are given by:

$$\vec{r}_{1c} = \vec{r}_c - \vec{r}_1 = \tfrac{1}{2}a\,\hat{i} + \tfrac{\sqrt{3}}{6}a\,\hat{j} - \vec{0} = \tfrac{1}{2}a\,\hat{i} + \tfrac{\sqrt{3}}{6}a\,\hat{j}$$

$$\vec{r}_{2c} = \vec{r}_c - \vec{r}_2 = \tfrac{1}{2}a\,\hat{i} + \tfrac{\sqrt{3}}{6}a\,\hat{j} - a\,\hat{i} = -\tfrac{1}{2}a\,\hat{i} + \tfrac{\sqrt{3}}{6}a\,\hat{j}$$

$$\vec{r}_{3c} = \vec{r}_c - \vec{r}_3 = \tfrac{1}{2}a\,\hat{i} + \tfrac{\sqrt{3}}{6}a\,\hat{j} - \tfrac{1}{2}a\,\hat{i} - \tfrac{\sqrt{3}}{2}a\,\hat{j} = -\tfrac{\sqrt{3}}{3}a\,\hat{j}$$

The magnitudes of these displacement vectors are all the same:

$$r_{1c} = r_{2c} = r_{3c} = \tfrac{\sqrt{3}}{3}a$$

The corresponding unit vectors are:

$$\hat{r}_{1c} = \frac{\vec{r}_{1c}}{r_{1c}} = \frac{\tfrac{1}{2}a\,\hat{i} + \tfrac{\sqrt{3}}{6}a\,\hat{j}}{\tfrac{\sqrt{3}}{3}a} = \tfrac{\sqrt{3}}{2}\,\hat{i} + \tfrac{1}{2}\,\hat{j}$$

$$\hat{r}_{2c} = \frac{\vec{r}_{2c}}{r_{2c}} = \frac{-\tfrac{1}{2}a\,\hat{i} + \tfrac{\sqrt{3}}{6}a\,\hat{j}}{\tfrac{\sqrt{3}}{3}a} = -\tfrac{\sqrt{3}}{2}\,\hat{i} + \tfrac{1}{2}\,\hat{j}$$

$$\hat{r}_{1c} = \frac{\vec{r}_{1c}}{r_{1c}} = \frac{-\tfrac{\sqrt{3}}{3}a\,\hat{j}}{\tfrac{\sqrt{3}}{3}a} = -\hat{j}$$

The electric field at the center due to each charge can now be determined:

$$\vec{E}_{1c} = \frac{1}{4\pi\varepsilon_0}\frac{-q}{r_{1c}^2}\hat{r}_{1c} = -\frac{1}{4\pi\varepsilon_0}\frac{q}{\left(\tfrac{\sqrt{3}}{3}a\right)^2}\left(\tfrac{\sqrt{3}}{2}\,\hat{i} + \tfrac{1}{2}\,\hat{j}\right) = -\frac{1}{4\pi\varepsilon_0}\frac{3q}{a^2}\left(\tfrac{\sqrt{3}}{2}\,\hat{i} + \tfrac{1}{2}\,\hat{j}\right)$$

$$\vec{E}_{2c} = \frac{1}{4\pi\varepsilon_0}\frac{q}{r_{2c}^2}\hat{r}_{2c} = \frac{1}{4\pi\varepsilon_0}\frac{q}{\left(\tfrac{\sqrt{3}}{3}a\right)^2}\left(-\tfrac{\sqrt{3}}{2}\,\hat{i} + \tfrac{1}{2}\,\hat{j}\right) = \frac{1}{4\pi\varepsilon_0}\frac{3q}{a^2}\left(-\tfrac{\sqrt{3}}{2}\,\hat{i} + \tfrac{1}{2}\,\hat{j}\right)$$

$$\vec{E}_{3c} = \frac{1}{4\pi\varepsilon_0}\frac{q}{r_{3c}^2}\hat{r}_{3c} = \frac{1}{4\pi\varepsilon_0}\frac{q}{\left(\tfrac{\sqrt{3}}{3}a\right)^2}\left(-\hat{j}\right) = -\frac{1}{4\pi\varepsilon_0}\frac{3q}{a^2}\hat{j}$$

The net electric field at the center is the sum of the electric fields from each point charge:

$$\vec{E} = \vec{E}_{1c} + \vec{E}_{2c} + \vec{E}_{3c} = \left[-\frac{1}{4\pi\varepsilon_0}\frac{3q}{a^2}\left(\tfrac{\sqrt{3}}{2}\,\hat{i} + \tfrac{1}{2}\,\hat{j}\right)\right] + \left[\frac{1}{4\pi\varepsilon_0}\frac{3q}{a^2}\left(-\tfrac{\sqrt{3}}{2}\,\hat{i} + \tfrac{1}{2}\,\hat{j}\right)\right] + \left[-\frac{1}{4\pi\varepsilon_0}\frac{3q}{a^2}\hat{j}\right] = -\frac{1}{4\pi\varepsilon_0}\frac{3q}{a^2}\left(\sqrt{3}\,\hat{i} + \hat{j}\right)$$

Section 21-7. Electric Field Calculations for Continuous Charge Distributions

In many cases, there are so many charges present that it is impossible to sum the electric field from each of the individual charges, but the charge distribution can be treated by a continuous charge density. For each infinitesimal charge element dq in the distribution of charge, there is an infinitesimal contribution to the net electric field, $d\vec{E}$, where

$$d\vec{E} = \frac{1}{4\pi\varepsilon_0}\frac{\hat{r}}{r^2}dq$$

The distance from the infinitesimal charge element to the point at which the field is being evaluated is r, and the unit vector in the direction from the infinitesimal charge element to the point the field is being evaluated is $\hat{\mathbf{r}}$. To determine the electric field at a given point, we integrate over all of the infinitesimal charge elements:

$$\vec{\mathbf{E}} = \int d\vec{\mathbf{E}} = \frac{1}{4\pi\varepsilon_0} \int \frac{\hat{\mathbf{r}}}{r^2} dq$$

Because the charge distribution is often a known function of position, we almost always change variables in the integral and write the charge density as a function of position. For a linear (or one-dimensional) density of charge, we write

$$dq = \lambda \, dl$$

where λ is the **linear charge density** (charge per unit length) and dl is an infinitesimal line element along the charge distribution. If the charge distribution is a surface charge density, we write

$$dq = \sigma \, dA$$

where σ is the **surface charge density** (charge per unit area) and dA is an infinitesimal area element of the charge distribution. If the charge distribution is three-dimensional, we write

$$dq = \rho \, dV$$

where ρ is the **volume charge density** (charge per unit volume) and dV is an infinitesimal volume element.

Example 21-7-A. Electric field due to a linear charge distribution. A linear charge density lies along the $+x$-axis between $x = 0$ and $x = L$. The charge density is a function of position and is given by

$$\lambda(x) = \lambda_0 \frac{x^2}{L^2}$$

Determine the electric field due to this charge distribution along the x-axis for all $x > L$.

Approach: To calculate the electric field, we need to determine the electric field due to a small segment dx, located between $x = 0$ and $x = L$. The electric field due to this segment is directed away from $x = 0$, along the x axis, assuming that $\lambda_0 > 0$. The total electric field is the vector sum due to all segments of the charge distribution. Since the contributions from all sections are pointing away from $x = 0$ and along the x axis, the net electric field is directed along the positive x axis and has a magnitude equal to the linear sum of the contributions of each section.

Solution: Consider a section of the charge distribution, located at x_{dist} and with a width dx_{dist}. The electric field at a point P, located at $x > L$, is equal to

$$d\vec{\mathbf{E}} = \frac{1}{4\pi\varepsilon_0} \frac{\lambda(x_{dist}) dx_{dist}}{(x - x_{dist})^2} \hat{\mathbf{i}} = \frac{1}{4\pi\varepsilon_0} \frac{\lambda_0 x_{dist}^2 dx_{dist}}{L^2 (x - x_{dist})^2} \hat{\mathbf{i}} = \frac{1}{4\pi\varepsilon_0} \frac{\lambda_0}{L^2} \frac{x_{dist}^2 dx_{dist}}{(x - x_{dist})^2} \hat{\mathbf{i}}$$

The total electric field can be obtained by integrating this expression over the region where the charge density is non-zero:

$$\vec{\mathbf{E}} = \int d\vec{\mathbf{E}} = \frac{1}{4\pi\varepsilon_0} \frac{\lambda_0}{L^2} \int_0^L \frac{x_{dist}^2 dx_{dist}}{(x - x_{dist})^2} \hat{\mathbf{i}} = \frac{1}{4\pi\varepsilon_0} \frac{\lambda_0}{L^2} \int_{x-L}^x \frac{(x - u)^2}{u^2} du \, \hat{\mathbf{i}} =$$

$$= \frac{1}{4\pi\varepsilon_0} \frac{\lambda_0}{L^2} \int_{x-L}^x \left(\frac{x^2}{u^2} - 2\frac{x}{u} + 1 \right) du \, \hat{\mathbf{i}} = \frac{1}{4\pi\varepsilon_0} \frac{\lambda_0}{L^2} \left[-\frac{x^2}{u} - 2x \ln u + u \right]_{x-L}^x \hat{\mathbf{i}} =$$

$$= \frac{1}{4\pi\varepsilon_0} \frac{\lambda_0}{L^2} \left[(-x - 2x \ln x + x) - \left(-\frac{x^2}{(x-L)} - 2x \ln(x-L) + (x-L) \right) \right] \hat{\mathbf{i}} =$$

$$= \frac{1}{4\pi\varepsilon_0} \frac{\lambda_0}{L^2} \left(2x \ln \frac{x-L}{x} + L + \frac{xL}{(x-L)} \right) \hat{\mathbf{i}}$$

The integral was simplified using the following substitution $u = x - x_{dist}$ and $du = -dx_{dist}$.

Section 21-8. Field Lines

A type of diagram that is used to help visualize the electric field is a diagram of **electric field lines**. An example of such a diagram is shown in the Figure on the right.

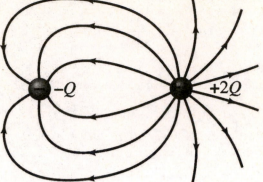

An electric field line is a path such that the electric field at each point along the path is tangent to the path at that point. An electric field line carries no information about the magnitude of the electric field, only about its direction. If a diagram of electric field lines is made, it is created according to the following rules:

1. Electric field lines only begin on positive charges and end on negative charges. The number of lines starting or ending on a given charge is proportional to the magnitude of the charge.
2. A given line has its tangent at a given point in the direction of the electric field at that point.
3. The magnitude of the electric field is proportional to the density of the electric field lines passing through a plane perpendicular to the direction of the electric field.

Section 21-9. Electric Fields and Conductors

Because charges move freely inside a conductor, there will be no electric field inside a conductor under static conditions. If there were an electric field, charges would continue to experience a force and move until the electric field is zero. The electric field at the surface just external to the conductor must be perpendicular to the surface under static conditions. If there was a component of the electric field parallel to the surface, the charges on the surface would move until the component parallel to the surface is zero.

Section 21-10. Motion of a Charged Particle in an Electric Field

The force on an object with charge q in an electric field $\vec{E}$ is given by

$$\vec{F} = q\vec{E}$$

Example 21-10-A. Motion of an electron in an electric field. An electron has a velocity $v_i = (3.99 \times 10^4\ \hat{i})$ m/s. It moves in an electric field $\vec{E} = (2.76 \times 10^2\ \hat{i})$ N/C. How far does the particle travel before its velocity reaches zero?

Approach: Assume that the electron starts at the origin of our coordinate system and is moving towards the right. The electric field points in the direction of the electric force on a particle with a positive charge. The electric force on the electron is thus pointing towards the left. The work done by the electric force is negative; using the work-energy theorem we can determine how far the electron travels before coming to rest.

Solution: The force on the electron with charge q is equal to

$$\vec{F} = q\vec{E}$$

The work done by this force when the electron moves a distance d along the positive x axis is equal to

$$W = \vec{F} \cdot \vec{d} = -qEd$$

The work done by the force lowers the kinetic energy of the electron. The electron will come to rest when the work done by the force is equal to the change in the kinetic energy of the electron:

$$-qEd = \Delta K = 0 - \frac{1}{2}mv^2 \quad \Rightarrow \quad d = \frac{-\frac{1}{2}mv^2}{qE} = -\frac{1}{2}\frac{\left(9.11 \times 10^{-31}\right)\left(3.99 \times 10^4\right)^2}{\left(-1.60 \times 10^{-19}\right)\left(2.76 \times 10^2\right)} = 1.64 \times 10^{-5} \text{ m}$$

Section 21-11. Electric Dipoles

A particular distribution of charge that is often a good approximation to many physical situations is the **electric dipole**. An electric dipole has two charges of equal magnitude but opposite sign, $+Q$ and $-Q$, separated by a displacement $\vec{\mathbf{d}}$. This displacement points from the location of the negative charge to the location of the positive charge. The **electric dipole moment** $\vec{\mathbf{p}}$ is a vector given by

$$\vec{\mathbf{p}} = Q\vec{\mathbf{d}}$$

If ℓ is the length of the displacement vector $\vec{\mathbf{d}}$, we can write the magnitude of the dipole moment is $Q\ell$. In an electric field, a dipole experiences a torque:

$$\vec{\tau} = \vec{\mathbf{p}} \times \vec{\mathbf{E}}$$

As the dipole is rotated by the electric field, work is done on the dipole by the torque. The potential energy associated with the orientation of the dipole moment in the electric field is given by:

$$U = -\vec{\mathbf{p}} \cdot \vec{\mathbf{E}}$$

A dipole in a uniform electric field feels no net electric force. In a uniform electric field, the positive charge and the negative charge of the dipole experience equal magnitude and opposite direction forces. An electric dipole in a non-uniform electric field can experience a net electric force because the electric field can be different at the locations of the positive and negative charges.

Example 21-11-A. Torque on an electric dipole. A rigid electric dipole consists of a charge $Q_1 = +6.32$ μC separated from a charge $Q_2 = -6.32$ μC by a distance of $d = 3.45$ cm. The dipole moment points in the direction of the unit vector $-\hat{\mathbf{i}}$. The dipole has a moment of inertia about the $\hat{\mathbf{k}}$ axis of 1.31×10^{-4} kg·m². The dipole is located in an electric field $\vec{\mathbf{E}} = (3.11 \times 10^1)(2\hat{\mathbf{i}} + \hat{\mathbf{j}})$ N/C. The torque on the dipole due to the electric field will cause the dipole to rotate. Calculate the torque on the dipole in its current orientation and calculate the rotational speed of the dipole when its dipole moment is pointing in the $+\hat{\mathbf{i}}$ direction.

Approach: The torque on the dipole can be calculated since the dipole moment and the electric field are known. Since the electric force is a conservative force, we can use the principle of conservation of energy to relate the change in the potential energy of the dipole to the change in its kinetic energy.

Solution: The magnitude of the dipole moment is equal to

$$p = Qd = \left(6.32 \times 10^{-6}\right)\left(3.45 \times 10^{-2}\right) = 2.18 \times 10^{-7} \text{ C} \cdot \text{m}$$

Since the current direction of the dipole moment is $-\hat{\mathbf{i}}$, we know that

$$\vec{\mathbf{p}}_i = -\left(2.18 \times 10^{-7}\,\hat{\mathbf{i}}\right)\text{C} \cdot \text{m}$$

The torque experienced by the dipole at its current orientation is equal to

$$\vec{\tau}_i = \vec{\mathbf{p}}_i \times \vec{\mathbf{E}} = \left(-2.18 \times 10^{-7}\,\hat{\mathbf{i}}\right) \times \left(3.11 \times 10^1\right)\left(2\hat{\mathbf{i}} + \hat{\mathbf{j}}\right) = -\left(6.78 \times 10^{-6}\right)\left(\hat{\mathbf{i}} \times \hat{\mathbf{j}}\right) = -\left(6.78 \times 10^{-6}\,\hat{\mathbf{k}}\right)\text{N} \cdot \text{m}$$

The energy of the dipole in its current orientation is equal to the sum of its kinetic energy, which is 0 since the dipole is initially at rest, and its potential energy:

$$E_i = K_i + U_i = -\vec{\mathbf{p}}_i \cdot \vec{\mathbf{E}} = -\left(-2.18 \times 10^{-7}\hat{\mathbf{i}}\right) \cdot \left(3.11 \times 10^1\left(2\hat{\mathbf{i}} + \hat{\mathbf{j}}\right)\right) = 1.36 \times 10^{-5} \text{ J}$$

The final energy of the dipole is equal to

$$E_f = K_f + U_f = K_f + \left(-\vec{\mathbf{p}}_f \cdot \vec{\mathbf{E}}\right) = K_f - \left(2.18 \times 10^{-7}\hat{\mathbf{i}}\right) \cdot \left(3.11 \times 10^1\left(2\hat{\mathbf{i}} + \hat{\mathbf{j}}\right)\right) = K_f - 1.36 \times 10^{-5} \text{ J} = E_i$$

The final kinetic energy of the dipole is thus equal to

$$K_f = E_i + 1.36 \times 10^{-5} = 2.72 \times 10^{-5} \text{ J}$$

The kinetic energy is entirely in the form of rotational kinetic energy. The angular velocity of the rotation motion is equal to

$$\omega_f = \sqrt{\frac{2K_f}{I}} = \sqrt{\frac{2\left(2.72 \times 10^{-5}\right)}{1.31 \times 10^{-4}}} = 0.644 \text{ rad/s}$$

Section 21-12. Electric Forces in Molecular Biology; DNA

The molecules inside a biological cell interact with one another via the electric force. The molecules of the strands of DNA have regions with net positive charges and regions with net negative charges. The two strands of DNA are attracted to each other by the electrostatic forces between regions of opposite charge. During replication of DNA, the electrostatic forces select the bases that attach to the old strands, ensuring that the new strands are an exact copy of the original strands.

Section 21-13. Photocopy Machines and Computer Printers Use Electrostatics

The principle of operation of a copy machine and a laser printer rely on the electric force for their operation. The image to be printed is mapped onto a selenium drum that is positively charged. The light image regions lose their charge while the dark regions remain charged. Negatively charged toner particles are attracted to the positively charged regions on the image drum. When the drum presses against the paper, which has a higher charge than the drum, the toner particles are transferred to the paper, and our page is ready.

Practice Quiz

1. A charged object A repels a charged object B. Charged object A also repels charged object C. What type of force exists between charged object B and charged object C?
    a) Object B will repel object C.
    b) Object B will attract object C.
    c) Object B will apply no force on object C.
    d) There is not enough information to determine the type of force of B on C.

2. Given two electrical charges separated by a distance d, will there always be a position, other than at infinity, at which a third charge will experience no net electrical force?
    a) There will always be such a position.
    b) There will only be such a position if the first two charges have opposite signs.
    c) There will only be such a position if the first two charges have the same sign.
    d) There will be cases for which no such position exists.

3. Electric field lines are:
    a) The direction of the force on a charged particle.
    b) The direction a charged particle follows in the electric field.
    c) Lines of constant magnitude of the electric field.
    d) Lines tangent to the direction of the electric field at each point.

4. Two surfaces, each of which are uncharged, are rubbed together. After rubbing the surfaces together, one has a charge of 3.4 μC and the other as a charge of –4.6 μC. The two surfaces only exchanged charged particles and no charges from other sources were lost or gained. Is this possible?
    a) No, it violates conservation of charge
    b) No, materials do not have that great of amount of charge to share
    c) No, the materials must each have the same charge as they started with
    d) Yes, the total charge can change in a process

5. At a distance of d from a point charge, the magnitude of the electric force on a charge q is F. What is the magnitude of the electric force if the charge is moved to a distance $3d$ from the other charge?
   a) $3F$
   b) $9F$
   c) $F/3$
   d) $F/9$

6. Suppose the convention on the sign of charges was reversed. That is, positive charges are called negative and negative charges are called positive. If Coulomb's law were written the same as it is now, what would happen to the prediction of electric forces?
   a) The direction of the predicted force would be opposite in direction to the actual force.
   b) The direction would be correct for like sign charges, but wrong for opposite sign charges.
   c) The direction would be correct for opposite sign charges, but wrong for like sign charges.
   d) The direction of the predicted force would be correct.

7. The gravitational constant in Newton's law of gravity is 20 orders of magnitude smaller than Coulomb's constant in SI units, but gravitational forces seem to dominate over electrical forces in most of everyday life. Why is that?
   a) Most objects have a relatively small net electrical charge.
   b) The distance between the masses is much smaller than the distance between the charges.
   c) Gravity is actually an electric force.
   d) Air is electrically conductive and the electric field is zero inside conductive media.

8. Which of the electric field line pictures is correct?

   a)   b)

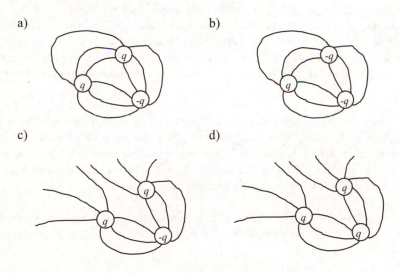

   c)   d)

9. Why is it difficult to accelerate a charged, thin flat object along a conductive surface using electrostatic fields?
   a) Flat objects are difficult to keep charged.
   b) Electrostatic fields near a conductive surface are always zero in magnitude.
   c) Electrostatic fields always increase the frictional force between the object and the surface.
   d) Electrostatic fields are perpendicular to conductive surfaces at the surface.

10. Can an electric field cause a translational acceleration of an electric dipole?
    a) No, it can only cause a torque on the electric dipole.
    b) Yes, all electric dipoles accelerate in the direction of the electric field.
    c) No, the electric force on the dipole is zero since its total charge is zero.
    d) Yes, if the electric field changes with position.

11. Three charges lie on a straight line. A 32 μC charge is 1.62 m to the east of a –6.73 μC charge and 3.64 m to the east of a +24.5 μC. What is the net electric force on the 32 μC charge?

12. A 5.6 µC charge is at the origin. A 3.5 µC charge is located at a position ($x = 1.00$ m, $y = 0.00$ m). A –2.3 µC charge is located at the position ($x = 0.00$ m, $y = 0.78$ m). What is the net electric force on the 5.6 µC charge?

13. What is the electric field at the position ($x = 0.00$ m, $y = 0.00$ m, $z = 1.00$ m) for the arrangement of three charges given in quiz problem 12?

14. A particle has an initial velocity ($245\ \hat{\mathbf{i}}$) m/s. The particle has a charge 4.76×10^{-4} C and a mass 3.78×10^{-3} kg. The particle is moving in a uniform electric field ($365\ \hat{\mathbf{i}}$) N/C. How far will the particle move in 3.87 s?

15. In a particular electric field, a 14.6 µC charge feels a force ($34\ \hat{\mathbf{i}} + 28\ \hat{\mathbf{j}}$) N. What is the torque on an electric dipole with electric dipole moment ($0.24\ \hat{\mathbf{i}} + 0.16\ \hat{\mathbf{k}}$) µC·m in this field?

Responses to Select End-of-Chapter Questions

1. Rub a glass rod with silk and use it to charge an electroscope. The electroscope will end up with a net positive charge. Bring the pocket comb close to the electroscope. If the electroscope leaves move farther apart, then the charge on the comb is positive, the same as the charge on the electroscope. If the leaves move together, then the charge on the comb is negative, opposite the charge on the electroscope.

7. Most of the electrons are strongly bound to nuclei in the metal ions. Only a few electrons per atom (usually one or two) are free to move about throughout the metal. These are called the "conduction electrons." The rest are bound more tightly to the nucleus and are not free to move. Furthermore, in the cases shown in Figures 21-7 and 21-8, not all of the conduction electrons will move. In Figure 21-7, electrons will move until the attractive force on the remaining conduction electrons due to the incoming charged rod is balanced by the repulsive force from electrons that have already gathered at the left end of the neutral rod. In Figure 21-8, conduction electrons will be repelled by the incoming rod and will leave the stationary rod through the ground connection until the repulsive force on the remaining conduction electrons due to the incoming charged rod is balanced by the attractive force from the net positive charge on the stationary rod.

13. When a charged ruler attracts small pieces of paper, the charge on the ruler causes a separation of charge in the paper. For example, if the ruler is negatively charged, it will force the electrons in the paper to the edge of the paper farthest from the ruler, leaving the near edge positively charged. If the paper touches the ruler, electrons will be transferred from the ruler to the paper, neutralizing the positive charge. This action leaves the paper with a net negative charge, which will cause it to be repelled by the negatively charged ruler.

19. Electric field lines can never cross because they give the direction of the electrostatic force on a positive test charge. If they were to cross, then the force on a test charge at a given location would be in more than one direction. This is not possible.

25. The motion of the electron in Example 21-16 is projectile motion. In the case of the gravitational force, the acceleration of the projectile is in the same direction as the field and has a value of g; in the case of an electron in an electric field, the direction of the acceleration of the electron and the field direction are opposite, and the value of the acceleration varies.

Solutions to Select End-of-Chapter Problems

1. Use Coulomb's law to calculate the magnitude of the force.

$$F = k\frac{Q_1 Q_2}{r^2} = \left(8.988 \times 10^9\ \text{N} \cdot \text{m}^2/\text{C}^2\right)\frac{\left(1.602 \times 10^{-19}\,\text{C}\right)\left(26 \times 1.602 \times 10^{-19}\,\text{C}\right)}{\left(1.5 \times 10^{-12}\,\text{m}\right)^2} = \boxed{2.7 \times 10^{-3}\,\text{N}}$$

7. Since the magnitude of the force is inversely proportional to the square of the separation distance, $F \propto \dfrac{1}{r^2}$, if the force is tripled, the distance has been reduced by a factor of $\sqrt{3}$.

$$r = \frac{r_0}{\sqrt{3}} = \frac{8.45\ \text{cm}}{\sqrt{3}} = \boxed{4.88\ \text{cm}}$$

13. The forces on each charge lie along a line connecting the charges. Let the variable d represent the length of a side of the triangle. Since the triangle is equilateral, each angle is $60°$. First calculate the magnitude of each individual force.

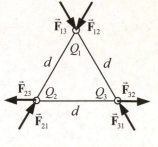

$$F_{12} = k\frac{|Q_1 Q_2|}{d^2} = \left(8.988 \times 10^9\ \text{N} \cdot \text{m}^2/\text{C}^2\right)\frac{\left(7.0 \times 10^{-6}\,\text{C}\right)\left(8.0 \times 10^{-6}\,\text{C}\right)}{\left(1.20\,\text{m}\right)^2}$$

$$= 0.3495\,\text{N}$$

$$F_{13} = k\frac{|Q_1 Q_3|}{d^2} = \left(8.988 \times 10^9\ \text{N} \cdot \text{m}^2/\text{C}^2\right)\frac{\left(7.0 \times 10^{-6}\,\text{C}\right)\left(6.0 \times 10^{-6}\,\text{C}\right)}{\left(1.20\,\text{m}\right)^2}$$

$$= 0.2622\,\text{N}$$

$$F_{23} = k\frac{|Q_2 Q_3|}{d^2} = \left(8.988 \times 10^9\ \text{N} \cdot \text{m}^2/\text{C}^2\right)\frac{\left(8.0 \times 10^{-6}\,\text{C}\right)\left(6.0 \times 10^{-6}\,\text{C}\right)}{\left(1.20\,\text{m}\right)^2} = 0.2996\,\text{N} = F_{32}$$

Now calculate the net force on each charge and the direction of that net force, using components.

$$F_{1x} = F_{12x} + F_{13x} = -\left(0.3495\,\text{N}\right)\cos 60° + \left(0.2622\,\text{N}\right)\cos 60° = -4.365 \times 10^{-2}\,\text{N}$$

$$F_{1y} = F_{12y} + F_{13y} = -\left(0.3495\,\text{N}\right)\sin 60° - \left(0.2622\,\text{N}\right)\sin 60° = -5.297 \times 10^{-1}\,\text{N}$$

$$F_1 = \sqrt{F_{1x}^2 + F_{1y}^2} = \boxed{0.53\,\text{N}} \qquad \theta_1 = \tan^{-1}\frac{F_{1y}}{F_{1x}} = \tan^{-1}\frac{-5.297 \times 10^{-1}\,\text{N}}{-4.365 \times 10^{-2}\,\text{N}} = \boxed{265°}$$

$$F_{2x} = F_{21x} + F_{23x} = \left(0.3495\,\text{N}\right)\cos 60° - \left(0.2996\,\text{N}\right) = -1.249 \times 10^{-1}\,\text{N}$$

$$F_{2y} = F_{21y} + F_{23y} = \left(0.3495\,\text{N}\right)\sin 60° + 0 = 3.027 \times 10^{-1}\,\text{N}$$

$$F_2 = \sqrt{F_{2x}^2 + F_{2y}^2} = \boxed{0.33\,\text{N}} \qquad \theta_2 = \tan^{-1}\frac{F_{2y}}{F_{2x}} = \tan^{-1}\frac{3.027 \times 10^{-1}\,\text{N}}{-1.249 \times 10^{-1}\,\text{N}} = \boxed{112°}$$

$$F_{3x} = F_{31x} + F_{32x} = -\left(0.2622\,\text{N}\right)\cos 60° + \left(0.2996\,\text{N}\right) = 1.685 \times 10^{-1}\,\text{N}$$

$$F_{3y} = F_{31y} + F_{32y} = \left(0.2622\,\text{N}\right)\sin 60° + 0 = 2.271 \times 10^{-1}\,\text{N}$$

$$F_3 = \sqrt{F_{3x}^2 + F_{3y}^2} = \boxed{0.26\,\text{N}} \qquad \theta_3 = \tan^{-1}\frac{F_{3y}}{F_{3x}} = \tan^{-1}\frac{2.271 \times 10^{-1}\,\text{N}}{1.685 \times 10^{-1}\,\text{N}} = \boxed{53°}$$

19. (*a*) The charge will experience a force that is always pointing towards the origin. In the diagram, there is a greater force of $\dfrac{Qq}{4\pi\varepsilon_0 \left(d-x\right)^2}$ to the left, and

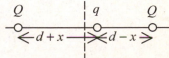

a lesser force of $\dfrac{Qq}{4\pi\varepsilon_0 \left(d+x\right)^2}$ to the right. So the net force is towards the

origin. The same would be true if the mass were to the left of the origin.

Calculate the net force.

$$F_{net} = \frac{Qq}{4\pi\varepsilon_0(d+x)^2} - \frac{Qq}{4\pi\varepsilon_0(d-x)^2} = \frac{Qq}{4\pi\varepsilon_0(d+x)^2(d-x)^2}\left[(d-x)^2 - (d+x)^2\right]$$

$$= \frac{-4Qqd}{4\pi\varepsilon_0(d+x)^2(d-x)^2}x = \frac{-Qqd}{\pi\varepsilon_0(d+x)^2(d-x)^2}x$$

We assume that $x \ll d$.

$$F_{net} = \frac{-Qqd}{\pi\varepsilon_0(d+x)^2(d-x)^2}x \approx \frac{-Qq}{\pi\varepsilon_0 d^3}x$$

This has the form of a simple harmonic oscillator, where the "spring constant" is $k_{elastic} = \frac{Qq}{\pi\varepsilon_0 d^3}$. The spring constant can be used to find the period. See Eq. 14-7b.

$$T = 2\pi\sqrt{\frac{m}{k_{elastic}}} = 2\pi\sqrt{\frac{m}{\frac{Qq}{\pi\varepsilon_0 d^3}}} = \boxed{2\pi\sqrt{\frac{m\pi\varepsilon_0 d^3}{Qq}}}$$

(b) Sodium has an atomic mass of 23.

$$T = 2\pi\sqrt{\frac{m\pi\varepsilon_0 d^3}{Qq}} = 2\pi\sqrt{\frac{(29)(1.66\times10^{-27}\,\text{kg})\pi(8.85\times10^{-12}\,\text{C}^2/\text{N}\cdot\text{m}^2)(3\times10^{-10}\,\text{m})^3}{(1.60\times10^{-19}\,\text{C})^2}}$$

$$= 2.4\times10^{-13}\,\text{s}\left(\frac{10^{12}\,\text{ps}}{1\,\text{s}}\right) = 0.24\,\text{ps} \approx \boxed{0.2\,\text{ps}}$$

25. Use the definition of the electric field, Eq. 21-3.

$$\vec{E} = \frac{\vec{F}}{q} = \frac{(7.22\times10^{-4}\,\text{N}\,\hat{j})}{4.20\times10^{-6}\,\text{C}} = \boxed{172\,\text{N/C}\,\hat{j}}$$

31. The field at the point in question is the vector sum of the two fields shown in Figure 21-56. Use the results of Example 21-11 to find the field of the long line of charge.

$$\vec{E}_{thread} = \frac{1}{2\pi\varepsilon_0}\frac{\lambda}{y}\hat{j} \;;\; \vec{E}_Q = \frac{1}{4\pi\varepsilon_0}\frac{|Q|}{d^2}(-\cos\theta\hat{i} - \sin\theta\hat{j}) \rightarrow$$

$$\vec{E} = \left(-\frac{1}{4\pi\varepsilon_0}\frac{|Q|}{d^2}\cos\theta\right)\hat{i} + \left(\frac{1}{2\pi\varepsilon_0}\frac{\lambda}{y} - \frac{1}{4\pi\varepsilon_0}\frac{|Q|}{d^2}\sin\theta\right)\hat{j}$$

$$d^2 = (0.070\,\text{m})^2 + (0.120\,\text{m})^2 = 0.0193\,\text{m}^2 \;;\; y = 0.070\,\text{m} \;;\; \theta = \tan^{-1}\frac{12.0\,\text{cm}}{7.0\,\text{cm}} = 59.7°$$

$$E_x = -\frac{1}{4\pi\varepsilon_0}\frac{|Q|}{d^2}\cos\theta = -(8.988\times10^9\,\text{N}\cdot\text{m}^2/\text{C}^2)\frac{(2.0\,\text{C})}{0.0193\,\text{m}^2}\cos59.7° = -4.699\times10^{11}\,\text{N/C}$$

$$E_y = \frac{1}{2\pi\varepsilon_0}\frac{\lambda}{y} - \frac{1}{4\pi\varepsilon_0}\frac{|Q|}{d^2}\sin\theta = \frac{1}{4\pi\varepsilon_0}\left(\frac{2\lambda}{y} - \frac{|Q|}{d^2}\sin\theta\right)$$

$$= \left(8.988\times10^9\,\text{N}\cdot\text{m}^2/\text{C}^2\right)\left[\frac{2(2.5\,\text{C/m})}{0.070\,\text{cm}} - \frac{(2.0\,\text{C})}{0.0193\,\text{m}^2}\sin 59.7°\right] = -1.622\times10^{11}\,\text{N/C}$$

$$\vec{E} = \boxed{\left(-4.7\times10^{11}\,\text{N/C}\right)\hat{i} + \left(-1.6\times10^{11}\,\text{N/C}\right)\hat{j}}$$

$$E = \sqrt{E_x^2 + E_y^2} = \sqrt{\left(-4.699\times10^{11}\,\text{N/C}\right)^2 + \left(-1.622\times10^{11}\,\text{N/C}\right)^2} = \boxed{5.0\times10^{11}\,\text{N/C}}$$

$$\theta_E = \tan^{-1}\frac{\left(-1.622\times10^{11}\,\text{N/C}\right)}{\left(-4.699\times10^{11}\,\text{N/C}\right)} = \boxed{199°}$$

37. Make use of Example 21-11. From that, we see that the electric field due to the line charge along the y axis is $\vec{E}_1 = \frac{1}{2\pi\varepsilon_0}\frac{\lambda}{x}\hat{i}$. In particular, the field due to that line of charge has no y dependence. In a similar fashion, the electric field due to the line charge along the x axis is $\vec{E}_2 = \frac{1}{2\pi\varepsilon_0}\frac{\lambda}{y}\hat{j}$. Then the total field at (x, y) is the vector sum of the two fields.

$$\vec{E} = \vec{E}_1 + \vec{E}_2 = \frac{1}{2\pi\varepsilon_0}\frac{\lambda}{x}\hat{i} + \frac{1}{2\pi\varepsilon_0}\frac{\lambda}{y}\hat{j} = \frac{\lambda}{2\pi\varepsilon_0}\left(\frac{1}{x}\hat{i} + \frac{1}{y}\hat{j}\right)$$

$$E = \frac{\lambda}{2\pi\varepsilon_0}\sqrt{\frac{1}{x^2} + \frac{1}{y^2}} = \boxed{\frac{\lambda}{2\pi\varepsilon_0 xy}\sqrt{x^2 + y^2}} \quad ; \quad \theta = \tan^{-1}\frac{E_y}{E_x} = \tan^{-1}\frac{\dfrac{1}{2\pi\varepsilon_0}\dfrac{\lambda}{y}}{\dfrac{1}{2\pi\varepsilon_0}\dfrac{\lambda}{x}} = \boxed{\tan^{-1}\frac{x}{y}}$$

43. (a) See the diagram. From the symmetry of the charges, we see that the net electric field points along the y axis.

$$\vec{E} = 2\frac{Q}{4\pi\varepsilon_0\left(\ell^2 + y^2\right)}\sin\theta\,\hat{j} = \boxed{\frac{Qy}{2\pi\varepsilon_0\left(\ell^2 + y^2\right)^{3/2}}\hat{j}}$$

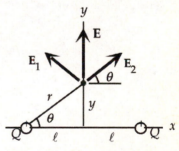

(b) To find the position where the magnitude is a maximum, set the first derivative with respect to y equal to 0, and solve for the y value.

$$E = \frac{Qy}{2\pi\varepsilon_0\left(\ell^2 + y^2\right)^{3/2}} \quad \rightarrow$$

$$\frac{dE}{dy} = \frac{Q}{2\pi\varepsilon_0\left(\ell^2 + y^2\right)^{3/2}} + \left(-\tfrac{3}{2}\right)\frac{Qy}{2\pi\varepsilon_0\left(\ell^2 + y^2\right)^{5/2}}(2y) = 0 \quad \rightarrow$$

$$\frac{1}{\left(\ell^2 + y^2\right)^{3/2}} = \frac{3y^2}{\left(\ell^2 + y^2\right)^{5/2}} \quad \rightarrow \quad y^2 = \tfrac{1}{2}\ell^2 \quad \rightarrow \quad y = \boxed{\pm\ell/\sqrt{2}}$$

This has to be a maximum, because the magnitude is positive, the field is 0 midway between the charges, and $E \rightarrow 0$ as $y \rightarrow \infty$.

49. Select a differential element of the arc which makes an angle
of θ with the x axis. The length of this element is $R d\theta$ and
the charge on that element is $dq = \lambda R d\theta$. The magnitude of
the field produced by that element is $dE = \dfrac{1}{4\pi\varepsilon_0}\dfrac{\lambda R d\theta}{R^2}$. From

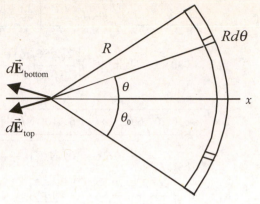

the diagram, considering pieces of the arc that are symmetric
with respect to the x axis, we see that the total field will only
have an x component. The vertical components of the field
due to symmetric portions of the arc will cancel each other.
So we have the following.

$$dE_{horizontal} = \frac{1}{4\pi\varepsilon_0}\frac{\lambda R d\theta}{R^2}\cos\theta$$

$$E_{horizontal} = \int_{-\theta_0}^{\theta_0}\frac{1}{4\pi\varepsilon_0}\cos\theta\frac{\lambda R d\theta}{R^2} = \frac{\lambda}{4\pi\varepsilon_0 R}\int_{-\theta_0}^{\theta_0}\cos\theta\, d\theta = \frac{\lambda}{4\pi\varepsilon_0 R}\left[\sin\theta_0 - \sin\left(-\theta_0\right)\right] = \frac{2\lambda\sin\theta_0}{4\pi\varepsilon_0 R}$$

The field points in the negative x direction, so $\boxed{E = -\dfrac{2\lambda\sin\theta_0}{4\pi\varepsilon_0 R}\hat{\mathbf{i}}}$.

55. Take Figure 21-28 and add the angle ϕ, measured from the $-z$ axis, as
indicated in the diagram. Consider an infinitesimal length of the ring $a\, d\phi$.

The charge on that infinitesimal length is $dq = \lambda\left(a d\phi\right) = \dfrac{Q}{\pi a}\left(a d\phi\right) = \dfrac{Q}{\pi}d\phi$.

The charge creates an infinitesimal electric field, $d\vec{\mathbf{E}}$, with magnitude

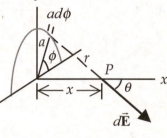

$dE = \dfrac{1}{4\pi\varepsilon_0}\dfrac{dq}{r^2} = \dfrac{1}{4\pi\varepsilon_0}\dfrac{\frac{Q}{\pi}d\phi}{x^2+a^2}$. From the symmetry of the figure, we see that

the z component of $d\vec{\mathbf{E}}$ will be cancelled by the z component due to the
piece of the ring that is on the opposite side of the y axis. The trigonometric relationships give $dE_x = dE\cos\theta$ and
$dE_y = -dE\sin\theta\sin\phi$. The factor of $\sin\phi$ can be justified by noting that $dE_y = 0$ when $\phi = 0$, and $dE_y = -dE\sin\theta$
when $\phi = \pi/2$.

$$dE_x = dE\cos\theta = \frac{Q}{4\pi^2\varepsilon_0}\frac{d\phi}{x^2+a^2}\frac{x}{\sqrt{x^2+a^2}} = \frac{Qx}{4\pi^2\varepsilon_0}\frac{d\phi}{\left(x^2+a^2\right)^{3/2}}$$

$$E_x = \frac{Qx}{4\pi^2\varepsilon_0\left(x^2+a^2\right)^{3/2}}\int_0^{\pi}d\phi = \boxed{\frac{Qx}{4\pi\varepsilon_0\left(x^2+a^2\right)^{3/2}}}$$

$$dE_y = -dE\sin\theta\sin\phi = -\frac{Q}{4\pi^2\varepsilon_0}\frac{d\phi}{x^2+a^2}\frac{a}{\sqrt{x^2+a^2}}\sin\phi = -\frac{Qa}{4\pi^2\varepsilon_0\left(x^2+a^2\right)^{3/2}}\sin\phi\, d\phi$$

$$E_y = -\frac{Qa}{4\pi^2\varepsilon_0\left(x^2+a^2\right)^{3/2}}\int_0^{\pi}\sin\phi\, d\phi = -\frac{Qa}{4\pi^2\varepsilon_0\left(x^2+a^2\right)^{3/2}}\left[\left(-\cos\pi\right)-\left(-\cos 0\right)\right] = \boxed{-\frac{2Qa}{4\pi^2\varepsilon_0\left(x^2+a^2\right)^{3/2}}}$$

We can write the electric field in vector notation.

$$\vec{\mathbf{E}} = \frac{Qx}{4\pi\varepsilon_0\left(x^2+a^2\right)^{3/2}}\hat{\mathbf{i}} - \frac{2Qa}{4\pi^2\varepsilon_0\left(x^2+a^2\right)^{3/2}}\hat{\mathbf{j}} = \boxed{\frac{Q}{4\pi\varepsilon_0\left(x^2+a^2\right)^{3/2}}\left(x\hat{\mathbf{i}} - \frac{2a}{\pi}\hat{\mathbf{j}}\right)}$$

61. (a) The field along the axis of the ring is given in Example 21-9, with the opposite sign because this ring is negatively charged. The force on the charge is the field times the charge q. Note that if x is positive, the force is to the left, and if x is negative, the force is to the right. Assume that $x \ll R$.

$$F = qE = \frac{q}{4\pi\varepsilon_0} \frac{(-Q)x}{\left(x^2 + R^2\right)^{3/2}} = \frac{-qQx}{4\pi\varepsilon_0} \frac{1}{\left(x^2 + R^2\right)^{3/2}} \approx \frac{-qQx}{4\pi\varepsilon_0 R^3}$$

This has the form of a simple harmonic oscillator, where the "spring constant" is $k_{elastic} = \dfrac{Qq}{4\pi\varepsilon_0 R^3}$.

(b)   The spring constant can be used to find the period. See Eq. 14-7b.

$$T = 2\pi\sqrt{\frac{m}{k_{elastic}}} = 2\pi\sqrt{\frac{m}{\dfrac{Qq}{4\pi\varepsilon_0 R^3}}} = 2\pi\sqrt{\frac{m4\pi\varepsilon_0 R^3}{Qq}} = 4\pi\sqrt{\frac{m\pi\varepsilon_0 R^3}{Qq}}$$

67. (a) Along the x axis the fields from the two charges are parallel so the magnitude is found as follows.

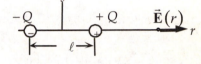

$$E_{net} = E_{+Q} + E_{-Q} =$$

$$= \frac{Q}{4\pi\varepsilon_0 \left(r - \tfrac{1}{2}\ell\right)^2} + \frac{(-Q)}{4\pi\varepsilon_0 \left(r + \tfrac{1}{2}\ell\right)^2} =$$

$$= \frac{Q\left[\left(r + \tfrac{1}{2}\ell\right)^2 - \left(r - \tfrac{1}{2}\ell\right)^2\right]}{4\pi\varepsilon_0 \left(r + \tfrac{1}{2}\ell\right)^2 \left(r - \tfrac{1}{2}\ell\right)^2} =$$

$$= \frac{Q(2r\ell)}{4\pi\varepsilon_0 \left(r + \tfrac{1}{2}\ell\right)^2 \left(r - \tfrac{1}{2}\ell\right)^2}$$

$$\approx \frac{Q(2r\ell)}{4\pi\varepsilon_0 r^4} = \frac{2Q\ell}{4\pi\varepsilon_0 r^3} = \boxed{\frac{1}{4\pi\varepsilon_0} \frac{2p}{r^3}}$$

The same result is obtained if the point is to the left of $-Q$.
(b) The electric field points in the $\boxed{\text{same direction as the dipole}}$ moment vector.

73. For the droplet to remain stationary, the magnitude of the electric force on the droplet must be the same as the weight of the droplet. The mass of the droplet is found from its volume times the density of water. Let n be the number of excess electrons on the water droplet.

$$F_E = |q|E = mg \rightarrow neE = \tfrac{4}{3}\pi r^3 \rho g \rightarrow$$

$$n = \frac{4\pi r^3 \rho g}{3eE} = \frac{4\pi\left(1.8\times10^{-5}\,\text{m}\right)^3\left(1.00\times10^3\,\text{kg/m}^3\right)\left(9.80\,\text{m/s}^2\right)}{3\left(1.602\times10^{-19}\,\text{C}\right)\left(150\,\text{N/C}\right)} =$$

$$= 9.96\times10^6 \approx \boxed{1.0\times10^7 \text{ electrons}}$$

79. The sphere will oscillate sinusoidally about the equilibrium point, with an amplitude of 5.0 cm. The angular frequency of the sphere is given by $\omega = \sqrt{k/m} = \sqrt{126\,\text{N/m}/0.650\,\text{kg}} = 13.92\,\text{rad/s}$. The distance of the sphere from the table is given by $r = \left[0.150 - 0.0500\cos(13.92t)\right]$m. Use this distance and the charge to give the electric field value at the tabletop.

That electric field will point upwards at all times, towards the negative sphere.

$$E = k\frac{|Q|}{r^2} = \frac{\left(8.988\times10^9\ \text{N}\cdot\text{m}^2/\text{C}^2\right)\left(3.00\times10^{-6}\ \text{C}\right)}{\left[0.150-0.0500\cos\left(13.92t\right)\right]^2\ \text{m}^2} =$$

$$= \frac{2.70\times10^4}{\left[0.150-0.0500\cos\left(13.92t\right)\right]^2}\ \text{N/C} =$$

$$= \boxed{\frac{1.08\times10^7}{\left[3.00-\cos\left(13.9t\right)\right]^2}\ \text{N/C, upwards}}$$

85. This is a constant acceleration situation, similar to projectile motion in a uniform gravitational field. Let the width of the plates be l, the vertical gap between the plates be h, and the initial velocity be v_0. Notice that the vertical motion has a maximum displacement of $h/2$. Let upwards be the positive vertical direction. We calculate the vertical acceleration produced by the electric field and the time t for the electron to cross the region of the field. We then use constant acceleration equations to solve for the angle.

$$F_y = ma_y = qE = -eE \quad\rightarrow\quad a_y = -\frac{eE}{m} \quad ; \quad \ell = v_0\cos\theta_0(t) \quad\rightarrow\quad t = \frac{\ell}{v_0\cos\theta_0}$$

$$v_{y_{\text{top}}} = v_{0y} + a_y t_{\text{top}} \quad\rightarrow\quad 0 = v_0\sin\theta_0 - \frac{eE}{m}\left(\frac{1}{2}\frac{\ell}{v_0\cos\theta_0}\right) \quad\rightarrow\quad v_0^2 = \frac{eE}{2m}\left(\frac{\ell}{\sin\theta_0\cos\theta_0}\right)$$

$$y_{\text{top}} = y_0 + v_{0y}t_{\text{top}} + \tfrac{1}{2}a_y t^2 \quad\rightarrow\quad \tfrac{1}{2}h = v_0\sin\theta_0\left(\tfrac{1}{2}\frac{\ell}{v_0\cos\theta_0}\right) - \tfrac{1}{2}\frac{eE}{m}\left(\tfrac{1}{2}\frac{\ell}{v_0\cos\theta_0}\right)^2 \quad\rightarrow$$

$$h = \ell\tan\theta_0 - \frac{eE\ell^2}{4m\cos^2\theta_0}\frac{1}{v_0^2} = \ell\tan\theta_0 - \frac{eE\ell^2}{4m\cos^2\theta_0}\frac{1}{\dfrac{eE}{2m}\left(\dfrac{\ell}{\sin\theta_0\cos\theta_0}\right)} = \ell\tan\theta_0 - \tfrac{1}{2}\ell\tan\theta_0$$

$$h = \tfrac{1}{2}\ell\tan\theta_0 \quad\rightarrow\quad \theta_0 = \tan^{-1}\frac{2h}{\ell} = \tan^{-1}\frac{2(1.0\,\text{cm})}{6.0\,\text{cm}} = \boxed{18°}$$

91. (*a*) The weight of the mass is only about 2 N. Since the tension in the string is more than that, there must be a downward electric force on the positive charge, which means that the electric field must be pointed $\boxed{\text{down}}$. Use the free-body diagram to write an expression for the magnitude of the electric field.

$$\sum F = F_T - mg - F_E = 0 \quad\rightarrow\quad F_E = QE = F_T - mg \quad\rightarrow$$

$$E = \frac{F_T - mg}{Q} = \frac{5.18\,\text{N} - (0.210\,\text{kg})(9.80\,\text{m/s}^2)}{3.40\times10^{-7}\,\text{C}} = \boxed{9.18\times10^6\ \text{N/C}}$$

(*b*) Use Eq. 21-7.

$$E = \frac{\sigma}{2\varepsilon_0} \quad\rightarrow\quad \sigma = 2E\varepsilon_0 = 2\left(9.18\times10^6\ \text{N/C}\right)\left(8.854\times10^{-12}\right) = \boxed{1.63\times10^{-4}\ \text{C/m}^2}$$

Chapter 22: Gauss's Law

Chapter Overview and Objectives

In this chapter, the concept of electric flux is presented. Gauss's Law is introduced and several applications of Gauss's law are discussed. Gauss's Law is equivalent to Coulomb's Law for situations involving static charge distributions.

After completing this chapter you should:
- Know the definition of electric flux.
- Be able to calculate the electric flux through a surface.
- Know Gauss's Law.
- Be able to calculate electric fields using Gauss's Law for certain symmetric charge distributions.

Summary of Equations

Electric flux through a planar area element in a uniform electric field:

$$\Phi_E = EA_\perp = EA\cos\theta \qquad\qquad \text{(Section 22-1)}$$

Electric flux through an area in an electric field: $\Phi_E = \int \vec{\mathbf{E}} \cdot d\vec{\mathbf{A}}$ (Section 22-1)

Gauss's Law: $\oint \vec{\mathbf{E}} \cdot d\vec{\mathbf{A}} = \dfrac{Q_{encl}}{\varepsilon_0}$ (Section 22-2)

Chapter Summary

Section 22-1. Electric Flux

The **electric flux**, Φ_E, passing through a planar surface of area A that lies in a plane perpendicular to a uniform electric field of magnitude E is defined as

$$\Phi_E = EA$$

If the plane of the area A is not perpendicular to the electric field, then the electric flux is given by

$$\Phi_E = EA_\perp = EA\cos\theta$$

where $A_\perp$ is the area of projection of the area A onto a plane perpendicular to the electric field and θ is the angle between the electric field and the direction normal to the area. If we define a vector $\vec{\mathbf{A}}$ that has magnitude A and a direction normal to the surface area, we can write the electric flux as

$$\Phi_E = \vec{\mathbf{E}} \cdot \vec{\mathbf{A}}$$

In the case of a non-uniform electric field and a general surface, we can calculate the flux through infinitesimal areas $d\vec{\mathbf{A}}$ and then integrate the flux over the entire surface:

$$\Phi_E = \int \vec{\mathbf{E}} \cdot d\vec{\mathbf{A}}$$

When the surface is a closed surface, a surface that separates space into an interior and an exterior region, we can write the integral as

$$\Phi_E = \oint \vec{\mathbf{E}} \cdot d\vec{\mathbf{A}}$$

When we evaluate the surface integral for a closed surface, we will use the convention that the area vector $d\vec{\mathbf{A}}$ always points outward. Electric field lines entering the volume enclosed by a closed surface make a negative contribution to the electric flux through that surface; electric field lines leaving the volume enclosed by a closed surface make a positive contribution to the electric flux through that surface.

Example 22-1-A. Calculating the electric flux. The electric field as a function of position along the *yz*-plane is given by

$$\vec{\mathbf{E}}(y,z) = E_0 \frac{yz}{L^2}\hat{\mathbf{i}} + E_1 \frac{y^2 z^2}{L^4}\hat{\mathbf{j}}$$

Determine the flux through a square with vertices (0,0,0), (0,*L*,0), (0,*L*,*L*), and (0,0,*L*).

Approach: The normal vector of the surface is directed parallel to the $\hat{\mathbf{i}}$ axis (see Figure). An expression of the electric flux through a surface element centered at (0, *y*, *z*) can be calculated easily, and the total electric flux can be obtained by integrating that expression with respect to *y* and *z*.

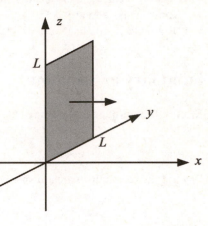

Solution: The flux $d\Phi_E$ through the surface dA, centered at (0, *y*, *z*) is equal to

$$d\Phi_E = \vec{\mathbf{E}} \cdot d\vec{\mathbf{A}} = \left(E_0 \frac{yz}{L^2}\hat{\mathbf{i}} + E_1 \frac{y^2 z^2}{L^4}\hat{\mathbf{j}} \right) \cdot \left(dy\,dz\,\hat{\mathbf{i}} \right) = E_0 \frac{yz}{L^2} dy\,dz$$

The total electric flux can be obtained by integrating the flux $d\Phi_E$ with respect to *y* and *z*:

$$\Phi_E = \int d\Phi_E = \int_{z=0}^{L} \int_{y=0}^{L} E_0 \frac{yz}{L^2} dy\,dz = \frac{E_0}{L^2} \int_{y=0}^{L} y\,dy \int_{z=0}^{L} z\,dz = \frac{1}{4} E_0 L^2$$

Section 22-2. Gauss's Law

The electric flux through a closed surface is proportional to the charge within the closed surface:

$$\oint \vec{\mathbf{E}} \cdot d\vec{\mathbf{A}} = \frac{Q_{encl}}{\varepsilon_0}$$

This relationship is called **Gauss's Law**. Gauss's Law is equivalent to Coulomb's Law for static charge distributions. Gauss's Law is more general than Coulomb's Law in that it is valid in situations where the electric field is produced by a non-static charge distribution.

Example 22-2-A. Calculating surface charge densities using Gauss's law. A spherical conductive shell of inner radius R_1 and outer radius R_2 has a charge Q at its center (see Figure). Determine the surface charge density on the inner and outer surfaces of the conductive shell and the electric field at the outer surface of the conductive shell.

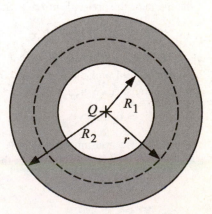

Approach: In this problem we use the fact that the electric field inside a conductor is 0 to determine the electric flux through a surface that is contained within the conductive shell. This allows us to determine the charge on the inner surface of the shell. Once the charge on the inner surface of the shell has been determined, we can determine the charge on the outer surface of the shell since the shell is neutral.

Solution: The flux through a sphere of radius *r* ($R_1 < r < R_2$) is 0 since the electric field inside the conductor is 0. The enclosed charge must thus be 0; this requires that the charge on the inner surface of the sphere is $-Q$.

The surface charge density of the inner surface is equal to

$$\sigma_1 = \frac{-Q}{4\pi R_1{}^2}$$

Since the shell is neutral, the charge on the outer surface must be $+Q$. The surface charge density of the outer surface is thus equal to

$$\sigma_2 = \frac{+Q}{4\pi R_2{}^2}$$

Section 22-3. Applications of Gauss's Law

To use Gauss's Law to determine an unknown electric field from a given charge distribution, it is necessary to choose a surface that has the correct symmetry. In many instances, we will try to use a gaussian surface for which the electric field is perpendicular to the surface and constant in magnitude across the surface. In this case, the electric flux through the surface will be equal to the product of the magnitude of the electric field and the area of the gaussian surface. Using the enclosed charge, the magnitude of the electric field can then be determined.

Example 22-3-A. Calculating the electric field using Gauss's law. A very long cylinder of charge has charge uniformly distributed along its length. The radial variation of charge density is given by

$$\rho(r) = \begin{cases} \rho_0(A-r)/A & \text{if } r < A \\ 0 & \text{if } r > A \end{cases}$$

Determine the electric field for $r < A$ and for $r > A$.

Approach: Due to the symmetry of the charge configuration, the electric field will be directed radially. The magnitude of the electric field only depends on r. If we choose a cylinder as our gaussian surface, the electric flux through this surface will be equal to the product of the magnitude of the electric field and the area of the curved surface of the cylinder. Since the electric field is parallel to the sides of the cylinder, the corresponding flux through these surfaces is zero.

Solution: Consider a cylinder of radius R and length L as our gaussian surface (see Figure). The magnitude of the electric field on the curved surface of the cylinder will be $E(R)$. The flux through this surface is thus equal to

$$\Phi_E = 2\pi R L E(R)$$

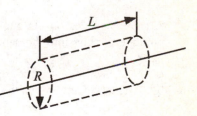

To determine the charge inside the cylinder, we integrate the charge density over the volume of the cylinder. If R is less than A, the charge inside the cylinder will be

$$Q_{inside} = \int \rho \, dV = \int_{z=0}^{L} \int_{r=0}^{R} \rho_0 \frac{A-r}{A} 2\pi r \, dr \, dz = 2\pi \rho_0 L \left(\frac{r^2}{2} - \frac{r^3}{3A} \right)\Bigg|_{r=0}^{R} = 2\pi \rho_0 L \left(\frac{R^2}{2} - \frac{R^3}{3A} \right)$$

The electric field for $R < A$ can now be determined by applying Gauss's Law:

$$\Phi_E = \frac{Q_{inside}}{\varepsilon_0} \quad \Rightarrow \quad 2\pi R L E(R) = \frac{2\pi \rho_0 L}{\varepsilon_0} \left(\frac{R^2}{2} - \frac{R^3}{3A} \right) \quad \Rightarrow \quad E(R) = \frac{\rho_0}{\varepsilon_0} \left(\frac{R}{2} - \frac{R^2}{3A} \right)$$

Once R becomes greater than the radius of the charge distribution, A, the charge inside the cylinder remains the same as for $R = A$:

$$Q_{inside} = 2\pi \rho_0 L \left(\frac{A^2}{2} - \frac{A^3}{3A} \right) = 2\pi \rho_0 L \left(\frac{A^2}{2} - \frac{A^2}{3} \right) = 2\pi \rho_0 L \left(\frac{A^2}{6} \right)$$

The electric field for $R > A$ can now be determined by applying Gauss's Law:

$$\Phi_E = \frac{Q_{inside}}{\varepsilon_0} \Rightarrow 2\pi RLE(R) = \frac{2\pi\rho_0 L}{6\varepsilon_0}A^2 \Rightarrow E(R) = \frac{\rho_0 A^2}{6\varepsilon_0 R}$$

Section 22-4. Experimental Basis of Gauss's and Coulomb's Law

Experimental tests of Gauss's Law and Coulomb's Law are in agreement with the laws to a very high precision. Any measured deviation from Coulomb's Law can be expressed in terms of the parameter δ:

$$F = \frac{1}{4\pi\varepsilon_0}\frac{Q_1 Q_2}{r^{2+\delta}}$$

The experimental value of δ is $(2.7 \pm 3.1) \times 10^{-16}$. The value of δ is in agreement with $\delta = 0$ which is consistent with a perfect inverse square law.

Practice Quiz

1. A charge of $+5\ \mu C$ is located at the origin. Through which of these closed surfaces is the electric flux the least?
   a) A cube with sides of 1.0 m that is centered at the origin
   b) A sphere with a radius 1.0 m that is centered at the origin
   c) A sphere with a radius of 0.8 m and a center that is located 0.5 m from the origin
   d) A sphere with a radius of 0.8 m and a center that is located 1.0 m from the origin

2. Suppose Coulomb's Law was $F = kQ_1Q_2/r^{2-\varepsilon}$ rather than $F = kQ_1Q_2/r^2$ where ε is a small positive number. If you take a gaussian surface that is a sphere around a positive point charge, how does the flux through the sphere depend on the radius of the sphere?
   a) The flux does not depend on the radius of the sphere.
   b) The flux increases as the radius of the sphere increases.
   c) The flux decreases as the radius of the sphere increases.
   d) There is insufficient information given to answer the question.

3. A cube has sides L. On one face of the cube, the electric field is uniform with magnitude E and has a direction pointing directly into the cube. The total charge on and within the cube is zero. Which statement is necessarily true?
   a) The electric field on the opposite face of the box has magnitude E and points directly out of the face.
   b) The total flux through the remaining five faces is out of the box and equal to EL^2.
   c) At least one other face of the cube must have an inward flux.
   d) None of the previous statements must be true.

4. Suppose our universe were squashed into an almost zero thickness two-dimensional surface and suppose Gauss's Law still held true. What would the distance dependence of Coulomb's law be in such a universe?
   a) $F \propto r$
   b) $F \propto 1/r$
   c) $F \propto 1/r^2$
   d) $F \propto 1/r^3$

5. The electric flux through a spherical surface is zero. Which condition is necessarily true?
   a) There net charge inside the sphere is zero.
   b) The electric field is zero everywhere within the sphere.
   c) The electric field is zero everywhere on the sphere.
   d) None of the above statements is necessarily true.

6. The electric flux through a given open surface is zero. What condition is certain to be true?
    a) The electric field is zero everywhere along the surface.
    b) There are no electric charges near the surface.
    c) If the electric flux is non-zero through some part of the surface, there is some other part of the surface through which the electric flux is non-zero.
    d) The electric field is zero everywhere.

7. In the diagram to the right is an arrangement of charges and the projection of the outline of some closed surfaces that are right cylinders with the bottom surface below the page and the top surface above the page. Through which of the closed surfaces is the flux zero?
    a) A
    b) B
    c) C
    d) D

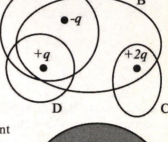

8. Consider a conductor with a net charge of +2 μC, surrounding a collection of point charges, as shown in the Figure. What is the total charge on the inner surface of the conductor?
    a) -1 μC
    b) +6 μC
    c) -2 μC
    d) -3 μC
    e) +2 μC
    f) 0 μC

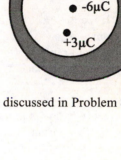

9. What is the total charge on the outer surface of the conductor for the charge distribution discussed in Problem 8?
    a) -1 μC
    b) +6 μC
    c) -2 μC
    d) -3 μC
    e) +2 μC
    f) 0 μC

10. Consider a point charge Q located at the origin of our coordinate system. A cube of sides L is located in this coordinate system such that its lower left corner coincides with the origin (see Figure). Through which side of the cube is the electric flux equal to 0?
    a) The side parallel to the xy plane, going through $(0, 0, L)$.
    b) The side parallel to the yz plane, going through $(L, 0, 0)$.
    c) The side parallel to the xz plane, going through $(0, L, 0)$.
    d) The side parallel to the xz plane, going through $(0, 0, 0)$.

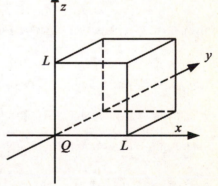

11. A uniform electric field of magnitude 30.4 μN/C passes through an equilateral triangle with 13.8-cm sides. The electric field is in a direction that makes a 37.5° angle with the perpendicular to the plane of the triangle. What is the electric flux through the triangle?

12. A cubical box with edges 3.2 cm long has an electric field of magnitude 0.46 N/C on one of its faces pointing directly into the cube. The box contains an arrangement of charge such that the electric field is zero on all other faces. What is the total charge within the box?

13. A distribution of charge is spherically symmetric. The charge density as a function of distance from the center of the distribution is given by

$$\rho(r) = \begin{cases} \rho_0 (R-r)^2 / R^2 & \text{if} \quad r < R \\ 0 & \text{if} \quad r > R \end{cases}$$

Determine the electric field for $r < R$ and for $r > R$.

14. An infinite slab of charge lies in the yz plane from $-L/2 < x < L/2$. On either side of the slab, the electric field is zero. Within the slab is an electric field given by

$$\vec{E}(x) = E_0 \left(1 - x^2\right)\hat{i}$$

What is the charge density in the slab as a function of x?

15. The electric field in a region of space is given by $\vec{E} = E_0 \cos(\pi y / 2L)\cos(\pi z / 2L)\hat{i}$. What is the electric flux through a square with corners $(0,0,0)$, $(0,L,0)$, $(0,L,L)$, $(0,0,L)$?

Responses to Select End-of-Chapter Questions

1. No. If the net electric flux through a surface is zero, then the net charge contained in the surface is zero. However, there may be charges both inside and outside the surface that affect the electric field at the surface. The electric field could point outward from the surface at some points and inward at others. Yes. If the electric field is zero for all points on the surface, then the net flux through the surface must be zero and no net charge is contained within the surface.

7. No. Gauss's law is most useful in cases of high symmetry, where a surface can be defined over which the electric field has a constant value and a constant relationship to the direction of the outward normal to the surface. Such a surface cannot be defined for an electric dipole.

13. Yes. The charge q will induce a charge $-q$ on the inside surface of the thin metal shell, leaving the outside surface with a charge $+q$. The charge Q outside the sphere will feel the same electric force as it would if the metal shell were not present.

Solutions to Select End-of-Chapter Problems

1. The electric flux of a uniform field is given by Eq. 22-1b.

(a) $\Phi_E = \vec{E} \cdot \vec{A} = EA\cos\theta = (580\,\text{N/C})\pi(0.13\,\text{m})^2 \cos 0 = \boxed{31\,\text{N·m}^2/\text{C}}$

(b) $\Phi_E = \vec{E} \cdot \vec{A} = EA\cos\theta = (580\,\text{N/C})\pi(0.13\,\text{m})^2 \cos 45° = \boxed{22\,\text{N·m}^2/\text{C}}$

(c) $\Phi_E = \vec{E} \cdot \vec{A} = EA\cos\theta = (580\,\text{N/C})\pi(0.13\,\text{m})^2 \cos 90° = \boxed{0}$

7. (a) Use Gauss's law to determine the electric flux.

$$\Phi_E = \frac{Q_{encl}}{\varepsilon_o} = \frac{-1.0 \times 10^{-6}\,\text{C}}{8.85 \times 10^{-12}\,\text{C}^2/\text{N·m}^2} = \boxed{-1.1 \times 10^5\,\text{N·m}^2/\text{C}}$$

(b) Since there is no charge enclosed by surface A_2, $\Phi_E = \boxed{0}$.

13. The electric field can be calculated by Eq. 21-4a, and that can be solved for the magnitude of the charge.

$$E = k\frac{Q}{r^2} \quad \rightarrow \quad Q = \frac{Er^2}{k} = \frac{\left(6.25\times10^2\,\text{N/C}\right)\left(3.50\times10^{-2}\,\text{m}\right)^2}{8.988\times10^9\,\text{N}\cdot\text{m}^2/\text{C}^2} = 8.52\times10^{-11}\,\text{C}$$

This corresponds to about 5×10^8 electrons. Since the field points toward the ball, the charge must be negative. Thus $Q = \boxed{-8.52\times10^{-11}\,\text{C}}$.

19. For points inside the nonconducting spheres, the electric field will be determined by the charge inside the spherical surface of radius r.

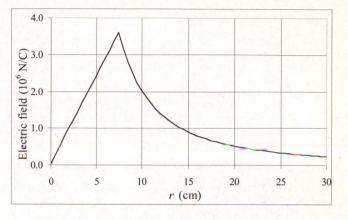

$$Q_{\text{encl}} = Q\left(\frac{\frac{4}{3}\pi r^3}{\frac{4}{3}\pi r_0^3}\right) = Q\left(\frac{r}{r_0}\right)^3$$

The electric field for $r \le r_0$ can be calculated from Gauss's law.

$$E(r \le r_0) = \frac{Q_{\text{encl}}}{4\pi\varepsilon_0 r^2}$$

$$= Q\left(\frac{r}{r_0}\right)^3 \frac{1}{4\pi\varepsilon_0 r^2} = \left(\frac{Q}{4\pi\varepsilon_0 r_0^3}\right)r$$

The electric field outside the sphere is calculated from Gauss's law with $Q_{\text{encl}} = Q$.

$$E\left(r \ge r_0\right) = \frac{Q_{\text{encl}}}{4\pi\varepsilon_0 r^2} = \frac{Q}{4\pi\varepsilon_0 r^2}$$

25. Example 22-7 gives the electric field from a positively charged plate as $E = \sigma/2\varepsilon_0$ with the field pointing away from the plate. The fields from the two plates will add, as shown in the figure.
(a) Between the plates the fields are equal in magnitude, but point in opposite directions.

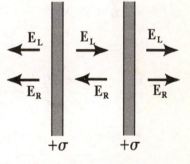

$$E_{\text{between}} = \frac{\sigma}{2\varepsilon_0} - \frac{\sigma}{2\varepsilon_0} = \boxed{0}$$

(b) Outside the two plates the fields are equal in magnitude and point in the same direction.

$$E_{\text{outside}} = \frac{\sigma}{2\varepsilon_0} + \frac{\sigma}{2\varepsilon_0} = \boxed{\frac{\sigma}{\varepsilon_0}}$$

(c) When the plates are conducting the charge lies on the surface of the plates. For nonconducting plates the same charge will be spread across the plate. This will not affect the electric field between or outside the two plates. It will, however, allow for a non-zero field inside each plate.

31. (*a*) Create a gaussian surface that just encloses the inner surface of the spherical shell. Since the electric field inside a conductor must be zero, Gauss's law requires that the enclosed charge be zero. The enclosed charge is the sum of the charge at the center and charge on the inner surface of the conductor.

$$Q_{enc} = q + Q_{inner} = 0$$

Therefore $Q_{inner} = \boxed{-q}$.

(*b*) The total charge on the conductor is the sum of the charges on the inner and outer surfaces.

$$Q = Q_{outer} + Q_{inner} \rightarrow Q_{outer} = Q - Q_{inner} = \boxed{Q + q}$$

(*c*) A gaussian surface of radius $r < r_1$ only encloses the center charge, q. The electric field will therefore be the field of the single charge.

$$\boxed{E(r < r_1) = \frac{q}{4\pi\varepsilon_0 r^2}}$$

(*d*) A gaussian surface of radius $r_1 < r < r_0$ is inside the conductor so $\boxed{E = 0}$.

(*e*) A gaussian surface of radius $r > r_0$ encloses the total charge $q + Q$. The electric field will then be the field from the sum of the two charges.

$$\boxed{E(r > r_0) = \frac{q + Q}{4\pi\varepsilon_0 r^2}}$$

37. (*a*) The final speed can be calculated from the work-energy theorem, where the work is the integral of the force on the electron between the two shells.

$$W = \int \vec{F} \cdot d\vec{r} = \tfrac{1}{2}mv^2 - \tfrac{1}{2}mv_0^2$$

Setting the force equal to the electric field times the charge on the electron, and inserting the electric field from Problem 36 gives the work done on the electron.

$$W = \int_{R_1}^{R_2} \frac{qQ}{2\pi\varepsilon_0 \ell R} dR = \frac{qQ}{2\pi\varepsilon_0 \ell} \ln\left(\frac{R_2}{R_1}\right)$$

$$= \frac{\left(-1.60\times10^{-19}\,\text{C}\right)\left(-0.88\,\mu\text{C}\right)}{2\pi\left(8.85\times10^{-12}\,\text{C}^2/\text{Nm}^2\right)\left(5.0\,\text{m}\right)} \ln\left(\frac{9.0\,\text{cm}}{6.5\,\text{cm}}\right) = 1.65\times10^{-16}\,\text{J}$$

Solve for the velocity from the work-energy theorem.

$$v = \sqrt{\frac{2W}{m}} = \sqrt{\frac{2\left(1.65\times10^{-16}\,\text{J}\right)}{9.1\times10^{-31}\,\text{kg}}} = \boxed{1.9\times10^7\,\text{m/s}}$$

(*b*) The electric force on the proton provides its centripetal acceleration.

$$F_c = \frac{mv^2}{R} = qE = \frac{|qQ|}{2\pi\varepsilon_0 \ell R}$$

The velocity can be solved for from the centripetal acceleration.

$$v = \sqrt{\frac{\left(1.60\times10^{-19}\,\text{C}\right)\left(0.88\,\mu\text{C}\right)}{2\pi\left(8.85\times10^{-12}\,\text{C}^2/\text{Nm}^2\right)\left(1.67\times10^{-27}\,\text{kg}\right)\left(5.0\,\text{m}\right)}} = \boxed{5.5\times10^5\,\text{m/s}}$$

Note that as long as the proton is between the two cylinders, the velocity is independent of the radius.

43. (*a*) Choose a cylindrical gaussian surface with the flat ends parallel to and equidistant from the slab. By symmetry the electric field must point perpendicularly away from the slab, resulting in no flux passing through the curved part of the gaussian cylinder. By symmetry the flux through each end of the cylinder must be equal with the electric field constant across the surface.

$$\oint \vec{E} \cdot d\vec{A} = 2EA$$

The charge enclosed by the surface is the charge density of the slab multiplied by the volume of the slab enclosed by the surface.

$$q_{enc} = \rho_E \left(Ad \right)$$

Gauss's law can then be solved for the electric field.

$$\oint \vec{E} \cdot d\vec{A} = 2EA = \frac{\rho_E A d}{\varepsilon_0} \quad \rightarrow \quad \boxed{E = \frac{\rho_E d}{2\varepsilon_0}}$$

Note that this electric field is independent of the distance from the slab.

(*b*) When the coordinate system of this problem is changed to axes parallel $\left(\hat{z} \right)$ and perpendicular $\left(\hat{r} \right)$ to the slab, it can easily be seen that the particle will hit the slab if the initial perpendicular velocity is sufficient for the particle to reach the slab before the acceleration decreases its velocity to zero. In the new coordinate system, the axes are rotated by 45°.

$$\vec{r}_0 = y_0 \cos 45° \hat{r} + y_0 \sin 45° \hat{z} = \frac{y_0}{\sqrt{2}} \hat{r} + \frac{y_0}{\sqrt{2}} \hat{z}$$

$$\vec{v}_0 = -v_0 \sin 45° \hat{r} + v_0 \cos 45° \hat{z} = -\frac{v_0}{\sqrt{2}} \hat{r} + \frac{v_0}{\sqrt{2}} \hat{z}$$

$$\vec{a} = qE / m \hat{r}$$

The perpendicular components are then inserted into Eq. 2-12c, with the final velocity equal to zero.

$$0 = v_{r0}^2 - 2a(r - r_0) = \frac{v_0^2}{2} - 2\frac{q}{m}\left(\frac{\rho_E d}{2\varepsilon_0} \right)\left(\frac{y_0}{\sqrt{2}} - 0 \right)$$

Solving for the velocity gives the minimum speed that the particle can have to reach the slab.

$$\boxed{v_0 \geq \sqrt{\frac{\sqrt{2} q \rho_E d y_0}{m \varepsilon_0}}}$$

49. The symmetry of the charge distribution allows the electric field inside the sphere to be calculated using Gauss's law with a concentric gaussian sphere of radius $r \leq r_0$. The enclosed charge will be found by integrating the charge density over the enclosed volume.

$$Q_{encl} = \int \rho_E dV = \int_0^r \rho_0 \left(\frac{r'}{r_0} \right) 4\pi r'^2 dr' = \frac{\rho_0 \pi r^4}{r_0}$$

The enclosed charge can be written in terms of the total charge by setting $r = r_0$ and solving for the charge density in terms of the total charge.

$$Q = \frac{\rho_0 \pi r_0^4}{r_0} = \rho_0 \pi r_0^3 \quad \rightarrow \quad \rho_0 = \frac{Q}{\pi r_0^3} \quad \rightarrow \quad Q_{encl}(r) = \frac{\rho_0 \pi r^4}{r_0} = Q \left(\frac{r}{r_0} \right)^4$$

The electric field is then found from Gauss's law

$$\oint \vec{E} \cdot d\vec{A} = \frac{Q_{encl}}{\varepsilon_0} \quad \rightarrow \quad E\left(4\pi r^2\right) = \frac{Q}{\varepsilon_0}\left(\frac{r}{r_0}\right)^4 \quad \rightarrow \quad \boxed{E = \frac{Q}{4\pi\varepsilon_0}\frac{r^2}{r_0^4}}$$

The electric field points $\boxed{\text{radially outward}}$ since the charge distribution is positive.

55. The flux through a gaussian surface depends only on the charge enclosed by the surface. For both of these spheres the two point charges are enclosed within the sphere. Therefore the flux is the same for both spheres.

$$\Phi = \frac{Q_{encl}}{\varepsilon_0} = \frac{\left(9.20 \times 10^{-9}\,\text{C}\right) + \left(-5.00 \times 10^{-9}\,\text{C}\right)}{8.85 \times 10^{-12}\,\text{C}^2/\text{N}\cdot\text{m}^2} = \boxed{475\,\text{N}\cdot\text{m}^2/\text{C}}$$

61. Consider this sphere as a combination of two spheres. Sphere 1 is a solid sphere of radius r_0 and charge density ρ_E centered at A and sphere 2 is a second sphere of radius $r_0/2$ and density $-\rho_E$ centered at C.

(a) The electric field at A will have zero contribution from sphere 1 due to its symmetry about point A. The electric field is then calculated by creating a gaussian surface centered at point C with radius $r_0/2$.

$$\oint \vec{E} \cdot d\vec{A} = \frac{q_{enc}}{\varepsilon_0} \quad \rightarrow \quad E \cdot 4\pi\left(\tfrac{1}{2}r_0\right)^2 = \frac{\left(-\rho_E\right)\frac{4}{3}\pi\left(\frac{1}{2}r_0\right)^3}{\varepsilon_0} \quad \rightarrow \quad \boxed{E = -\frac{\rho_E r_0}{6\varepsilon_0}}$$

Since the electric field points into the gaussian surface (negative) the electric field at point A $\boxed{\text{points to the right}}$.

(b) At point B the electric field will be the sum of the electric fields from each sphere. The electric field from sphere 1 is calculated using a gaussian surface of radius r_0 centered at A.

$$\oint \vec{E}_1 \cdot d\vec{A} = \frac{q_{enc}}{\varepsilon_0} \quad \rightarrow \quad E_1 \cdot 4\pi r_0^2 = \frac{\frac{4}{3}\pi r_0^3\left(\rho_E\right)}{\varepsilon_0} \quad \rightarrow \quad E_1 = \frac{\rho_E r_0}{3\varepsilon_0}$$

At point B the field from sphere 1 points toward the left. The electric field from sphere 2 is calculated using a gaussian surface centered at C of radius $3r_0/2$.

$$\oint \vec{E}_2 \cdot d\vec{A} = \frac{q_{enc}}{\varepsilon_0} \quad \rightarrow \quad E_2 \cdot 4\pi\left(\tfrac{3}{2}r_0\right)^2 = \frac{\left(-\rho_E\right)\frac{4}{3}\pi\left(\frac{1}{2}r_0\right)^3}{\varepsilon_0} \quad \rightarrow \quad E_2 = -\frac{\rho_E r_0}{54\varepsilon_0}$$

At point B, the electric field from sphere 2 points toward the right. The net electric field is the sum of these two fields. The net field $\boxed{\text{points to the left}}$.

$$E = E_1 + E_2 = \frac{\rho_E r_0}{3\varepsilon_0} + \frac{-\rho_E r_0}{54\varepsilon_0} = \boxed{\frac{17\rho_E r_0}{54\varepsilon_0}}.$$

67. The flux is the sum of six integrals, each of the form $\iint \vec{E} \cdot d\vec{A}$. Because the electric field has only x and y components, there will be no flux through the top or bottom surfaces. For the other faces, we choose a vertical strip of height a and width dy (for the front and back faces) or dx (for the left and right faces). See the diagram for an illustration of a strip on the front face. The total flux is then calculated, and used to determine the enclosed charge.

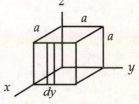

$$\Phi_{\substack{front \\ (x=a)}} = \int_0^a \left(E_{x0} e^{-\left(\frac{x+y}{a}\right)^2} \hat{\mathbf{i}} + E_{y0} e^{-\left(\frac{x+y}{a}\right)^2} \hat{\mathbf{j}} \right) \cdot a\,dy\,\hat{\mathbf{i}} = aE_{x0} \int_0^a e^{-\left(\frac{a+y}{a}\right)^2} dy$$

This integral does not have an analytic anti-derivative, and so must be integrated numerically. We approximate the integral by a sum: $\int_0^a e^{-\left(\frac{a+y}{a}\right)^2} dy \approx \sum_{i=1}^n e^{-\left(\frac{a+y_i}{a}\right)^2} \Delta y$. The region of integration is divided into n elements, and so $\Delta y = \frac{a-0}{n}$ and $y_i = i\Delta y$. We initially evaluate the sum for $n = 10$. Then we evaluate it for $n = 20$. If the two sums differ by no more than 2%, we take that as the value of the integral. If they differ by more than 2%, we choose a larger n, compute the sum, and compare that to the result for $n = 20$. We continue until a difference of 2% or less is reached. This integral, for $n = 100$ and $a = 1.0$ m, is 0.1335 m. So we have this intermediate result.

$$\Phi_{\substack{\text{front} \\ (x=a)}} = aE_{x0} \sum_{i=1}^n e^{-\left(\frac{a+y_i}{a}\right)^2} \Delta y = (1.0\,\text{m})(50\,\text{N/C})(0.1335\,\text{m}) = 6.675\,\text{N} \cdot \text{m}^2/\text{C}$$

Now do the integral over the back face.

$$\Phi_{\substack{\text{back} \\ (x=0)}} = \int_0^a \left(E_{x0} e^{-\left(\frac{x+y}{a}\right)^2} \hat{\mathbf{i}} + E_{y0} e^{-\left(\frac{x+y}{a}\right)^2} \hat{\mathbf{j}} \right) \cdot \left(-a\,dy\,\hat{\mathbf{i}} \right) = -aE_{x0} \int_0^a e^{-\left(\frac{y}{a}\right)^2} dy$$

We again get an integral that cannot be evaluated analytically. A similar process to that used for the front face is applied again, and so we make this approximation: $-aE_{x0} \int_0^a e^{-\left(\frac{y}{a}\right)^2} dy \approx -aE_{x0} \sum_{i=1}^n e^{-\left(\frac{y_i}{a}\right)^2} \Delta y$.

The numeric integration gives a value of 0.7405 m.

$$\Phi_{\substack{\text{back} \\ (x=0)}} = -aE_{x0} \sum_{i=1}^n e^{-\left(\frac{y_i}{a}\right)^2} \Delta y = -(1.0\,\text{m})(50\,\text{N/C})(0.7405\,\text{m}) = -37.025\,\text{N} \cdot \text{m}^2/\text{C}.$$

Now consider the right side.

$$\Phi_{\substack{\text{right} \\ (y=a)}} = \int_0^a \left(E_{x0} e^{-\left(\frac{x+y}{a}\right)^2} \hat{\mathbf{i}} + E_{y0} e^{-\left(\frac{x+y}{a}\right)^2} \hat{\mathbf{j}} \right) \cdot a\,dx\,\hat{\mathbf{j}} = aE_{y0} \int_0^a e^{-\left(\frac{x+a}{a}\right)^2} dx$$

Notice that the same integral needs to be evaluated as for the front side. All that has changed is the variable name. Thus we have the following.

$$\Phi_{\substack{\text{right} \\ (y=a)}} = aE_{y0} \int_0^a e^{-\left(\frac{x+a}{a}\right)^2} dx \approx (1.0\,\text{m})(25\,\text{N/C})(0.1335\,\text{m}) = 3.3375\,\text{N} \cdot \text{m}^2/\text{C}$$

Finally, do the left side, following the same process. The same integral arises as for the back face.

$$\Phi_{\substack{\text{left} \\ (y=0)}} = \int_0^a \left(E_{x0} e^{-\left(\frac{x+y}{a}\right)^2} \hat{\mathbf{i}} + E_{y0} e^{-\left(\frac{x+y}{a}\right)^2} \hat{\mathbf{j}} \right) \cdot \left(-a\,dx\,\hat{\mathbf{j}} \right) = -aE_{y0} \int_0^a e^{-\left(\frac{x}{a}\right)^2} dx$$

$$\approx -(1.0\,\text{m})(25\,\text{N/C})(0.7405\,\text{m}) = -18.5125\,\text{N} \cdot \text{m}^2/\text{C}$$

Sum to find the total flux, and multiply by ε_0 to find the enclosed charge.

$$\Phi_{\text{total}} = \Phi_{\text{front}} + \Phi_{\text{back}} + \Phi_{\text{right}} + \Phi_{\text{left}} + \Phi_{\text{top}} + \Phi_{\text{bottom}}$$

$$= (6.675 - 37.025 + 3.3375 - 18.5125)\,\text{N} \cdot \text{m}^2/\text{C} = -45.525\,\text{N} \cdot \text{m}^2/\text{C} \approx \boxed{-46\,\text{N} \cdot \text{m}^2/\text{C}}$$

$$Q_{\text{encl}} = \varepsilon_0 \Phi_{\text{total}} = (8.85 \times 10^{-12}\,\text{C}^2/\text{N} \cdot \text{m}^2)(-45.525\,\text{N} \cdot \text{m}^2/\text{C}) = \boxed{-4.0 \times 10^{-10}\,\text{C}}$$

Chapter 23: Electric Potential

Chapter Overview and Objectives

In this chapter, the concept of electric potential is introduced. The relationship between the electric potential and the work done in moving charged particles in electric fields, the relationship between electric fields and electric potential, and the electric potential of point charges and continuous charge distributions of are discussed. The concept of electric dipoles and their approximate potential energy are discussed.

After completing this chapter you should:
- Know the definition of electric potential and be able to relate the work done by electrical fields to electric potential.
- Know how to calculate the electric potential of a point charge.
- Be able to calculate the electric potential difference between different points in space for a collection of point charges.
- Be able to calculate the electric potential of a continuous charge distribution.
- Know both the integral and differential relationships between electric field and electric potential and be able to calculate electric field from electric potential and electric potential from electric field.
- Know what an electric dipole is and how to calculate its dipole moment.
- Know how to calculate the approximate electric potential of an electric dipole.

Summary of Equations

Definition of electric potential:
$$V_a = \frac{U_a}{q}$$
(Section 23-1)

Relationship between electric potential difference and work done by the electric field:
$$V_{ba} = V_b - V_a = -\frac{W_{ba}}{q}$$
(Section 23-1)

Integral relationship between electric potential and electric field:
$$V_b - V_a = -\int_a^b \vec{\mathbf{E}} \cdot d\vec{\mathbf{l}}$$
(Section 23-2)

Electric potential of a point charge:
$$V(r) = \frac{1}{4\pi\varepsilon_0} \frac{Q}{r}$$
(Section 23-3)

Electric potential due to a continuous charge distribution:
$$V_a = \frac{1}{4\pi\varepsilon_0} \int \frac{dq}{r} = \frac{1}{4\pi\varepsilon_0} \int \frac{\rho}{r} dV$$
(Section 23-4)

Definition of magnitude of dipole moment:
$$p = Q\ell$$
(Section 23-6)

Approximate electric potential of a dipole:
$$V(x,y,z) \approx \frac{1}{4\pi\varepsilon_0} \frac{p\cos\theta}{r^2}$$
(Section 23-6)

Obtaining the electric field from the potential:
$$E_x = -\frac{\partial V}{\partial x} \qquad E_y = -\frac{\partial V}{\partial y} \qquad E_z = -\frac{\partial V}{\partial z}$$
(Section 23-7)

Chapter Summary

Section 23-1. Electric Potential Energy and Potential Difference

As a charge q is moved through a static electric field, work is done by the electric force acting on the charge. Since the static electric force is a conservative force, the work done does not depend on the path taken but only on the initial and final positions. Because the static electric force is conservative, we can determine the potential energy U_a at point a. The **electric potential** at point a, V_a, is defined as the potential energy at point a per unit charge:

$$V_a = \frac{U_a}{q}$$

The SI unit for electric potential is J/C. A more commonly used unit for electric potential is the **volt** (V) defined as

$$1\,V = 1\,J/C$$

Since only changes in electric potential have a physical meaning, we can add a constant to the electric potential for all points in space and still have a correct potential energy function.

The work done by the electric field in moving a charge q from point a to point b, W_{ab}, is equal to the negative of the change in the potential energy. We can write this in terms of the electric potential at point a, V_a, and at point b, V_b, or the **potential difference** V_{ba}:

$$W_{ab} = -\Delta U_{ab} = -q(V_b - V_a) = -qV_{ba}$$

Example 23-1-A. Calculating the electric potential. The electric potential at a given position in space is $V_1 = 256$ V. A particle with mass of $m = 3.46 \times 10^{-8}$ kg and charge $q = -4.63$ μC starts at this location from rest and moves to another location where it has a speed $v_2 = 453$ m/s. What is the electric potential V_2 at the new location?

Approach: Since only conservative forces are involved, energy is conserved. The problem allows us to calculate the change in the kinetic energy of the particle, which must be opposite to the electrostatic potential energy. Since the initial electric potential is provided, we can determine the final electric potential.

Solution: Conservation of energy tells us that the change in the kinetic energy plus the change in the potential energy is equal to 0:

$$\Delta K + \Delta U = 0 \qquad \Rightarrow \qquad \frac{1}{2}mv_2^2 - \frac{1}{2}mv_1^2 + q(V_2 - V_1) = 0$$

This equation contains only one unknown parameter, V_2, which can now be determined:

$$V_2 = -\frac{\frac{1}{2}mv_2^2 - \frac{1}{2}mv_1^2}{q} + V_1 = -\frac{\frac{1}{2}\left(3.46 \times 10^{-8}\right)(453)^2 - 0}{-4.63 \times 10^{-6}} + 256 = 1.02 \times 10^3 \,V$$

Section 23-2. Relation between Electric Potential and Electric Field

The relationship between the change of the potential energy and the electrostatic force can be obtained from the definition of the work done by this force:

$$U_b - U_a = -W_{ba} = -\int_a^b \vec{F} \cdot d\vec{\ell}$$

If we divide both sides of this equation by the charge q we get a relationship between the change in the electric potential and the electric field:

$$\frac{U_b - U_a}{q} = -\frac{\int_a^b \vec{F} \cdot d\vec{\ell}}{q} \qquad \Rightarrow \qquad V_b - V_a = -\int_a^b \vec{E} \cdot d\vec{\ell}$$

In the case of a uniform electric field, this simplifies to

$$V_b - V_a = Ed$$

where E is the magnitude of the electric field and d is the distance between a and b in the direction of the electric field.

Example 23-2-A. Calculating the electric potential for a given electric field. The electric field in a region of space is given by $\vec{E}(x,y) = Axy\,\hat{\mathbf{i}} + Bx^2y\,\hat{\mathbf{j}}$. Determine the electric potential difference between $(0,0)$ and $(1,1)$.

Approach: To determine the electric potential difference we choose a convenient path to get from $(0,0)$ to $(1,1)$. Since the electric force is a conservative force, the force and field integrals only depend on the initial and final positions.

Solution: We know the electric potential difference between two points in space can be obtained from the path integral of the electric field:

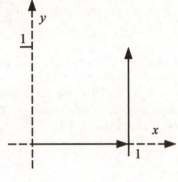

$$V_b - V_a = -\int_a^b \vec{E} \cdot d\vec{\ell}$$

Consider the path shown in the Figure, first moving along the x-axis from $x = 0$ to $x = 1$ and then along the line $x = 1$ from $y = 0$ to $y = 1$. Along the first segment of the path, $\vec{E} = 0$ and the path integral along this segment is thus 0. Along the second segment of the path, $d\vec{\ell}$ is in the $\hat{\mathbf{j}}$ direction, and we write $d\vec{\ell} = (dy)\hat{\mathbf{j}}$. The path integral becomes

$$V_b - V_a = -\int_a^b \vec{E} \cdot d\vec{\ell} = -\int_{y=0}^{1} \left(Ay\,\hat{\mathbf{i}} + By\,\hat{\mathbf{j}} \right) \cdot \hat{\mathbf{j}}\,dy =$$

$$= -\int_{y=0}^{1} By\,dy = -\frac{1}{2} By^2 \Big|_0^1 = -\frac{1}{2} B$$

Section 23-3. Electric Potential Due to Point Charges

It is relatively easy to derive the potential energy of a point charge. For a point charge Q, located at the origin of our coordinate system, the electric field is given by

$$\vec{E} = \frac{1}{4\pi\varepsilon_0} \frac{Q}{r^2} \hat{\mathbf{r}}$$

where r is the distance from the charge and $\hat{\mathbf{r}}$ is the unit vector that points from the charge to the point at which the electric field is being evaluated. The change in the electric potential between position $\vec{r}_a$ and position $\vec{r}_b$ is

$$V_b - V_a = -\int_{r_a}^{r_b} \vec{E} \cdot d\vec{\ell} = -\frac{1}{4\pi\varepsilon_0} \int_{r_a}^{r_b} \frac{Q}{r^2} \hat{\mathbf{r}} \cdot d\vec{\ell} =$$

$$= -\frac{1}{4\pi\varepsilon_0} \int_{r_a}^{r_b} \frac{Q}{r^2}\,dr = \frac{1}{4\pi\varepsilon_0} \left(\frac{Q}{r_b} - \frac{Q}{r_a} \right)$$

where dr is the change in the distance from the charge when moving along the path element $d\vec{\ell}$. The electric potential at position $\vec{r}$ is equal to

$$V(r) = \frac{1}{4\pi\varepsilon_0} \frac{Q}{r} + C$$

where C is an arbitrary constant potential. In general, the constant C is chosen such that the electric potential goes to zero when $\vec{r}$ goes to infinity. In this case, $C = 0$ and $V(r)$ becomes

$$V(r) = \frac{1}{4\pi\varepsilon_0} \frac{Q}{r}$$

Example 23-3-A. Calculating the change in the electric potential and the electric potential energy. Two charges are located a distance 1.00 m apart, as shown in the Figure on the right. If a 10.0 μC charge is moved from infinity to the position (1,1), what will be its change in electric potential energy?

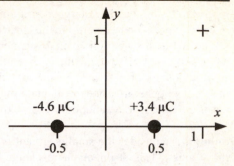

Approach: To solve this problem we use the principle of superposition. The electric potential at (1,1) is the sum of the electric potential at this point due to each of the two existing charges. Once we have determined the electric potential at (1,1) we can quickly determine the electric potential energy when the 10.0 μC charge is moved from infinity to this position.

Solution: The electric potential at an infinite distance from the two charges will be zero. The electric potential at the position (1, 1) will be the sum of the electric potentials due to each charge. The distance between the -4.6 μC charge and (1, 1) is equal to

$$r_1 = \sqrt{(x-x_1)^2 + (y-y_1)^2} = \sqrt{[1-(-0.5)]^2 + [1-0]^2} = 1.80 \, \text{m}$$

The distance between the +3.4 μC charge and (1,1) is equal to

$$r_2 = \sqrt{(x-x_2)^2 + (y-y_2)^2} = \sqrt{[1-(0.5)]^2 + [1-0.0]^2} = 1.12 \, \text{m}$$

The electric potential at (1, 1) is thus equal to

$$V = \frac{1}{4\pi\varepsilon_0}\frac{Q_1}{r_1} + \frac{1}{4\pi\varepsilon_0}\frac{Q_2}{r_2} = \frac{(9.0\times10^9)(-4.6\times10^{-6})}{1.80} + \frac{(9.0\times10^9)(3.4\times10^{-6})}{1.12} = 4.3\times10^3 \, \text{V}$$

The potential energy of the 10.0 μC charge located at (1,1) can now be determined:

$$U = QV = (10.0\times10^{-6} \, \text{C})(4.3\times10^3 \, \text{V}) = 4.3\times10^{-2} \, \text{J}$$

As the potential at an infinite distance is zero, the potential difference and the electric potential energy difference are the same as the values calculated for these quantities at the final location.

Section 23-4. Potential Due to Any Charge Distribution

Since the net electric field at a certain position due to a set of point charges is the vector sum of the electric fields due to the individual point charges, the net electric potential at this position due to this set of point charges will be the sum of the electric potentials due to the individual point charges. We write the potential at point a, V_a, due to n other charges as

$$V_a = \sum_{i=1}^{n} V_{ia} = \frac{1}{4\pi\varepsilon_0}\sum_{i=1}^{n}\left(\frac{Q_i}{r_{ia}}\right)$$

where i is an index that labels each of the n charges, V_{ia} is the electric potential at position a due to charge Q_i, and r_i is the distance from charge Q_i to point a. If the charge distribution is continuous, the sum is replaced by an integral:

$$V_a = \frac{1}{4\pi\varepsilon_0}\int\frac{dq}{r} = \frac{1}{4\pi\varepsilon_0}\int\frac{\rho}{r}dV$$

where r is the distance between charge dq and point a and ρ is the charge density in volume dV.

Example 23-4-A. Calculating V for a continuous charge distribution. Consider a linear charge distribution, located on the x-axis between $x = -L$ and $x = +L$. The linear charge density, as a function of position, is given by

$$\lambda(x) = \frac{\lambda_0}{L}|x|$$

Determine the electric potential as a function of the position along the y-axis.

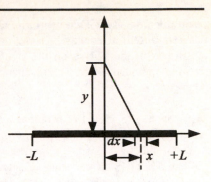

Approach: Since the charge distribution is continuous, we need to use the integral expression to determine the electric potential on the y axis. We will break up the charge distribution in small segments with length dx and integrate over the contributions of these segments to determine the total electric potential. Since the electric potential is a scalar, we do not have to worry about vector addition. This makes it in general easier to calculate the electric potential than to determine the electric field due to a continuous charge distribution.

Solution: Consider the small segment of length dx, located at a distance x from the y axis. This segment has a charge $dq = \lambda dx$. The distance between this segment and a point on the y axis is

$$r = \sqrt{x^2 + y^2}$$

The electric potential dV on the y axis due to this segment is thus equal to

$$dV = \frac{1}{4\pi\varepsilon_0} \frac{dq}{r} = \frac{1}{4\pi\varepsilon_0} \frac{\lambda dx}{\sqrt{x^2 + y^2}} = \frac{1}{4\pi\varepsilon_0} \frac{\lambda_0}{L} \frac{|x| dx}{\sqrt{x^2 + y^2}}$$

The total electric potential on the y axis can be found by integrating this expression between $x = -L$ and $x = L$:

$$V = \int dV = \frac{1}{4\pi\varepsilon_0} \frac{\lambda_0}{L} \int_{-L}^{L} \frac{|x| dx}{\sqrt{x^2 + y^2}} = \frac{1}{4\pi\varepsilon_0} \frac{\lambda_0}{L} \left\{ -\int_{-L}^{0} \frac{x dx}{\sqrt{x^2 + y^2}} + \int_{0}^{L} \frac{x dx}{\sqrt{x^2 + y^2}} \right\} =$$

$$= \frac{1}{4\pi\varepsilon_0} \frac{\lambda_0}{L} \left\{ -\sqrt{x^2 + y^2} \Big|_{-L}^{0} + \sqrt{x^2 + y^2} \Big|_{0}^{L} \right\} = \frac{1}{2\pi\varepsilon_0} \frac{\lambda_0}{L} \sqrt{L^2 + y^2}$$

Section 23-5. Equipotential Surfaces

Surfaces at the same electric potential are called **equipotential surfaces**. It is often useful to draw equipotential surfaces for a given arrangement of charges. The dashed curves in the Figure show the location of equipotential surfaces for a pair of opposite charges. The solid lines are the electric field lines. The electric field lines are always perpendicular to the equipotential surfaces. If a contour map of equipotential surfaces is drawn with equal potential differences between the different contour lines, the magnitude of the electric field at a certain position is proportional to the density of equipotential lines at that location.

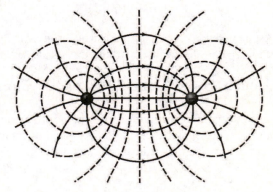

The surface of a conductor is an equipotential surface and the electric field always points perpendicular to the surface of a conductor when there is a static charge distribution.

Section 23-6. Electric Dipole Potential

A particular arrangement of charge that is often encountered in systems of interest or is often a good approximation to systems of interest is the **electric dipole**. An electric dipole is defined to be two charges of equal magnitude Q but opposite sign, separated by a distance ℓ, as shown in the Figure. The **dipole moment** $\vec{p}$ is a vector that points from the position of the negative charge to the position of the positive charge. The magnitude of the dipole moment is $p = Q\ell$.

The exact electric potential due to a dipole is easily calculated by using the principle of superposition. The electric potential at P due to charge Q is

$$V_Q = \frac{1}{4\pi\varepsilon_0} \frac{Q}{r}$$

The electric potential at P due to charge $-Q$ is

$$V_{-Q} = -\frac{1}{4\pi\varepsilon_0}\frac{Q}{r+\Delta r}$$

The total electric potential at P is the sum of the potentials due to charges Q and $-Q$:

$$V_P = V_Q + V_{-Q} = \frac{1}{4\pi\varepsilon_0}\frac{Q}{r} - \frac{1}{4\pi\varepsilon_0}\frac{Q}{r+\Delta r} = \frac{1}{4\pi\varepsilon_0}\frac{Q}{r}\left\{1 - \frac{r}{r+\Delta r}\right\} = \frac{1}{4\pi\varepsilon_0}\frac{Q}{r}\frac{\Delta r}{r+\Delta r}$$

When P is far from the charge distribution, we can approximate $r + \Delta r$ with r. The distance Δr can be expressed in terms of the angle θ and the distance ℓ between the charges: $\Delta r = \ell \cos\theta$. The electric potential at P is thus approximately equal to

$$V_P = \frac{1}{4\pi\varepsilon_0}\frac{Q}{r}\frac{\Delta r}{r+\Delta r} \approx \frac{1}{4\pi\varepsilon_0}\frac{Q}{r}\frac{\ell\cos\theta}{r} = \frac{1}{4\pi\varepsilon_0}\frac{p\cos\theta}{r^2}$$

Note that θ is the angle between the direction of the dipole moment and the position vector pointing from the dipole to P.

Example 23-6-A. Calculating the electric potential due to an electric dipole. An electric dipole with a dipole moment of magnitude 3.8×10^{-10} C·m is located at the origin of our coordinate system. Its dipole moment points in the $+\hat{\mathbf{i}}$ direction. Determine the difference in electric potential between a position on the x-axis at $x = 1.0$ m and a position on the x-axis at $x = -1.0$ m.

Approach: In this problem we use the expression for the electric potential discussed in this Section. The angle θ for a point on the x axis at $x = +1$ m is $0°$. The angle θ for a point on the x axis at $x = -1$ m is $180°$.

Solution: The electric potential difference between these two positions is equal to

$$\Delta V = V(x = 1\,\text{m}) - V(x = -1\,\text{m}) = \frac{1}{4\pi\varepsilon_0}\frac{p}{1^2}(\cos 0° - \cos 180°) = \frac{2}{4\pi\varepsilon_0}p =$$

$$= 2(9.0 \times 10^9)(3.8 \times 10^{-10}) = 6.8\text{ V}$$

Section 23-7. $\vec{\mathbf{E}}$ Determined from V

The change in the electrostatic potential dV across a displacement $d\vec{\ell}$ is related to the scalar product between the electric field and the displacement:

$$dV = -\vec{\mathbf{E}} \cdot d\vec{\ell} = -E_\ell d\ell$$

where E_ℓ is the component of $\vec{\mathbf{E}}$ in the direction of the displacement $d\vec{\ell}$. This relation can be rearranged as

$$E_\ell = -\frac{dV}{d\ell}$$

If we let $d\vec{\ell}$ be along the Cartesian coordinate axes, we see that

$$E_x = -\frac{\partial V}{\partial x} \qquad E_y = -\frac{\partial V}{\partial y} \qquad E_z = -\frac{\partial V}{\partial z}$$

where $\partial V/\partial x$ is the partial derivative of V with respect to x, $\partial V/\partial y$ is the partial derivative of V with respect to y, and $\partial V/\partial z$ is the partial derivative of V with respect to z.

These relations are important since in general it is much easier to first calculate the electrostatic potential associated with a charge distribution and use these relations to determine the corresponding electric field compared to calculating the electric field directly. Determining the electric potential requires scalar addition, while determining the electric field requires vector addition.

Example 23-7-A. Calculating the electric field if the electric potential is known. An electric potential is given by $V(x,y,z) = V_0 x^2 y z^3 / L^6$. Determine the electric field.

Approach: The problem provides us with the electric potential. By differentiating the electric potential with respect to x, y, and z, we can determine the x, y, and z components of the electric field using the relations discussed in this Section.

Solution: The Cartesian components of the electric field can be determined by taking the partial derivative of the electric potential function along the Cartesian axes:

$$E_x = -\frac{\partial V}{\partial x} = -\frac{\partial}{\partial x}\left(\frac{V_0}{L^6} x^2 y z^3\right) = -2\frac{V_0}{L^6} xyz^3$$

$$E_y = -\frac{\partial V}{\partial y} = -\frac{\partial}{\partial y}\left(\frac{V_0}{L^6} x^2 y z^3\right) = -\frac{V_0}{L^6} x^2 z^3$$

$$E_z = -\frac{\partial V}{\partial z} = -\frac{\partial}{\partial z}\left(\frac{V_0}{L^6} x^2 y z^3\right) = -3\frac{V_0}{L^6} x^2 y z^2$$

The corresponding electric field is thus equal to

$$\vec{E}(x,y,z) = -\frac{V_0}{L^6} xz^2 \left(2yz\,\hat{\mathbf{i}} + xz\,\hat{\mathbf{j}} + 3xy\,\hat{\mathbf{k}}\right)$$

Section 23-8. Electrostatic Potential Energy; the Electron Volt

The electrostatic potential energy of a system of point charges is the energy required to assemble the system. The electrostatic potential energy of a pair of point charges is

$$U = \frac{1}{4\pi\varepsilon_0} \frac{Q_1 Q_2}{r_{12}}$$

where Q_1 and Q_2 are the point charges and r_{12} is the distance between these point charges. If the system consists of more than two point charges we need to add the electrostatic energies associated with each pair of charges:

$$U = \frac{1}{4\pi\varepsilon_0} \sum_{i=1}^{N} \sum_{j=i+1}^{N} \frac{Q_i Q_j}{r_{ij}}$$

A convenient unit of energy for dealing with electrons, atoms, and molecules is the **electron volt** (eV). The electron volt is defined as the increase in the electrostatic energy of a particle with charge $|e|$ when it changes electric potential by 1 V:

$$1\,\text{eV} = \left(1.602 \times 10^{-19}\,\text{C}\right)\left(1\,\text{V}\right) = 1.602 \times 10^{-19}\,\text{CV} = 1.602 \times 10^{-19}\,\text{J}$$

Section 23-9. Cathode Ray Tube: TV and Computer Monitors, Oscilloscope

The **cathode ray tube (CRT)** is a device that accelerates electrons through a potential difference and allows the electrons to impinge on a screen coated with a fluorescent material that glows with visible light when the kinetic energy of the electrons is absorbed by the material. Thus, by moving the position where the electrons strike the screen with other electric and/or magnetic fields the electron beam can "write" on the screen and display information. Devices using this type of display device include television, computer monitors, and oscilloscopes.

Practice Quiz

1. A negatively charged particle is located at a position where the electric potential is 124 V and is traveling at a speed v. The particle moves to a position where the electric potential is 248 V. What can you say about the speed of the particle at this location?
    a) The speed will be $v/2$.
    b) The speed will be $2v$.
    c) The speed will be less than v.
    d) The speed will be greater than v.

2. The electric field is constant in magnitude and points toward the east. In which direction does the electric potential increase?
    a) East
    b) North
    c) West
    d) South

3. A conductor has no charge initially. Electrons are added one at a time to the conductor. The electrons are initially infinitely far from the conductor and at rest. How does the work done to add the second electron to the conductor compare to the work done to add the tenth electron to the conductor?
    a) The work done to add the second electron is less than the work done to add the tenth electron.
    b) The work done to add the second electron is the same as the work done to add the tenth electron.
    c) The work done to add the second electron is more than the work done to add the tenth electron.
    d) Not enough information is given to compare the work done to add the two electrons to the conductor.

4. A surface is an equipotential surface at a potential of zero. What is true about the electric field on the surface?
    a) The electric field on the surface is zero.
    b) The electric field on the surface is perpendicular to the surface.
    c) The electric field on the surface points parallel to the surface.
    d) None of the above statements is necessarily true about the electric field.

5. A particle has a kinetic energy of 346 eV. The particle moves to a position where the electric potential is 346 V higher. What is the kinetic energy of the particle at the new position?
    a) 0 eV
    b) 346 eV
    c) 692 eV
    d) Not enough information is given to determine the kinetic energy.

6. An electric dipole is placed in a uniform electric field. What orientation of the electric dipole minimizes its electric potential energy?
    a) Any orientation such that the line joining the charges is perpendicular to the electric field
    b) An orientation such that the displacement of the positive charge from the negative charge is in the direction of the electric field
    c) An orientation such that the displacement of the positive charge from the negative charge is in the opposite direction of the electric field
    d) The electric field is uniform, so all orientations of the dipole have the same electric potential energy.

7. Point a and point b are at the same electric potential. What can you conclude about the electric field between point a and point b?
    a) The electric field between point a and point b is zero everywhere.
    b) The electric field always points perpendicular to the line between point a and point b.
    c) Given a path from a to b, if the field has a component in the direction of the path in some region, it must have a component opposite to the path somewhere else.
    d) Nothing can be concluded about the electric field.

8. An electron is moved through a potential difference of +10 V. What is its change in electrical potential energy?
   a) +10eV
   b) −10 eV
   c) −1.602 × 10⁻¹⁸ eV
   d) −10 J

9. The electric field is uniform and has a magnitude of 100 V/m and a direction east. What is the difference in electric potential at any given position and a position 5.0 m to the north?
   a) +500 V
   b) −500 V
   c) 0 V
   d) 20 V

10. The electric field is uniform and has a magnitude of 100 V/m and a direction east. What is the difference in electric potential at any given position and a position 5.0 m to the west?
    a) +500 V
    b) −500 V
    c) 0 V
    d) −20 V

11. A charge of 24.6 μC is located at position 1.56 m from the origin on the +x-axis. A second charge of +32.7 μC is located 2.21 m from the origin on the +y-axis. How much work does it take to bring a charge of +4.11 μC from a position 5.63 m on the z-axis to the origin?

12. Determine the amount of work that is required to assemble an arrangement of charge that is four 1.22 μC charges each located on a different corner of a square that is 9.44 cm on a side if the charges are initially separated by an infinite distance.

13. An electron is accelerated from rest by a potential difference of +480 V. What are the kinetic energy and speed of the electron after being accelerated?

14. A dipole consists of charges +Q and −Q separated along the x-axis by a distance l as shown in the diagram. Determine the percent difference between the approximate dipole electric potential and the exact electric potential for a dipole at a positions $x = l$ and $x = 10l$.

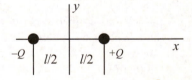

15. The electric potential in a region of space is given by $V(x, y, z) = V_0(xy/z^2 + y/x)$. Determine the components of the electric field in this region of space.

Responses to Select End-of-Chapter Questions

1. Not necessarily. If two points are at the same potential, then no *net* work is done in moving a charge from one point to the other, but work (both positive and negative) could be done at different parts of the path. No. It is possible that positive work was done over one part of the path, and negative work done over another part of the path, so that these two contributions to the net work sum to zero. In this case, a non-zero force would have to be exerted over both parts of the path.

7. (*a*) V at other points would be lower by 10 V. E would be unaffected, since E is the negative gradient of V, and a change in V by a constant value will not change the value of the gradient.
   (*b*) If V represents an absolute potential, then yes, the fact that the Earth carries a net charge would affect the value of V at the surface. If V represents a potential difference, then no, the net charge on the Earth would not affect the choice of V.

13. The charge density and the electric field strength will be greatest at the pointed ends of the football because the surface there has a smaller radius of curvature than the middle.

19. The electric potential energy of two unlike charges is negative. The electric potential energy of two like charges is positive. In the case of unlike charges, work must be done to separate the charges. In the case of like charges, work must be done to move the charges together.

Solutions to Select End-of-Chapter Problems

1. Energy is conserved, so the change in potential energy is the opposite of the change in kinetic energy. The change in potential energy is related to the change in potential.

$$\Delta U = q\Delta V = -\Delta K \quad \rightarrow$$

$$\Delta V = \frac{-\Delta K}{q} = \frac{K_{initial} - K_{final}}{q} = \frac{mv^2}{2q} = \frac{\left(9.11 \times 10^{-31}\,\text{kg}\right)\left(5.0 \times 10^5\,\text{m/s}\right)^2}{2\left(-1.60 \times 10^{-19}\,\text{C}\right)} = \boxed{-0.71\,\text{V}}$$

The final potential should be lower than the initial potential in order to stop the electron.

7. The maximum charge will produce an electric field that causes breakdown in the air. We use the same approach as in Examples 23-4 and 23-5.

$$V_{surface} = r_0 E_{breakdown} \quad \text{and} \quad V_{surface} = \frac{1}{4\pi\varepsilon_0}\frac{Q}{r_0} \quad \rightarrow$$

$$Q = 4\pi\varepsilon_0 r_0^2 E_{breakdown} = \left(\frac{1}{8.99 \times 10^9\,\text{N}\bullet\text{m}^2/\text{C}^2}\right)\left(0.065\,\text{m}\right)^2 \left(3 \times 10^6\,\text{V/m}\right) = \boxed{1.4 \times 10^{-6}\,\text{C}}$$

13. (a) The electric field at the surface of the Earth is the same as that of a point charge, $E = \dfrac{Q}{4\pi\varepsilon_0 r_0^2}$.

    The electric potential at the surface, relative to $V(\infty) = 0$ is given by Eq. 23-5. Writing this in terms of the electric field and radius of the earth gives the electric potential.

$$V = \frac{Q}{4\pi\varepsilon_0 r_0} = E r_0 = \left(-150\,\text{V/m}\right)\left(6.38 \times 10^6\,\text{m}\right) = \boxed{-0.96\,\text{GV}}$$

    (b) Part (a) demonstrated that the potential at the surface of the earth is 0.96 GV lower than the potential at infinity. Therefore if the potential at the surface of the Earth is taken to be zero, the potential at infinity must be $V(\infty) = \boxed{0.96\,\text{GV}}$. If the charge of the ionosphere is included in the calculation, the electric field outside the ionosphere is basically zero. The electric field between the earth and the ionosphere would remain the same. The electric potential, which would be the integral of the electric field from infinity to the surface of the earth, would reduce to the integral of the electric field from the ionosphere to the earth. This would result in a negative potential, but of a smaller magnitude.

19. (a) The electric field outside a charged, spherically symmetric volume is the same as that for a point charge of the same magnitude of charge. Integrating the electric field from infinity to the radius of interest will give the potential at that radius.

$$E\left(r \geq r_0\right) = \frac{Q}{4\pi\varepsilon_0 r^2} \quad ; \quad V\left(r \geq r_0\right) = -\int_{\infty}^{r} \frac{Q}{4\pi\varepsilon_0 r^2}\,dr = \frac{Q}{4\pi\varepsilon_0 r}\bigg|_{\infty}^{r} = \boxed{\frac{Q}{4\pi\varepsilon_0 r}}$$

    (b) Inside the sphere the electric field is obtained from Gauss's Law using the charge enclosed by a sphere of radius r.

$$4\pi r^2 E = \frac{Q\,\frac{4}{3}\pi r^3}{\varepsilon_0\,\frac{4}{3}\pi r_0^3} \quad \rightarrow \quad E\left(r < r_0\right) = \frac{Qr}{4\pi\varepsilon_0 r_0^3}$$

Integrating the electric field from the surface to $r < r_0$ gives the electric potential inside the sphere.

$$V\left(r < r_0\right) = V\left(r_0\right) - \int_{r_0}^{r} \frac{Qr}{4\pi\varepsilon_0 r_0^3}\, dr = \frac{Q}{4\pi\varepsilon_0 r_0} - \frac{Qr^2}{8\pi\varepsilon_0 r_0^3}\Bigg|_{r_0}^{r} = \boxed{\frac{Q}{8\pi\varepsilon_0 r_0}\left(3 - \frac{r^2}{r_0^2}\right)}$$

(c) To plot, we first calculate $V_0 = V\left(r = r_0\right) = \dfrac{Q}{4\pi\varepsilon_0 r_0}$ and $E_0 = E\left(r = r_0\right) = \dfrac{Q}{4\pi\varepsilon_0 r_0^2}$. Then we plot V/V_0 and E/E_0 as functions of r/r_0.

For $r < r_0$: $\quad V/V_0 = \dfrac{\dfrac{Q}{8\pi\varepsilon_0 r_0}\left(3 - \dfrac{r^2}{r_0^2}\right)}{\dfrac{Q}{4\pi\varepsilon_0 r_0}} = \tfrac{1}{2}\left(3 - \dfrac{r^2}{r_0^2}\right)$; $\quad E/E_0 = \dfrac{\dfrac{Qr}{4\pi\varepsilon_0 r_0^3}}{\dfrac{Q}{4\pi\varepsilon_0 r_0^2}} = \dfrac{r}{r_0}$

For $r > r_0$: $\quad V/V_0 = \dfrac{\dfrac{Q}{4\pi\varepsilon_0 r}}{\dfrac{Q}{4\pi\varepsilon_0 r_0}} = \dfrac{r_0}{r} = \left(r/r_0\right)^{-1}$; $\quad E/E_0 = \dfrac{\dfrac{Q}{4\pi\varepsilon_0 r^2}}{\dfrac{Q}{4\pi\varepsilon_0 r_0^2}} = \dfrac{r_0^2}{r^2} = \left(r/r_0\right)^{-2}$

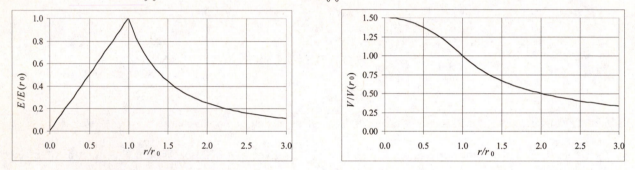

25. (a) The electric potential is given by Eq. 23-5.

$$V = \frac{1}{4\pi\varepsilon_0}\frac{Q}{r} = \left(8.99\times10^9\ \text{N}\cdot\text{m}^2/\text{C}^2\right)\frac{1.60\times10^{-19}\,\text{C}}{0.50\times10^{-10}\,\text{m}} = 28.77\ \text{V} \approx \boxed{29\ \text{V}}$$

(b) The potential energy of the electron is the charge of the electron times the electric potential due to the proton.

$$U = QV = \left(-1.60\times10^{-19}\,\text{C}\right)\left(28.77\ \text{V}\right) = \boxed{-4.6\times10^{-18}\ \text{J}}$$

31. By energy conservation, all of the initial potential energy will change to kinetic energy of the electron when the electron is far away. The other charge is fixed, and so has no kinetic energy. When the electron is far away, there is no potential energy.

$$E_{\text{initial}} = E_{\text{final}} \quad\rightarrow\quad U_{\text{initial}} = K_{\text{final}} \quad\rightarrow\quad \frac{(-e)(Q)}{4\pi\varepsilon_0 r} = \tfrac{1}{2}mv^2 \quad\rightarrow$$

$$v = \sqrt{\frac{2(-e)(Q)}{\left(4\pi\varepsilon_0\right)mr}} = \sqrt{\frac{2\left(8.99\times10^9\ \text{N}\cdot\text{m}^2/\text{C}^2\right)\left(-1.60\times10^{-19}\,\text{C}\right)\left(-1.25\times10^{-10}\,\text{C}\right)}{\left(9.11\times10^{-31}\,\text{kg}\right)\left(0.425\,\text{m}\right)}}$$

$$= \boxed{9.64\times10^5\ \text{m/s}}$$

37. The electric potential energy is the product of the point charge and the electric potential at the location of the charge. Since all points on the ring are equidistant from any point on the axis, the electric potential integral is simple.

$$U = qV = q\int \frac{dq}{4\pi\varepsilon_0 \sqrt{r^2 + x^2}} = \frac{q}{4\pi\varepsilon_0 \sqrt{r^2 + x^2}} \int dq = \frac{qQ}{4\pi\varepsilon_0 \sqrt{r^2 + x^2}}$$

Energy conservation is used to obtain a relationship between the potential and kinetic energies at the center of the loop and at a point 2.0 m along the axis from the center.

$$K_0 + U_0 = K + U$$

$$0 + \frac{qQ}{4\pi\varepsilon_0 \sqrt{r^2}} = \tfrac{1}{2}mv^2 + \frac{qQ}{4\pi\varepsilon_0 \sqrt{r^2 + x^2}}$$

This equation is solved to obtain the velocity at $x = 2.0$ m.

$$v = \sqrt{\frac{qQ}{2\pi\varepsilon_0 m}\left(\frac{1}{r} - \frac{1}{\sqrt{r^2 + x^2}}\right)}$$

$$= \sqrt{\frac{(3.0\,\mu C)(15.0\,\mu C)}{2\pi(8.85\times10^{-12}\,C^2/Nm^2)(7.5\times10^{-3}\,kg)}\left(\frac{1}{0.12\,m} - \frac{1}{\sqrt{(0.12\,m)^2 + (2.0\,m)^2}}\right)}$$

$$= \boxed{29 \text{ m/s}}$$

43. The electric field from a large plate is uniform with magnitude $E = \sigma/2\varepsilon_0$, with the field pointing away from the plate on both sides. Equation 23-4(a) can be integrated between two arbitrary points to calculate the potential difference between those points.

$$\Delta V = -\int_{x_0}^{x_1} \frac{\sigma}{2\varepsilon_0}\,dx = \frac{\sigma(x_0 - x_1)}{2\varepsilon_0}$$

Setting the change in voltage equal to 100 V and solving for $x_0 - x_1$ gives the distance between field lines.

$$x_0 - x_1 = \frac{2\varepsilon_0 \Delta V}{\sigma} = \frac{2(8.85\times10^{-12}\,C^2/Nm^2)(100\text{ V})}{0.75\times10^{-6}\,C/m^2} = 2.36\times10^{-3}\,m \approx \boxed{2\,mm}$$

49. The electric field between the plates is obtained from the negative derivative of the potential.

$$E = -\frac{dV}{dx} = -\frac{d}{dx}\left[(8.0 \text{ V/m})\,x + 5.0 \text{ V}\right] = -8.0 \text{ V/m}$$

The charge density on the plates (assumed to be conductors) is then calculated from the electric field between two large plates, $E = \sigma/\varepsilon_0$.

$$\sigma = E\varepsilon_0 = (8.0 \text{ V/m})(8.85\times10^{-12}\,C^2/Nm^2) = \boxed{7.1\times10^{-11}\,C/m^2}$$

The plate at the origin has the charge $-7.1\times10^{-11}\,C/m^2$ and the other plate, at a positive x, has charge $+7.1\times10^{-11}\,C/m^2$ so that the electric field points in the negative direction.

55. The gain of kinetic energy comes from a loss of potential energy due to conservation of energy, and the magnitude of the potential difference is the energy per unit charge. The helium nucleus has a charge of $2e$.

$$\Delta V = \frac{\Delta U}{q} = -\frac{\Delta K}{q} = -\frac{125\times10^3\,eV}{2e} = \boxed{-62.5\,kV}$$

The negative sign indicates that the helium nucleus had to go from a higher potential to a lower potential.

61. (*a*) The electron was accelerated through a potential difference of 1.33 kV (moving from low potential to high potential) in gaining 1.33 keV of kinetic energy. The proton is accelerated through the opposite potential difference as the electron, and has the exact opposite charge. Thus the proton gains the same kinetic energy, $\boxed{1.33 \text{ keV}}$.

(*b*) Both the proton and the electron have the same KE. Use that to find the ratio of the speeds.

$$\tfrac{1}{2}m_p v_p^2 = \tfrac{1}{2}m_e v_e^2 \quad \rightarrow \quad \frac{v_e}{v_p} = \sqrt{\frac{m_p}{m_e}} = \sqrt{\frac{1.67 \times 10^{-27}\,\text{kg}}{9.11 \times 10^{-31}\,\text{kg}}} = \boxed{42.8}$$

The lighter electron is moving about 43 times faster than the heavier proton.

67. Consider three parts to the electron's motion. First, during the horizontal acceleration phase, energy will be conserved and so the horizontal speed of the electron v_x can be found from the accelerating potential, V. Secondly, during the deflection phase, a vertical force will be applied by the uniform electric field which gives the electron an upward velocity, v_y. We assume that there is very little upward displacement during this time. Finally, after the electron leaves the region of electric field, it travels in a straight line to the top of the screen.

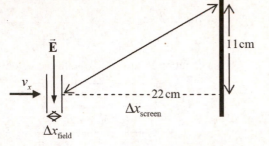

Acceleration:

$$U_{\text{initial}} = K_{\text{final}} \quad \rightarrow \quad eV = \tfrac{1}{2}mv_x^2 \quad \rightarrow \quad v_x = \sqrt{\frac{2eV}{m}}$$

Deflection:

time in field: $\Delta x_{\text{field}} = v_x t_{\text{field}} \quad \rightarrow \quad t_{\text{field}} = \dfrac{\Delta x_{\text{field}}}{v_x}$

$$F_y = eE = ma_y \quad \rightarrow \quad a_y = \frac{eE}{m} \quad v_y = v_0 + a_y t_{\text{field}} = 0 + \frac{eE\Delta x_{\text{field}}}{mv_x}$$

Screen:

$$\Delta x_{\text{screen}} = v_x t_{\text{screen}} \quad \rightarrow \quad t_{\text{screen}} = \frac{\Delta x_{\text{screen}}}{v_x} \qquad \Delta y_{\text{screen}} = v_y t_{\text{screen}} = v_y \frac{\Delta x_{\text{screen}}}{v_x}$$

$$\frac{\Delta y_{\text{screen}}}{\Delta x_{\text{screen}}} = \frac{v_y}{v_x} = \frac{\dfrac{eE\Delta x_{\text{field}}}{mv_x}}{v_x} = \frac{eE\Delta x_{\text{field}}}{mv_x^2} \quad \rightarrow$$

$$E = \frac{\Delta y_{\text{screen}} mv_x^2}{\Delta x_{\text{screen}} e\Delta x_{\text{field}}} = \frac{\Delta y_{\text{screen}} m \dfrac{2eV}{m}}{\Delta x_{\text{screen}} e\Delta x_{\text{field}}} = \frac{2V\Delta y_{\text{screen}}}{\Delta x_{\text{screen}} \Delta x_{\text{field}}} = \frac{2(7200\,\text{V})(0.11\,\text{m})}{(0.22\,\text{m})(0.028\,\text{m})}$$

$$= 2.57 \times 10^5 \,\text{V/m} \approx \boxed{2.6 \times 10^5 \,\text{V/m}}$$

As a check on our assumptions, we calculate the upward distance that the electron would move while in the electric field.

$$\Delta y = v_0 t_{\text{field}} + \tfrac{1}{2}a_y t_{\text{field}}^2 = 0 + \tfrac{1}{2}\left(\frac{eE}{m}\right)\left(\frac{\Delta x_{\text{field}}}{v_x}\right)^2 = \frac{eE(\Delta x_{\text{field}})^2}{2m\left(\dfrac{2eV}{m}\right)} = \frac{E(\Delta x_{\text{field}})^2}{4V} = 8.1 \times 10^{-3}\,\text{m}$$

This is about 7% of the total 11 cm vertical deflection, and so for an estimation, our approximation is acceptable.

73. The electric force on the electron must be the same magnitude as the weight of the electron. The magnitude of the electric force is the charge on the electron times the magnitude of the electric field. The electric field is the potential difference per meter: $E = V/d$.

$$F_E = mg \; ; \; F_E = |q|E = eV/d \; \rightarrow \; eV/d = mg \; \rightarrow$$

$$V = \frac{mgd}{e} = \frac{\left(9.11 \times 10^{-31} \text{kg}\right)\left(9.80 \text{ m/s}^2\right)\left(0.035 \text{m}\right)}{1.60 \times 10^{-19} \text{C}} = \boxed{2.0 \times 10^{-12} \text{V}}$$

Since it takes such a tiny voltage to balance gravity, the thousands of volts in a television set are more than enough (by many orders of magnitude) to move electrons upward against the force of gravity.

79. (*a*) The energy is related to the charge and the potential difference by Eq. 23-3.

$$\Delta U = q\Delta V \; \rightarrow \; \Delta V = \frac{\Delta U}{q} = \frac{4.8 \times 10^6 \text{J}}{4.0 \text{C}} = \boxed{1.2 \times 10^6 \text{V}}$$

(*b*) The energy (as heat energy) is used to raise the temperature of the water and boil it. Assume that room temperature is 20°C.

$$Q = mc\Delta T + mL_f \; \rightarrow$$

$$m = \frac{Q}{c\Delta T + L_f} = \frac{4.8 \times 10^6 \text{J}}{\left(4186 \frac{\text{J}}{\text{kg} \cdot \text{C}^\circ}\right)\left(80 \text{C}^\circ\right) + \left(22.6 \times 10^5 \frac{\text{J}}{\text{kg}}\right)} = \boxed{1.8 \text{kg}}$$

85. (*a*) The voltage at $x = 0.20$ m is obtained by inserting the given data directly into the voltage equation.

$$V\left(0.20 \text{m}\right) = \frac{B}{\left(x^2 + R^2\right)^2} = \frac{150 \text{ V} \cdot \text{m}^4}{\left[\left(0.20 \text{m}\right)^2 + \left(0.20 \text{m}\right)^2\right]^2} = \boxed{23 \text{ kV}}$$

(*b*) The electric field is the negative derivative of the potential.

$$\vec{\mathbf{E}}\left(x\right) = -\frac{d}{dx}\left[\frac{B}{\left(x^2 + R^2\right)^2}\right]\hat{\mathbf{i}} = \boxed{\frac{4Bx\,\hat{\mathbf{i}}}{\left(x^2 + R^2\right)^3}}$$

Since the voltage only depends on x the electric field points in the positive x direction.
(*c*) Inserting the given values in the equation of part (*b*) gives the electric field at $x = 0.20$ m

$$\vec{\mathbf{E}}(0.20 \text{ m}) = \frac{4\left(150 \text{ V} \cdot \text{m}^4\right)\left(0.20 \text{m}\right)\hat{\mathbf{i}}}{\left[\left(0.20 \text{m}\right)^2 + \left(0.20 \text{m}\right)^2\right]^3} = \boxed{2.3 \times 10^5 \text{ V/m}\,\hat{\mathbf{i}}}$$

Chapter 24: Capacitance, Dielectrics, Electric Energy Storage

Chapter Overview and Objectives

This chapter introduces the concept of capacitance. The calculation of the capacitance of arrangements of conductors and the effect of dielectric materials on the capacitance is discussed. The electrical potential energy stored in capacitors is studied.

After completing this chapter you should:
- Know what capacitance is.
- Know how to calculate the capacitance of parallel plates, concentric cylinders, and concentric spherical shells, with and without dielectric material between the conductors.
- Know how to determine the amount of electrical potential energy stored in a capacitor.
- Know how to calculate equivalent capacitances of serial and parallel networks of capacitances.

Summary of Equations

Definition of capacitance:
$$C = \frac{Q}{V}$$
(Section 24-1)

Capacitance of a parallel plate capacitor:
$$C = \frac{\varepsilon_0 A}{d}$$
(Section 24-2)

Capacitance of concentric cylinders:
$$C = \frac{2\pi\varepsilon_0 L}{\ln(R_a / R_b)}$$
(Section 24-2)

Capacitance of concentric spheres:
$$C = 4\pi\varepsilon_0 \frac{r_a r_b}{r_a - r_b}$$
(Section 24-2)

Capacitance of an isolated sphere:
$$C = 4\pi\varepsilon_0 r$$
(Section 24-2)

Equivalent capacitance of parallel network:
$$C_{eq} = \sum_{i=1}^{n} C_i$$
(Section 24-3)

Equivalent capacitance of series network:
$$\frac{1}{C_{eq}} = \sum_{i=1}^{n} \frac{1}{C_i}$$
(Section 24-3)

Electric potential energy of charged capacitor:
$$U = \tfrac{1}{2}QV = \tfrac{1}{2}CV^2 = \tfrac{1}{2}\frac{Q^2}{C}$$
(Section 24-4)

Energy density of an electric field:
$$u = \tfrac{1}{2}\varepsilon_0 E^2$$
(Section 24-4)

Capacitance of a capacitor with a dielectric:
$$C = KC_0$$
(Section 24-5)

Chapter Summary

Section 24-1. Capacitors

For a particular geometry of two conductors, a charge $+Q$ on one of the conductors and a charge $-Q$ on the other conductor will cause a potential difference V between the two conductors. The ratio of the charge Q to the potential difference V is a constant and called the **capacitance** C of the system of conductors:

$$C = \frac{Q}{V}$$

The SI unit of capacitance is the farad (F). One **farad** is one coulomb/volt.
A device constructed to provide capacitance to an electrical circuit is called a **capacitor**. The symbol of a capacitor in an electrical circuit is shown in the Figure.

Example 24-1-A. Measuring the capacitance of a capacitor. An initially uncharged capacitor has a current of $I_c = 3.64$ mA flow from one plate to another for a time $\Delta t = 23.9$ s. The resulting voltage across the capacitor is $V = 167$ V. What is the capacitance of the capacitor?

Approach: Since 1 A corresponds to 1 C/s, a knowledge of the current and the charging time allows us to determine the charge Q on the capacitor. Since we also know the voltage V across the capacitor, we can use this information to determine the capacitance $C = Q/V$.

Solution: The charge Q on the capacitor can be obtained by integrating the current with respect to time:

$$Q = \int_0^{\Delta t} I dt = I \int_0^{\Delta t} dt = I\Delta t = \left(3.64 \times 10^{-3}\right)(23.9) = 8.70 \times 10^{-2} \text{ C}$$

In this calculation, we have used the fact that the current is constant during the charging period. The capacitance of the capacitor can be determined from the charge Q and the voltage V:

$$C = \frac{Q}{V} = \frac{8.70 \times 10^{-2}}{167} = 5.21 \times 10^{-4} \text{ F} = 521\,\mu\text{F}$$

Section 24-2. Determination of Capacitance

For some very symmetric arrangements of conductors, we can determine the capacitance by using our knowledge of the electric field around the conductors. The known electric field can be used to determine the electric potential difference between the conductors. In this Section this approach is used to determine the capacitance of several arrangements of conductors.

- For **two conductive parallel plates** of equal area A, separated by a distance d, the capacitance C is

$$C = \frac{\varepsilon_0 A}{d}$$

- For **two concentric conductive circular cylinders** of length L, the capacitance C is

$$C = \frac{2\pi\varepsilon_0 L}{\ln(R_a / R_b)}$$

where R_a is the radius of the outer cylinder and R_b is the radius of the inner cylinder.
- For **two concentric conductive spherical shells**, the capacitance C is

$$C = 4\pi\varepsilon_0 \frac{r_a r_b}{r_a - r_b}$$

where r_a is the inner radius of the outer shell and r_b is the outer radius of the inner shell.

- A **single conductor** has a capacitance defined as the ratio of the charge on the conductor to the potential difference of conductor to the potential an infinite distance away. It is easy to show that the capacitance of an isolated conductive sphere or spherical shell of outer radius r is

$$C = 4\pi\varepsilon_0 r$$

Example 24-2-A. Capacitance of a coaxial cable. A coaxial cable has an inner wire of diameter 0.67 mm and an outer conductor of inside diameter 4.65 mm. What is the capacitance per length of cable?

Approach: Signal propagation through a coaxial cable depends on the capacitance of the cable per unit length. In this Section we already have determine the capacitance of a length L of concentric cylindrical conductors. The capacitance per unit length is the ratio of the capacitance divided by L.

Solution: The capacitance per unit length of the concentric cylindrical conductors is

$$\frac{C}{L} = \frac{2\pi\varepsilon_0}{\ln(R_a / R_b)} = \frac{2\pi\left(8.85 \times 10^{-12}\right)}{\ln(4.65/0.67)} = 1.31 \times 10^{-11}\,\text{F/m}$$

Example 24-2-B. Designing a capacitor. What is the minimum plate area of a parallel-plate capacitor that has a capacitance of 2.85 μF? The capacitor must be able to be charged to a voltage of $V_{max} = 1000$ V. The maximum electric field between the plates is $E_{max} = 1.00 \times 10^6$ V/m. At higher electric fields, the air between the plates will break down and become conductive.

Approach: The capacitor we need to design needs to have a specific capacitance. The capacitance is proportional to the area of the plates and inversely proportional to the distance between the plates. Both can be adjusted to obtain the requested capacitance. If we are interested in the minimum plate area, we want to minimize the distance between the plates. The minimum distance is determined by the maximum electric field between the plates for the maximum voltage V_{max}.

Solution: The electric field between the capacitor plates is equal to the ratio of the electric potential difference and the distance between the plates. The maximum electric field and the maximum voltage allows us to calculate the minimum separation between the capacitor plates:

$$d_{min} = \frac{V_{max}}{E_{max}} = \frac{1000}{1.00 \times 10^6} = 1.00 \times 10^{-3}\,\text{m}$$

The expression for the capacitance C of a parallel-plate capacitor can be used to determine the area of the plates that is required to obtain the required capacitance:

$$A = \frac{d_{min}C}{\varepsilon_0} = \frac{\left(1.00 \times 10^{-3}\right)\left(2.85 \times 10^{-6}\right)}{8.85 \times 10^{-12}} = 322\ \text{m}^2$$

This is quite a large area!

Section 24-3. Capacitors in Series and Parallel

Capacitors are **connected in parallel** if they are connected in such a way that the potential difference across each capacitor is identical. An example of three capacitors connected in parallel is shown in the diagram on the right. The conductors connecting the upper plates ensure that the upper plates are at the same electric potential. The conductors connecting the lower plates also ensure that the lower plates are at the same potential. When capacitors are connected in parallel, the equivalent capacitance of the circuit is the sum of the capacitances of each capacitor: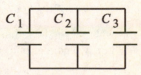

$$C_{eq} = C_1 + C_2 + C_3$$

If n capacitors are connected in parallel, the equivalent capacitance is

$$C_{eq} = \sum_{i=1}^{n} C_i$$

where C_i is the capacitance of the ith capacitor.

When capacitors are **connected in series**, the charge on each capacitance is the same. Three capacitors connected in series are shown in the diagram. The equivalent capacitance of this circuit is the reciprocal of the sum of the reciprocals of the individual capacitances:

$$\frac{1}{C_{eq}} = \frac{1}{C_1} + \frac{1}{C_2} + \frac{1}{C_3}$$

If n capacitors are connected in series, the equivalent capacitance can be obtained using the following relation

$$\frac{1}{C_{eq}} = \sum_{i=1}^{n} \frac{1}{C_i}$$

where C_i is the capacitance of the ith capacitor.

Example 24-3-A. Finding the equivalent capacitance of a capacitor network. Determine the equivalent capacitance of the capacitor network shown in the diagram.

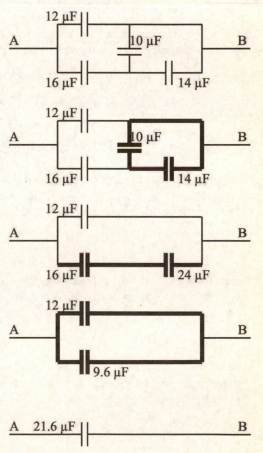

Approach: Since we know the equivalent capacitance of capacitors in parallel and in series, we need to examine the circuit in detail and determine which capacitor pairs are in parallel or in series and replace these by their equivalent capacitance. We repeat this procedure until we are left with a single equivalent capacitor.

Solution: We recognize that the 10 μF and the 14 μF capacitors are in parallel with each other. The equivalent capacitance is

$$C_{eq1} = 10\,\mu F + 14\,\mu F = 24\,\mu F$$

We redraw the network with the equivalent capacitance in place of the 10 μF and the 14 μF capacitors. We then recognize that the 16 μF and the 24 μF capacitors are in series. We calculate the equivalent capacitance of this series:

$$\frac{1}{C_{eq2}} = \frac{1}{16\,\mu F} + \frac{1}{24\,\mu F} \quad \Rightarrow \quad C_{eq2} = 9.6\,\mu F$$

We again redraw the circuit with this equivalent capacitance replacing the individual capacitors. We recognize that the 12 μF and the 9.6 μF capacitors are in parallel and calculate their equivalent capacitance:

$$C_{eq3} = 12\,\mu F + 9.6\,\mu F = 21.6\,\mu F$$

We have reduced the capacitor network to a single equivalent 21.6 μF capacitance.

Example 24-3-B. Calculating charges on and voltages across capacitors in a circuit. Determine the charge on each capacitor in the network in Example 24-3-A and the voltage across each capacitor if point A is at a potential 12 V higher than point B.

Approach: To determine the charges on and voltages across the capacitors in the network, we work backwards through the steps of reducing the circuit.

Solution: Examining the diagram shown as part of Example 24-3-A it can be determined immediately that the charge on the 12 μF capacitor is equal to

$$Q_{12} = C_{12}V_{12} = \left(12 \times 10^{-6}\right)(12) = 144\ \mu C$$

The total charge on C_{eq2} can also be calculated easily since there is 12 V across this capacitor. The charge across C_{eq2} is thus

$$Q_{eq2} = C_{eq2}V_{eq2} = \left(9.6 \times 10^{-6}\right)(12) = 115\ \mu C$$

The capacitor C_{eq2} is equivalent to a 16 μF capacitor in series with C_{eq1}. Since capacitors in series have the same charge we know the charge on the 16 μF capacitor is 115 μC, and the potential difference across this capacitor is

$$V_{16} = \frac{Q_{16}}{C_{16}} = \frac{115 \times 10^{-6}}{16 \times 10^{-6}} = 7.2\ \text{V}$$

Similarly, we calculate the voltage across C_{eq1}:

$$V_{eq1} = \frac{Q_{eq1}}{C_{eq1}} = \frac{115 \times 10^{-6}}{24 \times 10^{-6}} = 4.8\ \text{V}$$

Since C_{eq1} is the equivalent capacitance of two parallel capacitors, the voltage across each of these two capacitors is 4.8 V. The charge on the 10 μF and 14 μF capacitors is thus equal to

$$Q_{10} = C_{10}V_{10} = \left(10 \times 10^{-6}\right)(4.8) = 48\ \mu C$$
$$Q_{14} = C_{14}V_{14} = \left(14 \times 10^{-6}\right)(4.8) = 67\ \mu C$$

Section 24-4. Electric Energy Storage

Consider a charged capacitor with capacitance C and an electric potential difference V across its plates. The charge on the capacitor is Q, where $Q = CV$. The electric potential energy U of this charged capacitor can be determined by calculating the work that is required to add this charge to the capacitor. It is found that this potential energy is equal to

$$U = \tfrac{1}{2}QV = \tfrac{1}{2}CV^2 = \tfrac{1}{2}\frac{Q^2}{C}$$

Because the electric field between the plates of the capacitor is a well-defined function of the charge on the plates, we can write the electric potential energy stored in the capacitor in terms of the electric field in this region. The electric potential energy can be written as

$$U = \tfrac{1}{2}\varepsilon_0 E^2 Ad$$

where E is the magnitude of the electric field between the plates of the capacitor, A is the area of the plates, and d is the distance between the plates. The energy density u in the volume between the plates is the electric potential energy divided by the volume between the plates:

$$u = \frac{U}{Ad} = \tfrac{1}{2}\varepsilon_0 E^2$$

This expression is often applied to situations for which the electric field is known, but the charge distribution that created the electric field is unknown.

Section 24-5. Dielectrics

When an insulating material is placed between the plates of a capacitor, its capacitance is altered from its value without the insulator in place. The material that is inserted between the plates is called a **dielectric**. The capacitance of the capacitor with the dielectric in place is equal to

$$C = KC_0 = K\frac{\varepsilon_0 A}{d}$$

where C_0 is the capacitance of the capacitor without the dielectric material between the plates and K is called the **dielectric constant** of the material. Often, the dielectric constant and the free space permittivity are combined and called the **permittivity** ε of the dielectric material:

$$\varepsilon = K\varepsilon_0$$

The capacitance of the parallel plate capacitor is thus equal to

$$C = \frac{\varepsilon A}{d}$$

Example 24-5-A. Forces on dielectrics. Consider a charged, isolated parallel plate capacitor with plates of length L, width W, and separation distance d. When the region between the plates is empty, the capacitance of the capacitor is C_0. The capacitor is charged to a potential difference V. A dielectric material, with a dielectric constant K, is slowly inserted. Calculate the force on the dielectric, as function of x, by determining the electric potential energy of the capacitor as function of x.

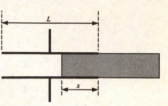

Approach: When the dielectric is inserted, we can determine the effective capacitance of the capacitor by treating it as a system of two capacitors, connected in parallel. One of these capacitors has a plate area of xW while the other one has an area of $(L - x)W$.

Solution: The capacitance of the capacitor with plate area $(L - x)W$ and no dielectric between the plates is equal to

$$C_1 = \varepsilon_0 \frac{(L-x)W}{d}$$

The capacitance of a capacitor with plate area xW, filled with the dielectric, is equal to

$$C_2 = K\varepsilon_0 \frac{x}{d}$$

The total effective capacitance of the capacitor is the sum of the individual capacitances since the capacitors are in parallel:

$$C_{eff} = C_1 + C_2 = \varepsilon_0 \frac{(L-x)W}{d} + K\varepsilon_0 \frac{xW}{d} = \varepsilon_0 \frac{WL}{d}\left(1 - \frac{1-K}{L}x\right) = C_0\left(1 - \frac{1-K}{L}x\right)$$

This expression is only valid for $0 < x < L$. The potential energy of the system is a function of x and is equal to:

$$U(x) = \frac{1}{2}\frac{Q^2}{C} = \frac{1}{2}\frac{Q^2}{C_0\left(1 - \frac{1-K}{L}x\right)} = \frac{1}{2}\frac{Q^2}{C_0}\frac{1}{\left(1 - \frac{1-K}{L}x\right)}$$

The force on the dielectric can be determined by differentiating the electric potential energy as function of x. All variables contained in the expression for U, except x, are constant and independent of x. For this reason we can determine the force F by differentiating U with respect to x:

$$F = -\frac{\partial U}{\partial x} = -\frac{1}{2}\frac{Q^2}{C_0\left(1 - \frac{1-K}{L}x\right)^2}\left(\frac{1-K}{L}\right) = \frac{1}{2}\frac{Q^2}{C_0}\left\{\frac{\left(\frac{K-1}{L}\right)}{\left(1 + \frac{K-1}{L}x\right)^2}\right\}$$

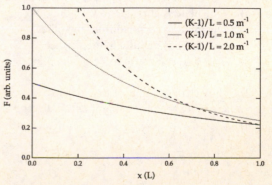

Since K is a number larger than 1, the force is positive and the dielectric is pulled into the gap. The force on the dielectric as function of x is shown in the Figure on the right for three different values of $(K-1)/L$. As the dielectric constant increases, the magnitude of the force increases, except for large values of x.

Section 24-6. Molecular Description of Dielectrics

A dielectric reduces the electric field in its interior because the applied electric field results in a change of the charge distribution inside the dielectric. Although the charges do not move freely inside the dielectric, the electrons in the atoms of the dielectric move slightly in a direction opposite to the direction of the applied electric field and the nuclei of the atoms of the dielectric move slightly in the direction of the applied electric field. The net effect of this movement of charge is a net positive charge density on the surface on one side of the material and a net negative surface charge density on the other surface. This charge distribution generates an electric field with a direction opposite to the direction of the applied electric field, and the net electric field in the region with the dielectric will be reduced. The electric field inside the dielectric, E_D, is reduced by a factor equal to the dielectric constant K compared to the applied electric field E_0:

$$E_D = \frac{E_0}{K}$$

Since the electric field inside the dielectric is the sum of the applied electric field and the induced electric field, the induced electric field, due to the polarization of the molecules of the dielectric, can be calculated:

$$E_D = E_0 + E_{induced} \quad \Rightarrow \quad E_{induced} = E_D - E_0 = -E_0\left(1 - \frac{1}{K}\right)$$

The minus sign indicates that the induced field points in a direction opposite to the direction of the applied field. The induced field is due to the induced surface charge density. We can use the induced electric field to determine the induced surface charge density:

$$E_{induced} = \frac{\sigma_{induced}}{\varepsilon_0}$$

Practice Quiz

1. If the electrostatic potential energy is given by QV as stated in Chapter 23, why is the potential energy stored in a capacitor $\frac{1}{2}QV$?
   a) Half of the energy is lost in charging up the capacitor.
   b) We divide by two because half of the charge is on each plate of the capacitor.
   c) The average voltage difference during the charge transfer is $\frac{1}{2}V$.
   d) The other half of the energy is stored in the electric field between the plates of the capacitor.

2. You have three capacitors of equal capacitance. How should they be connected together to reach the minimum capacitance of the combination of capacitors?
   a) Connect the three capacitors in parallel.
   b) Connect the three capacitors in series.
   c) Connect the first two capacitors in parallel and then in series with the third.
   d) Connect the first two capacitors in series and then in parallel with the third.

3. What is the direction of the electric force on the plates of a charged capacitor?
   a) One plate is attracted toward the other.
   b) One plate is repelled from the other.
   c) The force is parallel to the surface of the plates.
   d) There is no electric force on the plates.

4. The dielectric strength of dry air at room temperature and pressure is about 30,000 V/cm. What is the breakdown voltage of a capacitor with air as a dielectric and a gap of 1.00 mm?
   a) 300 V
   b) 3,000 V
   c) 30,000 V
   d) 300,000 V

5. You have a parallel plate capacitor with air between the plates. You have enough dielectric material with a relatively high dielectric constant to fill only half of the volume between the plates. To create the greatest capacitance, how should you fill the volume between the plates?
   a) Fill the gap between the plates completely over one half the area of the capacitor.
   b) Fill half the gap between the plates over the entire area of the capacitor.
   c) Don't fill the gap at all.
   d) It doesn't matter how you fill the gap, as long as half the volume is filled.

6. Suppose you have a capacitor with capacitance C charged to voltage V and a second capacitor with capacitance C charged to voltage $2V$. The two capacitors are connected in parallel. What will be the voltage across the two capacitors after being connected in parallel?
   a) The first capacitor will have voltage V and the second capacitor will have voltage $2V$.
   b) Both capacitors will be charged to voltage $3V$.
   c) Both capacitors will be charged to voltage $1.5V$.
   d) The first capacitor will have voltage $2V$ and the second capacitor will have voltage V.

7. Make an estimate of the capacitance of the human body based on the expression for the capacitance of an isolated sphere.
   a) 10 pF
   b) 10 nF
   c) 10 µF
   d) 10 mF

8. In order to double the capacitance of a parallel plate capacitor you can
   a) Double the area of its plates.
   b) Double the distance between its plates.
   c) Double the length of each of the edges of its plates.
   d) Any of the above.

9. In order to increase the capacitance of a concentric cylinder capacitor by a factor of 10 you can
   a) Make the outer conductor's inner radius a factor 10 larger.
   b) Make the inner conductor's outer radius a factor 10 larger.
   c) Make the outer conductor's inner radius a factor e^{10} larger.
   d) Make the ratio of the outer conductor's radius to the inner conductor's radius the tenth root of its current ratio.

10. Why does water have a relatively high dielectric constant?
    a) Its molecules have a large permanent dipole moment and are free to rotate.
    b) Its molecules are lower in mass than most diatomic molecules.
    c) It has a high specific heat.
    d) It has very low conductivity.

11. Determine the capacitance of two rectangular plates 24 cm × 36 cm separated by a distance of 0.02 cm.

12. Determine the equivalent capacitance between points A and B of the capacitor network shown.

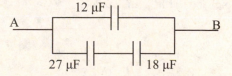

13. Determine the charge in each capacitor and the energy stored in each capacitor when the capacitor network in quiz Problem 12 is connected to a potential difference of 12 V.

14. A capacitor with no dielectric material between its plates has a capacitance C. If half of the volume between its plates is filled with a dielectric material with dielectric constant K_1 and half is filled with a dielectric material with dielectric constant K_2 as shown, what will the new capacitance be?

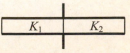

15. A capacitor with capacitance C is charged to a potential difference V and then disconnected from the potential difference source. A double layered dielectric with dielectric constant K_1 on one layer and dielectric K_2 on the other layer is placed in between the plates of the capacitor, as shown. How much work is done to insert the dielectric?

Responses to Select End-of-Chapter Questions

1. Yes. If the conductors have different shapes, then even if they have the same charge, they will have different charge densities and therefore different electric fields near the surface. There can be a potential difference between them. The definition of capacitance $C = Q/V$ cannot be used here because it is defined for the case where the charges on the two conductors of the capacitor are equal and opposite.

7. If a large copper sheet of thickness ℓ is inserted between the plates of a parallel-plate capacitor, the charge on the capacitor will appear on the large flat surfaces of the copper sheet, with the negative side of the copper facing the positive side of the capacitor. This arrangement can be considered to be two capacitors in series, each with a thickness of $\frac{1}{2}(d-\ell)$. The new net capacitance will be $C' = \varepsilon_0 A/(d-\ell)$, so the capacitance of the capacitor will be reduced.

13. (a) If the capacitor is isolated, Q remains constant, and $U = \frac{1}{2}\dfrac{Q^2}{C}$ becomes $U' = \frac{1}{2}\dfrac{Q^2}{KC}$ and the stored energy decreases.

    (b) If the capacitor remains connected to a battery so V does not change, $U = \frac{1}{2}CV^2$ becomes $U' = \frac{1}{2}KCV^2$, and the stored energy increases.

Solutions to Select End-of-Chapter Problems

1. The capacitance is found from Eq. 24-1.

$$Q = CV \quad \rightarrow \quad C = \frac{Q}{V} = \frac{2.8 \times 10^{-3}\,\text{C}}{930\,\text{V}} = 3.0 \times 10^{-6}\,\text{F} = \boxed{3.0\,\mu\text{F}}$$

7. The work to move the charge between the capacitor plates is $W = qV$, where V is the voltage difference between the plates, assuming that $q \ll Q$ so that the charge on the capacitor does not change appreciably. The charge is then found from Eq. 24-1. The assumption that $q \ll Q$ is justified.

$$W = qV = q\left(\frac{Q}{C}\right) \quad \rightarrow \quad Q = \frac{CW}{q} = \frac{(15\,\mu\text{F})(15\,\text{J})}{0.20\,\text{mC}} = \boxed{1.1\,\text{C}}$$

13. Inserting the potential at the surface of a spherical conductor into Eq. 24-1 gives the capacitance of a conducting sphere. Then inserting the radius of the Earth yields the Earth's capacitance.

$$C = \frac{Q}{V} = \frac{Q}{\left(Q/4\pi\varepsilon_0 r\right)} = 4\pi\varepsilon_0 r =$$

$$= 4\pi\left(8.85 \times 10^{-12}\,\text{F/m}\right)\left(6.38 \times 10^{6}\,\text{m}\right) = \boxed{7.10 \times 10^{-4}\,\text{F}}$$

19. (*a*) The distance between plates is obtained from Eq. 24-2.

$$C = \frac{\varepsilon_0 A}{x} \quad \rightarrow \quad x = \frac{\varepsilon_0 A}{C}$$

Inserting the maximum capacitance gives the minimum plate separation and the minimum capacitance gives the maximum plate separation.

$$x_{min} = \frac{\varepsilon_0 A}{C_{max}} = \frac{\left(8.85\,\text{pF/m}\right)\left(25 \times 10^{-6}\,\text{m}^2\right)}{1000.0 \times 10^{-12}\,\text{F}} = 0.22\,\mu\text{m}$$

$$x_{max} = \frac{\varepsilon_0 A}{C_{min}} = \frac{\left(8.85\,\text{pF/m}\right)\left(25 \times 10^{-6}\,\text{m}^2\right)}{1.0\,\text{pF}} = 0.22\,\text{mm} = 220\,\mu\text{m}$$

So $\boxed{0.22\,\mu\text{m} \leq x \leq 220\,\mu\text{m}}$.

(*b*) Differentiating the distance equation gives the approximate uncertainty in distance.

$$\Delta x \approx \frac{dx}{dC}\Delta C = \frac{d}{dC}\left[\frac{\varepsilon_0 A}{C}\right]\Delta C = -\frac{\varepsilon_0 A}{C^2}\Delta C \, .$$

The minus sign indicates that the capacitance increases as the plate separation decreases. Since only the magnitude is desired, the minus sign can be dropped. The uncertainty is finally written in terms of the plate separation using Eq. 24-2.

$$\Delta x \approx \frac{\varepsilon_0 A}{\left(\frac{\varepsilon_0 A}{x}\right)^2}\Delta C = \boxed{\frac{x^2 \Delta C}{\varepsilon_0 A}}$$

(*c*) The percent uncertainty in distance is obtained by dividing the uncertainty by the separation distance.

$$\frac{\Delta x_{min}}{x_{min}} \times 100\% = \frac{x_{min}\Delta C}{\varepsilon_0 A} \times 100\% = \frac{\left(0.22\,\mu\text{m}\right)\left(0.1\,\text{pF}\right)\left(100\%\right)}{\left(8.85\,\text{pF/m}\right)\left(25\,\text{mm}^2\right)} = \boxed{0.01\%}$$

$$\frac{\Delta x_{max}}{x_{max}} \times 100\% = \frac{x_{max}\Delta C}{\varepsilon_0 A} \times 100\% = \frac{\left(0.22\,\text{mm}\right)\left(0.1\,\text{pF}\right)\left(100\%\right)}{\left(8.85\,\text{pF/m}\right)\left(25\,\text{mm}^2\right)} = \boxed{10\%}$$

25. Capacitors in parallel add linearly, and so adding a capacitor in parallel will increase the net capacitance without removing the $5.0\,\mu\text{F}$ capacitor.

$$5.0\,\mu\text{F} + C = 16\,\mu\text{F} \quad \rightarrow \quad C = \boxed{11\,\mu\text{F connected in parallel}}$$

31. When the switch is down the initial charge on C_2 is calculated from Eq. 24-1.

$$Q_2 = C_2 V_0$$

When the switch is moved up, charge will flow from C_2 to C_1 until the voltage across the two capacitors is equal.

$$V = \frac{Q_2'}{C_2} = \frac{Q_1'}{C_1} \quad \rightarrow \quad Q_2' = Q_1'\frac{C_2}{C_1}$$

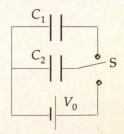

The sum of the charges on the two capacitors is equal to the initial charge on C_2.

$$Q_2 = Q_2' + Q_1' = Q_1' \frac{C_2}{C_1} + Q_1' = Q_1' \left(\frac{C_2 + C_1}{C_1} \right)$$

Inserting the initial charge in terms of the initial voltage gives the final charges.

$$Q_1' \left(\frac{C_2 + C_1}{C_1} \right) = C_2 V_0 \quad \rightarrow \quad \boxed{Q_1' = \frac{C_1 C_2}{C_2 + C_1} V_0} \ ; \ Q_2' = Q_1' \frac{C_2}{C_1} = \boxed{\frac{C_2^2}{C_2 + C} V_0}$$

37. (*a*) The series capacitors add reciprocally, and then the parallel combination is found by adding linearly.

$$C_{eq} = C_1 + \left(\frac{1}{C_2} + \frac{1}{C_3} \right)^{-1} = C_1 + \left(\frac{C_3}{C_2 C_3} + \frac{C_2}{C_2 C_3} \right)^{-1} = C_1 + \left(\frac{C_2 + C_3}{C_2 C_3} \right)^{-1} = \boxed{C_1 + \frac{C_2 C_3}{C_2 + C_3}}$$

(*b*) For each capacitor, the charge is found by multiplying the capacitance times the voltage. For C_1, the full 35.0 V is across the capacitance, so $Q_1 = C_1 V = \left(24.0 \times 10^{-6} \, \text{F} \right) \left(35.0 \, \text{V} \right) = \boxed{8.40 \times 10^{-4} \, \text{C}}$. The equivalent capacitance of the series combination of C_2 and C_3 has the full 35.0 V across it, and the charge on the series combination is the same as the charge on each of the individual capacitors.

$$C_{eq} = \left(\frac{1}{C} + \frac{1}{C/2} \right)^{-1} = \frac{C}{3} \quad Q_{eq} = C_{eq} V = \tfrac{1}{3} \left(24.0 \times 10^{-6} \, \text{F} \right) \left(35.0 \, \text{V} \right) = \boxed{2.80 \times 10^{-4} \, \text{C}} = Q_2 = Q_3$$

43. The energy stored is obtained from Eq. 24-5, with the capacitance of Eq. 24-2.

$$U = \frac{Q^2}{2C} = \frac{Q^2 d}{2 \varepsilon_0 A} = \frac{ \left(4.2 \times 10^{-4} \, \text{C} \right)^2 \left(0.0013 \, \text{m} \right) }{ 2 \left(8.85 \times 10^{-12} \, \text{C}^2/\text{N} \bullet \text{m}^2 \right) \left(0.080 \, \text{m} \right)^2 } = \boxed{2.0 \times 10^3 \, \text{J}}$$

49. (*a*) With the plate inserted, the capacitance is that of two series capacitors of plate separations $d_1 = x$ and $d_2 = d - \ell - x$.

$$C_i = \left[\frac{x}{\varepsilon_0 A} + \frac{d - x - \ell}{\varepsilon_0 A} \right]^{-1} = \frac{\varepsilon_0 A}{d - \ell}$$

With the plate removed the capacitance is obtained directly from Eq. 24-2.

$$C_f = \frac{\varepsilon_0 A}{d}$$

Since the voltage remains constant the energy of the capacitor will be given by Eq. 24-5 written in terms of voltage and capacitance. The work will be the change in energy as the plate is removed.

$$W = U_f - U_i = \tfrac{1}{2} \left(C_f - C_i \right) V^2 = \tfrac{1}{2} \left(\frac{\varepsilon_0 A}{d} - \frac{\varepsilon_0 A}{d - \ell} \right) V^2 = \boxed{- \frac{\varepsilon_0 A \ell V^2}{2d \left(d - \ell \right)}}$$

The net work done is negative. Although the person pulling the plate out must do work, charge is returned to the battery, resulting in a net negative work done.

(b) Since the charge now remains constant, the energy of the capacitor will be given by Eq. 24-5 written in terms of capacitance and charge.

$$W = \frac{Q^2}{2}\left(\frac{1}{C_f} - \frac{1}{C_i}\right) = \frac{Q^2}{2}\left(\frac{d}{\varepsilon_0 A} - \frac{d-\ell}{\varepsilon_0 A}\right) = \frac{Q^2 \ell}{2\varepsilon_0 A}$$

The original charge is $Q = CV_0 = \dfrac{\varepsilon_0 A}{d-\ell}V_0$ and so $W = \dfrac{\left(\dfrac{\varepsilon_0 A}{d-\ell}V_0\right)^2 \ell}{2\varepsilon_0 A} = \boxed{\dfrac{\varepsilon_0 A V_0^2 \ell}{2(d-\ell)^2}}$.

55. The change in energy of the capacitor is obtained from Eq. 24-5 in terms of the constant voltage and the capacitance.

$$\Delta U = U_f - U_i = \tfrac{1}{2}C_0 V^2 - \tfrac{1}{2}KC_0 V^2 = -\tfrac{1}{2}(K-1)C_0 V^2$$

The work done by the battery in maintaining a constant voltage is equal to the voltage multiplied by the change in charge, with the charge given by Eq. 24-1.

$$W_{battery} = V\left(Q_f - Q_i\right) = V\left(C_0 V - KC_0 V\right) = -(K-1)C_0 V^2$$

The work done in pulling the dielectric out of the capacitor is equal to the difference between the change in energy of the capacitor and the energy done by the battery.

$$W = \Delta U - W_{battery} = -\tfrac{1}{2}(K-1)C_0 V^2 + (K-1)C_0 V^2$$

$$= \tfrac{1}{2}(K-1)C_0 V^2 = (3.4-1)(8.8 \times 10^{-9}\,\text{F})(100\,\text{V})^2 = \boxed{1.1 \times 10^{-4}\,\text{J}}$$

61. The capacitor can be treated as two series capacitors with the same areas, but different plate separations and dielectrics. Substituting Eq. 24-8 into Eq. 24-4 gives the effective capacitance.

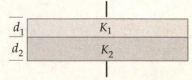

$$C = \left(\frac{1}{C_1} + \frac{1}{C_2}\right)^{-1} = \left(\frac{d_1}{K_1 A\varepsilon_0} + \frac{d_2}{K_2 A\varepsilon_0}\right)^{-1} = \boxed{\frac{A\varepsilon_0 K_1 K_2}{d_1 K_2 + d_2 K_1}}$$

67. The capacitance will be given by $C = Q/V$. When a charge Q is placed on one plate and a charge $-Q$ is placed on the other plate, an electric field will be set up between the two plates. The electric field in the air-filled region is just the

electric field between two charged plates, $E_0 = \dfrac{\sigma}{\varepsilon_0} = \dfrac{Q}{A\varepsilon_0}$. The electric field in the dielectric is equal to the electric

field in the air, divided by the dielectric constant: $E_D = \dfrac{E_0}{K} = \dfrac{Q}{KA\varepsilon_0}$.

The voltage drop between the two plates is obtained by integrating the electric field between the two plates. One plate is set at the origin with the dielectric touching this plate. The dielectric ends at $x = l$. The rest of the distance to $x = d$ is then air filled.

$$V = -\int_0^d \vec{\mathbf{E}} \cdot d\vec{\mathbf{x}} = \int_0^l \frac{Q\,dx}{KA\varepsilon_0} + \int_l^d \frac{Q\,dx}{A\varepsilon_0} = \frac{Q}{A\varepsilon_0}\left(\frac{l}{K} + (d-l)\right)$$

The capacitance is the ratio of the voltage to the charge.

$$C = \frac{Q}{V} = \frac{Q}{\dfrac{Q}{A\varepsilon_0}\left(\dfrac{l}{K} + (d-l)\right)} = \boxed{\dfrac{\varepsilon_0 A}{d-l+\dfrac{l}{K}}}$$

73. Since the capacitor is disconnected from the battery, the charge on it cannot change. The capacitance of the capacitor is increased by a factor of K, the dielectric constant.

$$Q = C_{initial}V_{initial} = C_{final}V_{final} \rightarrow V_{final} = V_{initial}\frac{C_{initial}}{C_{final}} = V_{initial}\frac{C_{initial}}{KC_{initial}} = (34.0\,\text{V})\frac{1}{2.2} = \boxed{15\,\text{V}}$$

79. The relative change in energy can be obtained by inserting Eq. 24-8 into Eq. 24-5.

$$\frac{U}{U_0} = \frac{\dfrac{Q^2}{2C}}{\dfrac{Q^2}{2C_0}} = \frac{C_0}{C} = \frac{\dfrac{A\varepsilon_0}{d}}{\dfrac{KA\varepsilon_0}{\left(\tfrac{1}{2}d\right)}} = \boxed{\dfrac{1}{2K}}$$

The dielectric is attracted to the capacitor. As such, the dielectric will gain kinetic energy as it enters the capacitor. An external force is necessary to stop the dielectric. The negative work done by this force results in the decrease in energy within the capacitor.

Since the charge remains constant, and the magnitude of the electric field depends on the charge, and not the separation distance, the electric field will not be affected by the change in distance between the plates. The electric field between the plates will be reduced by the dielectric constant, as given in Eq. 24-10.

$$\frac{E}{E_0} = \frac{E_0/K}{E_0} = \boxed{\frac{1}{K}}$$

85. (*a*) From the diagram, we see that one group of 4 plates is connected together, and the other group of 4 plates is connected together. This common grouping shows that the capacitors are connected $\boxed{\text{in parallel}}$.
(*b*) Since they are connected in parallel, the equivalent capacitance is the sum of the individual capacitances. The variable area will change the equivalent capacitance.

$$C_{eq} = 7C = 7\varepsilon_0\frac{A}{d}$$

$$C_{min} = 7\varepsilon_0\frac{A_{min}}{d} = 7(8.85\times10^{-12}\,\text{C}^2/\text{N}\cdot\text{m}^2)\frac{(2.0\times10^{-4}\,\text{m}^2)}{(1.6\times10^{-3}\,\text{m})} = 7.7\times10^{-12}\,\text{F}$$

$$C_{max} = 7\varepsilon_0\frac{A_{max}}{d} = 7(8.85\times10^{-12}\,\text{C}^2/\text{N}\cdot\text{m}^2)\frac{(9.0\times10^{-4}\,\text{m}^2)}{(1.6\times10^{-3}\,\text{m})} = 3.5\times10^{-11}\,\text{F}$$

And so the range is $\boxed{\text{from }7.7\,\text{pF to }35\,\text{pF}}$.

91. The force acting on one plate by the other plate is equal to the electric field produced by one charged plate multiplied by the charge on the second plate.

$$F = EQ = \left(\frac{Q}{2A\varepsilon_0}\right)Q = \frac{Q^2}{2A\varepsilon_0}$$

The force is attractive since the plates are oppositely charged. Since the force is constant, the work done in pulling the two plates apart by a distance x is just the force times distance.

$$W = Fx = \boxed{\frac{Q^2 x}{2A\varepsilon_0}}$$

The change in energy stored between the plates is obtained using Eq. 24-5.

$$W = \Delta U = \frac{Q^2}{2}\left(\frac{1}{C_2} - \frac{1}{C_1}\right) = \frac{Q^2}{2}\left(\frac{2x}{\varepsilon_0 A} - \frac{x}{\varepsilon_0 A}\right) = \boxed{\frac{Q^2 x}{2\varepsilon_0 A}}$$

The work done in pulling the plates apart is equal to the increase in energy between the plates.

97. (a) The initial capacitance is obtained directly from Eq. 24-8.

$$C_0 = \frac{K\varepsilon_0 A}{d} = \frac{3.7\left(8.85 \text{ pF/m}\right)\left(0.21\text{m}\right)\left(0.14\text{m}\right)}{0.030 \times 10^{-3}\text{m}} = \boxed{32\,\text{nF}}$$

(b) Maximum charge will occur when the electric field between the plates is equal to the dielectric strength. The charge will be equal to the capacitance multiplied by the maximum voltage, where the maximum voltage is the electric field times the separation distance of the plates.

$$Q_{max} = C_0 V = C_0 E d = \left(32\,\text{nF}\right)\left(15 \times 10^6 \text{ V/m}\right)\left(0.030 \times 10^{-3}\text{ m}\right)$$
$$= \boxed{14\,\mu\text{C}}$$

(c) The sheets of foil would be separated by sheets of paper with alternating sheets connected together on each side. This capacitor would consist of 100 sheets of paper with 101 sheets of foil.

$$t = 101 d_{\text{Al}} + 100 d_{\text{paper}} = 101\left(0.040 \text{ mm}\right) + 100\left(0.030 \text{ mm}\right)$$
$$= \boxed{7.0 \text{ mm}}$$

(d) Since the capacitors are in parallel, each capacitor has the same voltage which is equal to the total voltage. Therefore breakdown will occur when the voltage across a single capacitor provides an electric field across that capacitor equal to the dielectric strength.

$$V_{max} = E_{max} d = \left(15 \times 10^6 \text{ V/m}\right)\left(0.030 \times 10^{-3}\text{m}\right) = \boxed{450 \text{ V}}$$

Chapter 25: Electric Currents and Resistance

Chapter Overview and Objectives

In this chapter, electric current, electric circuits, and Ohm's Law are introduced. Topics related to circuits, such as power, resistivity, rms quantities, current density, and drift velocity, are also discussed.

After completing this chapter you should:
- Know that electric current is the rate of flow of electric charge.
- Know Ohm's Law.
- Be able to determine the resistance of a conductor given its cross-sectional area, length, and resistivity.
- Know how to relate resistivity and resistance at one temperature to the resistivity and resistance at another temperature.
- Know how to calculate power in both dc and ac resistive electrical circuits.
- Know what conventional current is.
- Know what rms quantities are and how they relate to peak values in sinusoidal functions.
- Know what current density is and how to calculate current from a known current density.
- Know what drift velocity is.

Summary of Equations

Definition of average current:
$$\bar{I} = \frac{\Delta Q}{\Delta t}$$
(Section 25-2)

Definition of instantaneous current:
$$I = \frac{dQ}{dt}$$
(Section 25-2)

Ohm's Law:
$$V = IR$$
(Section 25-3)

Resistance of a conductor:
$$R = \rho \frac{L}{A}$$
(Section 25-4)

Relationship between resistivity and conductivity:
$$\sigma = \frac{1}{\rho}$$
(Section 25-4)

Temperature dependence of resistivity:
$$\rho(T) = \rho(T_0)\left[1 + \alpha(T - T_0)\right]$$
(Section 25-4)

Power dissipated in a dc circuit:
$$P = IV = \frac{V^2}{R} = I^2 R$$
(Section 25-5)

Instantaneous power in a resistive ac circuit:
$$P = I_0 V_0 \sin^2 \omega t$$
(Section 25-7)

Average power in a resistive ac circuit:
$$\bar{P} = \frac{1}{2} I_0 V_0 = \frac{1}{2}\frac{V_0^2}{R} = \frac{1}{2} I_0^2 R = I_{RMS} V_{RMS}$$
(Section 25-7)

RMS voltage for an ac voltage with amplitude V_0:
$$V_{RMS} = \frac{V_0}{\sqrt{2}}$$
(Section 25-7)

RMS current for an ac current with amplitude I_0:
$$I_{RMS} = \frac{I_0}{\sqrt{2}}$$
(Section 25-7)

Definition of current density $\vec{j}$: $\qquad \vec{j} = \dfrac{\vec{I}}{A}$ (Section 25-8)

Relationship between $\vec{j}$ and the drift velocity $\vec{v}_d$: $\quad \vec{j} = nq\vec{v}_d$ (Section 25-8)

Relationship between $\vec{j}$ and the electric field $\vec{E}$: $\quad \vec{j} = \sigma\vec{E}$ or $\vec{E} = \rho\vec{j}$ (Section 25-8)

Chapter Summary

Section 25-1. The Electric Battery

When two dissimilar metal plates are immersed in an electrolyte, a potential difference develops between the plates. Such an arrangement is called an **electric cell** and the metal plates are called the **electrodes**. The potential difference between the electrodes depends on the type of metals that are used for the electrodes. A **battery** is either a single electric cell or a collection of electric cells connected either in series or in parallel. Each battery has two **terminals**, connected to the internal electrodes, to which external conductors can be connected.

Section 25-2. Electric Current

If a conductive path is connected between the terminals of a battery, electric charge will move along the conductive path. This flow of charge is called an **electric current**. The strength of the electric current is specified in terms of the amount of charge ΔQ that passes a given cross-section during a time interval Δt. The **time-averaged current** is given by

$$\overline{I} = \frac{\Delta Q}{\Delta t}$$

If the time interval Δt approaches zero, the time-averaged current approaches the **instantaneous current**:

$$I = \frac{dQ}{dt}$$

The unit of current is the **ampere** (A). One ampere corresponds to one Coulomb of charge passing through a given cross-section during a one second time interval:

$$1\,A = 1\,C/s$$

The direction of the current is defined as the direction from the region with the higher electric potential to the region with the lower electric potential. This corresponds to the direction in which positive charges will move. If the charge carriers in the material are electrons, they will move in a direction opposite to the direction in which the current flows.

Example 25-2-A. Calculating transferred charge. The current in a particular circuit as a function of time is given as

$$I(t) = 1.32 + 0.64t + 0.24t^2$$

where I is in amperes and t is in seconds. What is the total amount of charge transferred between time $t = 0$ and time $t = 10$ s?

Approach: Since the current I is defined as dQ/dt, we can determine the total charge transferred by integrating the current with respect to time between time $t = 0$ and time $t = 10$ s.

Solution: The total charge transferred is equal to

$$\Delta Q = \int_{Q_i}^{Q_f} dQ = \int_{t=0}^{t=10\,s} I\,dt = \int_0^{10} \left(1.32 + 0.64t + 0.24t^2\right)dt =$$
$$= \left[1.32t + 0.32t^2 + 0.080t^3\right]\Big|_0^{10} = 13.2 + 32 + 80 = 125\,C$$

The total charge transferred during these 10 seconds is thus 125 C.

Section 25-3. Ohm's Law: Resistance and Resistors

The electric current I that flows through a conductor is proportional to the potential difference V across the conductor. This relationship is known as **Ohm's Law**:

$$V = IR$$

The constant R is the **resistance** of the conductor. Resistance has the dimension of electric potential divided by current. The SI unit of resistance is the ohm (Ω), defined as one volt/ampere:

$$1\Omega = 1\,V/A$$

Ohm's Law is valid for a wide range of conductive materials and conditions. Resistance is represented in circuit diagrams by the symbol shown in the Figure on the right.

Example 25-3-A. Determining the resistance of a circuit. A circuit powered by a 9.0 V battery has a current of 17.0 mA flowing through it. What is the resistance of the circuit?

Approach: If both the current and the voltage in a circuit are specified, Ohm's law can be used to determine the resistance of the circuit.

Solution: Applying Ohm's law, the resistance of the circuit can be determined:

$$R = \frac{V}{I} = 5.3 \times 10^2\,\Omega$$

Section 25-4. Resistivity

The resistance of a conductor is proportional to the length of the conductor and inversely proportional to the cross-sectional area of the conductor:

$$R = \rho \frac{L}{A}$$

The constant of proportionality, ρ, is called the **resistivity** of the material. The resistivity depends on the type of material, its temperature, and pressure. Sometimes it is more useful to work with a quantity called **conductivity** σ. The conductivity is the inverse of resistivity:

$$\sigma = \frac{1}{\rho}$$

The resistivity depends on temperature. For metals, the resistivity has the following approximate linear dependence on the temperature T:

$$\rho(T) = \rho(T_0)\left[1 + \alpha(T - T_0)\right]$$

where $\rho(T)$ is the resistivity of the material at temperature T, $\rho(T_0)$ is the resistivity at temperature T_0, and α is the **linear temperature coefficient of resistivity**.

Example 25-4-A. Exploring the temperature dependence of the resistivity. Two round wires are made of the same material. The first wire has a diameter of 0.450 mm and a length of 23.4 cm. It has a resistance of 25.8 Ω at a temperature of 64.2°C. The second wire has a diameter of 0.750 mm and a length of 184 cm. It has a resistance of 64.8 Ω at a temperature of 33.2°C. What is the temperature coefficient of the resistivity of the material? What is the resistivity of this material at 20°C?

Approach: Although the wires have different dimensions and temperatures, sufficient information is provided to determine the resistivity as function of temperature. Assuming a linear relation between the resistivity and temperature, we can use the resistance at the two different temperatures to determine the resistivity at 20°C.

Solution: The relationship between resistance, resistivity, and geometry can be used to calculate the resistivity of the two samples at their given temperatures:

$$R_1 = \rho_{64.2} \frac{L_1}{A_1} \quad \Rightarrow \quad \rho_{64.2} = \frac{R_1 A_1}{L_1} = 1.75 \times 10^{-5} \, \Omega \cdot m$$

$$R_2 = \rho_{33.2} \frac{L_2}{A_2} \quad \Rightarrow \quad \rho_{33.2} = \frac{R_2 A_2}{L_2} = 1.56 \times 10^{-5} \, \Omega \cdot m$$

To determine the temperature coefficient of the resistivity, we relate the resistivity at the given temperatures in terms of the temperature coefficient of resistivity:

$$\rho_{64.2} = \rho_{33.2} \left[1 + \alpha \Delta T \right] = \rho_{33.2} \left[1 + \alpha (64.2 - 33.2) \right]$$

The only unknown is the linear temperature coefficient of resistivity α which can now be determined:

$$\alpha = \frac{(\rho_{64.2} / \rho_{33.2}) - 1}{\Delta T} = 3.9 \times 10^{-3} \, {}^{\circ}C^{-1}$$

The resistivity at 20°C can now be determined:

$$\rho_{20} = \rho_{33.2} \left[1 + \alpha (20 - 33.2) \right] = \left(1.56 \times 10^{-5} \right) \left[1 + \left(3.9 \times 10^{-3} \right)(20 - 33.2) \right] = 1.48 \times 10^{-5} \, \Omega \cdot m$$

Section 25-5. Electric Power

Consider a circuit in which a source of emf is connected to a resistive circuit element. When an electric charge dq moves across the electric potential difference, its electrical potential energy changes. The change in the potential energy dU of a charge dq moving through a potential difference V is equal to

$$dU = V \, dq$$

If we apply the law of conservation of energy or the work–energy theorem to this circuit, we must conclude that the resistive circuit element does negative work on the electric charge. If we divide the change in the potential energy dU by dt, we obtain

$$\frac{dU}{dt} = V \frac{dq}{dt} = VI$$

Since dU/dt is defined as the power P (see Section 8.8) we conclude that the power dissipated in the resistive circuit element is VI. The power dissipated in the resistive circuit is converted into other forms of energy in the material of the circuit elements.

Using Ohm's Law, we can rewrite the power in terms of either current or voltage and resistance:

$$P = IV = \frac{V^2}{R} = I^2 R$$

The unit of electrical energy supplied by electrical utility companies is the kilowatt-hour (kW-h). One kilowatt-hour is equal to the energy that corresponds to a power of 1 kW, supplied for 1 hour: 1 kW-h = 3.600×10^6 J.

Section 25-6. Power in Household Circuits

Electric power is supplied to customers at standardized voltages. The supplied voltage has a simple harmonic dependence on time. Loads or devices are connected to the supplied voltage in *parallel*. Each device connected to the supply thus has the same standardized voltage across its terminals. The total current flowing from the energy source is the sum of the currents flowing through each device.

To protect our homes from the possibility of fire due to overheated wires, household circuits include one or more fuses or circuit breakers that disconnect the circuit from the energy source when the current in the circuit exceeds a certain value.

Example 25-6-A. Calculating the proper rating of a circuit breaker. A motor that draws a maximum power of 2.6 horsepower is to be connected to a 120 V household circuit. What is the minimum current capacity of the circuit required to supply the motor, without tripping a circuit breaker? We must assume that the motor acts as a resistance under these circumstances.

Approach: This problem provides us with information about the maximum power that is used by the motor. The maximum power can be used to determine the maximum current that the motor will draw and thus the minimum requirement of the circuit breaker to be used. Note: since the power is provided in terms of horsepower, we need to convert it to units of W.

Solution: To convert the power to Watts, we use the following conversion factor: 1 hp = 746 W. The maximum power used by the motor is thus 1940 W. The current required to provide this power is equal to

$$I = \frac{P}{V} = \frac{1940}{120} = 16 \text{ A}$$

Standard household circuits have capacities of 15 A, 20 A, and 30 A. This motor thus requires a 20 A or 30 A household circuit.

Section 25-7. Alternating Current

A current that flows with a constant magnitude in a given direction is called a **direct current (dc)**. A current that varies with time, such as a current that has a simple harmonic time dependence, is called an **alternating current (ac)**. In linear circuits (circuits in which the current is proportional to the voltage) an ac current implies that an ac voltage is present. The ac voltage can be written as

$$V = V_0 \sin(2\pi ft) = V_0 \sin \omega t$$

where V_0 is called the **peak voltage** and f is the frequency of the simple harmonic function. In a resistive circuit, Ohm's law is satisfied at each instant in time. This implies that the current I is given by

$$I = \frac{V}{R} = \frac{V_0}{R} \sin(2\pi ft) = I_0 \sin \omega t$$

I_0 is called the **peak current** and is equal to the peak voltage divided by the resistance R.

The instantaneous power dissipated in the resistor of an ac circuit is equal to the instantaneous voltage multiplied by the instantaneous current:

$$P = IV = (I_0 \sin \omega t) \times (V_0 \sin \omega t) = I_0 V_0 \sin^2 \omega t$$

The instantaneous power can also be written as

$$P = I_0^2 R \sin^2 \omega t = \frac{V_0^2}{R} \sin^2 \omega t$$

In many situations, the time average power provides more useful information than the instantaneous power. The time average of $\sin^2 \omega t$, averaged over one period, is ½. The time-averaged power is equal to

$$\bar{P} = \frac{1}{T} \int_0^T I_0^2 R \sin^2 \omega t \, dt = \frac{1}{2} I_0^2 R$$

where T is the period of the simple harmonic function. This relation can also be written in terms of the peak voltage:

$$\bar{P} = \frac{1}{2} \frac{V_0^2}{R}$$

Commonly, the quantities **root-mean-square (rms) voltage** V_{RMS} and **root-mean-square current** I_{RMS} are used to describe the magnitude of ac voltages and currents. The root-mean-square voltage associated with an AC voltage with a peak voltage V_0 is equal to

$$V_{RMS} = \sqrt{\frac{1}{T} \int_0^T V_0^2 \sin^2 \omega t \, dt} = \frac{V_0}{\sqrt{2}}$$

The root-mean-square current associated with an AC current with a peak current I_0 is equal to

$$I_{RMS} = \sqrt{\frac{1}{T} \int_0^T I_0^2 \sin^2 \omega t \, dt} = \frac{I_0}{\sqrt{2}}$$

The average power can be expressed in terms of the root-mean-square voltage and current:

$$\bar{P} = I_{RMS} V_{RMS} = I_{RMS}^2 R = \frac{V_{RMS}^2}{R}$$

When we use the root-mean-square voltages and currents, the relationships between voltages and currents in AC circuits are similar to the corresponding relationships in DC circuits.

Example 25-7-A. Calculating properties of AC circuits. A resistive circuit with a resistance of 150 Ω is connected to a sinusoidal AC potential difference with a root-mean-square voltage of 120 V. Determine the average power in the circuit, the peak power, the root-mean-square current, and the peak current.

Approach: The problem provides information about the resistance R and the root-mean-square supply voltage. Based on this information, the relations discussed in this Section can be used to determine the requested quantities.

Solution: The average power in the circuit is equal to

$$\bar{P} = \frac{V_{rms}^2}{R} = 96 \, \text{W}$$

The peak power is given by the square of the peak voltage divided by the resistance. The peak voltage is $\sqrt{2}$ times as large as the root-mean-square voltage, and the peak power is thus 2 times as large as the average power:

$$P_P = \frac{V_0^2}{R} = \frac{\left(\sqrt{2} V_{rms}\right)^2}{R} = 2\frac{V_{rms}^2}{R} = 2\bar{P} = 192 \, \text{W}$$

The root-mean-square current can be determined from the root-mean-square voltage and the resistance, using Ohm's law:

$$I_{rms} = \frac{V_{rms}}{R} = 0.80 \, \text{A}$$

The peak current is $\sqrt{2}$ times as large as the root-mean-square current. We thus conclude that

$$I_P = \sqrt{2} I_{rms} = \sqrt{2}\left(0.80 \, \text{A}\right) = 1.13 \, \text{A}$$

Section 25-8. Microscopic View of Electric Current: Current Density and Drift Velocity

All conductors have a finite cross-sectional area A. The **current density** j in a conductor is the current I per cross-sectional area A:

$$j = \frac{I}{A}$$

In general, the current density will be a function of position, and must be written as

$$j = \frac{dI}{dA}$$

where dI is the infinitesimal current flowing through an infinitesimal cross-sectional area dA. The total current flowing through the wire can be found by integrating the current density across the cross-sectional area:

$$I = \int \vec{\mathbf{j}} \cdot d\vec{\mathbf{A}}$$

The direction of the vector $\vec{\mathbf{j}}$ is the direction of the current flow and the direction of $d\vec{\mathbf{A}}$ is perpendicular to the infinitesimal cross-sectional area element.

Although an exact description of microscopic current flow requires a quantum mechanical description, we can use a classical model of microscopic current flow that will allow us to understand many important aspects of currents in conductors. When an electric field is applied across a conductor, the charges initially accelerate, but they are eventually scattered into random directions by what we can picture as collisions with atoms or other moving charged particles. Equilibrium is reached between the acceleration and the scattering processes and the average velocity of the charged particles carrying the current reaches a time-independent value. This average velocity is called the **drift velocity**, $\vec{v}_d$. The magnitude of the drift velocity is much smaller than the average speed of the charged particles. The current density $\vec{j}$ is related to the drift velocity $\vec{v}_d$:

$$\vec{j} = nq\vec{v}_d$$

where n is the number of charged particles per unit volume and q is the charge of each particle. The current density is also proportional to the potential difference across the conductor and thus the electric field inside the conductor. The constant of proportionality is the conductivity σ or the inverse of the resistivity ρ:

$$\vec{j} = \sigma\vec{E} = \frac{1}{\rho}\vec{E}$$

This relation assumes that the local current is in the direction of the local electric field. This is typical of many materials, but there are some materials in which the current density is not parallel to the electric field.

Example 25-8-A. Calculating the resistivity of a material on the basis of its microscopic properties. A particular material has electrons as charge carriers and a carrier density of $1.96 \times 10^{19}/cm^3$. If the drift velocity is 3.84 cm/s when an electric field of 1.45 V/m exists within the conductor, what is the resistivity of the material?

Approach: The information provided in the problem can be used to determine the current density. This current density, combined with the known electric field, can be used to determine the resistivity.

Solution: The current density of the electrons is equal to

$$j = nqv_d$$

The resistivity can be obtained from the current density and the electric field:

$$\rho = \frac{E}{j}$$

Using the information provided in the problem we can now determine the resistivity:

$$\rho = \frac{E}{nqv_d} = 1.20 \times 10^{-5}\,\Omega \cdot m$$

Section 25-9. Superconductivity

The resistivity of some materials becomes 0 when their temperature drops below a material-dependent critical temperature. These materials are called **superconductors** and the condition of extremely low conductivity is called **superconductivity**. Superconductors have other unusual properties, such as their magnetic properties, which will be discussed in later chapters.

Section 25-10. Electric Conduction in the Nervous System

The human nervous system relies on electric signals to transmit information. The electric signals are transmitted by **neurons**. When a neuron is stimulated, it will conduct voltage pulses along its length. The shape of the voltage pulse is called the **action potential**; the voltage pulse travels along the neuron with a speed between 30 m/s and 150 m/s. The action potential is generated as a result of a rapid change of the permeability of the cell membrane.

Practice Quiz

1. The resistance of a piece of a wire depends on
   a) the length of the piece of material.
   b) the cross-sectional area of the piece of material.
   c) the resistivity of the piece of material.
   d) all of the above.

2. If the voltage across a fixed resistance is doubled, what happens to the power dissipated in the resistor?
   a) Power becomes twice as large.
   b) Power becomes four times as large.
   c) Power becomes half as large.
   d) Power becomes one fourth as large.

3. A resistor is connected across a fixed potential difference. The current is 2.00 A and the resistance is 4.00 Ω. If the resistance of the resistor is doubled, what will the power be that is dissipated in the new resistor?
   a) 32.0 W
   b) 16.0 W
   c) 8.00 W
   d) 4.00 W

4. If you take a 120 V 100 W light bulb and measure its resistance with an ohm meter, you measure a resistance of 40 Ω. If you calculate the power from a 120 V rms source with a 40 Ω load using $P=V^2/R$ you get a power of 360 W. Why is the light bulb only rated 100 W?
   a) The bulb rating only gives the power that goes into producing visible light, not into heat.
   b) The power calculation is incorrect because the rms voltage value was used.
   c) The resistance of the bulb is considerably higher at the operating temperature.
   d) The calculated power of 360 W is the peak power, the 100 W rating is the average power.

5. For a sinusoidal voltage with fixed amplitude, what happens to the rms voltage as the frequency is increased?
   a) When the frequency is increased, the rms value increases.
   b) When the frequency is increased, the rms value decreases.
   c) When the frequency is increased, the rms value remains the same.
   d) When the frequency is increased, the rms value could increase or decrease.

6. Two copper wires, each of the same length but different cross-sectional area, are connected across the same potential difference. Which statement is true?
   a) The current through each of the wires is the same.
   b) The current density in each wire is inversely proportional to its cross-sectional area.
   c) The current density in each of the wires is the same.
   d) The current in each of the wires is inversely proportional to its cross-sectional area.

7. Two wires of identical resistance, but made from different materials, are connected across the same potential difference at room temperature. The wires begin to warm up due to the electrical power dissipated in them. After current flows for some time, one of the wires is at a higher temperature than the other. Which wire is at the higher temperature?
   a) The wire with the greater temperature coefficient of resistivity.
   b) The wire with the lesser temperature coefficient of resistivity.
   c) The wire with the greater length.
   d) The wire with the greater cross-sectional area.

8. Four identically shaped cylinders are made from four different metals. Which of these four materials will produce the cylinder with the greatest resistance?
   a) Copper
   b) Platinum
   c) Gold
   d) Iron

9. A particular electrical circuit related quantity is measured in units of $kg \cdot m^2/(C^2 \cdot s)$. Which electrical circuit quantity is being measured?
    a) Voltage
    b) Current
    c) Resistance
    d) Power

10. The rms of a sinusoidal current in an ac circuit is 20.0 A. What is the peak current in the circuit?
    a) 14.1 A
    b) 28.2 A
    c) 20.0 A
    d) 40.0 A

11. A dc electrical source supplies a potential difference of 12.0 V. Loads with resistances 120 Ω, 240 Ω, and 300 Ω are connected in parallel to the source. What is the total current and power supplied by the source?

12. A lightning strike occurs between a potential difference of 3.8×10^7 V and an average current of 88 A flows for a time of 124 ms. If a power company charges 10 cents per kilowatt-hr, what would the amount of energy in the lightning strike be worth to the power company?

13. A time-varying voltage is a periodic triangular function as shown in the diagram. Determine the rms voltage of this voltage. The positive peaks are at V_0 and the negative peaks are at $-V_0$.

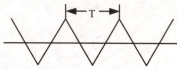

14. The change in resistance of a conductor with temperature can be used to construct a temperature measuring device. For a thermometer that needs to resolve a 0.5° C temperature difference, what must be the precision in measuring the resistance of a 1000 Ω resistor. Assume the conductor is made of copper.

15. Calculate the drift velocity of electrons in a copper wire with a cross-sectional area of 1.67×10^{-5} m² when it is carrying a current of 11.6 A.

Responses to Select End-of-Chapter Questions

1. A battery rating in ampere-hours gives the total amount of charge available in the battery.

7. If the emf in a circuit remains constant and the resistance in the circuit is increased, less current will flow, and the power dissipated in the circuit will decrease. Both power equations support this result. If the current in a circuit remains constant and the resistance is increased, then the emf must increase and the power dissipated in the circuit will increase. Both equations also support this result. There is no contradiction, because the voltage, current, and resistance are related to each other by V = IR.

13. The electric power transferred by the lines is P = IV. If the voltage across the transmission lines is large, then the current in the lines will be small. The power lost in the transmission lines is P = I²R. The power dissipated in the lines will be small, because I is small.

19. In the two wires described, the drift velocities of the electrons will be about the same, but the current density, and therefore the current, in the wire with twice as many free electrons per atom will be twice as large as in the other wire.

Solutions to Select End-of-Chapter Problems

1. Use the definition of current, Eq. 25-1a.

$$I = \frac{\Delta Q}{\Delta t} \rightarrow 1.30\,\text{A} = \frac{1.30\,\text{C}}{\text{s}} \times \frac{1\ \text{electron}}{1.60 \times 10^{-19}\text{C}} = \boxed{8.13 \times 10^{18}\ \text{electrons/s}}$$

7. Use Ohm's Law, Eq. 25-2a, to find the current. Then use the definition of current, Eq. 25-1a, to calculate the number of electrons per minute.

$$I = \frac{V}{R} = \frac{\Delta Q}{\Delta t} = \frac{4.5\,\text{V}}{1.6\,\Omega} = \frac{2.8\,\text{C}}{\text{s}} \times \frac{1\,\text{electron}}{1.60 \times 10^{-19}\,\text{C}} \times \frac{60\,\text{s}}{1\,\text{min}} = \boxed{1.1 \times 10^{21}\,\frac{\text{electrons}}{\text{minute}}}$$

13. Use Eq. 25-3 to calculate the resistances, with the area as $A = \pi r^2 = \pi d^2/4$.

$$R = \rho \frac{\ell}{A} = \rho \frac{4\ell}{\pi d^2}.$$

$$\frac{R_{Al}}{R_{Cu}} = \frac{\rho_{Al} \dfrac{4\ell_{Al}}{\pi d_{Al}^2}}{\rho_{Cu} \dfrac{4\ell_{Cu}}{\pi d_{Cu}^2}} = \frac{\rho_{Al}\ell_{Al}d_{Cu}^2}{\rho_{Cu}\ell_{Cu}d_{Al}^2} = \frac{\left(2.65 \times 10^{-8}\,\Omega\text{•m}\right)\left(10.0\,\text{m}\right)\left(1.8\,\text{mm}\right)^2}{\left(1.68 \times 10^{-8}\,\Omega\text{•m}\right)\left(20.0\,\text{m}\right)\left(2.0\,\text{mm}\right)^2} = \boxed{0.64}$$

19. Use Eq. 25-5 multiplied by ℓ/A so that it expresses resistances instead of resistivity.

$$R = R_0\left[1 + \alpha\left(T - T_0\right)\right] \rightarrow$$

$$T = T_0 + \frac{1}{\alpha}\left(\frac{R}{R_0} - 1\right) = 20°\text{C} + \frac{1}{0.0045\left(\text{C}°\right)^{-1}}\left(\frac{140\,\Omega}{12\,\Omega} - 1\right) = 2390°\text{C} \approx \boxed{2400°\text{C}}$$

25. The resistance depends on the length and area as $R = \rho\ell/A$. Cutting the wire and running the wires side by side will halve the length and double the area.

$$R_2 = \frac{\rho\left(\frac{1}{2}\ell\right)}{2A} = \frac{1}{4}\frac{\rho\ell}{A} = \boxed{\frac{1}{4}R_1}$$

31. Use Eq. 25-6 to find the power from the voltage and the current.

$$P = IV = \left(0.27\,\text{A}\right)\left(3.0\,\text{V}\right) = \boxed{0.81\,\text{W}}$$

37. (a) Use Eq. 25-6 to find the current.

$$P = IV \rightarrow I = \frac{P}{V} = \frac{95\,\text{W}}{115\,\text{V}} = \boxed{0.83\,\text{A}}$$

(b) Use Eq. 25-7b to find the resistance.

$$P = \frac{V^2}{R} \rightarrow R = \frac{V^2}{P} = \frac{\left(115\,\text{V}\right)^2}{95\,\text{W}} \approx \boxed{140\,\Omega}$$

43. Each bulb will draw an amount of current found from Eq. 25-6.

$$P = IV \rightarrow I_{bulb} = \frac{P}{V}$$

The number of bulbs to draw 15 A is the total current divided by the current per bulb.

$$I_{total} = nI_{bulb} = n\frac{P}{V} \rightarrow n = \frac{VI_{total}}{P} = \frac{\left(120\,\text{V}\right)\left(15\,\text{A}\right)}{75\,\text{W}} = \boxed{24\,\text{bulbs}}$$

49. Use Ohm's law and the relationship between peak and rms values.

$$I_{peak} = \sqrt{2}I_{rms} = \sqrt{2}\,\frac{V_{rms}}{R} = \sqrt{2}\,\frac{220\,\text{V}}{2700\,\Omega} = \boxed{0.12\,\text{A}}$$

55. (a) The average power used can be found from the resistance and the rms voltage by Eq. 25-10c.

$$\bar{P} = \frac{V_{rms}^2}{R} = \frac{(240\,\text{V})^2}{44\,\Omega} = 1309\,\text{W} \approx \boxed{1300\,\text{W}}$$

    (b) The maximum power is twice the average power, and the minimum power is 0.

$$P_{max} = 2\bar{P} = 2(1309\,\text{W}) \approx \boxed{2600\,\text{W}} \qquad P_{min} = \boxed{0\,\text{W}}$$

61. The speed is the change in position per unit time.

$$v = \frac{\Delta x}{\Delta t} = \frac{7.20 \times 10^{-2}\,\text{m} - 3.40 \times 10^{-2}\,\text{m}}{0.0063\,\text{s} - 0.0052\,\text{s}} = \boxed{35\,\text{m/s}}$$

Two measurements are needed because there may be a time delay from the stimulation of the nerve to the generation of the action potential.

67. From Eq. 25-2b, if $R = V/I$, then $G = I/V$

$$G = \frac{I}{V} = \frac{0.48\,\text{A}}{3.0\,\text{V}} = \boxed{0.16\,\text{S}}$$

73. (a) The resistance at the operating temperature can be calculated directly from Eq. 25-7.

$$P = \frac{V^2}{R} \rightarrow R = \frac{V^2}{P} = \frac{(120\,\text{V})^2}{75\,\text{W}} = \boxed{190\,\Omega}$$

    (b) The resistance at room temperature is found by converting Eq. 25-5 into an equation for resistances and solving for R_0.

$$R = R_0\left[1 + \alpha\left(T - T_0\right)\right]$$

$$R_0 = \frac{R}{\left[1 + \alpha\left(T - T_0\right)\right]} = \frac{192\,\Omega}{\left[1 + \left(0.0045\,\text{K}^{-1}\right)\left(3000\,\text{K} - 293\,\text{K}\right)\right]} = \boxed{15\,\Omega}$$

79. (a) Use Eq. 25-7b.

$$P = \frac{V^2}{R} \rightarrow R = \frac{V^2}{P} = \frac{(240\,\text{V})^2}{2800\,\text{W}} = 20.57\,\Omega \approx \boxed{21\,\Omega}$$

    (b) Only 75% of the heat from the oven is used to heat the water. Use Eq. 19-2.

$$0.75\left(P_{oven}\right)t = \text{Heat absorbed by water} = mc\Delta T \rightarrow$$

$$t = \frac{mc\Delta T}{0.75\left(P_{oven}\right)} = \frac{(0.120\,\text{L})\left(\frac{1\,\text{kg}}{1\,\text{L}}\right)(4186\,\text{J/kg·C°})(85\,\text{C°})}{0.75(2800\,\text{W})} = 20.33\,\text{s} \approx \boxed{20\,\text{s}}\,(2\text{ sig. fig.})$$

    (c) $\frac{11\,\text{cents}}{\text{kWh}}(2.8\,\text{kW})(20.33\,\text{s})\frac{1\,\text{h}}{3600\,\text{s}} = \boxed{0.17\,\text{cents}}$

85. Model the protons as moving in a continuous beam of cross-sectional area A. Then by Eq. 25-13, $I = neAv_d$, where we only consider the absolute value of the current. The variable n is the number of protons per unit volume, so $n = \dfrac{N}{A\ell}$, where N is the number of protons in the beam and ℓ is the circumference of the ring. The "drift" velocity in this case is the speed of light.

$$I = neAv_d = \frac{N}{A\ell} eAv_d = \frac{N}{\ell} ev_d \rightarrow$$

$$N = \frac{I\ell}{ev_d} = \frac{\left(11\times10^{-3}\right)\left(6300\,\text{m}\right)}{\left(1.60\times10^{-19}\,\text{C}\right)\left(3.00\times10^{8}\,\text{m/s}\right)} = \boxed{1.4\times10^{12}\ \text{protons}}$$

91. Eq. 25-3 can be used. The area to be used is the cross-sectional area of the pipe.

$$R = \frac{\rho\ell}{A} = \frac{\rho\ell}{\pi\left(r^2_{\text{outside}} - r^2_{\text{inside}}\right)} = \frac{\left(1.68\times10^{-8}\,\Omega\!\cdot\!\text{m}\right)\left(10.0\,\text{m}\right)}{\pi\left[\left(2.50\times10^{-2}\,\text{m}\right)^2 - \left(1.50\times10^{-2}\,\text{m}\right)^2\right]} = \boxed{1.34\times10^{-4}\,\Omega}$$

Chapter 26: DC Circuits

Chapter Overview and Objectives

In this chapter, DC circuits are discussed. Kirchhoff's circuit rules are introduced, and used to predict the current in the various circuit branches. The time dependence of the currents in RC circuits is determined. Finally, the construction of voltage and current meters from simple galvanometers is discussed.

After completing this chapter you should:

- Know what the terminal voltage of a battery is and how to calculate it from the emf and the internal resistance of the source.
- Know how to recognize when resistances are connected in series and in parallel.
- Know how to calculate equivalent resistances.
- Know Kirchhoff's current rule and Kirchhoff's voltage rule and how to apply these rules to solve for unknown currents and voltages in DC circuits.
- Know the time dependence of charge and current in an RC circuit with DC sources.
- Know how to calculate the time constant of an RC circuit.
- Know how to determine the resistances necessary to create ammeters and voltmeters from galvanometers.

Summary of Equations

Terminal voltage of a battery: $V_{ab} = \varepsilon - Ir$ (Section 26-1)

Equivalent resistance of resistances in series: $R_{eq} = \sum R_i$ (Section 26-2)

Equivalent resistance of resistances in parallel: $\dfrac{1}{R_{eq}} = \sum \dfrac{1}{R_i}$ (Section 26-2)

Charge on a capacitor in a RC circuit: $Q = C\varepsilon + (Q_0 - C\varepsilon)e^{-t/RC}$ (Section 26-5)

Current in an RC circuit: $I = -\dfrac{Q_0 - C\varepsilon}{RC}e^{-t/RC}$ (Section 26-5)

Time constant of an RC circuit: $\tau = RC$ (Section 26-5)

Series resistance of a voltmeter: $R_s = \dfrac{V_{fs}}{I_g} - R_g$ (Section 26-7)

Parallel or shunt resistance of a current meter: $R_p = \dfrac{I_g R_g}{I_{fs} - I_g}$ (Section 26-7)

Chapter Summary

Section 26-1. EMF and Terminal Voltage

The **electromotive force** or **emf** of a device that creates an electric potential difference is the electric potential difference across the two terminals that exists when no current flows from the source. When current does flow through the device, the device may have some **internal resistance** that causes a voltage drop internal to the device, so that the potential difference across the terminals falls below the emf of the source. The potential difference between the two terminals of the device is called the **terminal voltage**. The terminal voltage of the device is related to the emf ε, the internal resistance r, and the current I that flows from the terminals of the source

$$V_{ab} = \varepsilon - Ir$$

Example 26-1-A. Maximum power provided by a battery. Determine the maximum electrical power that can be provided by a battery with an emf ε and an internal resistance r. Also, determine the resistance R of the external circuit that causes this maximum power drain from the battery.

Approach: When the external resistance R decreases, the current increases. This results in an increase in the voltage drop across the internal resistor, and a decrease in the terminal voltage seen by the external circuit. By expressing the power provided by the battery as a function of R, we can determine for what external resistance R the power provided by the battery is maximized.

Solution: The power provided by the battery will be the product of the terminal voltage and the current. Using Ohm's law, we can replace the terminal voltage with the product of the current and the external resistance R. The power provided by the battery is thus equal to

$$P = VI = I^2 R$$

The current in the circuit will be equal to the ratio of the emf and the sum of the internal resistance r and the external resistance R:

$$I = \frac{\varepsilon}{r + R}$$

The power provided by the battery is thus equal to

$$P = \left(\frac{\varepsilon}{r + R}\right)^2 R$$

The emf and the internal resistance are fixed parameters, but the external resistance can be changed. If our goal is to obtain the maximum power from the battery, we need to determine which value of R optimizes P. At this value of R, the derivative of P will be zero:

$$\frac{dP}{dR} = \frac{d}{dR}\left\{\frac{\varepsilon^2}{(r+R)^2} R\right\} = -2\frac{\varepsilon^2}{(r+R)^3} R + \frac{\varepsilon^2}{(r+R)^2} = \frac{\varepsilon^2}{(r+R)^2}\left\{-\frac{2R}{r+R} + 1\right\} = \frac{\varepsilon^2}{(r+R)^2}\left\{\frac{r-R}{r+R}\right\} = 0$$

The derivative of P is zero when $R = r$. Putting this value of the external resistance into the expression of the power delivered by the battery, we obtain

$$P = \left(\frac{\varepsilon}{r + r}\right)^2 r = \frac{\varepsilon^2}{4r}$$

Section 26-2. Resistors in Series and in Parallel

Resistors can be connected in **series** or in **parallel**. When resistors are connected in series, an identical current flows through each resistor. Resistors that are connected in series have an equivalent resistance that is the sum of the resistances in the series network. The diagram shows three resistors connected in series. The total voltage across this resistor network if a current I is flowing through the resistors is equal to

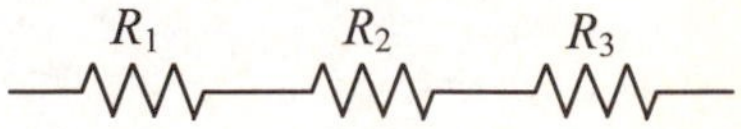

$$V = IR_1 + IR_2 + IR_3 = I\left(R_1 + R_2 + R_3\right) = IR_{eq}$$

The equivalent resistance of this series network is

$$R_{eq} = R_1 + R_2 + R_3$$

In general, for an arbitrary number of resistors connected in series, the equivalent resistance, R_{eq}, is

$$R_{eq} = \sum R_i$$

where R_i is the resistance of the ith resistor in the network.

When resistors are connected in parallel, the potential difference across each resistor is the same. The three resistors shown in the diagram are connected in parallel. The total current flowing through these resistors is equal to

$$I = \frac{V}{R_1} + \frac{V}{R_2} + \frac{V}{R_3} = V\left\{\frac{1}{R_1} + \frac{1}{R_2} + \frac{1}{R_3}\right\} = \frac{V}{R_{eq}}$$

The equivalent resistance of these three resistors is given by

$$\frac{1}{R_{eq}} = \frac{1}{R_1} + \frac{1}{R_2} + \frac{1}{R_3}$$

In general, the equivalent resistance of an arbitrary number of resistors connected in parallel can be obtained using the following relation:

$$\frac{1}{R_{eq}} = \sum \frac{1}{R_i}$$

Example 26-2-A. Current in a DC circuit. Determine the current through each resistor in the circuit shown in the diagram.

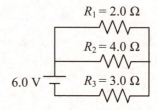

Approach: When we try to determine the currents in a DC circuit, we must carefully examine the various currents that flow in the circuit. A quick glance at the circuit diagram may suggest that this circuit contains 3 parallel resistors, but a closer examination shows that the sum of the currents flowing through resistors 1 and 2 is equal to the current flowing through resistor 3. Once we recognize this, we can determine the equivalent resistance of this circuit, and determine the overall current flowing through it. The division of the current through resistors 1 and 2 is related to the ratio of their resistances.

Solution: The 2.0 Ω and the 4.0 Ω resistors are in parallel. Their equivalent resistance can be calculated as follows

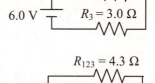

$$\frac{1}{R_{12}} = \frac{1}{R_1} + \frac{1}{R_2} = \frac{R_1 + R_2}{R_1 R_2} \quad \Rightarrow \quad R_{12} = \frac{R_1 R_2}{R_1 + R_2} = 1.33\,\Omega$$

The circuit can be redrawn with the equivalent resistance in place of the 2 Ω and the 4 Ω resistors. It is clear that the 1.33 Ω and the 3.0 Ω resistors in the equivalent circuit are in series. Their equivalent resistance is given by

$$R_{123} = R_{12} + R_3 = 1.33\,\Omega + 3.0\,\Omega = 4.33\,\Omega$$

Applying Ohm's law to this circuit, the current flowing from the battery through the equivalent resistance R_{123} can be determined

$$I = \frac{6.0\,\text{V}}{4.33\,\Omega} = 1.38\,\text{A}$$

This is also the current through R_{12}. The voltage drop across R_{12} is thus equal to

$$V_{R_{12}} = IR_{12} = (1.38\,\text{A})(1.33\,\Omega) = 1.85\,\text{V}$$

Ohm's law can now be used to determine the current through resistors R_1 and R_2:

$$I_{R_1} = \frac{V_{R_{12}}}{R_1} = \frac{1.85\,\text{V}}{2\,\Omega} = 0.92\,\text{A} \qquad I_{R_2} = \frac{V_{R_{12}}}{R_2} = \frac{1.85\,\text{V}}{4\,\Omega} = 0.46\,\text{A}$$

Example 26-2-B. Calculating equivalent resistance. Determine the equivalent resistance between points A and B of the resistance network shown in the diagram.

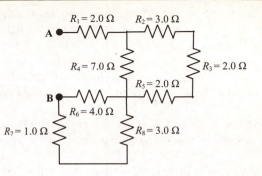

Approach: To determine the equivalent resistance of a circuit, we replace series and parallel sections of the circuit with their equivalent resistance and repeat this process until a single equivalent resistance remains.

Solution: The resistors R_2, R_3, and R_5 are in series with each other; R_7 and R_8 are also in series with each other. Their equivalent resistances are:

$$R_{235} = R_2 + R_3 + R_5 = 7.0\,\Omega$$
$$R_{78} = R_7 + R_8 = 4.0\,\Omega$$

The circuit can be redrawn with these equivalent resistances in place of the original individual resistors. The circuit diagram to the right shows that there are now two sets of parallel resistor networks: R_4 and R_{235} are in parallel and R_6 and R_{78} are in parallel. Their equivalent resistances can now be determined:

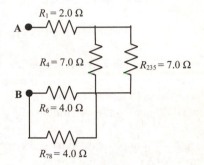

$$\frac{1}{R_{4235}} = \frac{1}{R_4} + \frac{1}{R_{235}} = \frac{R_4 + R_{235}}{R_4 R_{235}} \quad \Rightarrow \quad R_{4235} = \frac{R_4 R_{235}}{R_4 + R_{235}} = 3.5\,\Omega$$

$$\frac{1}{R_{678}} = \frac{1}{R_6} + \frac{1}{R_{78}} = \frac{R_6 + R_{78}}{R_6 R_{78}} \quad \Rightarrow \quad R_{678} = \frac{R_6 R_{78}}{R_6 + R_{78}} = 2.0\,\Omega$$

The equivalent circuit, obtained after replacing the parallel resistors with their equivalent resistors, shows a series network of three resistors. Their equivalent resistance is the sum of their individual resistances: with these equivalent resistors replacing the individual resistors:

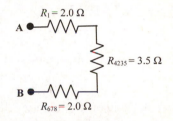

$$R_{eq} = R_1 + R_{4235} + R_{678} = 7.5\,\Omega$$

This is the equivalent resistance of the entire network between points A and B.

Section 26-3. Kirchhoff's Rules

We already know that electrical charge is conserved in any physical process. That fact, together with the fact that charge cannot accumulate at a given point in a conductor can be written as **Kirchhoff's current rule** or **Kirchhoff's node rule**:

>*The sum of the currents into any given point (or node) in an electric circuit is zero.*

We also know the electric potential is a well-defined function of position. If we move around through space and return to our starting point, the electric potential will be the same, and the change in the electric potential is thus equal to 0. For the same reason, the sum of the potential differences across the circuit elements of a closed path in an electric circuit must add up to zero. This is **Kirchhoff's voltage rule** or **Kirchhoff's loop rule**:

>*The sum of the changes in potential around any closed path in an electric circuit is zero.*

Kirchhoff's rules can be used to solve for currents and potential differences in circuits when there are interconnections of circuit elements that are not in series or in parallel. To make use of Kirchhoff's rules, there are some bookkeeping details that require careful attention.

1. For each unknown current we must define the direction that corresponds to a positive current. Our choice of the positive direction is arbitrary, but consistency must be maintained throughout the solution of the problem. It is recommended that you put arrows on the circuit diagram defining the positive direction for each current.

2. When Kirchhoff's current node rules are used, currents are treated as positive if they flow into a node and as negative if they flow out of that node. The directions of the currents have already been defined in step 1.

3. When Kirchhoff's voltage rules are applied, make sure a complete closed path through the circuit is used and components are traversed in a consistent direction. The potential change across a resistor is $-IR$ when it is traversed in the same direction as the current I, assuming $I > 0$; the potential change across the resistor is $+IR$ when it is

traversed in a direction opposite to the direction of the current I. The potential difference across a voltage source is positive if the source is traversed in a direction from its negative terminal to its positive terminal; the potential difference is negative if it is traversed in a direction from its positive terminal to its negative terminal.

4. To obtain a complete set of equations that can be solved for the unknown currents, each unknown current must appear in at least one Kirchhoff's current rule equation and at least one Kirchhoff's voltage rule equation.

These steps will be illustrated in the following example.

Example 26-3-A. Currents in a resistor network. Determine the currents through and the voltages across each of the resistors in the circuit shown in the diagram.

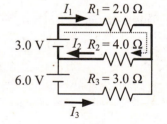

Approach: In this problem we need to follow the steps outlined above. It is critical that we keep track of the assumed direction of the currents, and define voltage drops that match the direction of these currents. In this particular circuit, there are 3 unknown currents and we thus need three equations to determine these currents.

Solution: We will follow the rules outlined above.

Step 1. The three currents we define to solve this problem are shown in the diagram on the right. The magnitude of the current flowing through the 3.0 V battery is equal to the magnitude of I_1. The magnitude of the current flowing through the 6.0 V battery is equal to the magnitude of I_3.

Step 2. We apply Kirchhoff's current node rule at point A. It is important that the signs of the currents are correct. The currents I_1 and I_3 are flowing toward A while current I_2 flows away from A. Based on this observation, Kirchhoff's current rule at A can be written as

$$I_1 - I_2 + I_3 = 0$$

Since all unknown currents appear in this equation, this equation will be the only Kirchhoff's current rule equation we need.

Step 3. We now need to use Kirchhoff's voltage rules to obtain two additional equations involving the unknown currents. To construct the first equation, consider the closed path, indicated by the heavy line in the diagram to the right, starting at the positive terminal of the 3.0 V battery. The first component we encounter is resistor R_1, across which we travel in the direction of the current I_1 and the potential change across this resistor is thus $-I_1R_1$. The next component is the resistor R_2. We travel across this resistor in the direction of the current I_2 and the potential change across this resistor is thus equal to $-I_2R_2$. Finally, we cross the 3.0 V potential source, which is crossed in the direction from its negative terminal to its positive terminal, and the corresponding potential change is thus equal to +3.0 V. Adding the potential differences around this closed path, we obtain the first voltage rule equation:

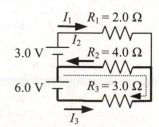

$$-I_1R_1 - I_2R_2 + 3.0 = 0$$

In order to obtain an equation that includes the current I_3 we consider the closed path indicated by the bold line in the diagram to the right. We traverse the path clockwise as indicated by the dashed arrow, starting at the positive terminal of the 6.0 V battery. The first component we encounter is resistor R_2, across which we travel in a direction opposite to the direction of current I_2. The potential change across this resistor is thus equal to $+I_2R_2$. The next component is the resistor R_3, which we travel across in a direction opposite to current I_3. The potential change across this resistor is thus equal to $+I_3R_3$. Finally, we cross the 6.0 V potential source, which is crossed in the direction from its negative terminal to its positive terminal, and the corresponding potential change is thus equal to +6.0 V. Adding the potential differences around this closed path, we obtain the second voltage rule equation:

$$+I_2R_2 + I_3R_3 + 6.0 = 0$$

We now have three independent linear equations for the three unknown currents that can be solved.

Consider the voltage-rule equations. These equations can be rewritten as

$$I_1 = \frac{-I_2 R_2 + 3.0}{R_1}$$

$$I_3 = \frac{-6.0 - I_2 R_2}{R_3}$$

Substituting these expressions for the currents I_1 and I_3 in the current-rule equation we obtain:

$$I_1 - I_2 + I_3 = \frac{-I_2 R_2 + 3.0}{R_1} - I_2 + \frac{-6.0 - I_2 R_2}{R_3} = \frac{-I_2 R_2 R_3 + 3.0 R_3 - I_2 R_1 R_3 - 6.0 R_1 - I_2 R_1 R_2}{R_1 R_3} = 0$$

This equation can be solved for current 2:

$$\frac{-I_2 (R_1 R_2 + R_1 R_3 + R_2 R_3) + 3.0 R_3 - 6.0 R_1}{R_1 R_3} = 0 \quad \Rightarrow \quad I_2 = \frac{3.0 R_3 - 6.0 R_1}{(R_1 R_2 + R_1 R_3 + R_2 R_3)} = -0.115 \,\text{A}$$

The negative sign means that current I_2 flows in the direction opposite to what we defined as the positive current direction. Using this value of current I_2 we can now determine the currents I_1 and I_3:

$$I_1 = \frac{3.0 - I_2 R_2}{R_1} = 1.73 \,\text{A}$$

$$I_3 = \frac{-6.0 - I_2 R_2}{R_3} = -1.85 \,\text{A}$$

Section 26-4. Series and Parallel EMFs; Battery Charging.

Sources of emfs can be arranged in series or in parallel.

When we place sources of emf in series, the total voltage provided by the system is the sum of the voltages provided by each source. The maximum current that can flow through the system is equal to the maximum current that can flow through each source. The bulb in a flashlight may require 3 V to produce light; this voltage can be generated by placing 2 1.5 V batteries in series.

Sources of emf are placed in parallel when large currents are required. In such an arrangement, the voltage across the system is equal to the voltage across each source, and the maximum current that can be provided is the sum of the maximum currents that can be provided by each source.

Section 26-5. Circuits Containing Resistor and Capacitor (*RC* Circuits)

If we apply Kirchhoff's voltage rule to a circuit consisting of a source of emf, a resistor, and a capacitor in series, as shown in the diagram, we get an equation of the form

$$\varepsilon - IR - \frac{Q}{C} = 0$$

To obtain this equation, we have assumed that a positive current is a current flowing in the direction shown in the diagram and that the charge Q is the charge on the upper plate of the capacitor. In this equation, we recognize that I and Q are functions of time and are related in the following way:

$$I = \frac{dQ}{dt}$$

Substituting this expression for I into Kirchhoff's voltage rule equation for this circuit gives:

$$\varepsilon - \frac{dQ}{dt} R - \frac{Q}{C} = 0$$

The most general solution to this equation is

$$Q = C\varepsilon + Ae^{-t/RC} = C\varepsilon + Ae^{-t/\tau}$$

where $\tau = RC$ is called the time constant of the circuit. The constant A is determined by the value of the charge on the capacitor at the time $t = 0$. If Q_0 is the charge on the capacitor at time $t = 0$, then

$$Q_0 = C\varepsilon + A \quad \Rightarrow \quad A = Q_0 - C\varepsilon$$

In this case, the solution can be written as

$$Q = C\varepsilon + \left(Q_0 - C\varepsilon\right)e^{-t/RC}$$

The current flowing in the circuit can be obtained by differentiating the charge Q with respect to time:

$$I = \frac{dQ}{dt} = \frac{d}{dt}\left[C\varepsilon + \left(Q_0 - C\varepsilon\right)e^{-t/RC}\right] = -\frac{Q_0 - C\varepsilon}{RC}e^{-t/RC}$$

The voltage on the capacitor is the charge on the capacitor divided by the capacitance:

$$V_C = \varepsilon + \left(Q_0/C - \varepsilon\right)e^{-t/RC}$$

Example 26-5-A. Discharging a capacitor. For the RC circuit shown to the right, determine the charge on the capacitor as a function of time if the switch is closed at time $t = 0$. The capacitor has an initial voltage across it of $\varepsilon_0 = 12.0$ V before the switch is closed with the upper plate positively charged. Determine the time at which the voltage across the capacitor is equal to 5.0 V. Determine the current in the circuit at time $t = 0$. Assume $R = 1.5$ kΩ, $C = 2.0$ μF, and $\varepsilon = 3.0$ V.

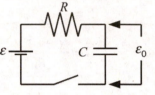

Approach: The time dependence of the voltage across the capacitor is given by the equation derived in this Section. To answer the questions, we need to determine the appropriate initial conditions.

Solution: The voltage across the capacitor as a function of time is given by the following equation

$$V_C(t) = \varepsilon + \left(Q_0/C - \varepsilon\right)e^{-t/RC}$$

In this equation, Q_0/C is the voltage across the capacitor at time $t = 0$. In terms of the variables provided, the voltage across the capacitor is thus equal to

$$V_C(t) = \varepsilon + \left(\varepsilon_0 - \varepsilon\right)e^{-t/RC}$$

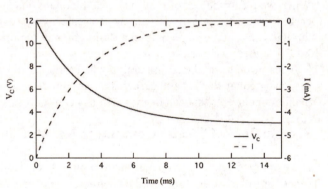

This function is shown by the solid curve in the graph on the right.

To determine the time at which the voltage across the capacitor is $V_5 = 5.0$ V, we solve the following equation for time t:

$$V_5 = \varepsilon + \left(\varepsilon_0 - \varepsilon\right)e^{-t/RC}$$

The time t is equal to

$$t = -RC \ln\left(\frac{V_5 - \varepsilon}{\varepsilon_0 - \varepsilon}\right) = 4.5 \times 10^{-3}\,\text{s}$$

The current at time t can be determined using Kirchhoff's voltage loop rule around the circuit:

$$\varepsilon - I(t)R - V_C(t) = 0$$

where we have defined the current flow across the resistor as positive when the current flows from left to right in the circuit diagram.

Using the known time dependence of the voltage across the capacitor, this equation can be used to determine the time dependence of the current:

$$I(t) = \frac{\varepsilon - V_C(t)}{R} = \frac{\varepsilon - \left(\varepsilon + (\varepsilon_0 - \varepsilon)e^{-t/RC}\right)}{R} = -\frac{(\varepsilon_0 - \varepsilon)e^{-t/RC}}{R}$$

The minus sign tells us that the current flows from right to left. The current as function of time is also shown in the graph. We see that as the rate of change in the voltage across the capacitor decreases (with increasing time), so does the magnitude of the current through the circuit. The current at $t = 0$ can now be determined:

$$I(0) = -\frac{(\varepsilon_0 - \varepsilon)e^{-0/RC}}{R} = -\frac{(12-3)}{1.5 \times 10^3} = -6\,\text{mA}$$

Section 26-6. Electric Hazards

Electric currents passing through the human body present dangers through two different mechanisms. First, large currents passing through tissue deposit energy into the tissue, causing the tissue's internal energy and temperature to rise. If the temperature rise is great enough, a burn is the result. Second, and more dangerously, currents flowing through the body cause potential differences across neurons that cause the neurons to fire uncontrollably. If the neurons of the heart fire uncontrollably, the heart fibrillates. A heart that is fibrillating is unable to pump blood.

The health effects of electric currents depends on the magnitude of the current and the path the current follows through the body. Current above 10 mA can cause serious health effects and/or death.

It is impossible to maintain an exactly zero potential and infinite resistance between people and equipment that uses electrical power. This means that in most circumstances, making contact with electrically powered equipment causes a current to flow through the human body. Safety design of electrically powered equipment ensures that this leakage current from the equipment stays at magnitudes far below that which would cause injury or death in properly used equipment.

Section 26-7. DC Ammeters and Voltmeters

Devices used to measure currents are called **ammeters**. Devices used to measure voltages are called **voltmeters**. Devices used to measure resistances are called **ohmmeters**. A single instrument that can operate as more than one of these devices is called a **multimeter**. All of these devices can be constructed from a basic instrument that can measure either current or voltage. There are two basic devices in common use today:

- A **galvanometer** is a meter that has an indicator with deflection that is proportional to the current passing through it.
- A **digital voltmeter** is a device that displays a numerical voltage reading on a display that is proportional to the voltage across its terminals.

A galvanometer has a deflection that is proportional to the current flowing through it. The current required to deflect the indicator to its maximum deflection is called the full scale current I_g of the galvanometer. The galvanometer has a finite resistance R_g. If we want to construct a voltmeter with a full-scale reading V_{fs} from a galvanometer, we must add a series resistance R_s to the galvanometer so that when V_{fs} is applied across the series network of the galvanometer and the series resistance, the current will be I_g. The equivalent resistance of the galvanometer and the series resistance is $R = R_g + R_s$. Using Ohm's law, we conclude that the resistance R must be equal to the desired full-scale voltage divided by the galvanometer current required for a full deflection of the indicator:

$$R = \frac{V_{fs}}{I_g} \quad \Rightarrow \quad R_s = \frac{V_{fs}}{I_g} - R_g$$

To construct a current meter with a full-scale current reading I_{fs} we must add a resistance R_p in parallel to the galvanometer. This resistance is often called a **shunt resistance** because it shunts a fraction of the current around the meter. To determine the resistance needed we must ensure that the desired full-scale current through the network of the galvanometer and the shunt resistance causes the current through the galvanometer to be that which causes full-scale deflection. The easiest way to do this analysis is by using the fact that the voltage across a parallel network is the same across each component resistance. The voltage across the galvanometer when its deflection is maximum is

$$V_g = I_g R_g$$

This must be the voltage across the shunt resistance when the remainder of the current, $I_P = I_{fs} - I_g$, flows through it:

$$V_g = I_p R_p \quad \Rightarrow \quad R_p = \frac{V_g}{I_p} = \frac{I_g R_g}{I_{fs} - I_g}$$

Example 26-7-A. Converting a galvanometer into a voltmeter. A galvanometer reaches its full-scale deflection when a current $I_g = 50$ μA passes through it. The galvanometer has a resistance $R_g = 100$ Ω. How can a voltmeter with a full-scale deflection of $V_{fs} = 1.00$ V be made?

Approach: The potential drop across the galvanometer when it reaches full-scale deflection is $I_g R_g = 0.5$ mV. In order for the voltmeter to have a full-scale deflection at 1.00 V, a resistor must be added in series with the galvanometer such that a current of 50 μA flows through the galvanometer and the resistor when 1.00 V is applied across its terminal.

Solution: If we add a resistor R_S in series to the galvanometer, the total resistance will be $R_S + R_g$. The voltage across this resistor chain when a current I_g is flowing through it must be $V_{fs} = 1.00$ V. The requirement is thus

$$I_g \left(R_S + R_g \right) = V_{fs} \quad \Rightarrow \quad R_S = \frac{V_{fs}}{I_g} - R_g = 1.99 \times 10^4 \ \Omega$$

Practice Quiz

1. Consider Example 26-1-A in this study guide, where the maximum electrical power out of the battery was calculated to occur when the external resistance of the circuit is equal to the internal resistance of the battery. Why doesn't decreasing the external resistance further, causing an increase in current, increase the electrical power out of the battery?
    a) Decreasing the resistance causes a decrease in the current drawn from the battery.
    b) Decreasing the resistance causes chemical energy to be transformed into electrical energy at a greater rate, but more of the electrical energy goes into heating the internal resistance of the battery.
    c) Decreasing the resistance decreases the voltage drop across the internal resistance of the battery.
    d) The electrical power out of the battery remains the same regardless of the external resistance.

2. Given three equal resistances with resistance R, how can they be interconnected so their equivalent resistance is $\frac{2}{3}R$ between the right and left ends of the network?

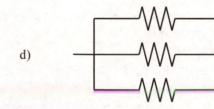

3. A network of resistors, all with resistance R, is connected such that it can be reduced, using equivalent parallel and/or series resistance, to a single resistor. The equivalent resistance is greater than R. What must be true about the network of resistors?
    a) There must be at least two resistors in series.
    b) There must be at least two resistors in parallel.
    c) A series equivalent must be used at some point in reducing the network to its equivalent resistance.
    d) A parallel equivalent must be used at some point in reducing the network to its equivalent resistance.

4. If in an arbitrary network of batteries and resistances, the voltage of each battery is doubled, what is true about the currents in the circuit?
    a) They will all increase by a factor 2^N, where N is the number of batteries.
    b) They will all increase by a factor of two.
    c) How the currents change will depend on the details of the circuit.
    d) The currents will not change.

5. A parallel network of resistances has an equivalent resistance that
    a) is always greater than any of the component resistances.
    b) is always less than any of the component resistances.
    c) is always less than the greatest component resistance, but greater than the least component resistance.
    d) can be less than, equal to, or greater than any of the component resistances.

6. A series network of resistances has an equivalent resistance that
    a) is always greater than any of the component resistances.
    b) is always less than any of the component resistances.
    c) is always less than the greatest component resistance, but greater than the least component resistance.
    d) can be less than, equal to, or greater than any of the component resistances.

7. The time constant of an RC circuit is the time for the charge on the capacitor
    a) to go from its initial to its final value.
    b) to change from its initial value by a fraction $1/e$ of the charge difference between its initial and its final values.
    c) to go to zero.
    d) to not change.

8. How can the time constant of an RC circuit be doubled?
    a) Double the voltage.
    b) Double the capacitance.
    c) Double both the resistance and the capacitance simultaneously.
    d) All of the above.

9. What is the minimum full-scale voltage possible for a voltmeter made from a galvanometer with a full deflection current I_g and a resistance R_g?
    a) There is no minimum full-scale voltage.
    b) I_g/R_g
    c) $I_g R_g$
    d) R_g/I_g

10. The current in a series RC circuit after many time constants will
    a) approach zero.
    b) approach infinity.
    c) approach ε/R.
    d) become opposite in direction to its initial direction.

11. A minimum of 7.4 V is required to power a radio. If the power is supplied by a battery with a 9.0 V emf and the circuit draws 0.18 A when the terminal voltage is 7.4 V, what is the maximum internal resistance the battery can have in order to be able to power the radio?

12. Determine the equivalent resistance between points A and B of the resistance network shown in the diagram to the right.

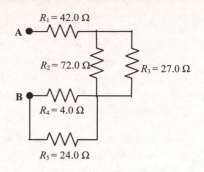

13. Determine the current through each of the resistors in the circuit shown in the diagram below.

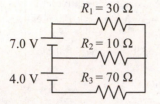

14. The capacitor in the circuit shown in the diagram to the right is initially uncharged. The switch is closed at time $t = 0$. At what time is the voltage across the capacitor equal to 8.0 V?

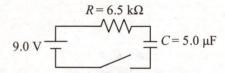

15. Determine the shunt resistance necessary to make an ammeter with a full-scale deflection of 1.00 A from a galvanometer with a full-scale deflection of 100 µA and a resistance of 250 Ω.

Responses to Select End-of-Chapter Questions

1. Even though the bird's feet are at high potential with respect to the ground, there is very little potential difference between them, because they are close together on the wire. The resistance of the bird is much greater than the resistance of the wire between the bird's feet. These two resistances are in parallel, so very little current will pass through the bird as it perches on the wire. When you put a metal ladder up against a power line, you provide a direct connection between the high potential line and ground. The ladder will have a large potential difference between its top and bottom. A person standing on the ladder will also have a large potential difference between his or her hands and feet. Even if the person's resistance is large, the potential difference will be great enough to produce a current through the person's body large enough to cause substantial damage or death.

7. The battery has to supply less power when the two resistors are connected in series than it has to supply when only one resistor is connected. $P = IV = \dfrac{V^2}{R}$, so if V is constant and R increases, the power decreases.

13. Put the battery in a circuit in series with a very large resistor and measure the terminal voltage. With a large resistance, the current in the circuit will be small, and the potential across the battery will be mainly due to the emf. Next put the battery in parallel with the large resistor (or in series with a small resistor) and measure the terminal voltage and the current in the circuit. You will have enough information to use the equation $V_{\text{terminal}} = \text{emf} - Ir$ to determine the internal resistance r.

19. If you use an ammeter where you need to use a voltmeter, you will short the branch of the circuit. Too much current will pass through the ammeter and you will either blow the fuse on the ammeter or burn out its coil.

Solutions to Select End-of-Chapter Problems

1. See Figure 26-2 for a circuit diagram for this problem. Using the same analysis as in Example 26-1, the current in the circuit is $I = \dfrac{\varepsilon}{R+r}$. Use Eq. 26-1 to calculate the terminal voltage.

(a) $V_{ab} = \varepsilon - Ir = \varepsilon - \left(\dfrac{\varepsilon}{R+r}\right)r = \dfrac{\varepsilon(R+r) - \varepsilon r}{R+r} = \varepsilon\dfrac{R}{R+r} = (6.00\,\text{V})\dfrac{81.0\,\Omega}{(81.0+0.900)\,\Omega} = \boxed{5.93\,\text{V}}$

(b) $V_{ab} = \varepsilon\dfrac{R}{R+r} = (6.00\,\text{V})\dfrac{810\,\Omega}{(810+0.900)\,\Omega} = \boxed{5.99\,\text{V}}$

7. (*a*) The maximum resistance is made by combining the resistors in series.

$$R_{eq} = R_1 + R_2 + R_3 = 680\,\Omega + 720\,\Omega + 1200\,\Omega = \boxed{2.60\,\text{k}\Omega}$$

(*b*) The minimum resistance is made by combining the resistors in parallel.

$$\frac{1}{R_{eq}} = \frac{1}{R_1} + \frac{1}{R_2} + \frac{1}{R_3} \;\rightarrow\; R_{eq} = \left(\frac{1}{R_1} + \frac{1}{R_2} + \frac{1}{R_3}\right)^{-1} = \left(\frac{1}{680\,\Omega} + \frac{1}{720\,\Omega} + \frac{1}{1200\,\Omega}\right)^{-1} = \boxed{270\,\Omega}$$

13. We model the resistance of the long leads as a single resistor r. Since the bulbs are in parallel, the total current is the sum of the current in each bulb, and so $I = 8I_R$. The voltage drop across the long leads is $V_{leads} = Ir = 8I_R r = 8(0.24\,\text{A})(1.4\,\Omega) = 2.688\,\text{V}$. Thus the voltage across each of the parallel resistors is $V_R = V_{tot} - V_{leads} = 110\,\text{V} - 2.688\,\text{V} = 107.3\,\text{V}$. Since we have the current through each resistor, and the voltage across each resistor, we calculate the resistance using Ohm's law.

$$V_R = I_R R \;\rightarrow\; R = \frac{V_R}{I_R} = \frac{107.3\,\text{V}}{0.24\,\text{A}} = 447.1\,\Omega = \boxed{450\,\Omega}$$

The total power delivered is $P = V_{tot}I$, and the "wasted" power is $I^2 r$. The fraction wasted is the ratio of those powers.

$$\text{fraction wasted} = \frac{I^2 r}{IV_{tot}} = \frac{Ir}{V_{tot}} = \frac{8(0.24\,\text{A})(1.4\,\Omega)}{110\,\text{V}} = \boxed{0.024}$$

So about 2.5% of the power is wasted.

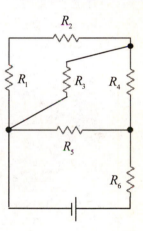

19. The resistors have been numbered in the accompanying diagram to help in the analysis. R_1 and R_2 are in series with an equivalent resistance of $R_{12} = R + R = 2R$. This combination is in parallel with R_3, with an equivalent resistance of $R_{123} = \left(\frac{1}{R} + \frac{1}{2R}\right)^{-1} = \frac{2}{3}R$. This combination is in series with R_4, with an equivalent resistance of $R_{1234} = \frac{2}{3}R + R = \frac{5}{3}R$. This combination is in parallel with R_5, with an equivalent resistance of $R_{12345} = \left(\frac{1}{R} + \frac{3}{5R}\right)^{-1} = \frac{5}{8}R$. Finally, this combination is in series with R_6, and we calculate the final equivalent resistance. $R_{eq} = \frac{5}{8}R + R = \boxed{\frac{13}{8}R}$

25. (*a*) Note that adding resistors in series always results in a larger resistance, and adding resistors in parallel always results in a smaller resistance. Closing the switch adds another resistor in parallel with R_3 and R_4, which lowers the net resistance of the parallel portion of the circuit, and thus lowers the equivalent resistance of the circuit. That means that more current will be delivered by the battery. Since R_1 is in series with the battery, its voltage will increase. Because of that increase, the voltage across R_3 and R_4 must decrease so that the total voltage drops around the loop are equal to the battery voltage. Since there was no voltage across R_2 until the switch was closed, its voltage will increase. To summarize:

$$\boxed{V_1 \text{ and } V_2 \text{ increase} \;;\; V_3 \text{ and } V_4 \text{ decrease}}$$

(*b*) By Ohm's law, the current is proportional to the voltage for a fixed resistance. Thus

$$\boxed{I_1 \text{ and } I_2 \text{ increase} \;;\; I_3 \text{ and } I_4 \text{ decrease}}$$

(*c*) Since the battery voltage does not change and the current delivered by the battery increases, the power delivered by the battery, found by multiplying the voltage of the battery by the current delivered, $\boxed{\text{increases}}$.

(*d*) Before the switch is closed, the equivalent resistance is R_3 and R_4 in parallel, combined with R_1 in series.

$$R_{eq} = R_1 + \left(\frac{1}{R_3} + \frac{1}{R_4} \right)^{-1} = 125\,\Omega + \left(\frac{2}{125\,\Omega} \right)^{-1} = 187.5\,\Omega$$

The current delivered by the battery is the same as the current through R_1.

$$I_{total} = \frac{V_{battery}}{R_{eq}} = \frac{22.0\,\text{V}}{187.5\,\Omega} = 0.1173\,\text{A} = I_1$$

The voltage across R_1 is found by Ohm's law.

$$V_1 = IR_1 = (0.1173\,\text{A})(125\,\Omega) = 14.66\,\text{V}$$

The voltage across the parallel resistors is the battery voltage less the voltage across R_1.

$$V_p = V_{battery} - V_1 = 22.0\,\text{V} - 14.66\,\text{V} = 7.34\,\text{V}$$

The current through each of the parallel resistors is found from Ohm's law.

$$I_3 = \frac{V_p}{R_2} = \frac{7.34\,\text{V}}{125\,\Omega} = 0.0587\,\text{A} = I_4$$

Notice that the current through each of the parallel resistors is half of the total current, within the limits of significant figures. The currents before closing the switch are as follows.

$$\boxed{I_1 = 0.117\,\text{A} \quad I_3 = I_4 = 0.059\,\text{A}}$$

After the switch is closed, the equivalent resistance is R_2, R_3, and R_4 in parallel, combined with R_1 in series. Do a similar analysis.

$$R_{eq} = R_1 + \left(\frac{1}{R_2} + \frac{1}{R_3} + \frac{1}{R_4} \right)^{-1} = 125\,\Omega + \left(\frac{3}{125\,\Omega} \right)^{-1} = 166.7\,\Omega$$

$$I_{total} = \frac{V_{battery}}{R_{eq}} = \frac{22.0\,\text{V}}{166.7\,\Omega} = 0.1320\,\text{A} = I_1 \qquad V_1 = IR_1 = (0.1320\,\text{A})(125\,\Omega) = 16.5\,\text{V}$$

$$V_p = V_{battery} - V_1 = 22.0\,\text{V} - 16.5\,\text{V} = 5.5\,\text{V} \qquad I_2 = \frac{V_p}{R_2} = \frac{5.5\,\text{V}}{125\,\Omega} = 0.044\,\text{A} = I_3 = I_4$$

Notice that the current through each of the parallel resistors is one third of the total current, within the limits of significant figures. The currents after closing the switch are as follows.

$$\boxed{I_1 = 0.132\,\text{A} \quad I_2 = I_3 = I_4 = 0.044\,\text{A}}$$

$\boxed{\text{Yes}}$, the predictions made in part (*b*) are all confirmed.

31. This circuit is identical to Example 26-9 and Figure 26-13 except for the numeric values. So we may copy the same equations as developed in that Example, but using the current values.

Eq. (*a*): $I_3 = I_1 + I_2$ 　　　　　　　　　　Eq. (*b*): $-34I_1 + 45 - 48I_3 = 0$

Eq. (*c*): $-34I_1 + 19I_2 - 75 = 0$ 　　　　　Eq. (*d*): $I_2 = \dfrac{75 + 34I_1}{19} = 3.95 + 1.79I_1$

Eq. (*e*): $I_3 = \dfrac{45 - 34I_1}{48} = 0.938 - 0.708I_1$

$$I_3 = I_1 + I_2 \quad \rightarrow \quad 0.938 - 0.708I_1 = I_1 + 3.95 + 1.79I_1 \quad \rightarrow \quad I_1 = -0.861\,\text{A}$$

$$I_2 = 3.95 + 1.79I_1 = 2.41\,\text{A} \; ; \; I_3 = 0.938 - 0.708I_1 = 1.55\,\text{A}$$

(*a*) To find the potential difference between points a and d, start at point a and add each individual potential difference until reaching point d. The simplest way to do this is along the top branch.

$$V_{ad} = V_d - V_a = -I_1(34\,\Omega) = -(-0.861\,\text{A})(34\,\Omega) = 29.27\,\text{V} \approx \boxed{29\,\text{V}}$$

Slight differences will be obtained in the final answer depending on the branch used, due to rounding. For example, using the bottom branch, we get the following.

$$V_{ad} = V_d - V_a = \varepsilon_1 - I_2(19\,\Omega) = 75\,\text{V} - (2.41\,\text{A})(19\,\Omega) = 29.21\,\text{V} \approx 29\,\text{V}$$

(*b*) For the 75-V battery, the terminal voltage is the potential difference from point g to point e. For the 45-V battery, the terminal voltage is the potential difference from point d to point b.

75 V battery: $V_{\text{terminal}} = \varepsilon_1 - I_2 r = 75\,\text{V} - (2.41\,\text{A})(1.0\,\Omega) = \boxed{73\,\text{V}}$

45 V battery: $V_{\text{terminal}} = \varepsilon_2 - I_3 r = 45\,\text{V} - (1.55\,\text{A})(1.0\,\Omega) = \boxed{43\,\text{V}}$

37. This problem is the same as Problem 36, except the total resistance in the top branch is now $23\,\Omega$ instead of $35\,\Omega$. We simply reproduce the adjusted equations here without the prose.

$$I_1 = I_2 + I_3$$

$$12.0\,\text{V} - I_2(12\,\Omega) + 12.0\,\text{V} - I_1(23\,\Omega) = 0 \quad \rightarrow \quad 24 = 23I_1 + 12I_2$$

$$12.0\,\text{V} - I_2(12\,\Omega) - 6.0\,\text{V} + I_3(34\,\Omega) = 0 \quad \rightarrow \quad 6 = 12I_2 - 34I_3$$

$$24 = 23I_1 + 12I_2 = 23(I_2 + I_3) + 12I_2 = 35I_2 + 23I_3$$

$$6 = 12I_2 - 34I_3 \quad \rightarrow \quad I_2 = \frac{6 + 34I_3}{12} \quad ; \quad 24 = 35I_2 + 23I_3 = 35\left(\frac{6 + 34I_3}{12}\right) + 23I_3 \quad \rightarrow$$

$$I_3 = 0.0532\,\text{A} \quad ; \quad I_2 = \frac{6 + 34I_3}{12} = 0.6508\,\text{A} \quad ; \quad I_1 = I_2 + I_3 = 0.704\,\text{A} \approx \boxed{0.70\,\text{A}}$$

43. We estimate the time between cycles of the wipers to be from 1 second to 15 seconds. We take these times as the time constant of the RC combination.

$$\tau = RC \quad \rightarrow \quad R_{1s} = \frac{\tau}{C} = \frac{1\,\text{s}}{1\times10^{-6}\,\text{F}} = 10^6\,\Omega \quad ; \quad R_{1s} = \frac{\tau}{C} = \frac{15\,\text{s}}{1\times10^{-6}\,\text{F}} = 15\times10^6\,\Omega$$

So we estimate the range of resistance to be

$$V_{\text{full}} = I_G(r_G + R) \quad \rightarrow$$

$$R = \frac{V_{\text{full}}}{I_G} - r_G = \frac{1.00\,\text{V}}{2.222\times10^{-5}\,\text{A}} - 20.0\,\Omega = 44985\,\Omega \approx \boxed{45\,\text{k}\Omega \text{ in series}}$$

49. (a) At t = 0, the capacitor is uncharged and so there is no voltage difference across it. The capacitor is a "short," and so a simpler circuit can be drawn just by eliminating the capacitor. In this simpler circuit, the two resistors on the right are in parallel with each other, and then in series with the resistor by the switch. The current through the resistor by the switch splits equally when it reaches the junction of the parallel resistors.

$$R_{eq} = R + \left(\frac{1}{R} + \frac{1}{R}\right)^{-1} = \tfrac{3}{2}R \quad \rightarrow \quad I_1 = \frac{\varepsilon}{R_{eq}} = \frac{\varepsilon}{\tfrac{3}{2}R} = \boxed{\frac{2\varepsilon}{3R}} \quad ; \quad I_2 = I_3 = \tfrac{1}{2}I_1 = \boxed{\frac{\varepsilon}{3R}}$$

(b) At $t = \infty$, the capacitor will be fully charged and there will be no current in the branch containing the capacitor, and so a simpler circuit can be drawn by eliminating that branch. In this simpler circuit, the two resistors are in series, and they both have the same current.

$$R_{eq} = R + R = 2R \quad \rightarrow \quad I_1 = I_2 = \frac{\varepsilon}{R_{eq}} = \boxed{\frac{\varepsilon}{2R}} \quad ; \quad I_3 = \boxed{0}$$

(c) At $t = \infty$, since there is no current through the branch containing the capacitor, there is no potential drop across that resistor. Therefore the voltage difference across the capacitor equals the voltage difference across the resistor through which I_2 flows.

$$V_C = V_{R_2} = I_2 R = \left(\frac{\varepsilon}{2R} \right) R = \boxed{\tfrac{1}{2} \varepsilon}$$

55. (a) The current for full-scale deflection of the galvanometer is

$$I_G = \frac{1}{\text{sensitivity}} = \frac{1}{45,000 \, \Omega/V} = 2.222 \times 10^{-5} \, \text{A}$$

To make an ammeter, a shunt resistor must be placed in parallel with the galvanometer. The voltage across the shunt resistor must be the voltage across the galvanometer. The total current is to be 2.0 A. See Figure 26-28 for a circuit diagram.

$$I_G r_G = I_s R_s \quad \rightarrow \quad R_s = \frac{I_G}{I_s} r_G = \frac{I_G}{I_{full} - I_G} r_G = \frac{2.222 \times 10^{-5} \, \text{A}}{2.0 \, \text{A} - 2.222 \times 10^{-5} \, \text{A}} (20.0 \, \Omega)$$

$$= 2.222 \times 10^{-4} \, \Omega \approx \boxed{2.2 \times 10^{-4} \, \Omega \text{ in parallel}}$$

(b) To make a voltmeter, a resistor must be placed in series with the galvanometer, so that the desired full scale voltage corresponds to the full scale current of the galvanometer. See Figure 26-29 for a circuit diagram. The total current must be the full-scale deflection current.

$$V_{full} = I_G \left(r_G + R \right) \quad \rightarrow$$

$$R = \frac{V_{full}}{I_G} - r_G = \frac{1.00 \, \text{V}}{2.222 \times 10^{-5} \, \text{A}} - 20.0 \, \Omega = 44985 \, \Omega \approx \boxed{45 \, \text{k}\Omega \text{ in series}}$$

61. Find the equivalent resistance for the entire circuit, and then find the current drawn from the source. That current will be the ammeter reading. The ammeter and voltmeter symbols in the diagram below are each assumed to have resistance.

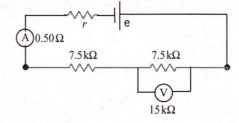

$$R_{eq} = 1.0 \, \Omega + 0.50 \, \Omega + 7500 \, \Omega + \frac{(7500 \, \Omega)(15000 \, \Omega)}{(7500 \, \Omega + 15000 \, \Omega)} =$$

$$= 12501.5 \, \Omega \approx 12500 \, \Omega \quad ;$$

$$I_{source} = \frac{\varepsilon}{R_{eq}} = \frac{12.0 \, \text{V}}{12500 \, \Omega} = \boxed{9.60 \times 10^{-4} \, \text{A}}$$

The voltmeter reading will be the source current times the equivalent resistance of the resistor–voltmeter combination.

$$V_{meter} = I_{source} R_{eq} = \left(9.60 \times 10^{-4} \, \text{A} \right) \frac{(7500 \, \Omega)(15000 \, \Omega)}{(7500 \, \Omega + 15000 \, \Omega)} = \boxed{4.8 \, \text{V}}$$

67. The voltage drop across the two wires is the 3.0 A current times their total resistance.

$$V_{wires} = IR_{wires} = (3.0 \text{ A})(0.0065 \, \Omega/\text{m})(130 \text{ m}) R_p = 2.535 \text{ V} \approx \boxed{2.5 \text{ V}}$$

Thus the voltage applied to the apparatus is

$$V = V_{source} - V_{wires} = 120 \text{ V} - 2.535 \text{ V} = 117.465 \text{ V} \approx \boxed{117 \text{ V}}$$

73. Divide the power by the required voltage to determine the current drawn by the hearing aid.

$$I = \frac{P}{V} = \frac{2.5 \text{ W}}{4.0 \text{ V}} = 0.625 \text{ A}$$

Use Eq. 26-1 to calculate the terminal voltage across the three batteries for mercury and dry cells.

$$V_{Hg} = 3(\varepsilon - Ir) = 3\left[1.35 \text{ V} - (0.625 \text{ A})(0.030 \, \Omega)\right] = 3.99 \text{ V}$$

$$V_D = 3(\varepsilon - Ir) = 3\left[1.50 \text{ V} - (0.625 \text{ A})(0.35 \, \Omega)\right] = 3.84 \text{ V}$$

The terminal voltage of the mercury cell batteries is closer to the required 4.0 V than the voltage from the dry cell.

79. The current in the circuit can be found from the resistance and the power dissipated. Then the product of that current and the equivalent resistance is equal to the battery voltage.

$$P = I^2 R \quad \rightarrow \quad I = \sqrt{\frac{P_{33}}{R_{33}}} = \sqrt{\frac{0.80 \text{ W}}{33 \, \Omega}} = 0.1557 \text{ A}$$

$$R_{eq} = 33 \, \Omega + \left(\frac{1}{68 \, \Omega} + \frac{1}{75 \, \Omega}\right)^{-1} = 68.66 \, \Omega \quad V = IR_{eq} = (0.1557 \text{ A})(68.66 \, \Omega) = 10.69 \text{ V} \approx \boxed{11 \text{ V}}$$

85. (a) When the galvanometer gives a null reading, no current is passing through the galvanometer or the emf that is being measured. All of the current is flowing through the slide wire resistance. Application of the loop rule to the lower loop gives $\varepsilon - IR = 0$, since there is no current through the emf to cause voltage drop across any internal resistance. The amount of current flowing through the slide wire resistor will be the same no matter what emf is used since no current is flowing through the lower loop. Apply this relationship to the two emfs.

$$\varepsilon_x - IR_x = 0 \; ; \; \varepsilon_s - IR_s = 0 \quad \rightarrow \quad ; \; I = \frac{\varepsilon_x}{R_x} = \frac{\varepsilon_s}{R_s} \quad \rightarrow \quad \boxed{\varepsilon_x = \left(\frac{R_x}{R_s}\right)\varepsilon_s}$$

(b) Use the equation derived above. We use the fact that the resistance is proportional to the length of the wire, by Eq. 25-3, $R = \rho \ell / A$.

$$\varepsilon_x = \left(\frac{R_x}{R_s}\right)\varepsilon_s = \left(\frac{\rho \dfrac{\ell_x}{A}}{\rho \dfrac{\ell_s}{A}}\right)\varepsilon_s = \left(\frac{\ell_x}{\ell_s}\right)\varepsilon_s = \left(\frac{45.8 \text{ cm}}{33.6 \text{ cm}}\right)(1.0182 \text{ V}) = \boxed{1.39 \text{ V}}$$

(c) If there is current in the galvanometer, then the voltage between points A and C is uncertainty by the voltage drop across the galvanometer, which is $V_G = I_G R_G = (0.012 \times 10^{-3} \text{A})(35 \, \Omega) = \boxed{4.2 \times 10^{-4} \text{V}}$. The uncertainty might of course be more than this, due to uncertainties compounding from having to measure distance for both the standard emf and the unknown emf. Measuring the distances also has some uncertainty associated with it.

(d) Using this null method means that the (unknown) internal resistance of the unknown emf does not enter into the calculation. No current passes through the unknown emf, and so there is no voltage drop across that internal resistance.

91. We represent the $10.00\text{-}M\Omega$ resistor by R_{10}, and the resistance of the voltmeter as R_V. In the first configuration, we find the equivalent resistance R_{eqA}, the current in the circuit I_A, and the voltage drop across R.

$$R_{eqA} = R + \frac{R_{10}R_V}{R_{10}+R_V} \quad ; \quad I_A = \frac{\varepsilon}{R_{eqA}} \quad ; \quad V_R = I_A R = \varepsilon - V_A \quad \rightarrow \quad \varepsilon\frac{R}{R_{eqA}} = \varepsilon - V_A$$

In the second configuration, we find the equivalent resistance R_{eqB}, the current in the circuit I_B, and the voltage drop across R_{10}.

$$R_{eqB} = R_{10} + \frac{RR_V}{R+R_V} \quad ; \quad I_B = \frac{\varepsilon}{R_{eqB}} \quad ; \quad V_{R_{10}} = I_B R_{10} = \varepsilon - V_B \quad \rightarrow \quad \varepsilon\frac{R_{10}}{R_{eqB}} = \varepsilon - V_B$$

We now have two equations in the two unknowns of R and R_V. We solve the second equation for R_V and substitute that into the first equation. We are leaving out much of the algebra in this solution.

$$\varepsilon\frac{R}{R_{eqA}} = \varepsilon\frac{R}{R + \dfrac{R_{10}R_V}{R_{10}+R_V}} = \varepsilon - V_A \quad ;$$

$$\varepsilon\frac{R_{10}}{R_{eqB}} = \varepsilon\frac{R_{10}}{R_{10} + \dfrac{RR_V}{R+R_V}} = \varepsilon - V_B \quad \rightarrow \quad R_V = \frac{V_B R_{10} R}{\left(\varepsilon R - V_B R_{10} - V_B R\right)}$$

$$\varepsilon - V_A = \varepsilon\frac{R}{R + \dfrac{R_{10}R_V}{R_{10}+R_V}} = \varepsilon\frac{R}{R + \dfrac{R_{10}\left[\dfrac{V_B R_{10} R}{\left(\varepsilon R - V_B R_{10} - V_B R\right)}\right]}{R_{10} + \left[\dfrac{V_B R_{10} R}{\left(\varepsilon R - V_B R_{10} - V_B R\right)}\right]}} \quad \rightarrow$$

$$R = \frac{V_B}{V_A}R_{10} = \frac{7.317\,\text{V}}{0.366\,\text{V}}\left(10.00\,\text{M}\Omega\right) = 199.92\,\text{M}\Omega \approx \boxed{200\,\text{M}\Omega} \quad (3 \text{ sig. fig.})$$

Chapter 27: Magnetism

Chapter Overview and Objectives

This chapter introduces magnetic fields, magnetic forces on currents, and magnetic forces on charged particles. The magnetic force is used to determine the torque on a current loop, to predict the trajectory of a charged particle in a magnetic field, and to explain the Hall effect. The concept of the magnetic dipole moment is introduced, and the torque on and the potential energy of a magnetic dipole in a magnetic field are determined.

After completing this chapter you should:
- Know how to calculate the magnetic force on a current.
- Know how to calculate the magnetic force on a charged particle.
- Know the right-hand rule to determine the magnetic force on a current or a charged particle.
- Know how to determine the cyclotron period and the frequency of a particle in a magnetic field.
- Know how to calculate the torque due to a magnetic field on a current loop or magnetic dipole.
- Know how to calculate the magnetic dipole moment of a current loop.
- Know how to calculate the potential energy of a magnetic dipole moment in a magnetic field.
- Know what the Hall emf and Hall field are and be able to calculate these quantities.

Summary of Equations

Force on a current in a magnetic field: $\vec{\mathbf{F}} = I\vec{\ell} \times \vec{\mathbf{B}}$ (Section 27-3)

Magnetic force on an infinitesimal length of current: $d\vec{\mathbf{F}} = I\,d\vec{\ell} \times \vec{\mathbf{B}}$ (Section 27-3)

Magnitude of the magnetic force on a current: $F = I\ell B \sin\theta$ (Section 27-3)

Magnetic force on a charged particle in a magnetic field: $\vec{\mathbf{F}} = q\vec{\mathbf{v}} \times \vec{\mathbf{B}}$ (Section 27-4)

Magnitude of the magnetic force on a charged particle: $F = qvB \sin\theta$ (Section 27-4)

Cyclotron period: $T = \dfrac{2\pi m}{qB}$ (Section 27-4)

Cyclotron frequency: $f = \dfrac{qB}{2\pi m}$ (Section 27-4)

Magnitude of the torque on a coil: $\tau = NIAB \sin\theta$ (Section 27-5)

Torque on a coil: $\vec{\tau} = NI\vec{\mathbf{A}} \times \vec{\mathbf{B}} = \vec{\mu} \times \vec{\mathbf{B}}$ (Section 27-5)

Magnetic dipole moment of a coil: $\vec{\mu} = NI\vec{\mathbf{A}}$ (Section 27-5)

Potential energy of a magnetic dipole in a magnetic field: $U = -\vec{\mu} \cdot \vec{\mathbf{B}}$ (Section 27-5)

Hall emf: $\varepsilon_H = v_d B d$ (Section 27-8)

Determination of particle mass from spectrometer parameters: $m = \dfrac{qBr}{v}$ (Section 27-9)

Chapter Summary

Section 27-1. Magnets and Magnetic Fields

A magnet has two regions called **poles**. One region is called the **north pole**; the other region is called the **south pole**. The north pole of a magnet attracts the south poles of other magnets. The north pole of a magnet repels the north poles of other magnets; the south pole of one magnet repels the south poles of other magnets. The north pole of a magnet derives its name from the fact that it points toward the geographic north pole of the earth. The earth thus acts like a magnet with its south pole pointing toward the geographic north pole.

It is useful to introduce the concept of **magnetic field** when we discuss magnetic forces; this is similar to the introduction of the electric field when we discussed electrostatic forces. **Magnetic field lines** are drawn parallel to the magnetic field in analogy to electric field lines. A major difference between magnetic field lines and electric field lines is that magnetic field lines are continuous closed curves without a beginning or an end. Electric field lines on the other hand begin on positive charges and end on negative charges.

Section 27-2. Electric Currents Produce Magnetic Fields

Electric currents create magnetic fields. The direction of the magnetic field created by a current is given by the **right-hand rule**. If you grasp a wire carrying a current with your right hand with your thumb extended in the direction of the current flowing through the wire, your fingers wrap around the wire in the same direction as the magnetic field lines.

Section 27-3. Force on an Electric Current in a Magnetic Field; Definition of $\vec{B}$

A magnetic field exerts a force on a current-carrying wire. The force exerted on the wire by the magnetic field is always in a direction that is perpendicular to both the direction of the current flowing through the wire and the direction of the magnetic field. The direction of the force is given by the **right-hand rule**. We can write the relationship between the force on the wire, $\vec{F}$, the current I, and the uniform magnetic field $\vec{B}$, in terms of the vector cross product:

$$\vec{F} = I\vec{\ell} \times \vec{B}$$

where $\vec{\ell}$ is a vector with a magnitude equal to the length of the wire and pointing in the direction of the current flow. If the magnetic field is not uniform, we can look at the force on a small length $d\vec{\ell}$ of the current:

$$d\vec{F} = I\,d\vec{\ell} \times \vec{B}$$

where $d\vec{F}$ is the infinitesimal force on the segment of current of length $d\vec{\ell}$ and $\vec{B}$ is the magnetic field at that location. The vector $d\vec{\ell}$ has a magnitude equal to $d\ell$ and points in the direction of the current. The magnitude of the magnetic force F on a wire the length ℓ through which a current I is flowing depends on the magnitude of the magnetic field $\vec{B}$ in the following way

$$F = I\ell B \sin\theta$$

where θ is the angle between the direction of the current and the direction of the magnetic field.

The SI unit of the magnetic field is the tesla (T). One tesla is equal to 1 N/(Am). In cgs units, the unit of magnetic field is called the gauss (G). One tesla is 10,000 gauss.

Example 27-3-A. The magnetic force on a current loop. The circuit shown in the Figure on the right carries a current $I = 5.22$ A that flows clockwise around the circuit. A magnetic field points downward into the page with a magnitude $B = 3.55$ T. Each long segment, segments 1 and 6, is 1.000 m in length and each short segment, segments 2, 3, 4, and 5, is 0.500 m in length. What is the magnetic force on each segment of the circuit and what is the total magnetic force on the circuit?

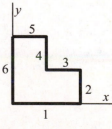

Approach: The current flows in the xy plane and the magnetic field is directed into the page. Using the properties of the vector product between the direction of the current and the direction of the magnetic field, we conclude that the direction of the magnetic force acting on those segments that are parallel to the x axis will be directed parallel or anti parallel to the y axis and the direction of the magnetic force acting on those segments that are parallel to the y axis will be directed parallel or anti parallel to the x axis.

Solution: The current in each segment of the loop is the same. The total length of current loop through which the current is flowing in the positive x direction is the same as the length of the current loop through which the current is flowing in the negative x direction. The corresponding magnetic forces, whose magnitudes are proportional to the magnitude of the current and the length of the current loop, cancel each other. The same conclusion can be drawn for the segments of the current loop that are parallel or anti parallel to the y axis. The net force is thus equal to 0.

The same conclusion can be obtained by determining the force separately for each of the 6 segments shown in the Figure, and using vector addition to determine the total magnetic force. The force on segment 1 is

$$\vec{F}_1 = I\,\vec{\ell}_1 \times \vec{B} = (5.22)(-1.000)\hat{i} \times (-3.55)\hat{k} = -18.5\,\hat{j}$$

Similarly, we calculate the forces on each of the other segments:

$$\vec{F}_2 = I\,\vec{\ell}_2 \times \vec{B} = (5.22)(-0.500)\hat{j} \times (-3.55)\hat{k} = 9.27\,\hat{i}$$

$$\vec{F}_3 = I\,\vec{\ell}_3 \times \vec{B} = (5.22)(0.500)\hat{i} \times (-3.55)\hat{k} = 9.27\,\hat{j}$$

$$\vec{F}_4 = I\,\vec{\ell}_4 \times \vec{B} = (5.22)(-0.500)\hat{j} \times (-3.55)\hat{k} = 9.27\,\hat{i}$$

$$\vec{F}_5 = I\,\vec{\ell}_5 \times \vec{B} = (5.22)(0.500)\hat{i} \times (-3.55)\hat{k} = 9.27\,\hat{j}$$

$$\vec{F}_6 = I\,\vec{\ell}_6 \times \vec{B} = (5.22)(1.000)\hat{j} \times (-3.55)\hat{k} = -18.5\,\hat{i}$$

As expected, the vector sum of these forces is zero.

Section 27-4. Force on an Electric Charge Moving in a Magnetic Field

The magnetic force $\vec{F}$ on a particle of charge q moving with a velocity $\vec{v}$ in a magnetic field $\vec{B}$ is given by

$$\vec{F} = q\vec{v} \times \vec{B}$$

The direction of the force is perpendicular to the direction of $\vec{v}$ and the direction of $\vec{B}$ and is given by the right-hand rule. Note that reversing the sign of the charge, will reverse the direction of the force. The magnitude of the magnetic force on the charged particle can be written as

$$F = qvB\sin\theta$$

where θ is the angle between the direction of the velocity and the direction of the magnetic field.

A particle moving in a uniform magnetic field with its velocity directed perpendicular to the direction of the field moves in a circular path. The period T of the circular motion is independent of the speed of the particle and is given by

$$T = \frac{2\pi m}{qB}$$

where m is the mass of the particle, q is its charge, and B is the magnitude of the magnetic field. The frequency of the circular motion is called the **cyclotron frequency** and is equal to

$$f = \frac{1}{T} = \frac{qB}{2\pi m}$$

Example 27-4-A. Net force on charged particles in the Earth's magnetic field. The magnetic field of the Earth is approximately pointing horizontally toward the north and has a magnitude $B = 0.500$ T. Consider a charged particle moving with a speed $v = 300$ m/s. What is the minimum charge-to-mass ratio of this particle that can result in a net force due to the magnetic and gravitational fields that is equal to 0? What must be the direction of this velocity for this to be true? How does this compare to the charge-to-mass ratio of an electron?

Approach: The magnetic force acting on the charged particle depends on the vector product between the velocity and the magnetic field. The magnitude of the magnetic force will have a maximum value when the velocity and the field are at right angles with respect to each other. In order to cancel the gravitational force, the magnetic force must be directed upwards in the vertical direction. Using the properties of the vector product and the known direction of the Earth's magnetic field, we conclude that the charged particle must be moving towards the East if the particle has a positive charge or towards the West if the particle has a negative charge.

Solution: Consider a positively charge particle moving towards the East with a velocity v. If the net force on this particle is zero, the magnitude of the magnetic force on this particle must be equal to its weight:

$$qvB = mg$$

The charge-to-mass ratio of the particle is thus equal to

$$\frac{q}{m} = \frac{g}{vB}$$

If the particle is not directed towards the East, the velocity in this equation must be the Easterly component of the velocity of the particle, and the q/m ratio will increase. The minimum charge to mass ratio is thus equal to

$$\left(\frac{q}{m}\right)_{min} = \frac{g}{vB} = \frac{9.81}{(300)(0.500)} = 0.0654\,\text{C/kg}$$

The charge-to-mass ratio of the electron is

$$\left(\frac{q}{m}\right)_{electron} = \frac{1.602 \times 10^{-19}}{9.11 \times 10^{-31}} = 1.76 \times 10^{11}\,\text{C/kg}$$

Since this ratio is larger than the minimum ratio required for a zero net force, there will be a range of directions and velocities for which the net force on the electron is equal to 0.

Example 27-4-B. Calculating the magnetic force. What is the magnetic force on a particle with a charge $q = 3.77$ µC, moving with a velocity $\vec{v} = (3.04\,\hat{i} + 4.76\,\hat{j} + 3.11\,\hat{k})$ m/s moving in a magnetic field $\vec{B} = (1.23\,\hat{i} - 2.11\,\hat{k})$ T?

Approach: To determine the magnetic force we need to evaluate the vector product between the velocity of the charged particle and the magnetic field.

Solution: The magnetic force on the charged particle is equal to

$$\vec{F} = q\vec{v} \times \vec{B} = (3.77 \times 10^{-6})(3.04\,\hat{i} + 4.76\,\hat{j} + 3.11\,\hat{k}) \times (1.23\,\hat{i} - 2.11\,\hat{k}) =$$

$$= (3.77 \times 10^{-6})\{3.04(-2.11\,\hat{i} \times \hat{k}) + 4.76(1.23\,\hat{j} \times \hat{i} - 2.11\,\hat{j} \times \hat{k}) + 3.11(1.23\,\hat{k} \times \hat{i})\} =$$

$$= -3.79 \times 10^{-5}\hat{i} + 3.86 \times 10^{-5}\hat{j} - 2.21 \times 10^{-5}\hat{k}$$

Section 27-5. Torque on a Current Loop; Magnetic Dipole Moment

The net magnetic force on a current loop in a magnetic field is zero. Since the force is distributed around the loop, the net torque on the current loop is in general not zero. For a planar circuit loop, the magnitude of the torque τ on the loop is given by

$$\tau = NIAB \sin\theta$$

where N is the number of turns in the coil, I is the current flowing in the coil, A is the area of the coil, and B is the magnitude of the magnetic field. This can be written in vector form to help clarify the direction of the torque:

$$\vec{\tau} = NI\vec{A} \times \vec{B}$$

where the vector $\vec{A}$ is a vector with a magnitude equal to the area of the loops of the coil and a direction given by a right-hand rule. To use the right-hand rule to determine the direction of the vector $\vec{A}$, place your hand over the coil so that your fingers curl around in the direction that the current flows around the loops of the coil. Then your extended thumb points in the direction of the $\vec{A}$ vector.

The magnitude of the **magnetic dipole moment** $\vec{\mu}$ of a coil is defined as

$$\mu = NIA$$

The direction of the magnetic dipole moment is the same as the direction of the area vector $\vec{A}$. The torque on a magnetic dipole is equal to

$$\vec{\tau} = \vec{\mu} \times \vec{B}$$

The potential energy associated with the torque on the magnetic dipole is equal to:

$$U = -\vec{\mu} \cdot \vec{B}$$

Since the potential energy depends on the angle between the magnetic dipole moment and the magnetic field, we conclude that work is done when the magnetic dipole is rotated in a magnetic field.

Example 27-5-A. Torque on a circular coil. A current $I = 3.76$ A flows through a circular coil with a diameter $D = 1.34$ cm and $N = 24$ turns of wire. The coil lies in the xy-plane and the current flows counterclockwise around the coil when viewed from above. The magnetic field at the location of the coil is equal to $1.78(3\,\hat{\mathbf{i}} + 4\,\hat{\mathbf{k}})$ T. What is the torque on the coil?

Approach: To calculate the torque on the coil, we first determine its magnetic moment and then use the vector product between this magnetic moment and the magnetic field to calculate the torque on the coil.

Solution: Because the coil lies in the xy-plane and the current flows in the counterclockwise direction, the area vector $\vec{A}$ points in the positive $\hat{\mathbf{k}}$ direction. The area vector $\vec{A}$ is thus given by

$$\vec{A} = \pi \left(\frac{D}{2}\right)^2 \hat{\mathbf{k}} = 1.41 \times 10^{-4}\,\hat{\mathbf{k}}$$

The torque on the coil can now be calculated

$$\vec{\tau} = NI\vec{A} \times \vec{B} = (24)(3.76)\left\{\left(1.41 \times 10^{-4}\,\hat{\mathbf{k}}\right) \times \left[1.78\left(3\hat{\mathbf{i}} + 4\hat{\mathbf{k}}\right)\right]\right\} = 6.79 \times 10^{-2}\,\hat{\mathbf{j}}\ \text{Nm}$$

Section 27-6. Applications: Galvanometers, Motors, Loudspeakers

There are several practical applications of the magnetic torque on a current loop. **Galvanometers** are electric meters that balance the magnetic torque on a current loop against the torque of a torsional spring. Since the magnetic torque is proportional to the current and the spring torque is proportional to the rotation of the coil, the angle of rotation of the coil is proportional to the current flowing through it. An indicator attached to the coil rotates through an angle proportional to the current, giving an indication on a scale of the magnitude of the current. The shaft of an **electric motor** rotates as a result of the torque on a current-carrying coil that is located within a magnetic field. Some means of switching current directions through the coil of the motor is provided to keep the torque acting in the same direction for any orientation of the coils.

Section 27-7. Discovery and Properties of the Electron

By measuring the deflection of the path of electrons in a magnetic field, J. J. Thomson was able to measure the charge-to-mass ratio of the electron. Robert A. Millikan was able to determine the charge on the electron using his oil-drop experiment. These two results can be combined to provide information on the mass of the electron.

Section 27-8. The Hall Effect

When a current-carrying conductor is held statically in a magnetic field and the current flowing through it has a component perpendicular to the direction of the magnetic field, a voltage difference develops across the conductor due to the deflection of the charge carriers by the magnetic field. A dynamic equilibrium is reached when the electric field due to this potential difference creates an electrostatic force that balances the magnetic force on the charge. This effect is called the **Hall effect**. The potential difference across the wire is called the **Hall emf** or **Hall voltage**, ε_H. When the direction of the current is perpendicular to the direction of the magnetic field, the Hall emf is given by

$$\varepsilon_H = v_d B \ell$$

where v_d is the drift velocity of the charge carriers in the conductor and ℓ is the dimension of the wire transverse to the direction of the current. If we know the drift velocity of the charge carriers, we can use the Hall effect to measure the magnetic field strength. The Hall effect can also be used to determine the drift velocity or the density of charged carriers in a conductor if we know the magnetic field in which our sample is located.

Example 27-8-A. Calculating the Hall voltage. What is the Hall emf across a copper wire with a diameter $D = 2.00$ mm that carries a current $I = 5.78$ A, perpendicular to a magnetic field of magnitude $B = 6.61$ T?

Approach: The Hall voltage depends on the drift velocity of the charge carriers and this velocity must be determined first. Using this velocity, the magnetic field, and the wire dimensions provided in the problem, we can determine the Hall voltage using the theory developed in this Section.

Solution: Consider a 1-m long segment of wire. Its volume V is equal to $\pi(D/2)^2 = \pi D^2/4$. Assuming that each copper atom contributes one electron to the carrier concentration we can determine the number of charge carriers N in this segment:

$$N = \frac{\rho V}{m} = \frac{\pi \rho D^2}{4m}$$

where m is the atomic mass of a copper atom ($m = 1.06 \times 10^{-25}$ kg) and ρ is the density of copper, $\rho = 8.9 \times 10^3$ kg/m^3. The concentration of charge carriers n will thus be equal to

$$n = \frac{N}{V} = \frac{\rho}{m}$$

The drift velocity v_d of the electrons in the copper is equal to

$$v_d = \frac{I}{Ane} = \frac{I}{\left(\pi \dfrac{D^2}{4}\right)\left(\dfrac{\rho}{m}\right)(e)} = 1.37 \times 10^{-4} \text{ m/s}$$

The Hall emf can now be calculated:

$$\varepsilon_H = v_d B \ell = \left(1.37 \times 10^{-4}\right)(6.61)\left(2 \times 10^{-3}\right) = 1.81 \,\mu\text{V}$$

Section 27-9. Mass Spectrometer

A mass spectrometer is a device that measures the mass of atoms, molecules, or other small particles. Such a measurement can be made using the trajectory of charged particles in a magnetic field. A charged particle traveling in a magnetic field follows a circular path in the plane perpendicular to the direction of the magnetic field. The radius r of this path is given by

$$r = \frac{mv}{qB}$$

where m is the mass of the particle, B is the magnitude of the magnetic field, q is the charge of the particle, and v is the speed of the particle perpendicular to the magnetic field. The measured radius r, combined with the known speed v and the strength of the magnetic field, can now be used to determine the mass of the particle:

$$m = \frac{qBr}{v}$$

Example 27-9-A. Measuring the mass of an ion. A particular ion is formed by removing its two outermost electrons. The ion is accelerated from rest by a $\Delta V = -300$ V potential difference. It then enters a magnetic field with magnitude B = 2.24 T, directed perpendicular to the direction of its velocity, in which it travels along a trajectory with a radius of curvature equal to $r = 12.7$ cm. What is the mass of the ion?

Approach: To determine the mass of the ion using the data provided by a mass spectrometer, we need to know its radius of curvature, the strength of the magnetic field, and the speed of the ion. The speed of the ion can be adjusted by changing the electrostatic potential that is used to accelerate it.

Solution: The velocity of the ion when it enters the mass spectrometer can be determined by calculating the change in its kinetic energy on the basis of the change of its electrostatic potential energy:

$$\Delta K = -\Delta U = -q\Delta V$$

Assuming that the ion starts from rest, its final speed can be determined by solving the following equation for v_f:

$$\tfrac{1}{2} m v_f^2 = -q\Delta V$$

The speed of the ion when it enters the spectrometer is thus equal to

$$v_f = \sqrt{\frac{-2q\Delta V}{m}}$$

Inserting this velocity into the expression for the mass of the particle we get

$$m = \frac{qBr}{\sqrt{\dfrac{-2q\Delta V}{m}}} = \frac{qBr}{\sqrt{-2q\Delta V}}\sqrt{m}$$

By squaring both sides of this equation we obtain

$$m^2 = \frac{(qBr)^2}{(-2q\Delta V)}\, m = -\frac{qB^2 r^2}{2\Delta V}\, m$$

The mass of the particle is thus equal to

$$m = -\frac{qB^2 r^2}{2\Delta V} = -\frac{2\left(1.60\times10^{-19}\right)(2.24)^2(0.127)^2}{2(-300)} = 4.32\times10^{-23}\ \text{kg}$$

Practice Quiz

1. A negative charge is moving into this page and a magnetic field points to the right parallel to this page. Which direction is the magnetic force on the charge?
   a) Toward the left.
   b) Toward the top of the page.
   c) Toward the bottom of the page.
   d) Out of the page.

2. A loop is partially inserted into a region with a magnetic field pointing into the page as shown in the diagram. Current flows around the loop in a counterclockwise direction. What is the direction of the magnetic force on the loop?
   a) Toward the top of the page.
   b) Into the page.
   c) Toward the right.
   d) Toward the left.

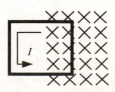

3. If you know the charge and velocity of a particle and the magnetic force acting on that particle, is this sufficient information to determine the strength and the direction of the magnetic field?
   a) Yes.
   b) No, you also need to know the mass of the object.
   c) Yes, unless the velocity is zero.
   d) No, only the component of magnetic field perpendicular to the velocity can be determined.

4. A particle moves in a uniform magnetic field from point A to point C, as shown in the picture. The path length from A to C is L. How much work is done by the magnetic field $\vec{B}$, shown in the diagram, as the particle moves from A to C?
   a) $qvBL \sin \theta$
   b) $qvBL \cos \theta$
   c) $-qvBL \cos \theta$
   d) zero

5. A loop, located in a uniform magnetic field, has a constant current I flowing through it. What is the best description of the most general type of motion the loop will undergo due to the interaction with the magnetic field if it is released from rest?
   a) The loop will spin with constant angular speed.
   b) The loop will spin with constant angular velocity.
   c) The loop will rotate back and forth.
   d) The loop will translationally accelerate with constant acceleration.

6. For a given length of wire, what shape of loop will have the greatest torque for a given current and magnetic field?
   a) A square.
   b) A rectangle with the long sides twice as long as the short sides.
   c) A rectangle with the long sides the square root of two longer than the short sides.
   d) A circle.

7. A power line runs east-west in a place where there is no magnetic dip. It carries a DC current to the west. In what direction does the power line deflect, compared to the direction in which it would hang when no current is flowing through it?
   a) North.
   b) South.
   c) Up.
   d) Down.

8. In the situation shown in the diagram, which direction is the Hall field in the conductor if electrons are the charge carriers of the current?
   a) Toward the top of the page.
   b) Toward the bottom of the page.
   c) Out of the page.
   d) Into the page.

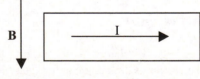

9. Two conductors are made of identical materials. They are in the same uniform magnetic field and have identical currents passing through them. The length of the first conductor in the direction perpendicular to the current and the magnetic field is twice the length of the second conductor in the direction perpendicular to the current and the magnetic field. How do the Hall fields compare in the two conductors?
   a) The Hall field of the first conductor is twice the Hall field in the second conductor.
   b) The Hall field of the second conductor is twice the Hall field in the first conductor.
   c) The Hall field in the first conductor is the same as the Hall field in the second conductor.
   d) There is insufficient information given to compare the Hall fields.

10. Two conductors have identical geometry. They are in the same uniform magnetic field and have identical currents passing through them. The density of charge carriers in the first conductor is twice the density of charge carriers in the second conductor. How do the Hall emfs compare in the two conductors?
    a) The Hall emf of the first conductor is twice the Hall emf in the second conductor.
    b) The Hall emf of the second conductor is twice the Hall emf in the first conductor.
    c) The Hall emf in the first conductor is the same as the Hall emf in the second conductor.
    d) There is insufficient information given to compare the Hall emfs.

11. A current-carrying wire lies along the $\hat{\mathbf{i}}$ direction. What is the force per unit length of wire on the wire if it carries 3.56 A of current in the $\hat{\mathbf{i}}$ direction in a magnetic field $\vec{\mathbf{B}} = 1.65\left(\hat{\mathbf{i}} + 1.34\hat{\mathbf{j}}\right)$ T ?

12. Determine the acceleration of an electron with a velocity $\vec{\mathbf{v}} = 7.72\left(3.42\hat{\mathbf{i}} + 4.34\hat{\mathbf{j}}\right)$ m/s , moving through a magnetic field $\vec{\mathbf{B}} = 2.75\left(-2.21\hat{\mathbf{i}} + 3.67\hat{\mathbf{k}}\right)$ T .

13. An electron moves through an electric field $\vec{\mathbf{E}} = E_0\hat{\mathbf{j}}$ and a magnetic field $\vec{\mathbf{B}} = B_0\hat{\mathbf{k}}$. What must be the velocity of the electrons if they move through these fields with no net acceleration?

14. A mass spectrometer is designed with a magnetic field of 1.43 T and particles are injected with a velocity of 380 m/s. Determine the difference in radii of curvature of the trajectories for singly ionized carbon-12 and carbon-14 atoms.

15. What is the concentration of charge carriers in a piece of material if a 2 mm wide ribbon wire with a thickness of 0.5 mm that carries a 1.69 A current in the direction perpendicular to a 4.76 T magnetic field generates a 0.761 V Hall emf?

Responses to Select End-of-Chapter Questions

1. The compass needle aligns itself with the local magnetic field of the Earth, and the Earth's magnetic field lines are not always parallel to the surface of the Earth.

7. Positive particle in the upper left: force is downward toward the wire. Negative particle in the upper right: force is to the left. Positive particle in the lower right: force is to the left. Negative particle in the lower left: force is upward toward the wire.

13. The particle will move in an elongating helical path in the direction of the electric field (for a positive charge). The radius of the helix will remain constant.

19. Use a small current-carrying coil or solenoid for the compass needle.

Solutions to Select End-of-Chapter Problems

1. (a) Use Eq. 27-1 to calculate the force with an angle of 90° and a length of 1 meter.

$$F = I\ell B\sin\theta \quad \rightarrow \quad \frac{F}{\ell} = IB\sin\theta = (9.40\text{ A})(0.90\text{ T})\sin 90.0° = \boxed{8.5\text{ N/m}}$$

    (b) $\dfrac{F}{\ell} = IB\sin\theta = (9.40\text{ A})(0.90\text{ T})\sin 35.0° = \boxed{4.9\text{ N/m}}$

7. We find the force using Eq. 27-3, where the vector length is broken down into two parts: the portion along the z-axis and the portion along the line $y=2x$.

$$\vec{L}_1 = -0.250\text{ m }\hat{\mathbf{k}} \qquad \vec{L}_2 = 0.250\text{ m}\left(\frac{\hat{\mathbf{i}} + 2\hat{\mathbf{j}}}{\sqrt{5}}\right)$$

$$\vec{F} = I\vec{L} \times \vec{B} = I\left(\vec{L}_1 + \vec{L}_2\right) \times \vec{B} = \left(20.0\ \text{A}\right)\left(0.250\ \text{m}\right)\left(-\hat{k} + \frac{\hat{i} + 2\hat{j}}{\sqrt{5}}\right) \times \left(0.318\hat{i}\ \text{T}\right)$$

$$= \left(1.59\ \text{N}\right)\left(-\hat{k} \times \hat{i} + \frac{2}{\sqrt{5}}\hat{j} \times \hat{i}\right) = -\left(1.59\hat{j} + 1.42\hat{k}\right)\text{N}$$

$$F = \left|\vec{F}\right| = \sqrt{1.59^2 + 1.42^2}\ \text{N} = \boxed{2.13\ \text{N}}$$

$$\theta = \tan^{-1}\left(\frac{-1.42\ \text{N}}{-1.59\ \text{N}}\right) = \boxed{41.8^\circ \text{ below the negative y-axis}}$$

13. The maximum magnetic force as given in Eq. 27-5b can be used since the velocity is perpendicular to the magnetic field.

$$F_{\text{max}} = qvB = \left(1.60 \times 10^{-19}\ \text{C}\right)\left(8.75 \times 10^5\ \text{m/s}\right)\left(0.45\ \text{T}\right) = \boxed{6.3 \times 10^{-14}\ \text{N}}$$

By the right-hand rule, the force must be directed to the $\boxed{\text{North}}$.

19. (a) The velocity of the ion can be found using energy conservation. The electrical potential energy of the ion becomes kinetic energy as it is accelerated. Then, since the ion is moving perpendicular to the magnetic field, the magnetic force will be a maximum. That force will cause the ion to move in a circular path.

$$E_{\text{initial}} = E_{\text{final}} \quad \rightarrow \quad qV = \tfrac{1}{2}mv^2 \quad \rightarrow \quad v = \sqrt{\frac{2qV}{m}}$$

$$F_{\text{max}} = qvB = m\frac{v^2}{r} \quad \rightarrow$$

$$r = \frac{mv}{qB} = \frac{m\sqrt{\dfrac{2qV}{m}}}{qB} = \frac{1}{B}\sqrt{\frac{2mV}{q}} = \frac{1}{0.340\ \text{T}}\sqrt{\frac{2\left(6.6 \times 10^{-27}\ \text{kg}\right)\left(2700\ \text{V}\right)}{2\left(1.60 \times 10^{-19}\ \text{C}\right)}} = \boxed{3.1 \times 10^{-2}\ \text{m}}$$

(b) The period can be found from the speed and the radius. Use the expressions for the radius and the speed from above.

$$v = \frac{2\pi r}{T} \quad \rightarrow \quad T = \frac{2\pi r}{v} = \frac{2\pi \dfrac{1}{B}\sqrt{\dfrac{2mV}{q}}}{\sqrt{\dfrac{2qV}{m}}} = \frac{2\pi m}{qB} = \frac{2\pi\left(6.6 \times 10^{-27}\ \text{kg}\right)}{2\left(1.60 \times 10^{-19}\ \text{C}\right)\left(0.340\ \text{T}\right)} = \boxed{3.8 \times 10^{-7}\ \text{s}}$$

25. The total force on the proton is given by the Lorentz equation, Eq. 27-7.

$$\vec{F}_{\text{B}} = q\left(\vec{E} + \vec{v} \times \vec{B}\right) = e\left[\left(3.0\hat{i} - 4.2\hat{j}\right) \times 10^3\ \text{V/m} + \begin{vmatrix} \hat{i} & \hat{j} & \hat{k} \\ 6.0 \times 10^3\ \text{m/s} & 3.0 \times 10^3\ \text{m/s} & -5.0 \times 10^3\ \text{m/s} \\ 0.45\ \text{T} & 0.38\ \text{T} & 0 \end{vmatrix}\right]$$

$$= \left(1.60 \times 10^{-19}\ \text{C}\right)\left[\left(3.0\hat{i} - 4.2\hat{j}\right) + \left(1.9\hat{i} - 2.25\hat{j} + 0.93\hat{k}\right)\right] \times 10^3\ \text{N/C}$$

$$= \left(1.60 \times 10^{-19}\ \text{C}\right)\left[\left(4.9\hat{i} - 6.45\hat{j} + 0.93\hat{k}\right)\right] \times 10^3\ \text{N/C}$$

$$= \left(7.84 \times 10^{-16}\hat{i} - 1.03 \times 10^{-15}\hat{j} + 1.49 \times 10^{-16}\hat{k}\right)\text{N/C}$$

$$= \boxed{\left[\left(0.78\hat{i} - 1.0\hat{j} + 0.15\hat{k}\right)\right] \times 10^{-15}\ \text{N}}$$

31. The magnetic force will produce centripetal acceleration. Use that relationship to calculate the speed. The radius of the Earth is $6.38 \times 10^6 \, \text{km}$, and the altitude is added to that.

$$F_B = qvB = m\frac{v^2}{r} \quad \rightarrow \quad v = \frac{qrB}{m} = \frac{\left(1.60 \times 10^{-19} \, \text{C}\right)\left(6.385 \times 10^6 \, \text{m}\right)\left(0.50 \times 10^{-4} \, \text{T}\right)}{238\left(1.66 \times 10^{-27} \, \text{kg}\right)} = \boxed{1.3 \times 10^8 \, \text{m/s}}$$

Compare the size of the magnetic force to the force of gravity on the ion.

$$\frac{F_B}{F_g} = \frac{qvB}{mg} = \frac{\left(1.60 \times 10^{-19} \, \text{C}\right)\left(1.3 \times 10^8 \, \text{m/s}\right)\left(0.50 \times 10^{-4} \, \text{T}\right)}{238\left(1.66 \times 10^{-27} \, \text{kg}\right)\left(9.80 \, \text{m/s}^2\right)} = 2.3 \times 10^8$$

$\boxed{\text{Yes}}$, we may ignore gravity. The magnetic force is more than 200 million times larger than gravity.

37. (a) The torque is given by Eq. 27-9. The angle is the angle between the B-field and the perpendicular to the coil face.

$$\tau = NIAB\sin\theta = 12\left(7.10 \, \text{A}\right)\left[\pi\left(\frac{0.180 \, \text{m}}{2}\right)^2\right]\left(5.50 \times 10^{-5} \, \text{T}\right)\sin 24° = \boxed{4.85 \times 10^{-5} \, \text{m} \cdot \text{N}}$$

(b) In Example 27-11 it is stated that if the coil is free to turn, it will rotate toward the orientation so that the angle is 0. In this case, that means the $\boxed{\text{north}}$ edge of the coil will rise, so that a perpendicular to its face will be parallel with the Earth's magnetic field.

43. From the galvanometer discussion in Section 27-6, the amount of deflection is proportional to the ratio of the current and the spring constant: $\phi \propto \frac{I}{k}$. Thus if the spring constant decreases by 15%, the current can decrease by 15% to produce the same deflection. The new current will be 85% of the original current.

$$I_{\text{final}} = 0.85 I_{\text{initial}} = 0.85\left(46 \mu\text{A}\right) = \boxed{39 \mu\text{A}}$$

49. We find the magnetic field using Eq. 27-14, with the drift velocity given by Eq. 25-13. To determine the electron density we divide the density of sodium by its atomic weight. This gives the number of moles of sodium per cubic meter. Multiplying the result by Avogadro's number gives the number of sodium atoms per cubic meter. Since there is one free electron per atom, this is also the density of free electrons.

$$B = \frac{\varepsilon_H}{v_d d} = \frac{\varepsilon_H}{\left(\dfrac{I}{ne(td)}\right)d} = \frac{\varepsilon_H net}{I} = \frac{\varepsilon_H et}{I}\left(\frac{\rho N_A}{m_A}\right) =$$

$$= \frac{\left(1.86 \times 10^{-6} \, \text{V}\right)\left(1.60 \times 10^{-19} \, \text{C}\right)\left(1.30 \times 10^{-3} \, \text{m}\right)}{12.0 \, \text{A}} \frac{\left(0.971\right)\left(1000 \, \text{kg/m}^3\right)\left(6.022 \times 10^{23} \, \text{e/mole}\right)}{0.02299 \, \text{kg/mole}} =$$

$$= \boxed{0.820 \, \text{T}}$$

55. Since the particle is undeflected in the crossed fields, its speed is given by Eq. 27-8. Without the electric field, the particle will travel in a circle due to the magnetic force. Using the centripetal acceleration, we can calculate the mass of the particle. Also, the charge must be an integer multiple of the fundamental charge.

$$qvB = \frac{mv^2}{r} \quad \rightarrow$$

$$m = \frac{qBr}{v} = \frac{qBr}{\left(E/B\right)} = \frac{neB^2r}{E} = \frac{n\left(1.60 \times 10^{-19} \, \text{C}\right)\left(0.034 \, \text{T}\right)^2\left(0.027 \, \text{m}\right)}{1.5 \times 10^3 \, \text{V/m}} = n\left(3.3 \times 10^{-27} \, \text{kg}\right) \approx n\left(2.0 \, \text{u}\right)$$

The particle has an atomic mass of a multiple of 2.0 u. The simplest two cases are that it could be a hydrogen-2 nucleus (called a deuteron), or a helium-4 nucleus (called an alpha particle): $\boxed{{}^2_1\text{H}, {}^4_2\text{He}}$.

61. The magnetic force must be equal in magnitude to the weight of the electron.

$$mg = qvB \quad \rightarrow \quad v = \frac{mg}{qB} = \frac{\left(9.11 \times 10^{-31} \text{kg}\right)\left(9.80 \text{ m/s}^2\right)}{\left(1.60 \times 10^{-19} \text{C}\right)\left(0.50 \times 10^{-4} \text{T}\right)} = \boxed{1.1 \times 10^{-6} \text{ m/s}}$$

The magnetic force must point upwards, and so by the right-hand rule and the negative charge of the electron, the electron must be moving $\boxed{\text{west}}$.

67. The protons will follow a circular path as they move through the region of magnetic field, with a radius of curvature given in Example 27-7 as $r = \dfrac{mv}{qB}$.

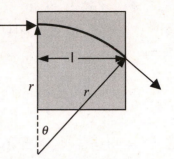

Fast-moving protons will have a radius of curvature that is too large and so they will exit above the second tube. Likewise, slow-moving protons will have a radius of curvature that is too small and so $\boxed{\text{they will exit below the second tube}}$. Since the exit velocity is perpendicular to the radius line from the center of curvature, the bending angle can be calculated.

$$\sin\theta = \frac{\ell}{r} \quad \rightarrow$$

$$\theta = \sin^{-1}\frac{\ell}{r} = \sin^{-1}\frac{\ell qB}{mv} = \sin^{-1}\frac{\left(5.0 \times 10^{-2} \text{m}\right)\left(1.60 \times 10^{-19} \text{C}\right)\left(0.38 \text{ T}\right)}{\left(1.67 \times 10^{-27} \text{kg}\right)\left(0.85 \times 10^{7} \text{ m/s}\right)} = \sin^{-1} 0.214 = \boxed{12°}$$

73. The speed of the proton can be calculated based on the radius of curvature of the (almost) circular motion. From that the kinetic energy can be calculated.

$$qvB = \frac{mv^2}{r} \quad \rightarrow \quad v = \frac{qBr}{m} \qquad K = \tfrac{1}{2}mv^2 = \tfrac{1}{2}m\left(\frac{qBr}{m}\right)^2 = \frac{q^2B^2r^2}{2m}$$

$$\Delta K = \frac{q^2B^2}{2m}\left(r_2^2 - r_1^2\right) = \frac{\left(1.60 \times 10^{-19} \text{C}\right)^2\left(0.018 \text{ T}\right)^2}{2\left(1.67 \times 10^{-27} \text{kg}\right)}\left[\left(8.5 \times 10^{-3}\text{m}\right)^2 - \left(10.0 \times 10^{-3}\text{m}\right)^2\right]$$

$$= \boxed{-6.9 \times 10^{-20} \text{J}} \text{ or } -0.43 \text{eV}$$

Chapter 28: Sources of Magnetic Field

Chapter Overview and Objectives

This chapter introduces different sources of magnetic fields. The calculation of the magnetic field for simple arrangements of currents is presented. The magnetic properties of materials are discussed.

After completing study of this chapter you should:
- Know how to calculate the magnetic field in the vicinity of a long straight wire.
- Know how to calculate the force between two long straight current-carrying wires.
- Know Ampere's law and how to use it to calculate magnetic fields for symmetrical current distributions.
- Know how to calculate the magnetic field inside a solenoid.
- Know how to calculate the magnetic field inside a toroidal coil.
- Know the Biot-Savart law and how it can be used to calculate magnetic fields.
- Know what the properties of ferromagnetic, paramagnetic, and diamagnetic materials are.

Summary of Equations

Magnitude of magnetic field around a straight wire:

$$B = \frac{\mu_0}{2\pi}\frac{I}{r}$$
(Section 28-1)

Force per unit length between two parallel wires:

$$\frac{F}{l} = \frac{\mu_0}{2\pi}\frac{I_1 I_2}{d}$$
(Section 28-2)

Ampere's law:

$$\oint \vec{B} \cdot d\vec{\ell} = \mu_0 I_{encl}$$
(Section 28-4)

Magnetic field inside a solenoid:

$$B = \mu_0 n I$$
(Section 28-5)

Magnetic field inside a toroid:

$$B = \frac{\mu_0 N I}{2\pi r}$$
(Section 28-5)

Biot-Savart law:

$$d\vec{B} = \frac{\mu_0 I}{4\pi}\frac{d\vec{\ell} \times \hat{\mathbf{r}}}{r^2}$$
(Section 28-6)

$$\vec{B} = \int \frac{\mu_0}{4\pi}\frac{I\,d\vec{\ell} \times \hat{\mathbf{r}}}{r^2}$$
(Section 28-6)

Chapter Summary

Section 28-1. Magnetic Field Due to a Straight Wire

The magnetic field created by a long straight wire carrying a current I has a magnitude

$$B = \frac{\mu_0}{2\pi}\frac{I}{r}$$

where r is the distance from the axis of the wire and $\mu_0 = 4\pi \times 10^{-7}$ Tm/A is the **permeability of free space**. The direction of the magnetic field is tangent to the circle of radius r, concentric with the wire. The right-hand rule specifies the direction of the magnetic field along the circle's tangent. Grasp the wire with your right hand with your thumb pointing in the direction of the current. The magnetic field points in the same direction around the circle as your fingers curl around the wire.

Example 28-1-A. Magnetic field due to two current-carrying wires. Two long wires run parallel to each other and are separated by a distance $d = 1.5$ cm. Wire 1 carries a current of $I_1 = 5.6$ A out of the page; wire 2 carries a current of $I_2 = 3.9$ A in the opposite direction. What is the magnetic field at a point P, located a distance $d = 1.5$ cm above wire 2?

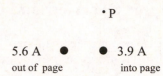

Approach: This problem is solved using the principle of superposition. The magnetic field at P is the vector sum of the magnetic field generated by the wires. To solve this problem, a coordinate system with its origin located at the position of wire 1 is used. The x axis is defined by the location of the two wires; wire 2 is located at $(d, 0, 0)$. The y axis is parallel to the line between wire 2 and point P; point P is located at $(d, d, 0)$. The z axis is perpendicular to the page, upwards.

Solution: The position vector of point P with respect to wire 1 is

$$\vec{r}_1 = d\,\hat{i} + d\,\hat{j} = d\sqrt{2}\left(\frac{1}{2}\sqrt{2}\,\hat{i} + \frac{1}{2}\sqrt{2}\,\hat{j}\right) = d\sqrt{2}\,\hat{r}_1$$

The position vector of point P with respect to wire 2 is

$$\vec{r}_2 = d\,\hat{j} = d\,\hat{r}_2$$

The magnetic field at P is the vector sum of the magnetic field due to both wires:

$$\vec{B} = \frac{\mu_0}{2\pi}\frac{\vec{I}_1 \times \hat{r}_1}{r_1} + \frac{\mu_0}{2\pi}\frac{\vec{I}_2 \times \hat{r}_2}{r_2} = \frac{\mu_0}{2\pi}\left[\frac{I_1\,\hat{k}\times\left(\frac{1}{2}\sqrt{2}\,\hat{i} + \frac{1}{2}\sqrt{2}\,\hat{j}\right)}{d\sqrt{2}} + \frac{-I_2\,\hat{k}\times\hat{j}}{d}\right] =$$

$$= \frac{\mu_0}{2\pi}\frac{1}{d}\left[\frac{1}{2}I_1\left(\hat{j}-\hat{i}\right) + I_2\,\hat{i}\right] = \frac{\mu_0}{2\pi}\frac{1}{d}\left[\left(I_2 - \frac{1}{2}I_1\right)\hat{i} + \frac{1}{2}I_1\,\hat{j}\right] = 1.33\times10^{-5}\left[1.1\,\hat{i} + 2.8\,\hat{j}\right]$$

Section 28-2. Force Between Two Parallel Wires

The magnitude of the magnetic force per unit length between two long straight parallel current-carrying wires is equal to

$$\frac{F}{\ell} = \frac{\mu_0}{2\pi}\frac{I_1 I_2}{d}$$

where I_1 is the current in one wire, I_2 is the current in the other wire, and d is the distance between the two wires. This equation is correct if the wires are circular in cross-section, have a uniform current density, and the radius of the wires is small compared to the distance d between the wires.

If the currents in the two wires flow in the same direction, the force is attractive. If the currents flow in opposite directions, the force is repulsive.

Example 28-2-A. Calculating the force between two parallel wires. Determine the net force per unit length on the wire 2 of Example 28-1-A.

Approach: Since all currents and dimensions are specified in Example 28-1-A, the calculation of the force per unit length only involves a direct substitution of the currents and distance in the force per unit length formula discussed in this section.

Solution: The currents in the two wires are 3.9 A and 5.6 A, and their separation is 0.015 m. The force per unit length on wire 2 is equal to

$$\frac{F}{l} = \frac{\mu_0}{2\pi}\frac{I_1 I_2}{d} = \frac{4\pi \times 10^{-7}}{2\pi}\frac{(3.9)(5.6)}{(0.015)} = 2.9 \times 10^{-4}\ \text{N}$$

Since the currents in the two wires flow in opposite directions, the force between the wires is repulsive.

Section 28-3. Definitions of the Ampere and the Coulomb

The modern definition of an ampere is in terms of the magnitude of the magnetic force between two parallel current-carrying wires. One **ampere** is defined as the current that must flow through each of two parallel wires, one meter apart, such that the magnitude of the magnetic force per unit length between the wires is $2\pi \times 10^{-7}$ N/m. The modern definition of the **coulomb** is the amount of charge that is transported in one second by a current of one ampere.

Section 28-4. Ampère's Law

Consider a closed path in space. We can move around the path with a large number of small displacements $d\vec{\ell}$. The vector $d\vec{\ell}$ is directed tangentially to the path, in the direction in which we traverse it. **Ampere's law** states that the integral around the closed path of the scalar product of the magnetic field $\vec{B}$ with $d\vec{\ell}$ is equal to μ_0 times the current that is enclosed by the loop:

$$\oint \vec{B} \cdot d\vec{\ell} = \mu_0 I_{encl}$$

Ampere's law, like Gauss's law for electric fields, can be used to determine the magnitude of the magnetic field if the current distributions have a certain degree of symmetry.

Example 28-4-A. Calculating the magnetic field within a conductor. A long straight wire has a circular cross-section of radius R. The current density j is a function of the distance r from the center of the wire and is given by

$$j(r) = j_0 r$$

Determine the magnitude of the magnetic field within the wire.

Approach: Due to the radial symmetry, we expect that the magnetic field lines will be circular, centered on the center of the wire. Since the magnitude of the magnetic field will only depend on r, the path integral of the magnetic field around circular paths, centered on the center of the wire, is easy to calculate. Applying Ampere's law, we can use this path integral to determine the magnitude of the magnetic field.

Solution: Consider a circular path of radius r. The path integral of the magnetic field around this path is equal to

$$\oint \vec{B} \cdot d\vec{\ell} = 2\pi B(r)$$

To determine the current enclosed by this path we integrate the current density over the circular cross-section of the wire enclosed by the path:

$$I_{enclosed} = \int_0^r j(r')\,dA = \int_0^r (j_0 r')(2\pi r'\,dr') = 2\pi j_0 \int_0^r r'^2\,dr = \frac{2\pi}{3}j_0 r^3$$

Using Ampere's law, we can now determine the magnitude of the magnetic field

$$2\pi B(r) = \mu_0 \frac{2\pi}{3}j_0 r^3 \quad \Rightarrow \quad B(r) = \frac{1}{3}\mu_0 j_0 r^3$$

Section 28-5. Magnetic Field of a Solenoid and a Toroid

A **solenoid** is a relatively long coil with many loops of wire, uniformly wound around a cylindrical surface, centered on the axis of the solenoid. The magnetic field inside the solenoid, far from its ends, is uniform and has a magnitude

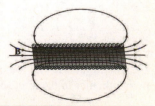

$$B = \frac{\mu_0 NI}{\ell} = \mu_0 nI$$

where N is the total number of turns, ℓ is the length of the solenoid, n is the number of turns per unit length, and I is the current flowing through the coil. The direction of the magnetic field inside the solenoid is parallel to the cylindrical axis and its direction can be determined using the right-hand rule.

A toroidal coil is shaped like a solenoid but bent around in a circle so that the two ends are located next to each other. The magnetic field of a tightly and uniformly wound toroidal coil will be zero outside the toroid. Inside the toroid, the magnetic field will have a magnitude

$$B = \frac{\mu_0 N I}{2\pi r}$$

where N is the number of turns, I is the current, and r is the distance from the center of toroid.

Example 28-5-A. Designing a solenoid. You must design a solenoid that can generate a magnetic field of magnitude B. You have a length L of wire that can carry a maximum current I. What is the maximum radius of the solenoid that will use this length of wire and have a volume V?

Approach: The magnetic field inside a solenoid depends on the length of the solenoid S, the number of turns N, and the current flowing through the turns I. In this problem, the number of turns is constrained by the radius of the solenoid and the total length of the wire available. The radius also constrains the length of the solenoid since the volume must be V.

Solution: The volume V of the solenoid, its radius R, and its length S must satisfy the following relation:

$$V = \pi R^2 S$$

The number of turns of the solenoid, N, will be the length of the wire L divided by the circumference of each turn:

$$N = \frac{L}{2\pi R}$$

The magnetic field inside the solenoid is given by

$$B = \mu_0 n I = \frac{\mu_0 N I}{S} = \frac{\mu_0 L I}{2\pi R S}$$

Instead of expressing the magnetic field in terms of R and S we can rewrite the expression in terms of R and V:

$$B = \frac{\mu_0 L I}{2\pi R \left(\dfrac{V}{\pi R^2}\right)} = \frac{1}{2}\mu_0 L I \frac{R}{V}$$

The radius of the solenoid is thus equal to

$$R = \frac{2B}{\mu_0 L I} V$$

Section 28-6. Biot-Savart Law

The Biot-Savart law relates the distribution of currents in space to the magnetic field at a particular point. An infinitesimal current element $d\vec{\ell}$ contributes $d\vec{B}$ to the total magnetic field at that point

$$d\vec{B} = \frac{\mu_0 I}{4\pi} \frac{d\vec{\ell} \times \hat{\mathbf{r}}}{r^2}$$

To determine the total magnetic field, the contributions from each infinitesimal current element $d\vec{\ell}$ must be added together:

$$\vec{B} = \int \frac{\mu_0}{4\pi} \frac{I\, d\vec{\ell} \times \hat{\mathbf{r}}}{r^2}$$

The magnitude of the magnetic field of a circular current loop with radius R and current I along the axis of the loop, perpendicular to the plane of the loop, is

$$B = \frac{\mu_0 I R^2}{2\left(R^2 + x^2\right)^{\frac{3}{2}}} = \frac{\mu_0 \mu}{2\pi \left(R^2 + x^2\right)^{\frac{3}{2}}}$$

where μ is the magnetic moment of the current loop and x is the distance from the plane of the current loop to the point at which the magnetic field is evaluated. If $x \gg R$, then the magnitude of the magnetic field can be approximated by

$$B = \frac{\mu_0 I R^2}{2\pi x^3 \left(\frac{R^2}{x^2} + 1\right)^{\frac{3}{2}}} \approx \frac{\mu_0 \mu}{2\pi x^3}$$

Section 28-7. Magnetic Materials–Ferromagnetism

Many atoms have magnetic dipole moments because of the motion of the electrons within the atoms and because the electrons themselves have magnetic moments. In most materials, these magnetic dipole moments are oriented in random directions and the net magnetic field due to the magnetic dipoles is negligibly small. In a certain class of materials, the so-called **ferromagnetic** materials, it is energetically favorable for the magnetic dipole moments of some of the electrons to be aligned with each other. The magnetic field due to the aligned magnetic dipoles can be relatively large. The regions where the magnetic dipoles are aligned are called **domains**. The alignment of the magnetic dipoles is lost when the material is heated to a temperature above the **Curie temperature**.

Section 28-8. Electromagnets and Solenoids

Electromagnets and solenoids have many practical applications. Electromagnets are used in electric motors and devices that need to produce variable magnetic fields. When superconductors are used in electromagnets, very high magnetic fields can be generated.
In certain applications, solenoids are used to move soft-iron rods partially inserted in their central region. Changes in the solenoidal magnetic field generate magnetic forces on the rod and can induce motion. Applications include the operation of a door bell and the starter of a car engine.

Section 28-9. Magnetic Fields in Magnetic Materials; Hysteresis

The magnetic field in magnetic materials can be resolved into two components when a magnetic material is present. One component of the field is due to the magnetic fields from sources outside the material and the free currents inside the material; this component is labeled as $\vec{B}_0$. The other component is the magnetic field due to the magnetic dipole moments of the magnetic material and is labeled $\vec{B}_M$. The total magnetic field is the vector sum of these two fields:

$$\vec{B} = \vec{B}_0 + \vec{B}_M$$

In a ferromagnetic material, $\vec{B}_M$ can be much larger than $\vec{B}_0$. When we insert ferromagnetic material into a solenoid, the magnetic field inside the solenoid can be written as

$$\vec{B} = \mu n I$$

where μ is called the **magnetic permeability** of the ferromagnetic material. The value of μ is not a constant; it depends on the external magnetic field $\vec{B}_0$. At a certain external field, nearly all magnetic domains of the material are aligned and the material approaches **saturation**.
$\vec{B}_M$ can be non-zero even if $\vec{B}_0$ is zero. This property is called **retentivity** and is necessary for a material to form a permanent magnet. The value of $\vec{B}_M$ is not only a function of the current value of $\vec{B}_0$, but also depends on past values of $\vec{B}_0$. This property is called **hysteresis**.

Section 28-10. Paramagnetism and Diamagnetism

In materials that are non-ferromagnetic, $\vec{B}_M$ is small compared to $\vec{B}_0$. The total magnetic field is the sum of the two component fields. For these types of materials, it is a good approximation to assume that the total magnetic field is proportional to $\vec{B}_0$:

$$\vec{B} = \mu \vec{B}_0 = K_m \mu_0 \vec{B}_0$$

where $K_m = \mu/\mu_0$ is called the **relative magnetic permeability** of the material. The expression of the total magnetic field can also be written in terms of the **magnetic susceptibility** χ_m of the material. The magnetic susceptibility is defined as

$$\chi_m = K_m - 1$$

A **paramagnetic material** has a relative magnetic permeability greater than one and a magnetic susceptibility less than one. A **diamagnetic material** has a relative magnetic permeability less than one and a magnetic susceptibility greater than one. In a paramagnetic material, the magnetic field within the material is slightly greater in magnitude than the applied magnetic field. In a diamagnetic material, the magnetic field is slightly less in magnitude than the applied magnetic field.

Practice Quiz

1. The magnetic force per unit length between two non-parallel wires, with a distance of closest approach equal to d and an angle between currents equal to $\theta < 90°$, will be
   a) equal to $\dfrac{\mu_0}{2\pi}\dfrac{I_1 I_2}{d}$.
   b) equal to $\dfrac{\mu_0}{2\pi}\dfrac{I_1 I_2}{d}\sin\theta$.
   c) equal to $\dfrac{\mu_0}{2\pi}\dfrac{I_1 I_2}{d}\cos\theta$.
   d) less than $\dfrac{\mu_0}{2\pi}\dfrac{I_1 I_2}{d}\cos\theta$.

2. You are located north of a wire carrying an upward vertical current. What is the direction of the magnetic field due to this current at your location?
   a) North.
   b) South.
   c) East.
   d) West.

3. You walk around a vertical metal conduit carrying a magnetic compass. As you walk completely around the conduit in a clockwise direction, the compass always points in the direction you are moving. What can you conclude?
   a) There is a permanent magnet inside the conduit.
   b) The conduit has a net electrical charge at its surface.
   c) The compass must be broken.
   d) There is a current flowing through the conduit.

4. To double the magnetic field inside a solenoid you can
   a) double the length of the solenoid.
   b) double the diameter of the solenoid.
   c) double the square of the radius of the solenoid.
   d) double the current in the solenoid.

5. The magnitude of the magnetic field a distance d from a current-carrying wire is B. What is the magnitude of the magnetic field a distance $2d$ from the wire?
   a) $2B$
   b) $B/2$
   c) $4B$
   d) $B/4$

6. A segment of a wire of length L carries a current I in the direction shown in the diagram. Point P is located on the same line as the wire segment, a distance L from the end of the segment. What is the magnitude of the magnetic field at point P due to the current in the wire segment?
   a) $\mu_0 I/L$
   b) $\mu_0 I/2L$
   c) $\mu_0 I/3L$
   d) 0

7. A type of superconducting material always has zero magnetic field within it when it is in a superconducting state. What is the magnetic susceptibility of such a material?
   a) 0
   b) 1
   c) −1
   d) infinite

8. If you look into the north pole end of a solenoid, which way does the current travel around the solenoid?
   a) Clockwise.
   b) Counterclockwise.
   c) East.
   d) West.

9. A permanent magnet is brought near an unknown type of material. No matter which pole of the permanent magnet is brought near the material, the material is repelled by the permanent magnet. What is the type of this material?
   a) Diamagnetic material.
   b) Paramagnetic material.
   c) Ferromagnetic material.
   d) Non-magnetic material.

10. Which of the following properties of a material is necessary for a permanent magnet to be constructed from the material?
    a) Diamagnetism.
    b) Paramagnetism.
    c) Ferromagnetism.
    d) All of the above.

11. Determine the magnitude and direction of the magnetic field at a position 0.16 m East of a long vertical wire carrying an upward current of 4.8 A.

12. A square loop of wire with sides of length L carries a current I in the clockwise direction as shown in the diagram. A long wire lies parallel to the left side of the square and is a distance L away from the square. The long wire also carries a current I toward the top of the page as shown in the diagram. What is the magnetic force on the loop?

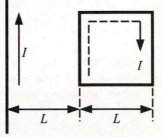

13. A solenoid is to be designed so that the magnetic field within the solenoid has a magnitude of 0.28 T. The diameter of the coil needs to be 8.8 cm and the length needs to be 12 cm to accommodate the material that is to be placed within the solenoid. How many turns of wire that can carry a maximum current of 3.0 A must be used?

14. A long straight coaxial cable carries a current of 3.0 A in one direction on its inner conductor and a current of 5.0 A in the opposite direction on its outer conductor. The inner conductor has a diameter of 0.50 mm and the outer conductor has a diameter of 6.0 mm. Determine the magnitude of the magnetic field in the space between the two conductors and in the space outside the outer conductor.

15. Determine the magnetic field at the center of a square loop of wire that carries a current I.

Responses to Select End-of-Chapter Questions

1. Alternating currents will have little effect on the compass needle, due to the rapid change of the direction of the current and of the magnetic field surrounding it. Direct currents will deflect a compass needle. The deflection depends on the magnitude and direction of the current and the distance from the current to the compass. The effect on the compass decreases with increasing distance from the wire.

7. Inside the cavity $\vec{\mathbf{B}} = 0$ since the geometry is cylindrical and no current is enclosed.

13. The Biot-Savart law and Coulomb's law are both inverse-square in the radius and both contain a proportionality constant. Coulomb's law describes a central force; the Biot-Savart law involves a cross product of vectors and so cannot describe a central force.

19. An unmagnetized nail has randomly oriented domains and will not generate an external magnetic field. Therefore, it will not attract an unmagnetized paper clip, which also has randomly oriented domains. When one end of the nail is in contact with a magnet, some of the domains in the nail align, producing an external magnetic field and turning the nail into a magnet. The magnetic nail will cause some of the domains in the paper clip to align, and it will be attracted to the nail.

25. (*a*) The magnetization curve for a paramagnetic substance is a straight line with slope slightly greater than 1. It passes through the origin; there is no hysteresis.
(*b*) The magnetization curve for a diamagnetic substance is a straight line with slope slightly less than 1. It passes through the origin; there is no hysteresis.
The magnetization curve for a ferromagnetic substance is a hysteresis curve (see Figure 28-29).

Solutions to Select End-of-Chapter Problems

1. We assume the jumper cable is a long straight wire, and use Eq. 28-1.

$$B_{cable} = \frac{\mu_0 I}{2\pi r} = \frac{\left(4\pi \times 10^{-7} \text{ T} \cdot \text{m/A}\right)(65 \text{A})}{2\pi (0.035 \text{ m})} = 3.714 \times 10^{-4} \text{ T} \approx \boxed{3.7 \times 10^{-4} \text{ T}}$$

Compare this to the Earth's field of 0.5×10^{-4} T.

$$B_{cable}/B_{Earth} = \frac{3.714 \times 10^{-4} \text{ T}}{5.0 \times 10^{-5} \text{ T}} = 7.43$$

so $\boxed{\text{the field of the cable is over 7 times that of the Earth.}}$

7. Since the magnetic field from a current carrying wire circles the wire, the individual field at point P from each wire is perpendicular to the radial line from that wire to point P. We define $\vec{\mathbf{B}}_1$ as the field from the top wire, and $\vec{\mathbf{B}}_2$ as the field from the bottom wire. We use Eq. 28-1 to calculate the magnitude of each individual field.

$$B_1 = \frac{\mu_0 I}{2\pi r_1} = \frac{\left(4\pi \times 10^{-7} \text{ T} \cdot \text{m/A}\right)(35 \text{ A})}{2\pi (0.060 \text{ m})} = 1.17 \times 10^{-4} \text{ T}$$

$$B_2 = \frac{\mu_0 I}{2\pi r_2} = \frac{\left(4\pi \times 10^{-7} \text{ T} \cdot \text{m/A}\right)(35 \text{ A})}{2\pi (0.100 \text{ m})} = 7.00 \times 10^{-5} \text{ T}$$

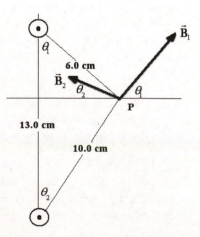

We use the law of cosines to determine the angle that the radial line from each wire to point P makes with the vertical. Since the field is perpendicular to the radial line, this is the same angle that the magnetic fields make with the horizontal.

$$\theta_1 = \cos^{-1}\left(\frac{(0.060 \text{ m})^2 + (0.130 \text{ m})^2 - (0.100 \text{ m})^2}{2(0.060 \text{ m})(0.130 \text{ m})}\right) = 47.7°$$

$$\theta_2 = \cos^{-1}\left(\frac{(0.100 \text{ m})^2 + (0.130 \text{ m})^2 - (0.060 \text{ m})^2}{2(0.100 \text{ m})(0.130 \text{ m})}\right) = 26.3°$$

Using the magnitudes and angles of each magnetic field we calculate the horizontal and vertical components, add the vectors, and calculate the resultant magnetic field and angle.

$$B_{net\,x} = B_1\cos(\theta_1) - B_2\cos\theta_2 = (1.174\times10^{-4} \text{ T})\cos47.7° - (7.00\times10^{-5} \text{ T})\cos26.3° = 1.626\times10^{-5}\text{T}$$

$$B_{net\,y} = B_1\sin(\theta_1) + B_2\sin\theta_1 = (1.17\times10^{-4} \text{ T})\sin47.7° + (7.00\times10^{-5} \text{ T})\sin26.3° = 1.18\times10^{-4}\text{T}$$

$$B = \sqrt{B_{net\,x}^2 + B_{net\,y}^2} = \sqrt{(1.626\times10^{-5} \text{ T})^2 + (1.18\times10^{-4} \text{ T})^2} = 1.19\times10^{-4} \text{ T}$$

$$\theta = \tan^{-1}\frac{B_{net\,y}}{B_{net\,y}} = \tan^{-1}\frac{1.18\times10^{-4} \text{ T}}{1.626\times10^{-5} \text{ T}} = 82.2°$$

$$\vec{B} = 1.19\times10^{-4} \text{ T @ } 82.2° \approx \boxed{1.2\times10^{-4}\text{T @ } 82°}$$

13. Use the right hand rule to determine the direction of the magnetic field from each wire. Remembering that the magnetic field is inversely proportional to the distance from the wire, qualitatively add the magnetic field vectors. The magnetic field at point #2 is zero.

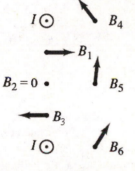

19. The left wire will cause a field on the x axis that points in the y direction, and the right wire will cause a field on the x axis that points in the negative y direction. The distance from the left wire to a point on the x axis is x, and the distance from the right wire is $d - x$.

$$\vec{B}_{net} = \frac{\mu_0 I}{2\pi x}\hat{j} - \frac{\mu_0 I}{2\pi(d-x)}\hat{j} = \boxed{\frac{\mu_0 I}{2\pi}\left(\frac{1}{x} - \frac{1}{d-x}\right)\hat{j}} = \boxed{\frac{\mu_0 I}{2\pi}\left(\frac{d-2x}{x(d-x)}\right)\hat{j}}$$

25. Use Eq. 28-4 for the field inside a solenoid.

$$B = \frac{\mu_0 I N}{\ell} \;\rightarrow\; I = \frac{B\ell}{\mu_0 N} = \frac{(0.385\times10^{-3} \text{ T})(0.400 \text{ m})}{(4\pi\times10^{-7} \text{ T·m/A})(765)} = \boxed{0.160 \text{ A}}$$

31. Because of the cylindrical symmetry, the magnetic fields will be circular. In each case, we can determine the magnetic field using Ampere's law with concentric loops. The current densities in the wires are given by the total current divided by the cross-sectional area.

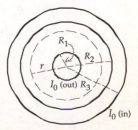

$$J_{inner} = \frac{I_0}{\pi R_1^2} \qquad J_{outer} = -\frac{I_0}{\pi(R_3^2 - R_2^2)}$$

(*a*) Inside the inner wire the enclosed current is determined by the current density of the inner wire.

$$\oint \vec{B} \cdot d\vec{s} = \mu_0 I_{encl} = \mu_0 \left(J_{inner} \pi R^2 \right)$$

$$B(2\pi R) = \mu_0 \frac{I_0 \pi R^2}{\pi R_1^2} \rightarrow \boxed{B = \frac{\mu_0 I_0 R}{2\pi R_1^2}}$$

(*b*) Between the wires the current enclosed is the current on the inner wire.

$$\oint \vec{B} \cdot d\vec{s} = \mu_0 I_{encl} \rightarrow B(2\pi R) = \mu_0 I_0 \rightarrow \boxed{B = \frac{\mu_0 I_0}{2\pi R}}$$

(*c*) Inside the outer wire the current enclosed is the current from the inner wire and a portion of the current from the outer wire.

$$\oint \vec{B} \cdot d\vec{s} = \mu_0 I_{encl} = \mu_0 \left[I_0 + J_{outer} \pi \left(R^2 - R_2^2 \right) \right]$$

$$B(2\pi r) = \mu_0 \left[I_0 - I_0 \frac{\pi \left(R^2 - R_2^2 \right)}{\pi \left(R_3^2 - R_2^2 \right)} \right] \rightarrow$$

$$\boxed{B = \frac{\mu_0 I_0}{2\pi R} \frac{\left(R_3^2 - R^2 \right)}{\left(R_3^2 - R_2^2 \right)}}$$

(*d*) Outside the outer wire the net current enclosed is zero.

$$\oint \vec{B} \cdot d\vec{s} = \mu_0 I_{encl} = 0 \rightarrow B(2\pi R) = 0 \rightarrow \boxed{B = 0}$$

(*e*) See the graph on the right.

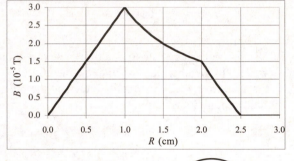

37. (*a*) The magnetic field at point C can be obtained using the Biot-Savart law (Eq. 28-5, integrated over the current). First break the loop into four sections: 1) the upper semi-circle, 2) the lower semi-circle, 3) the right straight segment, and 4) the left straight segment. The two straight segments do not contribute to the magnetic field as the point C is in the same direction that the current is flowing. Therefore, along these segments $\hat{r}$ and $d\hat{\ell}$ are parallel and $d\hat{\ell} \times \hat{r} = 0$. For the upper segment, each infinitesimal line segment is perpendicular to the constant magnitude radial vector, so the magnetic field points downward with constant magnitude.

$$\vec{B}_{upper} = \int \frac{\mu_0 I}{4\pi} \frac{d\hat{\ell} \times \hat{r}}{r^2} = \frac{\mu_0 I}{4\pi} \frac{-\hat{k}}{R_1^2} \left(\pi R_1 \right) = -\frac{\mu_0 I}{4 R_1} \hat{k}$$

Along the lower segment, each infinitesimal line segment is also perpendicular to the constant radial vector.

$$\vec{B}_{lower} = \int \frac{\mu_0 I}{4\pi} \frac{d\hat{\ell} \times \hat{r}}{r^2} = \frac{\mu_0 I}{4\pi} \frac{-\hat{k}}{R_2^2} \left(\pi R_2 \right) = -\frac{\mu_0 I}{4 R_2} \hat{k}$$

Adding the two contributions yields the total magnetic field.

$$\vec{B} = \vec{B}_{upper} + \vec{B}_{lower} = -\frac{\mu_0 I}{4 R_1} \hat{k} - \frac{\mu_0 I}{4 R_2} \hat{k} = \boxed{-\frac{\mu_0 I}{4} \left(\frac{1}{R_1} + \frac{1}{R_2} \right) \hat{k}}$$

(*b*) The magnetic moment is the product of the area and the current. The area is the sum of the two half circles. By the right-hand-rule, curling your fingers in the direction of the current, the thumb points into the page, so the magnetic moment is in the $-\hat{k}$ direction.

$$\vec{\mu} = -\left(\frac{\pi R_1^2}{2} + \frac{\pi R_2^2}{2} \right) I\hat{k} = \boxed{\frac{-\pi I}{2} \left(R_1^2 + R_2^2 \right) \hat{k}}$$

43. (*a*) The angle subtended by one side of a polygon, θ, from the center point P is 2π divided by the number of sides, n. The length of the side L and the distance from the point to the center of the side, D, are obtained from trigonometric relations.

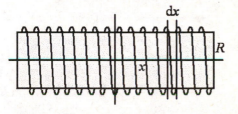

$$L = 2R\sin(\theta/2) = 2R\sin(\pi/n)$$
$$D = R\cos(\theta/2) = R\cos(\pi/n)$$

The magnetic field contribution from each side can be found using the result of Problem 40.

$$B = \frac{\mu_0 I}{2\pi D}\frac{L}{\left(L^2 + 4D^2\right)^{\frac{1}{2}}} = \frac{\mu_0 I}{2\pi\left(R\cos(\pi/n)\right)}\frac{2R\sin(\pi/n)}{\left(\left(2R\sin(\pi/n)\right)^2 + 4\left(R\cos(\pi/n)\right)^2\right)^{\frac{1}{2}}}$$

$$= \frac{\mu_0 I}{2\pi R}\tan(\pi/n)$$

The contributions from each segment add, so the total magnetic field is n times the field from one side.

$$\boxed{B_{\text{total}} = \frac{\mu_0 In}{2\pi R}\tan(\pi/n)}$$

(*b*) In the limit of large n, π/n, becomes very small, so $\tan(\pi/n) \approx \pi/n$.

$$B_{\text{total}} = \frac{\mu_0 In}{2\pi R}\frac{\pi}{n} = \boxed{\frac{\mu_0 I}{2R}}$$

This is the magnetic field at the center of a circle.

49. The magnetic field of a long, thin torus is the same as the field given by a long solenoid, as in Eq. 28-9.

$$B = \mu nI = (2200)\left(4\pi\times 10^{-7}\text{ Tm/A}\right)\left(285\text{ m}^{-1}\right)(3.0\text{ A}) =$$
$$= \boxed{2.4\text{ T}}$$

55. (*a*) Use Eq. 28-1 to calculate the field due to a long straight wire.

$$B_{\text{A at B}} = \frac{\mu_0 I_A}{2\pi r_{\text{A to B}}} = \frac{\left(4\pi\times 10^{-7}\text{ T}\cdot\text{m/A}\right)(2.0\text{ A})}{2\pi(0.15\text{ m})} = 2.667\times 10^{-6}\text{ T} \approx \boxed{2.7\times 10^{-6}\text{ T}}$$

(*b*) $$B_{\text{B at A}} = \frac{\mu_0 I_B}{2\pi r_{\text{B to A}}} = \frac{\left(4\pi\times 10^{-7}\text{ T}\cdot\text{m/A}\right)(4.0\text{ A})}{2\pi(0.15\text{ m})} = 5.333\times 10^{-6}\text{ T} \approx \boxed{5.3\times 10^{-6}\text{ T}}$$

(*c*) The two fields are not equal and opposite. Each individual field is due to a single wire, and has no dependence on the other wire. The magnitude of current in the second wire has nothing to do with the value of the field caused by the first wire.

(*d*) Use Eq. 28-2 to calculate the force due to one wire on another. The forces are attractive since the currents are in the same direction.

$$\frac{F_{\text{on A due to B}}}{\ell_A} = \frac{F_{\text{on B due to A}}}{\ell_B} = \frac{\mu_0}{2\pi}\frac{I_A I_B}{d_{\text{A to B}}} = \frac{\left(4\pi\times 10^{-7}\text{ T}\cdot\text{m/A}\right)}{2\pi}\frac{(2.0\text{ A})(4.0\text{ A})}{(0.15\text{ m})}$$

$$= 1.067\times 10^{-5}\text{ N/m} \approx \boxed{1.1\times 10^{-5}\text{ N/m}}$$

These two forces per unit length are equal and opposite because they are a Newton's third law pair of forces.

61. (*a*) Choose $x = 0$ at the center of one coil. The center of the other coil will then be at $x = R$. Since the currents flow in the same direction in both coils, the right-hand-rule shows that the magnetic fields from the two coils will point in the same direction along the axis. The magnetic field from a current loop was found in Example 28-12. Adding the two magnetic fields together yields the total field.

$$B(x) = \boxed{\frac{\mu_0 NIR^2}{2\left[R^2 + x^2\right]^{3/2}} + \frac{\mu_0 NIR^2}{2\left[R^2 + (x-R)^2\right]^{3/2}}}$$

(*b*) Evaluate the derivative of the magnetic field at $x = \frac{1}{2}R$.

$$\frac{dB}{dx} = -\frac{3\mu_0 NIR^2 x}{2\left[R^2 + x^2\right]^{5/2}} - \frac{3\mu_0 NIR^2 (x-R)}{2\left[R^2 + (x-R)^2\right]^{5/2}} = -\frac{3\mu_0 NIR^3}{4\left[R^2 + R^2/4\right]^{5/2}} - \frac{-3\mu_0 NIR^3}{4\left[R^2 + R^2/4\right]^{5/2}} = \boxed{0}$$

Evaluate the second derivative of the magnetic field at $x = \frac{1}{2}R$.

$$\frac{d^2 B}{dx^2} = -\frac{3\mu_0 NIR^2}{2\left[R^2 + x^2\right]^{5/2}} + \frac{15\mu_0 NIR^2 x^2}{2\left[R^2 + x^2\right]^{7/2}} - \frac{3\mu_0 NIR^2}{2\left[R^2 + (x-R)^2\right]^{5/2}} + \frac{15\mu_0 NIR^2 (x-R)^2}{2\left[R^2 + (x-R)^2\right]^{7/2}}$$

$$= -\frac{3\mu_0 NIR^2}{2\left[5R^2/4\right]^{5/2}} + \frac{15\mu_0 NIR^4}{8\left[5R^2/4\right]^{7/2}} - \frac{3\mu_0 NIR^2}{2\left[5R^2/4\right]^{5/2}} + \frac{15\mu_0 NIR^4}{8\left[5R^2/4\right]^{7/2}}$$

$$= \frac{\mu_0 NIR^2}{\left[5R^2/4\right]^{5/2}}\left(-\frac{3}{2} + \frac{15\cdot4}{8\cdot5} - \frac{3}{2} + \frac{15\cdot4}{8\cdot5}\right) = \boxed{0}$$

Therefore, at the midpoint $\dfrac{dB}{dx} = 0$ and $\dfrac{d^2 B}{dx^2} = 0$.

(*c*) We insert the given data into the magnetic field equation to calculate the field at the midpoint.

$$B\left(\tfrac{1}{2}R\right) = \frac{\mu_0 NIR^2}{2\left[R^2 + \left(\tfrac{1}{2}R\right)^2\right]^{3/2}} + \frac{\mu_0 NIR^2}{2\left[R^2 + \left(\tfrac{1}{2}R\right)^2\right]^{3/2}} = \frac{\mu_0 NIR^2}{\left[R^2 + \left(\tfrac{1}{2}R\right)^2\right]^{3/2}}$$

$$= \frac{\left(4\pi \times 10^{-7}\ \text{T} \cdot \text{m/A}\right)(250)(2.0\ \text{A})(0.10\ \text{m})^2}{\left[(0.10\ \text{m})^2 + (0.05\ \text{m})^2\right]^{3/2}} = \boxed{4.5\ \text{mT}}$$

67. The wire can be broken down into five segments: the two long wires, the left vertical segment, the right vertical segment, and the top horizontal segment. Since the current in the two long wires either flow radially toward or away from the point P, they will not contribute to the magnetic field. The magnetic field from the top horizontal segment points into the page and is obtained from the solution to Problem 40.

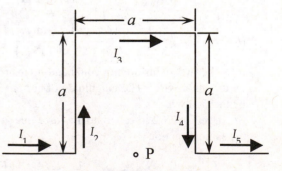

$$B_{top} = \frac{\mu_0 I}{2\pi a} \frac{a}{\left(a^2 + 4a^2\right)^{\frac{1}{2}}} = \frac{\mu_0 I}{2\pi a\sqrt{5}}$$

The magnetic fields from the two vertical segments both point into the page with magnitudes obtained from the solution to Problem 41.

$$B_{vert} = \frac{\mu_0 I}{4\pi (a/2)} \frac{a}{\left(a^2 + (a/2)^2\right)^{\frac{1}{2}}} = \frac{\mu_0 I}{\pi a\sqrt{5}}$$

Summing the magnetic fields from all the segments yields the net field.

$$B = B_{top} + 2B_{vert} = \frac{\mu_0 I}{2\pi a\sqrt{5}} + 2\frac{\mu_0 I}{\pi a\sqrt{5}} = \boxed{\frac{\mu_0 I\sqrt{5}}{2\pi a}}\text{, into the page.}$$

73. (*a*) Set $x = 0$ at the midpoint on the axis between the two loops. Since the loops are a distance R apart, the center of one loop will be at $x = -\frac{1}{2}R$ and the center of the other at $x = \frac{1}{2}R$. The currents in the loops flow in opposite directions, so by the right-hand-rule the magnetic fields from the two wires will subtract from each other. The magnitude of each field can be obtained from Example 28-12.

$$B(x) = \frac{\mu_0 NIR^2}{2\left[R^2 + \left(\frac{1}{2}R - x\right)^2\right]^{3/2}} - \frac{\mu_0 NIR^2}{2\left[R^2 + \left(\frac{1}{2}R + x\right)^2\right]^{3/2}}$$

Factoring out $\frac{1}{8}R^3$ from each of the denominators yields the desired equation.

$$B(x) = \frac{4\mu_0 NI}{R\left[4 + \left(1 - 2x/R\right)^2\right]^{3/2}} - \frac{4\mu_0 NI}{R\left[4 + \left(1 + 2x/R\right)^2\right]^{3/2}}$$

$$\boxed{= \frac{4\mu_0 NI}{R}\left\{\left[4 + \left(1 - \frac{2x}{R}\right)^2\right]^{-3/2} - \left[4 + \left(1 + \frac{2x}{R}\right)^2\right]^{-3/2}\right\}}$$

(*b*) For small values of x, we can use the approximation $\left(1 \pm \dfrac{2x}{R}\right)^2 \approx 1 \pm \dfrac{4x}{R}$.

$$B(x) = \frac{4\mu_0 NI}{R}\left\{\left[4 + 1 - \frac{4x}{R}\right]^{-3/2} - \left[4 + 1 + \frac{4x}{R}\right]^{-3/2}\right\}$$

$$= \frac{4\mu_0 NI}{5R\sqrt{5}}\left\{\left[1 - \frac{4x}{5R}\right]^{-3/2} - \left[1 + \frac{4x}{5R}\right]^{-3/2}\right\}$$

Again we can use the expansion for small deviations $\left(1 \pm \dfrac{4x}{5R}\right)^{-3/2} \approx 1 \mp \dfrac{6x}{5R}$

$$B(x) = \frac{4\mu_0 NI}{5R\sqrt{5}}\left[\left(1 + \frac{6x}{5R}\right) - \left(1 - \frac{6x}{5R}\right)\right] = \boxed{\frac{48\mu_0 NIx}{25R^2\sqrt{5}}}$$

This magnetic field has the expected linear dependence on x with a coefficient of $C = 48\mu_0 NI / \left(25R^2\sqrt{5}\right)$.

(*c*) Set C equal to 0.15 T/m and solve for the current.

$$I = \frac{25CR^2\sqrt{5}}{48\mu_0 N} = \frac{25\left(0.15 \text{ T/m}\right)\left(0.04 \text{ m}\right)^2 \sqrt{5}}{48\left(4\pi \times 10^{-7} \text{ T}\cdot\text{m/A}\right)\left(150\right)} = \boxed{1.5 \text{ A}}$$

Chapter 29: Electromagnetic Induction and Faraday's Law

Chapter Overview and Objectives

This chapter introduces Faraday's law, Lenz's law, and the concept of magnetic flux. Several implications and applications of Faraday's law are discussed.

After completing this chapter you should:
- Know the definition of magnetic flux and be able to calculate the magnetic flux for a given situation.
- Know Faraday's law and Lenz's law.
- Know how to calculate the emf induced in a moving conductor.
- Know what generators, dynamos, and alternators are.
- Know what a transformer is and how to relate primary and secondary voltages and currents.

Summary of Equations

Magnetic flux for a uniform magnetic field:
$$\Phi_B = BA\cos\theta \qquad \text{(Section 29-2)}$$

General definition of magnetic flux:
$$\Phi_B = \int \vec{B} \cdot d\vec{A} \qquad \text{(Section 29-2)}$$

Faraday's law:
$$\varepsilon = -\frac{d\Phi_B}{dt} \qquad \text{(Section 29-2)}$$

Emf induced in a moving conductor:
$$\varepsilon = vLB \qquad \text{(Section 29-3)}$$

Emf induced in generator coil:
$$\varepsilon = NBA\omega\sin(\omega t + \phi) \qquad \text{(Section 29-4)}$$

Relationship between voltages in a transformer:
$$\frac{V_s}{V_p} = \frac{N_s}{N_p} \qquad \text{(Section 29-6)}$$

Relationship between currents in a transformer:
$$\frac{I_s}{I_p} = \frac{N_p}{N_s} \qquad \text{(Section 29-6)}$$

Faraday's law written in terms of electric field:
$$\oint \vec{E} \cdot d\vec{\ell} = -\frac{d\Phi_B}{dt} \qquad \text{(Section 29-7)}$$

Chapter Summary

Section 29-1. Induced EMF

A coil of wire will have an **induced emf** if the magnetic field within the coil is changing with time. The induced emf can generate an electric current. This current is called the **induced current**.

Section 29-2. Faraday's Law of Induction; Lenz's Law

In a uniform magnetic field of magnitude B, the **magnetic flux** Φ_B through an area A is defined as

$$\Phi_B = BA\cos\theta$$

where θ is the angle between the normal of the plane and the direction of the magnetic field. The magnetic flux can also be written in vector notation:

$$\Phi_B = \vec{\mathbf{B}} \cdot \vec{\mathbf{A}}$$

If the magnetic field is not uniform, the magnetic flux must be determined using the surface integral across the surface area:

$$\Phi_B = \int \vec{\mathbf{B}} \cdot d\vec{\mathbf{A}}$$

The SI unit of magnetic flux is the **weber** (Wb). One Wb is equal to one T·m^2.

Faraday's law of induction states that the induced emf in a loop is equal to the negative of the time derivative of the flux through that loop:

$$\varepsilon = -\frac{d\Phi_B}{dt}$$

The induced emf in the coil can generate a current that will create a magnetic field that opposes the change in flux that is taking place. This is called **Lenz's law**.

Example 29-2-A. Calculating the magnetic flux associated with a non-uniform magnetic field. A magnetic field is given by the expression

$$\vec{\mathbf{B}}(x,y,z) = B_0 \frac{(y+z)}{L}\hat{\mathbf{i}} + B_0 \frac{(x-z)}{L}\hat{\mathbf{j}}$$

Determine the flux through the triangle, located in the yz plane (see Figure).

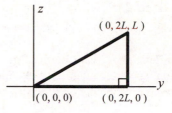

Approach: In order to calculate the magnetic flux, we need to consider the flux through a small area element at (y, z) with a width dy and height dz. The total flux through the triangle is the integral over the entire surface area.

Solution: The magnetic flux through a surface element of area $dy\,dz$, located at (y, z), is equal to

$$d\Phi_B = \vec{\mathbf{B}} \cdot d\vec{\mathbf{A}} = \left(B_0 \frac{(y+z)}{L}\hat{\mathbf{i}} + B_0 \frac{(x-z)}{L}\hat{\mathbf{j}} \right) \cdot \left(dz\,dy\,\hat{\mathbf{i}} \right) = B_0 \frac{(y+z)}{L} dz\,dy$$

We have assumed that the surface area vector of the element is directed along the positive x axis.

Now consider all surface elements between y and $y+dy$. They will make the following contribution to the magnetic flux:

$$d\Phi_B = \int_0^{y/2} \vec{\mathbf{B}} \cdot d\vec{\mathbf{A}} = \int_0^{y/2} \left(B_0 \frac{(y+z)}{L} dy \right) dz = \frac{B_0}{L}\left(yz + \frac{1}{2}z^2 \right) dy \bigg|_0^{y/2} = \frac{B_0}{L}\left(\frac{y^2}{2} + \frac{1}{8}y^2 \right) dy = \frac{5}{8}\frac{B_0}{L}y^2 dy$$

The total magnetic flux through the surface area can now be determined by integrating y between $y = 0$ and $y = 2L$:

$$\Phi_B = \int_0^{2L} d\Phi_B = \int_0^{2L} \frac{5}{8}\frac{B_0}{L}y^2 dy = \frac{5}{24}\frac{B_0}{L}y^3 \bigg|_0^{2L} = \frac{5}{3}\frac{B_0}{L}L^3 = \frac{5}{3}B_0 L^2$$

Example 29-2-B. Changing the magnetic flux by changing area. A square loop of wire with sides of length L is located in a uniform magnetic field, pointing into the page, as shown in the diagram on the right. The magnitude of the magnetic field is B. At time $t = 0$, the square is stretched in such a way that it remains a rectangle, but two opposite sides grow at the rate v and the other two sides shrink at the same rate v so that the perimeter of the rectangle remains constant. Determine the induced emf in the loop as a function of time.

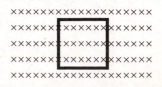

Approach: Since the magnetic flux is proportional to the surface area, the rate of change of the magnetic flux is proportional to the rate of change of the surface area of the loop. The induced emf is thus also proportional to the rate of change of the surface area.

Solution: At time t, the length of the loop will be $L + vt$ and its width will be $L - vt$. The area of the loop will be

$$A(t) = (L + vt)(L - vt) = L^2 - v^2 t^2$$

The magnetic flux through the loop is

$$\Phi_B(t) = BA(t) = B\left(L^2 - v^2 t^2\right)$$

Here we have used the fact that the surface area vector and the magnetic field vector are pointing in the same direction and the scalar product between them is equal to the product of their magnitudes.

The induced emf in the loop is proportional to the rate of change of the magnetic flux:

$$\varepsilon = -\frac{d\Phi_B}{dt} = -\frac{d}{dt}\left[B\left(L^2 - v^2 t^2\right)\right] = 2Bv^2 t$$

Section 29-3. EMF Induced in a Moving Conductor

If a conductor of length L moves with a velocity $\vec{v}$ perpendicular to its length and perpendicular to a magnetic field $\vec{B}$, the induced emf ε along its length will be equal to

$$\varepsilon = vLB$$

where v is the speed of the conductor and B is the magnitude of the magnetic field.

Example 29-3-A. Emf induced in a rolling metal ball. A metal ball of diameter 1.00 inches rolls from east to west in a region where the magnetic field points toward the north and has a magnitude of 1.3×10^{-2} T. The emf induced across the diameter of the ball is 12.6 µV. How fast is the ball moving?

Approach: The conductor moves with a speed v in a direction perpendicular to the direction of the magnetic field. The induced emf will be induced across the ball in a direction perpendicular to the direction of motion and perpendicular to the direction of the magnetic field. Based on the directions provided in the problem, the emf will be induced between the top and the bottom of the ball.

Solution: The dimension of the ball perpendicular to both the direction of its velocity and the direction of the magnetic field is equal to its diameter. The induced emf is equal to the product of the diameter, the speed, and the magnitude of the magnetic field:

$$\varepsilon = BLv$$

The problem provides the induced emf, the dimension L, and the magnetic field strength B. The only unknown, v, can thus be determined:

$$v = \frac{\varepsilon}{BL} = \frac{12.6 \times 10^{-6}}{\left(1.3 \times 10^{-2}\right)\left(1.00 \times 0.0254\right)} = 3.8 \times 10^{-2}\ \text{m/s}$$

Using the right-hand rule we can determine that the force on a negative charge carrier will be directed from the bottom of the ball to the top of the ball.

Section 29-4. Electric Generators

An application of Faraday's law is the conversion of mechanical energy into electrical energy. Generators, dynamos, and alternators are devices that utilize this principle. Mechanical work is used to rotate a coil within a magnetic field of magnitude B. If the coil consists of N loops of area A and is rotated at an angular speed ω, the induced emf ε will be equal to

$$\varepsilon = NBA\omega \sin(\omega t + \phi)$$

where ϕ is a phase angle that depends on the orientation of the coil at time $t = 0$.

Example 29-4-A. Designing an electric generator. Design an electric generator that provides an output voltage $V_{rms} =$ 120 V with a frequency $f = 60$ Hz. The coil to be used has $N = 60$ turns and an area $A = 0.063$ m^2. What must be the magnitude of the magnetic field inside the generator in order to create the desired output voltage?

Approach: The induced emf has a frequency that is equal to the frequency with which the coil is rotated. The magnitude of the induced voltage, which is directly related to the rms voltage, depends on the number of turns N, the angular frequency ω, and the yet to-be-determined magnetic field. The voltage requirement thus defines the required magnetic field.

Solution: The output voltage of the generator can be described by the following function:

$$V(t) = NBA\omega \sin(\omega t + \phi)$$

The amplitude of the output voltage of the generator is $NBA\omega$ which is √2 times the root-mean-square output voltage:

$$\sqrt{2}V_{rms} = NBA\omega$$

The angular frequency ω is fixed by the requirement that the frequency of the output voltage has to be 60 Hz. The corresponding angular frequency is

$$\omega = 2\pi f = 3.8 \times 10^2 \text{ s}^{-1}$$

The only unknown in the expression for the root-mean-square voltage is the magnetic field, which can now be determined:

$$B = \frac{\sqrt{2}V_{rms}}{NA\omega} = \frac{\sqrt{2}(120)}{(60)(0.063)(3.8 \times 10^2)} = 0.12 \text{ T}$$

Section 29-5. Back EMF and Counter Torque; Eddy Currents

When a motor operates, it turns a coil within a magnetic field. As a consequence, the magnetic flux intercepted by the coil will change and an emf is induced. The induced emf opposes the applied emf and reduces the current in the coil supplied by the external emf source. This opposing emf is called a **back emf** or **counter emf**.

When an electric generator supplies power, the current flowing through its coil will induce an emf that will oppose the motion. The result is a torque, called **counter torque**, which tries to slow down the rotation of the generator coil. In order to keep the generator turning at the desired rate, defined by the required frequency of the generated power, the applied torque must be increased.

When a conductor moves through a magnetic field or is within a changing magnetic field, currents are produced within the conductor. These currents are called **eddy currents**.

Section 29-6. Transformers and Transmission of Power

A **transformer** contains two coils that are located in close proximity and oriented in such a way that the magnetic field from the current in one coil causes a flux in the other coil. One of the coils is called the **primary coil** with N_p turns; the other coil is called the **secondary coil** with N_s turns. When an AC voltage of amplitude V_p is applied to the primary coil of the transformer, the magnetic flux through the secondary coil will induce an emf that results in an AC voltage of amplitude V_s across the secondary coil. If the transformer is constructed in such a way that the magnetic flux through the primary and secondary coils are identical and the internal resistance of the coils is negligible, then

$$\frac{V_s}{V_p} = \frac{N_s}{N_p}$$

The ratio of the currents in the coils is inversely proportional to the ratio of the number of turns:

$$\frac{I_s}{I_p} = \frac{N_p}{N_s}$$

These relationships apply to both peak values and rms values, as long as you are consistent within a given application of the relationships.

Example 29-6-A. Designing a transformer. A transformer is used to convert a voltage $V_{in,rms} = 120$ V to a voltage $V_{out,rms} = 5000$ V. The primary coil of the transformer has $N_p = 180$ turns of wire. The secondary circuit has a resistance of $R_s = 180$ kΩ. Determine the required number of turns in the secondary coil, the rms current in the primary and secondary circuits, and the power transformed by the transformer.

Approach: The problem provides us with information about the primary and secondary voltages and this information fixes the ratio of the number of turns. Based on the resistance of the secondary circuit, we can determine the current in the secondary circuit. The relation between the ratio of currents and the ratio of the number of turns can then be used to determine the current in the primary circuit. Based on the known currents and voltages in the primary and secondary circuits, we can determine the power dissipated in these circuits.

Solution: Using the ratio of voltages we can determine the ratio of the number of turns in the coils. Based on the known number of the turns of the primary coil we can calculate the number of turns of the secondary coil:

$$\frac{V_{out,rms}}{V_{in,rms}} = \frac{N_s}{N_p} \quad \Rightarrow \quad N_s = \frac{V_{out,rms}}{V_{in,rms}} N_p = \frac{5000}{120} 180 = 7500$$

The secondary current can be found using Ohm's law:

$$I_{out,rms} = \frac{V_{out,rms}}{R_s} = \frac{5000}{180 \times 10^3} = 2.8 \times 10^{-2} \text{ A}$$

Since the secondary voltage is the rms voltage, the secondary current determined here is the rms current. The primary current in the transformer can now be determined:

$$I_{in,rms} = \frac{N_s}{N_p} I_{out,rms} = \frac{7500}{180} \left(2.8 \times 10^{-2}\right) = 1.2 \text{ A}$$

The average power dissipated is the product of the rms voltage and the rms current in either the primary or the secondary circuit:

$$\bar{P} = V_{in,rms} I_{in,rms} = (120)(1.2) = 140 \text{ W} = V_{out,rms} I_{out,rms} = (5000)\left(2.8 \times 10^{-2}\right)$$

Section 29-7. A Changing Magnetic Flux Produces an Electric Field

A changing magnetic flux through any closed path in space will induce an emf around that path. By writing the emf as a path integral of the electrical field, we obtain the following more general form of Faraday's law:

$$\oint \vec{E} \cdot d\vec{\ell} = -\frac{d\Phi_B}{dt}$$

Example 29-7-A. Required magnetic flux to generate a time-independent electric field. The electric field is integrated over a circular path of radius $r = 16.2$ cm. At each point along the path, the electric field points outward from the tangent to the circle (as shown in the diagram) at an angle $\theta = 30°$. The magnitude of the electric field is $E = 20$ V/m at every point on the circle. What is the rate of change of magnetic flux through the circle?

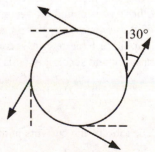

Approach: Faraday's law tells us that the rate of change of magnetic flux is related to the path integral of the electric field around the circular path.

Solution: To apply Faraday's law we need to calculate the path integral of the electric field around the circular path. Consider moving along the path in a counterclockwise direction. At every point along the path, the electric field will make an angle of 30° with respect to the direction of the path element.

The scalar product between the electric field and the displacement is thus equal to:

$$\vec{E} \cdot d\vec{\ell} = E\, d\ell \cos\theta = \frac{1}{2}\sqrt{3}\, E\, d\ell$$

Since the magnitude of the electric field and the angle between the electric field and the displacement vector are constant along the path, the scalar product $\vec{E} \cdot d\vec{\ell}$ will be constant along the path. The path integral of $\vec{E} \cdot d\vec{\ell}$ will thus be equal to

$$\oint \vec{E} \cdot d\vec{\ell} = E\cos\theta \oint dl = \frac{1}{2}\sqrt{3}\, E(2\pi r) = 18\,V$$

Since the path integral is positive, the electric field can induce a current flowing in the counterclockwise direction (the same direction used to evaluate the path integral). This current will produce a magnetic field that points out of the page. Using Faraday's law, we can determine the rate of change of the magnetic flux through the area enclosed by the circle:

$$\frac{d\Phi_B}{dt} = -\oint \vec{E} \cdot d\vec{\ell} = -18\,V = -18\,\frac{T\cdot m^2}{s}$$

Faraday's law indicates that the flux of the magnetic field that induced the electric field is decreasing. The direction of the induced current is thus consistent with Lenz's law since the magnetic field generated by the induced current counters the decrease in the magnetic flux.

Section 29-8. Applications of Induction: Sound Systems, Computer Memory, the Seismograph

There are many applications of magnetic induction, including microphones, magnetic recording, and seismographs.
In microphones, the motion of a membrane to which a coil is attached and which is placed in the vicinity of a magnet, induces a current that can be amplified or recorded.
Reading and writing data on magnetic media, such as tapes and disks, requires magnetic induction between the read/write head and the medium. When the read/write head writes data, it converts an electric signal to a magnetic field that is used to change the orientation of an element of the medium; when data are read, changes in the magnetic flux at the location of the read/write head are converted into an electric signal for processing by the computer.
Other applications that rely on magnetic induction are credit card swipe machines, seismographs, and ground fault circuit interrupters. All these applications rely on the induction of currents when external magnetic fields are changing.

Practice Quiz

1. A loop of area A lies flat on the ground. A uniform magnetic field of magnitude B is directed horizontally and points to the north. The magnetic field decreases uniformly to zero during a time interval t. What is the emf induced in the loop?
   a) $-BA/t$
   b) $+BA/t$
   c) $-Bat$
   d) zero

2. A loop of area A lies flat on the ground. A uniform magnetic field of magnitude B points vertically upward. The magnetic field decreases uniformly to zero during a time interval t. What is the magnitude of the emf induced in the loop?
   a) BA/t
   b) B/At
   c) A/Bt
   d) zero

3. What is the direction of the current induced in the loop described in Question 2, as viewed looking down at the loop from above?
   a) Clockwise.
   b) Counterclockwise.
   c) Depends on the resistance of the loop.
   d) No emf is induced, so there is no direction.

4. If the angular speed of a generator is doubled, what happens to the amplitude of the output voltage?
   a) Doubles
   b) Decreases by 50%
   c) Amplitude doesn't change, only the frequency changes
   d) Quadruples

5. Considering the emf, when is an electric motor operating most efficiently? This means, when is the motor converting the highest fraction of its input electrical energy into mechanical work?
   a) When the motor is at rest
   b) When the motor is turning as slow as it goes without stalling
   c) When the motor is turning at full speed
   d) The motor is always converting energy to mechanical energy at the same efficiency.

6. Consider a transformer in which the flux in the secondary coil is less than the flux through the primary coil. For this device we can write the relationship between secondary and primary voltage as

$$\frac{V_s}{V_p} = \alpha \frac{N_s}{N_p}$$

   where α is a number. What do you know about α?
   a) α is less than one.
   b) α is greater than one.
   c) α must be equal to one.
   d) α must be negative.

7. An arrow is shot from a bow with a speed v in a direction perpendicular to the direction of a magnetic field of magnitude B. The arrow's shaft has a length L and a diameter D. What is the magnitude of the emf induced in the shaft of the arrow as it moves through the magnetic field?
   a) BLv
   b) BDv
   c) $Bv(L + D)/2$
   d) zero

8. An AC voltage of 10.0 V is applied across the primary coil of a transformer that has 300 turns. What is the voltage across the secondary coil of the transformer if it has 150 turns?
   a) 20.0 V
   b) 5.0 V
   c) 1500 V
   d) zero

9. What is the current in the primary circuit of the transformer in Question 8, if there are no voltage sources in the secondary circuit of the transformer, the resistance of the secondary circuit is 20 Ω, and the resistance of the primary circuit is 0 Ω?
   a) 0.125 A
   b) 0.50 A
   c) 0.25 A
   d) 0.75 A

10. A coil of wire is located in a magnetic field pointing perpendicular to and into the page, as shown in the diagram. The coil is being pulled from the magnetic field as shown by the arrow. What is the direction of induced current in the coil?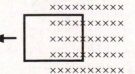
    a) Clockwise
    b) Counterclockwise
    c) It depends on how fast the coil is pulled from the magnetic field.
    d) No induced current flows because there is no induced emf.

11. A circular loop of wire of diameter 15.6 cm rests in a uniform magnetic field of magnitude 1.19 T that is directed at an angle of 32° with respect to the normal of the plane of the loop. The resistance of the loop is 3.8 Ω. The magnitude of the magnetic field goes to zero uniformly in 2.78 s. What is the current induced in the loop while the magnetic field is changing?

12. A magnetic field is given by

$$\vec{\mathbf{B}}(x,y,z) = x\,\hat{\mathbf{i}} + y\,\hat{\mathbf{j}} - 2(x+z)\,\hat{\mathbf{k}}$$

Calculate the magnetic flux through the square that lies in the xy plane with vertices at (0,0), (0,1), (1,1), and (1,0).

13. A generator is made from a coil with 120 turns of wire that are rectangular and have a width of 12.8 cm and a length of 21.6 cm. The coil turns in a uniform magnetic field of magnitude 1.93 T. At what frequency must the coil turn to produce an output voltage with an amplitude of 10.0 V?

14. An engineer wants to design a transformer that has a primary voltage of 120 V and a secondary voltage of 48 V. If the transformer has 300 turns in the primary circuit, how many turns are needed in the secondary circuit?

15. An engineer designs a transformer that requires a primary voltage of 120 V rms and the secondary circuit has a resistance of 10.0 Ω with a power requirement of 360 W. What ratio of secondary turns to primary turns must be used and what currents must the primary and secondary coils be able to safely carry?

Responses to Select End-of-Chapter Questions

1. Using coils with many (N) turns increases the values of the quantities to be experimentally measured, because the induced emf and therefore the induced current are proportional to N.

7. Yes, a current will be induced in the second coil. It will start when the battery is disconnected from the first coil and stop when the current falls to zero in the first coil. The current in the second loop will be clockwise.

13. (*a*) Yes. If a rapidly changing magnetic field exists outside, then currents will be induced in the metal sheet. These currents will create magnetic fields which will partially cancel the external fields.
(*b*) Yes. Since the metal sheet is permeable, it will partially shield the interior from the exterior static magnetic field; some of the magnetic field lines will travel through the metal sheet.
(*c*) The superconducting sheet will shield the interior from magnetic fields.

19. At the moment shown in Figure 29-15, the armature is rotating clockwise and so the current in length *b* of the wire loop on the armature is directed outward. (Use the right-hand rule: the field is north to south and the wire is moving with a component downward, therefore force on positive charge carriers is out.) This current is increasing, because as the wire moves down, the downward component of the velocity increases. As the current increases, the flux through the loop also increases, and therefore there is an induced emf to oppose this change. The induced emf opposes the current flowing in section *b* of the wire, and therefore creates a counter-torque.

25. As the bar moves in the magnetic field, induced eddy currents are created in the bar. The magnetic field exerts a force on these currents that opposes the motion of the bar. (See Figure 29-21.)

Solutions to Select End-of-Chapter Problems

1. The average induced emf is given by Eq. 29-2b.

$$\varepsilon = -N\frac{d\Phi_B}{dt} = -N\frac{\Delta\Phi_B}{\Delta t} = -2\frac{38\,\text{Wb} - (-58\,\text{Wb})}{0.42\,\text{s}} = \boxed{-460\,\text{V}}$$

7. (a) When the plane of the loop is perpendicular to the field lines, the flux is given by the maximum of Eq. 29-1a.

$$\Phi_B = BA = B\pi r^2 = (0.50\,\text{T})\pi(0.080\,\text{m})^2 = \boxed{1.0\times10^{-2}\,\text{Wb}}$$

(b) The angle is $\theta = \boxed{55^\circ}$

(c) Use Eq. 29-1a.

$$\Phi_B = BA\cos\theta = B\pi r^2 = (0.50\,\text{T})\pi(0.080\,\text{m})^2\cos55^\circ = \boxed{5.8\times10^{-3}\,\text{Wb}}$$

13. (a) Use Eq. 29-2a to calculate the emf induced in the ring, where the flux is the magnetic field multiplied by the area of the ring. Then using Eq. 25-7, calculate the average power dissipated in the ring as it is moved away. The thermal energy is the average power times the time.

$$e = -\frac{\Delta\Phi_B}{\Delta t} = -\frac{\Delta BA}{\Delta t} = -\frac{\Delta B\left(\frac{1}{4}\pi d^2\right)}{\Delta t}$$

$$Q = P\Delta t = \left(\frac{e^2}{R}\right)\Delta t = \left(\frac{\Delta B\left(\frac{1}{4}\pi d^2\right)}{\Delta t}\right)^2\left(\frac{\Delta t}{R}\right) = \frac{(\Delta B)^2\,\pi^2 d^4}{16R\Delta t}$$

$$= \frac{(0.80\,\text{T})^2\,\pi^2(0.015\,\text{m})^4}{16(55\times10^{-6}\,\Omega)(45\times10^{-3}\,\text{s})} = 8.075\times10^{-3}\,\text{J} \approx \boxed{8.1\,\text{mJ}}$$

(b) The temperature change is calculated from the thermal energy using Eq. 19-2.

$$\Delta T = \frac{Q}{mc} = \frac{8.075\times10^{-3}\,\text{J}}{(15\times10^{-3}\,\text{kg})(129\,\text{J/kg}\cdot\text{°C})} = \boxed{4.2\times10^{-3}\,\text{C}^\circ}$$

19. The energy dissipated in the process is the power dissipated by the resistor, times the elapsed time that the current flows. The average induced emf is given by the "difference" version of Eq. 29-2a.

$$\varepsilon = -\frac{\Delta\Phi_B}{\Delta t} \quad ; \quad P = \frac{\varepsilon^2}{R} \quad ;$$

$$E = P\Delta t = \frac{\varepsilon^2}{R}\Delta t = \left(\frac{\Delta\Phi_B}{\Delta t}\right)^2\frac{\Delta t}{R} = \frac{A^2(\Delta B)^2}{R\Delta t} = \frac{\left[\pi(0.125\text{m})^2\right]^2(0.40\,\text{T})^2}{(150\,\Omega)(0.12\,\text{s})} = \boxed{2.1\times10^{-5}\,\text{J}}$$

25. (a) The magnetic field a distance r from the wire is perpendicular to the wire and given by Eq. 28-1. Integrating this magnetic field over the area of the loop gives the flux through the loop.

$$\Phi_B = \int BdA = \int_b^{b+a}\frac{\mu_0 I}{2\pi r}adr = \boxed{\frac{\mu_0 Ia}{2\pi}\ln\left(1+\frac{a}{b}\right)}$$

(b) Since the loop is being pulled away, $v = \dfrac{db}{dt}$. Differentiate the magnetic flux with respect to time to calculate the emf in the loop.

$$\varepsilon = -\frac{d\Phi_B}{dt} = -\frac{d}{dt}\left[\frac{\mu_0 Ia}{2\pi}\ln\left(1+\frac{a}{b}\right)\right] = -\frac{\mu_0 Ia}{2\pi}\frac{d}{db}\left[\ln\left(1+\frac{a}{b}\right)\right]\frac{db}{dt} = \boxed{\frac{\mu_0 Ia^2 v}{2\pi b(b+a)}}$$

Note that this is the emf at the instant the loop is a distance b from the wire. The value of b is changing with time.

(c) Since the magnetic field at the loop points into the page, and the flux is decreasing, the induced current will create a downward magnetic field inside the loop. The current in the loop then flows $\boxed{\text{clockwise}}$.

(*d*) The power dissipated in the loop as it is pulled away is related to the emf and resistance by Eq. 25-7b. This power is provided by the force pulling the loop away. We calculate this force from the power using Eq. 8-21. As in part (*b*), the value of *b* is changing with time.

$$F = \frac{P}{v} = \frac{\varepsilon^2}{Rv} = \boxed{\frac{\mu_0^2 I^2 a^4 v}{4\pi^2 R b^2 (b+a)^2}}$$

31. The rod will descend at its terminal velocity when the magnitudes of the magnetic force (found in Example 29-8) and the gravitational force are equal. We set these two forces equal and solve for the terminal velocity.

$$\frac{B^2 \ell^2 v_t}{R} = mg \;\rightarrow$$

$$v_t = \frac{mgR}{B^2 \ell^2} = \frac{(3.6 \times 10^{-3} \text{ kg})(9.80 \text{ m/s}^2)(0.0013 \text{ }\Omega)}{(0.060 \text{ T})^2 (0.18 \text{ m})^2} = \boxed{0.39 \text{ m/s}}$$

37. We find the number of turns from Eq. 29-4. The factor multiplying the sine term is the peak output voltage.

$$\varepsilon_{\text{peak}} = NB\omega A \;\rightarrow\; N = \frac{\varepsilon_{\text{peak}}}{B\omega A} = \frac{24.0 \text{ V}}{(0.420 \text{ T})(2\pi \text{ rad/rev})(60 \text{ rev/s})(0.0515 \text{ m})^2} = \boxed{57.2 \text{ loops}}$$

43. The back emf is proportional to the rotation speed (Eq. 29-4). Thus if the motor is running at half speed, the back emf is half the original value, or 54 V. Find the new current from writing a loop equation for the motor circuit, from Figure 29-20.

$$\varepsilon - \varepsilon_{\text{back}} - IR = 0 \;\rightarrow\; I = \frac{\varepsilon - \varepsilon_{\text{back}}}{R} = \frac{120 \text{ V} - 54 \text{ V}}{5.0 \text{ }\Omega} = \boxed{13 \text{ A}}$$

49. (*a*) We assume 100% efficiency, and find the input voltage from $P = IV$.

$$P = I_P V_P \;\rightarrow\; V_P = \frac{P}{I_P} = \frac{75 \text{ W}}{22 \text{ A}} = 3.409 \text{ V}$$

Since $V_P < V_S$, this is a $\boxed{\text{step-up}}$ transformer.

(*b*) $\dfrac{V_S}{V_P} = \dfrac{12 \text{ V}}{3.409 \text{ V}} = \boxed{3.5}$

55. (*a*) The increasing downward magnetic field creates a circular electric field along the electron path. This field applies an electric force to the electron causing it to accelerate.
(*b*) For the electrons to move in a circle, the magnetic force must provide a centripetal acceleration. With the magnetic field pointing downward, the right-hand rule requires the electrons travel in a $\boxed{\text{clockwise}}$ direction for the force to point inward.
(*c*) For the electrons to accelerate, the electric field must point in the counterclockwise direction. A current in this field would create an upward magnetic flux. So by Lenz's law, the downward magnetic field must be $\boxed{\text{increasing}}$.
(*d*) For the electrons to move in a circle and accelerate, the field must be pointing downward and increasing in magnitude. For a sinusoidal wave, the field is downward half of the time and upward the other half. For the half that it is downward its magnitude is decreasing half of the time and increasing the other half. Therefore, the magnetic field is pointing downward and increasing for only one fourth of every cycle.

61. The charge on the capacitor can be written in terms of the voltage across the battery and the capacitance using Eq. 24-1. When fully charged the voltage across the capacitor will equal the emf of the loop, which we calculate using Eq. 29-2b.

$$Q = CV = C \frac{d\Phi_B}{dt} = CA \frac{dB}{dt} = (5.0 \times 10^{-12} \text{ F})(12 \text{ m}^2)(10 \text{ T/s}) = \boxed{0.60 \text{ nC}}$$

67. The induced current in the coil is the induced emf divided by the resistance. The induced emf is found from the changing flux by Eq. 29-2a. The magnetic field of the solenoid, which causes the flux, is given by Eq. 28-4. For the area used in Eq. 29-2a, the cross-sectional area of the solenoid (not the coil) must be used, because all of the magnetic flux is inside the solenoid.

$$I = \frac{e_{\text{ind}}}{R} \qquad |e_{\text{ind}}| = N_{\text{coil}} \frac{d\Phi}{dt} = N_{\text{coil}} A_{\text{sol}} \frac{dB_{\text{sol}}}{dt} \qquad B_{\text{sol}} = \mu_0 \frac{N_{\text{sol}} I_{\text{sol}}}{l_{\text{sol}}}$$

$$I = \frac{N_{\text{coil}} A_{\text{sol}} \mu_0 \frac{N_{\text{sol}}}{l_{\text{sol}}} \frac{dI_{\text{sol}}}{dt}}{R} = \frac{N_{\text{coil}} A_{\text{sol}} \mu_0}{R} \frac{N_{\text{sol}}}{l_{\text{sol}}} \frac{dI_{\text{sol}}}{dt}$$

$$= \frac{(150 \text{ turns}) \pi (0.050 \text{ m})^2 (4\pi \times 10^{-7} \text{ T·m/A})}{12 \, \Omega} \frac{(230 \text{ turns})}{(0.01 \text{ m})} \frac{2.0 \text{ A}}{0.10 \text{ s}} = \boxed{5.7 \times 10^{-2} \text{ A}}$$

As the current in the solenoid increases, a magnetic field from right to left is created in the solenoid and the loop. The induced current will flow in such a direction as to oppose that field, and so must flow from $\boxed{\text{left to right}}$ through the resistor.

73. The total emf across the rod is the integral of the differential emf across each small segment of the rod. For each differential segment, dr, the differential emf is given by the differential version of Eq. 29-3. The velocity is the angular speed multiplied by the radius. The figure is a top view of the spinning rod.

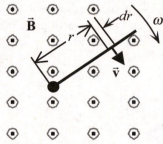

$$d\varepsilon = Bvd\ell = B\omega r dr \quad \rightarrow \quad \varepsilon = \int d\varepsilon = \int_0^\ell B\omega r dr = \boxed{\tfrac{1}{2} B\omega \ell^2}$$

79. (a) As the loop falls out of the magnetic field, the flux through the loop decreases with time creating an induced emf in the loop. The current in the loop is equal to the emf divided by the resistance, which can be written in terms of the resistivity using Eq. 25-3.

$$I = \frac{\varepsilon}{R} = \left(\frac{\pi d^2 / 4}{\rho 4\ell} \right) \frac{d\Phi_B}{dt} = \left(\frac{\pi d^2}{16\rho\ell} \right) B \frac{dA}{dt} = \frac{\pi d^2}{16\rho\ell} B\ell v$$

This current induces a force on the three sides of the loop in the magnetic field. The forces on the two vertical sides are equal and opposite and therefore cancel.

$$F = I\ell B = \frac{\pi d^2}{16\rho\ell} B\ell v\ell B = \boxed{\frac{\pi d^2 B^2 \ell v}{16\rho}}$$

By Lenz's law this force is upward to slow the decrease in flux.

(b) Terminal speed will occur when the gravitational force is equal to the magnetic force.

$$F_g = \rho_m \left(4\pi\ell \frac{d^2}{4} \right) g = \frac{\pi d^2 B^2 \ell v_T}{16\rho} \rightarrow v_T = \boxed{\frac{16\rho\rho_m g}{B^2}}$$

(c) We calculate the terminal velocity using the given magnetic field, the density of copper from Table 13-1, and the resistivity of copper from Table 25-1

$$v_T = \frac{16(8.9 \times 10^3 \text{ kg/m}^3)(1.68 \times 10^{-8} \, \Omega\text{m})(9.80 \text{ m/s}^2)}{(0.80 \text{ T})^2} = \boxed{3.7 \text{ cm/s}}$$

Chapter 30: Inductance, Electromagnetic Oscillations, and AC Circuits

Chapter Overview and Objectives

This chapter introduces the concept of mutual and self-inductance. Faraday's law is rewritten in terms of inductance. The energy density of a magnetic field is discussed. Solutions to the time dependence of charge and current in LR, LC, and LRC circuits are developed. The behavior of ac circuits with resistors, inductors, and capacitors is discussed. The resonances of LRC series circuits, impedance matching, and three-phase power are presented.

After completing this chapter you should:
- Know what mutual-inductance and self-inductance are.
- Know how mutual and self-inductance are related to Faraday's law.
- Know how to calculate induced emfs in electronic circuits.
- Know how to calculate the energy density of a magnetic field.
- Know what the characteristics of LR, LC, and LRC circuits are.
- Know how to calculate the time constant of an LR circuit.
- Know how to calculate the natural frequencies of LC and LRC circuits.
- Know how to relate ac current to ac voltage across resistors, capacitors, and inductors.
- Know how to calculate reactance of inductors and capacitors.
- Know that current lags voltage in an inductor and that current leads voltage in a capacitor.
- Know how to calculate the impedance of an LRC series circuit.
- Know how to determine the resonant frequency of an LRC series circuit.
- Know what impedance matching is.
- Know what three-phase power is.

Summary of Equations

Definition of mutual inductance:	$$M_{21} = \frac{N_2 \Phi_{21}}{I_1}$$	(Section 30-1)
Emf induced in a secondary coil:	$$\varepsilon_2 = -M_{21} \frac{dI_1}{dt}$$	(Section 30-1)
Definition of self-inductance:	$$L = \frac{N \Phi_B}{I}$$	(Section 30-2)
Emf induced by self-inductance:	$$\varepsilon = -L \frac{dI}{dt}$$	(Section 30-2)
Power into an inductor:	$$P = I\varepsilon = LI \frac{dI}{dt}$$	(Section 30-3)
Energy stored in an inductor:	$$U = \frac{1}{2} LI^2$$	(Section 30-3)
Energy density in a magnetic field:	$$u = \frac{1}{2} \frac{B^2}{\mu_0}$$	(Section 30-3)

Time constant of an LR circuit:	$\tau = \dfrac{L}{R}$	(Section 30-4)
Time dependence of current in an LR circuit:	$I(t) = \dfrac{V_0}{R}\left(1 - e^{-Rt/L}\right) = \dfrac{V_0}{R}\left(1 - e^{-t/\tau}\right)$	(Section 30-4)
Natural frequency of oscillation of an LC circuit:	$\omega = \sqrt{\dfrac{1}{LC}}$	(Section 30-5)
Time dependence of charge in an LC circuit:	$Q(t) = Q_0 \cos(\omega t + \phi)$	(Section 30-5)
Natural frequency of an LRC series circuit:	$\omega = \sqrt{\dfrac{1}{LC} - \dfrac{R^2}{4L^2}}$	(Section 30-6)
Time dependence of charge in series LRC circuit:	$Q(t) = Q_0 e^{-\frac{R}{2L}t} \cos(\omega t + \phi)$	(Section 30-6)
Definition of inductive reactance:	$X_L = \omega L$	(Section 30-7)
Definition of capacitive reactance:	$X_C = \dfrac{1}{\omega C}$	(Section 30-7)
Impedance of an LRC series circuit:	$Z = \sqrt{R^2 + (X_L - X_C)^2}$	(Section 30-8)
Average power dissipated in an LRC series circuit:	$\overline{P} = I_{rms}^2 Z \cos\phi = I_{rms} V_{rms} \cos\phi$	(Section 30-8)
Resonant frequency of an LRC series circuit:	$f_0 = \dfrac{\omega_0}{2\pi} = \dfrac{1}{2\pi\sqrt{LC}}$	(Section 30-9)
Impedance matching:	$R_{load} = R_{source}$	(Section 30-10)

Chapter Summary

Section 30-1. Mutual Inductance

When two coils of wire are relatively close to each other, a changing current in coil 1 creates a changing magnetic field through coil 2. The changing flux through coil 2 induces an emf in coil 2. The induced emf ε_2 in coil 2 is related to the rate of change of the magnetic flux Φ_{21} through each turn of coil 2 and the number of turns of coil 2, N_2:

$$\varepsilon_2 = -N_2 \frac{d\Phi_{21}}{dt}$$

For a given current in coil 1, the flux through coil 2 depends on the geometry of the two coils and their position and orientation relative to each other. The flux through coil 2 will be proportional to the current in coil 1. All geometrical factors can be lumped together into a constant called the **mutual inductance** M_{21} of the coils, and the flux through each turn of coil 2 can be written as

$$\Phi_{21} = \frac{M_{21} I_1}{N_2}$$

The induced emf in coil 2 can be written in terms of the mutual inductance of the two coils and the rate of change of the current in coil 1:

$$\varepsilon_2 = -M_{21}\frac{dI_1}{dt}$$

A changing current in coil 2 will also cause an induced emf in coil 1:

$$\varepsilon_1 = -M_{12}\frac{dI_2}{dt}$$

The mutual inductances M_{12} and M_{21} are the same and usually the subscripts on M are dropped. The dimension of mutual inductance is voltage × time / current. The SI unit of inductance is the Henry (H). One Henry is one volt second per ampere (1 H = 1 V·s/A).

Example 30-1-A. Calculating the mutual inductance. Determine the mutual inductance of two small single-turn wire loops of radius r, separated by a distance x along a line joining the centers of the loops, perpendicular to the plane of both loops. Assume the separation distance is much greater than the radii of the loops.

Approach: To calculate the mutual inductance between the two coils, we calculate the magnetic field at the location of coil 2 due to a current I_1 in coil 1. The magnetic field is used to determine the magnetic flux through coil 2. This information can then be used to determine the mutual inductance between the two coils.

Solution: Since the distance between the two coils is large compared to their radii, we can use the approximation derived in Chapter 28, equation (28-7b), to find the magnetic field of a small circular loop far from the loop along a line that passes through the center of the loop. The magnetic field is given by

$$\vec{B} = \frac{\mu_0}{2\pi}\frac{N_1 I_1 A_1}{x^3}\hat{n} = \frac{\mu_0}{2\pi}\frac{I_1 \pi r^2}{x^3}\hat{n} = \frac{1}{2}\mu_0\frac{I_1 r^2}{x^3}\hat{n}$$

where $\hat{n}$ is the unit vector that points along the line joining the centers of the coils. Because coil 2 is small compared to the distance between the loops, it is reasonable to assume that the magnetic field is uniform across the area of coil 2. The magnetic flux through coil 2 is thus equal to

$$\Phi_{21} = \vec{B}\cdot\vec{A} = \left(\frac{1}{2}\mu_0\frac{I_1 r^2}{x^3}\right)(\pi r^2) = \frac{1}{2}\mu_0\pi\frac{I_1 r^4}{x^3}$$

The magnetic flux through coil 2 is thus proportional to the current in coil 1, and the constant of proportionality is related to the mutual inductance between the coils:

$$M_{21} = \frac{\Phi_{21}N_2}{I_1} = \frac{\frac{1}{2}\mu_0\pi\frac{I_1 r^4}{x^3}}{I_1} = \frac{1}{2}\mu_0\pi\frac{r^4}{x^3}$$

Section 30-2. Self-Inductance

A changing current through a coil causes a change of the magnetic flux intercepted by that coil. According to Faraday's law, the changing magnetic flux induces an emf in the coil. This phenomenon is called self inductance. The **self-inductance** L of a coil of wire is defined as:

$$L = \frac{N\Phi_B}{I}$$

where N is the number of turns of the coil, Φ_B is the magnetic flux through the coil, and I is the current in the coil. The induced emf can be written in terms of the self-inductance L and the rate of change of the current through the coil dI/dt:

$$\varepsilon = -L\frac{dI}{dt}$$

The minus sign indicates that the induced emf is in a direction that creates a current that opposes the change in the current that induces the emf. The dimension of self-inductance is the same as the dimension of mutual inductance. The symbol used for an **inductor** in an electronic circuit diagram is shown in the Figure on the right.

The inductance of a circuit element can be minimized by using **noninductive windings** in which wire is wound in opposite senses so that the net magnetic flux is negligible.

Example 30-2-A. Calculating the induced emf in an inductor. The current through an inductor with inductance L as a function of time is $I_0 t^2$. What is the induced emf as function of time?

Approach: The problem provides us with information about the time dependence of the current and the self-inductance of the inductor. This information is sufficient to determine the time dependence of the induced emf.

Solution: The induced emf is given by

$$\varepsilon = -L\frac{dI}{dt} = -L\frac{d\left(I_0 t^2\right)}{dt} = -2LI_0 t$$

Section 30-3. Energy Stored in a Magnetic Field

Consider an inductor with negligible resistance. When a constant current I is flowing through the inductor, the induced emf will be zero and no power is dissipated. If the current is changing, an emf will be induced and energy must be supplied in order to maintain a flow of current. The power P that must be supplied to the inductor is equal to

$$P = I\varepsilon = LI\frac{dI}{dt}$$

The power P is positive when the magnitude of the current is increasing and negative when the magnitude of the current is decreasing. The power added to the inductor increases its potential energy U:

$$U = \frac{1}{2}LI^2 = \frac{1}{2}\left(\frac{N\Phi_B}{I}\right)I^2 = \frac{1}{2}(NAB)I = \frac{1}{2}(NI)AB = \frac{1}{2}\left(\frac{B\ell}{\mu_0}\right)AB = \frac{1}{2}\frac{B^2}{\mu_0}A\ell$$

In the previous equation, the potential energy has been rewritten in terms of the magnetic field inside the inductor and its geometrical properties. The potential energy of the inductor can also be expressed in terms of the **energy density** u, which is defined as the ratio of the potential energy and the volume of the inductor:

$$u = \frac{U}{A\ell} = \frac{1}{2}\frac{B^2}{\mu_0}$$

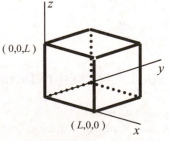

Example 30-3-A. Calculating the magnetic potential energy. Calculate the total magnetic field energy within the cube shown in the diagram if the magnetic field is given by

$$B(x,y,z) = B_0\frac{x}{L}i - \frac{1}{2}B_0\frac{y}{L}j - \frac{1}{2}B_0\frac{z}{L}k$$

Approach: This problem illustrates how to calculate the potential energy associated with non-uniform magnetic fields. In order to determine the energy density, we need to start with calculating the energy density for a small volume element and then integrate this expression over all volume elements contained within the cube.

Solution: Consider a volume element located at (x, y, z) with dimensions (dx, dy, dz). If we make the dimensions of this element small enough, we can assume that the magnetic field is constant over its volume. In this limit, the energy contained within this volume element is equal to

$$dU = udV = \frac{1}{2}\frac{B_x^2 + B_y^2 + B_z^2}{\mu_0}(dx\,dy\,dz) = \frac{B_0^2}{2\mu_0 L^2}\left(x^2 + \frac{1}{4}y^2 + \frac{1}{4}z^2\right)dx\,dy\,dz$$

To determine the total energy contained within the cube, we integrate this expression across the cube. This integration can be carried out in steps. For example, we can first integrate with respect to z:

$$dU = \frac{B_0^2}{2\mu_0 L^2} \int_0^L \left\{ \left(x^2 + \frac{1}{4}y^2 + \frac{1}{4}z^2 \right) dx\, dy \right\} dz = \frac{B_0^2}{2\mu_0 L^2} \left\{ \left(x^2 z + \frac{1}{4}y^2 z + \frac{1}{12}z^3 \right) dx\, dy \right\} \Big|_0^L =$$

$$= \frac{B_0^2}{2\mu_0 L} \left(x^2 + \frac{1}{4}y^2 + \frac{1}{12}L^2 \right) dx\, dy$$

Next we integrate with respect to y:

$$dU = \frac{B_0^2}{2\mu_0 L} \int_0^L \left\{ \left(x^2 + \frac{1}{4}y^2 + \frac{1}{12}L^2 \right) dx \right\} dy = \frac{B_0^2}{2\mu_0 L} \left\{ \left(x^2 y + \frac{1}{12}y^3 + \frac{1}{12}L^2 y \right) dx \right\} \Big|_0^L =$$

$$= \frac{B_0^2}{2\mu_0} \left(x^2 + \frac{1}{12}L^2 + \frac{1}{12}L^2 \right) dx = \frac{B_0^2}{2\mu_0} \left(x^2 + \frac{1}{6}L^2 \right) dx$$

The total energy contained in the magnetic field can now be determined by integrating with respect to x:

$$U = \frac{B_0^2}{2\mu_0} \int_0^L \left(x^2 + \frac{1}{6}L^2 \right) dx = \frac{B_0^2}{2\mu_0} \left(\frac{1}{3}x^3 + \frac{1}{6}L^2 x \right) \Big|_0^L = \frac{B_0^2 L^3}{4\mu_0}$$

Example 30-3-B. Doing work with the energy stored in an inductor. Consider the possibility to store electrical energy in the magnetic field of a large inductor made of superconducting ($R = 0$) material. How long could you operate a 100-W light bulb with the energy stored in a 10 H inductor that carries a current $I = 40$ A?

Approach: The total energy stored in the magnetic field can be calculated on the basis of the current I and the inductance L. Assuming this energy is completely dissipated in the light bulb, which uses energy at a rate of 100 J/s, we can easily determine how long we can operate the bulb.

Solution: The energy stored in the inductor is equal to

$$U = \frac{1}{2}LI^2 = \frac{1}{2}(10)(40)^2 = 8{,}000 \text{ J}$$

Since the rate of energy consumption is known, we can determine how long we can operate the light bulb:

$$t = \frac{U}{P} = \frac{8{,}000}{100} = 80 \text{ s}$$

Section 30-4. *LR* Circuits

An *LR* series circuit is shown in the diagram on the right. Consider that at time $t = 0$, the switch is closed. The current I flowing in the circuit will be time dependent. At any given time, Kirchhoff's voltage rule around the circuit must be satisfied and results in the following requirement:

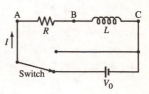

$$V_0 - L\frac{dI}{dt} - IR = 0$$

The general solution of this equation is:

$$I(t) = \frac{V_0}{R} + I_1 e^{-Lt/R} = \frac{V_0}{R} + I_1 e^{-t/\tau}$$

where I_1 is a constant that depends on the current at time $t = 0$ and the **time constant** $\tau = L/R$. If the current at time $t = 0$ is 0, the constant I_1 is equal to

$$I_1 = I(0) - \frac{V_0}{R} = -\frac{V_0}{R}$$

and the current in the circuit is equal to

$$I(t) = \frac{V_0}{R}\left(1 - e^{-t/\tau}\right)$$

Example 30-4-A. Time-dependent current in an *LR* series circuit. An *LR* series circuit utilizes a voltage source that provides $V_0 = 5.0$ V, an $R = 45$ Ω resistor, an $L = 34$ mH inductor, and a switch. The switch is initially open and the current is 0. How long after the switch is closed will the current in the circuit reach $I_{90} = 90$ mA?

Approach: To solve this problem we use the general solution of the current in an *LR* series circuit with the appropriate initial conditions. The problem asks us to determine the time at which the current in the circuit reaches the specified value. This requires us to solve one equation with one unknown.

Solution: The current in the *LR* series circuit has the following time dependence if the current at time $t = 0$ is 0:

$$I(t) = \frac{V_0}{R}\left(1 - e^{-Rt/L}\right)$$

The current as function of time is shown in the Figure on the right. To determine the time at which the current reaches a value I_{90} we need to solve the following equation:

$$I_{90} = \frac{V_0}{R}\left(1 - e^{-Rt_{90}/L}\right)$$

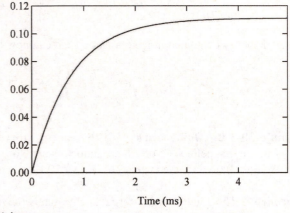

The time at which the current reaches a value of 90 mA (0.09 A) is

$$t_{90} = -\frac{L}{R}\ln\left(1 - \frac{I_{90}R}{V_0}\right) = -\frac{3.4 \times 10^{-2}}{45}\ln\left(1 - \frac{\left(90 \times 10^{-3}\right)\left(45\right)}{5.0}\right) = 1.3 \times 10^{-3}\,\text{s} = 1.3\,\text{ms}$$

Section 30-5. *LC* Circuits and Electromagnetic Oscillations

An *LC* series circuit has a capacitor C and an inductor L connected in series, as shown in the diagram on the right. The capacitor is initially charged. At time $t = 0$ the switch is closed, the capacitor starts to discharge, and a current I starts to flow in the clockwise direction. Applying Kirchhoff's voltage rule to this circuit we get

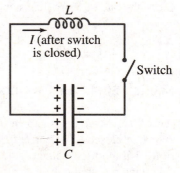

$$-L\frac{dI}{dt} + \frac{Q}{C} = 0$$

where Q is the charge on the capacitor. A current flow in the clockwise direction will reduce the charge on the capacitor. The relation between the current I and the charge Q on the capacitor can be written as

$$I = -\frac{dQ}{dt}$$

Using this relation, Kirchhoff's voltage rule can be rewritten as

$$L\frac{d^2Q}{dt^2} + \frac{Q}{C} = 0 \quad \Rightarrow \quad \frac{d^2Q}{dt^2} + \frac{Q}{LC} = 0$$

The general solution of this equation depends on time t and is given by

$$Q(t) = Q_0 \cos\left(\omega t + \phi\right)$$

The constants Q_0 and ϕ are defined by the initial capacitor charge and the initial current. The angular frequency ω is fixed by the circuit components and is equal to

$$\omega = \sqrt{\frac{1}{LC}}$$

The current in the circuit can now be calculated:

$$I(t) = -\frac{dQ}{dt} = -\frac{d}{dt}\left[Q_0 \cos(\omega t + \phi)\right] = \omega Q_0 \sin(\omega t + \phi) = I_0 \sin(\omega t + \phi)$$

Example 30-5-A. Charge on a capacitor in an *LC* circuit. Determine the charge on the capacitor of an *LC* series circuit as a function of time. The inductor has an inductance $L = 365$ mH and the capacitor has a capacitance $C = 534$ nF. At time $t_1 = 1.76$ ms, the charge on the capacitor is $Q_1 = 0.365$ μC and the current in the circuit is $I_1 = 2.45$ mA.

Approach: In this section we have derived the charge and current in an *LC* series circuit as function of time. In this problem, we determine the constants Q_0 and ϕ by requiring that the conditions at time $t = 1.76$ ms are satisfied.

Solution: The charge and current in an *LC* circuit are given by the following expressions:

$$Q(t) = Q_{max} \cos(\omega t + \phi)$$

$$I(t) = \omega Q_{max} \sin(\omega t + \phi)$$

The angular frequency ω is equal to

$$\omega = \sqrt{\frac{1}{LC}} = \sqrt{\frac{1}{(365 \times 10^{-3})(534 \times 10^{-9})}} = 2.27 \times 10^3 \, \text{s}^{-1}$$

The boundary conditions require:

$$Q_1 = Q_{max} \cos(\omega t_1 + \phi)$$

$$I_1 = \omega Q_{max} \sin(\omega t_1 + \phi) \quad \Rightarrow \quad \frac{I_1}{\omega} = Q_{max} \sin(\omega t_1 + \phi)$$

By squaring each equation and adding them we obtain one equation with one unknown:

$$Q_1^2 + \left(\frac{I_1}{\omega}\right)^2 = Q_{max}^2 \cos^2(\omega t_1 + \phi) + Q_{max}^2 \sin^2(\omega t_1 + \phi) = Q_{max}^2 \quad \Rightarrow \quad Q_{max} = \sqrt{Q_1^2 + \left(\frac{I_1}{\omega}\right)^2} = \pm 1.14 \, \mu\text{C}$$

Although there are two possible values of Q_{max}, we can limit ourselves to the positive solution. The negative solution is equivalent to adding π rad to the phase angle.
The boundary condition for the charge on the capacitor can be rewritten as

$$\left(\frac{Q_1}{Q_{max}}\right) = \cos(\omega t_1 + \phi) \quad \Rightarrow \quad \phi = \text{acos}\left(\frac{Q_1}{Q_{max}}\right) - \omega t_1$$

The acos function in general has two solutions since there are two angles that have the same cosine. As a result, there are two solutions for ϕ:

$$\phi = \text{acos}\left(\frac{0.365 \times 10^{-6}}{1.14 \times 10^{-6}}\right) - (2.27 \times 10^3)(1.76 \times 10^{-3}) = \pm 1.24 - 4.00 = \begin{cases} -5.24 \text{ rad} \\ -2.76 \text{ rad} \end{cases}$$

The two different phase angles result in different currents. This is illustrated in the Figure on the next page, which shows the current in the circuit as function of time for the two possible phase angles.

As can be seen in the Figure, the two possible solutions for ϕ correspond to two different currents at time t_1:

$\phi = -5.24$ rad: $I(t_1) = \omega Q_{max} \sin(-1.24) < 0$

$\phi = -2.76$ rad: $I(t_1) = \omega Q_{max} \sin(+1.24) > 0$

Since the current at time t_1 is positive, the phase angle must be equal to -2.76 rad.

The charge on the capacitor as function of time is thus equal to

$$Q(t) = (1.14 \times 10^{-6}) \cos(2.27 \times 10^{-3} t - 2.76)$$

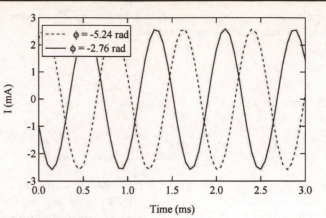

Section 30-6. *LC* Oscillations with Resistance (*LRC* Circuit)

An *LRC* series circuit is shown in the circuit diagram. Consider the current I flowing around this circuit in the clockwise direction. According Kirchhoff's voltage rule the total voltage drop around the circuit must be zero:

$$-L\frac{dI}{dt} - IR + \frac{Q}{C} = 0$$

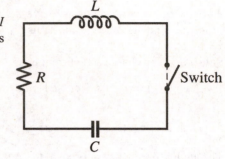

The current I is related to the charge Q on the capacitor:

$$I = -\frac{dQ}{dt}$$

Using this relation, we can rewrite Kirchhoff's voltage rule in terms of just the charge on the capacitor:

$$L\frac{d^2Q}{dt^2} + \frac{dQ}{dt}R + \frac{Q}{C} = 0$$

This equation has the same form as the equation that describes a damped harmonic oscillator, as discussed in Chapter 14. Similar to the case of the damped harmonic oscillator, the *LRC* circuit has three distinct solutions, depending on the values of R, C, and L:

$R^2 < \frac{4L}{C}$ underdamped

$R^2 = \frac{4L}{C}$ critically damped

$R^2 > \frac{4L}{C}$ overdamped

The solution for the underdamped case is

$$Q(t) = Q_0 e^{-\frac{R}{2L}t} \cos(\omega t + \phi)$$

where

$$\omega = \sqrt{\frac{1}{LC} - \frac{R^2}{4L^2}}$$

The values of the constants Q_0 and ϕ depend on the initial charge on the capacitor and the initial current.

Section 30-7. AC Circuits with AC Sources

An **ac circuit** is an electrical circuit that has ac current or voltage sources in the circuit. The **ac source** is characterized by its frequency f and its peak voltage (for a voltage source) or its peak current (for a current source).

In describing ac circuits, we will assume that the current has the following time dependence

$$I = I_0 \cos\left(2\pi ft\right) = I_0 \cos\omega t$$

where $\omega = 2\pi f$.

Before discussing more complicated ac circuits, we start with discussing basic circuits in which either a resistor, a capacitor, or an inductor is connected to the ac source.

- **AC resistor circuit**.

  In a resistor circuit, there is a simple relation between the current in the circuit and the voltage from the source:

  $$V(t) = I(t)R = I_0 R\cos\omega t = V_0 \cos\omega t$$

  where $V_0 = I_0 R$ is the amplitude of the ac voltage. We note that the phase angles of the voltage and the current are identical: **the current and the voltage are in phase**. The average electrical power dissipated in the resistor is

  $$\overline{P} = \tfrac{1}{2}I_0 V_0 = I_{rms}^2 R = V_{rms}^2 / R$$

- **AC inductor circuit**.

  In an inductor circuit, Kirchhoff's voltage rule can be used to relate the voltage across the power source to the rate of change of the current through the inductor:

  $$V - L\frac{dI}{dt} = 0 \quad \Rightarrow \quad V = L\frac{dI}{dt} = -\omega I_0 L\sin\omega t = \omega I_0 L\cos\left(\omega t + \frac{1}{2}\pi\right) = V_0 \cos\left(\omega t + \frac{1}{2}\pi\right)$$

  where $V_0 = \omega L I_0$ is the amplitude of the ac voltage. We note that the phase of the voltage is $\pi/2$ radians (90°) greater than the current: **the current lags the voltage across the inductor by 90°**.

  The relation between the amplitude of the voltage and the amplitude of the current looks similar to Ohm's law except that ωL replaces the resistance R. The quantity ωL is called the **inductive reactance X_L**:

  $$X_L = \omega L$$

  The inductive reactance increases with increasing frequency.

- **AC capacitor circuit**.

  In a capacitor circuit, Kirchhoff's voltage rule can be used to relate the voltage across the power source to the charge on the capacitor:

  $$V - \frac{Q}{C} = 0 \qquad V = \frac{Q}{C}$$

  Since the current I is the time derivative of the charge on the capacitor we can rewrite this relation in the following way:

  $$I = \frac{dQ}{dt} = C\frac{dV}{dt} \quad \Rightarrow \quad dV = \frac{1}{C}I\,dt \quad \Rightarrow \quad V = \frac{1}{C}\int I\,dt = \frac{1}{C}\int I_0 \cos\omega t\,dt = \frac{I_0}{\omega C}\sin\omega t = \frac{I_0}{\omega C}\cos\left(\omega t - \frac{1}{2}\pi\right)$$

  The amplitude of the voltage of the ac source is thus equal to

  $$V_0 = \frac{1}{\omega C}I_0$$

  We note that the phase of the voltage is $\pi/2$ radians (90°) less than the phase of the current: **the current leads the voltage across the capacitor by 90°**.

  The relation between the amplitude of the voltage and the amplitude of the current looks similar to Ohm's law except that $(\omega C)^{-1}$ replaces the resistance R. The quantity $(\omega C)^{-1}$ is called the **capacitive reactance X_C**:

  $$X_C = \frac{1}{\omega C}$$

  The capacitive reactance decreases with increasing frequency.

Example 30-7-A. Properties of an inductor. An inductor is made from a solenoid with a length $\ell = 10.0$ cm and a diameter $d = 2.88$ cm. When the solenoid is connected to a voltage source with an amplitude $V_0 = 30.0$ V and a frequency $f = 400$ Hz, a current of amplitude $I_0 = 10.6$ mA flows in the circuit. Determine the number of turns in the inductor.

Approach: The amplitude of the current in the circuit depends on the amplitude and frequency of the applied voltage and the inductance of the inductor. The information provided can be used to determine the inductance of the inductor. Since the length and diameter of the inductor are provided, we can use the inductance to determine the number of turns.

Solution: The amplitude of the voltage across the inductor is related to the current amplitude in the circuit in the following way:

$$V_0 = I_0 \omega L$$

Based on the information provided in the problem, we can determine the inductance L

$$L = \frac{V_0}{\omega I_0}$$

The inductance L of the solenoid can also be determined on the basis of its geometrical properties:

$$L = \frac{\mu_0 N^2 A}{\ell}$$

where A is the cross-sectional area of the solenoid, ℓ is its length, and N is its number of turns. Using these two expressions for the inductance of the solenoid we conclude that

$$\frac{V_0}{\omega I_0} = \frac{\mu_0 N^2 A}{\ell} \quad \Rightarrow \quad N = \sqrt{\frac{V_0 \ell}{\mu_0 \omega I_0 A}} = \sqrt{\frac{(30.0)(0.01)}{\left(4\pi \times 10^{-7}\right)\left(2\pi \times 400\right)\left(10.6 \times 10^{-3}\right)\left(\frac{1}{4}\pi [0.0288]^2\right)}} = 3700$$

Example 30-7-B. Measuring the capacitance of a circuit by determining its capacitance reactance. The rms current in a capacitive circuit with a voltage source of rms voltage $V_{rms} = 12.0$ V and frequency $f = 1.20$ kHz is $I_{rms} = 2.78$ mA. What is the capacitance of the capacitor in the circuit?

Approach: To measure the capacitance of a circuit element, we have to measure its capacitance reactance using an ac voltage source. The reactance will depend on both the capacitance of the circuit and the frequency of the ac source. If the frequency of the ac voltage source is known, a measurement of the rms values of both the current and the voltage is sufficient to determine the capacitance of the circuit.

Solution: The known rms current and voltage can be used to determine the capacitance reactance:

$$X_C = \frac{V_{rms}}{I_{rms}} = \frac{1}{\omega C}$$

The capacitance C can now be determined:

$$C = \frac{1}{\omega X_C} = \frac{I_{rms}}{(2\pi f)V_{rms}} = \frac{\left(2.78 \times 10^{-3}\right)}{2\pi \left(1.20 \times 10^3\right)(12.0)} = 3.07 \times 10^{-8} \text{ F} = 30.7 \text{ nF}$$

Section 30-8. *LRC* Series AC Circuit

A series *LRC* circuit has an inductor, a resistor, and a capacitor in series, as shown in the diagram. Assume that the current in the circuit has the form $I(t) = I_0 \cos\omega t$. Using Kirchhoff's voltage rule, we know that the sum of the voltage drops across each circuit element must add up to the voltage provided by the ac source: $V = V_L + V_R + V_C$

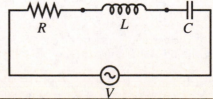

Using the expression of the current $I(t)$ we can write the voltage across each circuit element as a function of time:

$$V = L\frac{dI}{dt} + RI + \frac{1}{C}\int I dt = \omega L I_0 \cos(\omega t + \pi/2) + R I_0 \cos(\omega t) + \frac{1}{\omega C}\cos(\omega t - \pi/2)$$

The sum of several cosine functions that have the same frequency but different amplitudes and phases can always be rewritten, using trigonometric identities, as a single cosine function with the same frequency and some amplitude and phase. After manipulating the expression for V for the LRC circuit we obtain

$$V = \sqrt{R^2 + \left(\omega L - \frac{1}{\omega C}\right)^2}\ I_0 \sin[\omega t + \phi] = \sqrt{R^2 + (X_L - X_C)^2}\ I_0 \sin[\omega t + \phi]$$

where

$$\phi = \arctan\left(\frac{X_L - X_C}{R}\right)$$

The coefficient of I_0 is called the **impedance** of the circuit and is denoted by Z:

$$Z = \sqrt{R^2 + (X_L - X_C)^2}$$

The average power dissipated in the circuit is

$$\overline{P} = I_{rms}^2 Z \cos\phi = I_{rms} V_{rms} \cos\phi$$

The $\cos\phi$ factor in the average power expression is called the **power factor**.

Example 30-8-A. Studying the response of an _LRC_ series circuit. An LRC series circuit has an inductor with an inductance of $L = 34$ mH, a resistor with a resistance of $R = 2.8$ kΩ, and a capacitor with a capacitance of $C = 20$ nF. The voltage source has a voltage with an rms voltage of $V_{rms} = 4.8$ V and a frequency of $f = 36$ kHz. What is the resonant frequency of the circuit? What is the impedance of the circuit at the frequency of the voltage source? What is the rms current in the circuit? What is the power delivered to the circuit?

Approach: The best approach to this problem is to start with calculating the impedance of the circuit. Using the impedance, we can determine the power factor of the circuit and its rms current.

Solution: The impedance at the frequency of the voltage source is

$$Z = \sqrt{R^2 + \left(\omega L - \frac{1}{\omega C}\right)^2} = \sqrt{(2.8\times10^3)^2 + \left((2\pi\times36\times10^3)(34\times10^{-3}) - \frac{1}{(2\pi\times36\times10^3)(20\times10^{-9})}\right)^2} =$$
$$= 8.0\times10^3\ \Omega$$

The rms current in the circuit will be the rms voltage of the ac source divided by the impedance:

$$I_{rms} = \frac{V_{rms}}{Z} = \frac{4.8\text{ V}}{8.0\times10^3\ \Omega} = 6.0\times10^{-4}\text{ A}$$

To calculate the power dissipated in the circuit, we need to calculate the phase angle ϕ:

$$\phi = \arctan\left(\frac{X_L - X_C}{R}\right) = \arctan\left(\frac{(2\pi\times36\times10^3)(34\times10^{-3}) - \frac{1}{(2\pi\times36\times10^3)(20\times10^{-9})}}{2.8\times10^3}\right) = 69°$$

The average power dissipated in the circuit is equal to

$$\overline{P} = I_{rms} V_{rms} \cos\phi = (6.0\times10^{-4})(4.8)\cos(69°) = 1.0\times10^{-3}\text{ W}$$

Section 30-9. Resonance in AC Circuits

For a voltage source of fixed amplitude in an *LRC* series circuit, the amplitude of the current I_0 depends on the amplitude of the voltage source V_0 and the frequency ω:

$$I_0 = \frac{V_0}{Z} = \frac{V_0}{\sqrt{R^2 + \left(\omega L - \dfrac{1}{\omega C}\right)^2}}$$

The amplitude of the current will reach a maximum value V_0/R when the angular frequency ω satisfies the following relation:

$$\omega L - \frac{1}{\omega C} = 0$$

The angular frequency at which this happens, ω_0, is equal to

$$\omega_0 = \frac{1}{\sqrt{LC}}$$

When the ac voltage source has this angular frequency, the circuit is in resonance. The **resonant frequency** of the circuit is

$$f_0 = \frac{\omega_0}{2\pi} = \frac{1}{2\pi\sqrt{LC}}$$

Example 30-9-A. Operating an *LRC* series circuit at resonance. Determine the resonant frequency of a *LRC* series circuit that has an inductance $L = 66.8$ mH, a resistance $R = 2.82\ \Omega$, and a capacitance $C = 92.6$ nF. Determine the rms current in this *LRC* series circuit when the source operates at the resonant frequency with an rms voltage of 12.0 V. Determine the rms voltage across the resistor, the inductor, and the capacitor with this source.

Approach: The first step in solving this problem is finding the resonance frequency. When the system operates at resonance, the rms current is equal to the ratio of the rms voltage and the resistance R. Using the rms current, the rms voltage across each of the circuit elements can be determined.

Solution: The resonant frequency of the circuit is

$$f_0 = \frac{\omega_0}{2\pi} = \frac{1}{2\pi\sqrt{LC}} = \frac{1}{2\pi\sqrt{(66.8\times10^{-3})(92.6\times10^{-9})}} = 2.02\times10^3\ \text{Hz}$$

The impedance at the resonant frequency of the circuit is just the resistance of the circuit: $Z = 2.82\ \Omega$. The rms current can be calculated from the rms voltage and the impedance of the circuit:

$$I_{rms} = \frac{V_{rms}}{Z} = \frac{12.0}{2.82} = 4.26\ \text{A}$$

The rms current is the same for each component of the circuit. We calculate the rms voltage across each circuit element using its impedance:

Resistor: $V_{rms} = I_{rms}R = (4.26)(2.82) = 12.0\ \text{V}$

Inductor: $V_{rms} = I_{rms}X_L = I_{rms}\omega_0 L = (4.26)(2\pi\times2.02\times10^3)(66.8\times10^{-3}) = 3.62\times10^3$

Capacitor: $V_{rms} = I_{rms}X_C = \dfrac{I_{rms}}{\omega_0 C} = \dfrac{(4.26)}{(2\pi\times2.02\times10^3)(92.6\times10^{-9})} = 3.62\times10^3\ \text{V}$

Note that the voltages across the inductor and capacitor are huge, but since they are 180° out of phase, they cancel each other.

Section 30-10. Impedance Matching

A source of electrical power can usually be modeled as an ideal voltage source in series with an impedance. The source is often connected to a load that has its own impedance, as shown in the diagram on the right. The total impedance is the sum of the source impedance and the load impedance. In some instances, we want to transfer the maximum possible power from the source and it is reasonable to ask what the load impedance should be so that the power transferred to the load is maximized. We can easily determine the power delivered to the load in the case where the impedance is due only to a resistance:

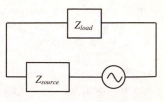

$$\bar{P} = I_{rms}^2 R_{load} = \left(\frac{V_{rms\,source}}{R_{total}}\right)^2 R_{load} = \left(\frac{V_{rms\,source}}{R_{source} + R_{load}}\right)^2 R_{load}$$

To determine for what external load the power transferred reaches a maximum value, we take the derivative of the power with respect to R_{load} and require that it is equal to zero:

$$\frac{d\bar{P}}{dR_{load}} = -2\frac{V_{rms\,source}^2}{\left(R_{source} + R_{load}\right)^3} R_{load} + \left(\frac{V_{rms\,source}}{R_{source} + R_{load}}\right)^2 = \left(\frac{V_{rms\,source}}{R_{source} + R_{load}}\right)^2 \left\{\frac{-2R_{load}}{R_{source} + R_{load}} + 1\right\} =$$

$$= \left(\frac{V_{rms\,source}}{R_{source} + R_{load}}\right)^2 \left\{\frac{R_{source} - R_{load}}{R_{source} + R_{load}}\right\} = 0$$

The power thus has a maximum value when

$$R_{load} = R_{source}$$

Adjusting the load impedance to equal the source impedance so that maximum power is transferred is called **impedance matching**.

Section 30-11. Three-Phase AC

A three-phase voltage source supplies three voltages of the same amplitude and frequency relative to a common potential (usually ground potential). The three voltages differ in phase by 120° ($2\pi/3$ radians) from each other:

$$V_1 = V_0 \sin \omega t$$

$$V_2 = V_0 \sin(\omega t + 2\pi/3)$$

$$V_3 = V_0 \sin(\omega t + 4\pi/3)$$

The biggest advantage to three-phase voltage sources over single-phase voltage sources is that the power is delivered at a constant rate if the load is balanced. If the load on each phase is a resistor R, connected to the common potential, then the total power delivered as function of time is

$$P = P_1 + P_2 + P_3 = \frac{V_1^2}{R} + \frac{V_2^2}{R} + \frac{V_3^2}{R} = \frac{[V_0 \sin \omega t]^2}{R} + \frac{[V_0 \sin(\omega t + 2\pi/3)]^2}{R} + \frac{[V_0 \sin(\omega t + 4\pi/3)]^2}{R}$$

We can use the trigonometric identity $\sin a + b = \sin a \cos b + \sin b \cos a$ to rewrite the second two terms in the expression:

$$P = \frac{V_0^2}{R}\left[\sin^2 \omega t + (\sin \omega t \cos 2\pi/3 + \sin 2\pi/3 \cos \omega t)^2 + (\sin \omega t \cos 4\pi/3 + \sin 4\pi/3 \cos \omega t)^2\right] =$$

$$= \frac{V_0^2}{R}\left[\sin^2 \omega t + \left(-\tfrac{1}{2}\sin \omega t + \tfrac{\sqrt{3}}{2}\cos \omega t\right)^2 + \left(-\tfrac{1}{2}\sin \omega t - \tfrac{\sqrt{3}}{2}\cos \omega t\right)^2\right] =$$

$$= \frac{V_0^2}{R}\left[\sin^2 \omega t + \left(\tfrac{1}{4}\sin^2 \omega t - \tfrac{\sqrt{3}}{2}\sin \omega t \cos \omega t + \tfrac{3}{4}\cos \omega t\right) + \left(\tfrac{1}{4}\sin^2 \omega t + \tfrac{\sqrt{3}}{2}\sin \omega t \cos \omega t + \tfrac{3}{4}\cos \omega t\right)\right] =$$

$$= \frac{V_0^2}{R}\left(\tfrac{3}{2}\sin^2 \omega t + \tfrac{3}{2}\cos^2 \omega t\right) = \frac{3V_0^2}{2R}$$

Note that there is no time dependence in the power delivered by the power source.

Example 30-11-A. Calculating the power delivered by a three-phase voltage source. Determine the power delivered by a three-phase voltage source where the amplitude of each source is V_0 relative to a common potential and resistors of resistance R connected between each pair of phases makes the load on the source as shown in the diagram.

Approach: Since the voltages across the resistors differ, the currents flowing through them will also differ. We will assume that the voltages V_1, V_2, and V_3 are given by the expression listed at the start of this section. As a result, we know the voltage across each resistor and thus the current flowing through it. Using the known voltage across and the current through each resistor, we can determine the power dissipated in the circuit.

Solution: The voltage drop across each resistor is the difference between two of the phase voltages:

$$V_{R_A} = V_2 - V_1 = V_0 \sin(\omega t + 2\pi/3) - V_0 \sin \omega t$$

$$V_{R_B} = V_3 - V_2 = V_0 \sin(\omega t + 4\pi/3) - V_0 \sin(\omega t + 2\pi/3)$$

$$V_{R_C} = V_1 - V_3 = V_0 \sin \omega t - V_0 \sin(\omega t + 4\pi/3)$$

The total power in the circuit is

$$P = \frac{V_{R_A}^2}{R_A} + \frac{V_{R_B}^2}{R_B} + \frac{V_{R_C}^2}{R_C}$$

$$= \frac{\left[V_0 \sin(\omega t + 2\pi/3) - V_0 \sin \omega t\right]^2}{R} + \frac{\left[V_0 \sin(\omega t + 4\pi/3) - V_0 \sin(\omega t + 2\pi/3)\right]^2}{R}$$

$$+ \frac{\left[V_0 \sin \omega t - V_0 \sin(\omega t + 4\pi/3)\right]^2}{R}$$

Using the trigonometric identities that were used previously and expanding the binomial squares, we can write this as

$$P = \frac{V_0^2}{R}\left\{ \left[\tfrac{9}{4}\sin^2 \omega t - \tfrac{3\sqrt{3}}{2}\sin \omega t \cos \omega t + \tfrac{3}{4}\cos^2 \omega t\right] \right.$$

$$\left. + \left[3\cos^2 \omega t\right] + \left[\tfrac{9}{4}\sin^2 \omega t + \tfrac{3\sqrt{3}}{2}\sin \omega t \cos \omega t + \tfrac{3}{4}\cos^2 \omega t\right]\right\}$$

$$= \frac{V_0^2}{R}\left\{\tfrac{9}{2}\sin^2 \omega t + \tfrac{9}{2}\cos^2 \omega t\right\} = \frac{9V_0^2}{2R}$$

Again, we see that this circuit, called a delta connected three-phase circuit also has constant power.

Practice Quiz

1. Two circuits are located side by side, as shown in the Figure on the right. What is the direction of current flow induced in the right-hand circuit when the switch in the left-hand circuit is closed?
   a) Clockwise.
   b) Counterclockwise.
   c) No current flow is induced in the coil on the right.
   d) It depends on how quickly the switch is closed.

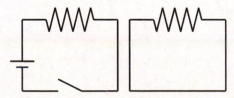

2. In the diagram for Question 1, if the switch has been closed for some time so that the current in the left-hand circuit is constant, what will be the direction of the induced current in the right-hand circuit?
   a) Clockwise.
   b) Counterclockwise.
   c) No current flow is induced in the coil on the right.
   d) It depends on how long the current has been constant.

3. To help eliminate unwanted emfs induced in a circuit due to changing currents elsewhere, you should
   a) make the area enclosed by the conductors of the circuit as large as possible.
   b) make the area enclosed by the conductors of the circuit as small as possible.
   c) place the circuit as close as possible to the other currents.
   d) only use ac current in your circuit.

4. The natural frequency of a damped harmonic oscillator is often approximated by the natural frequency of the undamped harmonic oscillator. When is this a good approximation?
   a) It is always a good approximation.
   b) The smaller the resistance in the circuit, the better the approximation is.
   c) The larger the resistance in the circuit, the better the approximation is.
   d) It is a good approximation whenever you are solving a homework or test problem.

5. An LR circuit without a battery has a current I at time $t = 0$. At time $t = t_1$, the circuit has a current $I/2$. At what time does the circuit have a current $I/4$?
   a) $1.5\, t_1$
   b) $2.0\, t_1$
   c) $4.0\, t_1$
   d) The circuit will never have a current $I/4$.

6. An electric circuit component is of an unknown type of resistor, inductor, or capacitor. When connected to an ac source, the amplitude of the current increases with increasing frequency for a fixed voltage amplitude. What type of component is this?
   a) Resistor.
   b) Inductor.
   c) Capacitor.
   d) Can't tell from the information given.

7. An electric circuit component is of an unknown type of resistor, inductor, or capacitor. When connected to an ac source of voltage with a time dependence $V(t) = V_0 \cos(\omega t + 3\pi/7)$, the current is $I(t) = I_0 \sin(\omega t + 3\pi/7)$. What type of component is this?
   a) Resistor.
   b) Inductor.
   c) Capacitor.
   d) Can't tell from the information given.

8. An amplifier is designed to power some loudspeakers that have an impedance of 8 Ω. Four of these speakers are connected in parallel to the amplifier. What should the output impedance be of the amplifier so that maximum power is transferred to the loudspeakers?
   a) 2 Ω
   b) 8 Ω
   c) 32 Ω
   d) 64 Ω

9. What is one advantage of three-phase power?
   a) The power delivered to a balanced load is constant in time.
   b) The total voltage delivered to a balanced load is constant in time.
   c) The total current delivered to a balanced load is constant in time.
   d) If one phase of the supply fails, equipment will function properly on the two remaining phases.

10. If you lower the load resistance below the voltage source resistance, the current from the voltage source increases, but according to our maximum power condition, the power to the load is smaller. The voltage source is supplying greater power because voltage times current has increased. Where is the additional power going?
    a) The power disappears.
    b) The power from the source actually decreases because the power factor drops.
    c) The power goes into heating the wires that connect the source to the load.
    d) The power goes into the source resistance.

11. In an LR series circuit with a resistance of 123 Ω, it takes the current 32.6 ms after closing the switch to reach 50.0% of its maximum value. If the resistance is changed to 394 Ω, how long would it take the current to reach 98% of its maximum value?

12. You want to design an AM radio receiver to pick up a station operating at a frequency of 1040 kHz. You have an inductor that is a solenoid with a diameter of 5.78 cm, a length of 14.6 cm, and 240 turns. What value of capacitance do you need so that the natural frequency of an LC circuit made from these components is 1040 kHz?

13. An LRC series circuit has a frequency 340 kHz and loses 3.78% of its energy on each cycle. The resistor in the circuit has a resistance of 28.5 Ω. What are the values of the inductance and the capacitance of the circuit?

14. Determine the resonant frequency of an LRC series circuit that has an inductance of 10.8 mH, a resistance of 5.82 Ω, and a capacitance of 45.6 nF. Determine the rms current in this LRC series circuit when the source operates at the resonant frequency with an rms voltage of 12.0 V.

15. Determine the power delivered by a three-phase power source in which each phase voltage has an amplitude of 280 V and the load consists of a 32.6 Ω resistance connected between each pair of voltages.

Responses to Select End-of-Chapter Questions

1. (*a*) For the maximum value of the mutual inductance, place the coils close together, face to face, on the same axis.
   (*b*) For the least possible mutual inductance, place the coils with their faces perpendicular to each other.

7. A circuit with a large inductive time constant is resistant to changes in the current. When a switch is opened, the inductor continues to force the current to flow. A large charge can build up on the switch, and may be able to ionize a path for itself across a small air gap, creating a spark.

13. Yes. When ω approaches zero, X_L approaches zero, and X_C becomes infinitely large. This is consistent with what happens in an ac circuit connected to a dc power supply. For the dc case, ω is zero and X_L will be zero because there is no changing current to cause an induced emf. X_C will be infinitely large, because steady direct current cannot flow across a capacitor once it is charged.

19. In an LRC circuit, the current and the voltage in the circuit both oscillate. The energy stored in the circuit also oscillates and is alternately stored in the magnetic field of the inductor and the electric field of the capacitor.

Solutions to Select End-of-Chapter Problems

1. (*a*) The mutual inductance is found in Example 30-1.

$$M = \frac{\mu N_1 N_2 A}{l} = \frac{1850 \left(4\pi \times 10^{-7}\ \text{T·m/A}\right)(225)(115)\pi \left(0.0200\ \text{m}\right)^2}{2.44\ \text{m}} = \boxed{3.10 \times 10^{-2}\ \text{H}}$$

   (*b*) The emf induced in the second coil can be found from Eq. 30-3b.

$$\varepsilon_2 = -M\frac{dI_1}{dt} = -M\frac{\Delta I_1}{\Delta t} = \left(-3.10 \times 10^{-2}\ \text{H}\right)\frac{(-12.0\ \text{A})}{0.0980\ \text{ms}} = \boxed{3.79\ \text{V}}$$

7. Because the current is increasing, the emf is negative. We find the self-inductance from Eq. 30-5.

$$\varepsilon = -L\frac{dI}{dt} = -L\frac{\Delta I}{\Delta t} \quad \rightarrow \quad L = -e\frac{\Delta t}{\Delta I} = -(-2.50\,\text{V})\frac{0.0120\,\text{s}}{\left[0.0250\,\text{A} - (-0.0280\,\text{A})\right]} = \boxed{0.566\,\text{H}}$$

13. We use Eq. 30-4 to calculate the self-inductance, where the flux is the integral of the magnetic field over a cross-section of the toroid. The magnetic field inside the toroid was calculated in Example 28-10.

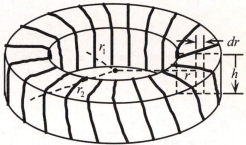

$$L = \frac{N}{I}\Phi_B = \frac{N}{I}\int_{r_1}^{r_2}\frac{\mu_0 NI}{2\pi r}h\,dr = \boxed{\frac{\mu_0 N^2 h}{2\pi}\ln\left(\frac{r_2}{r_1}\right)}$$

19. We use Eq. 30-7 to calculate the energy density in the toroid, with the magnetic field calculated in Example 28-10. We integrate the energy density over the volume of the toroid to obtain the total energy stored in the toroid. Since the energy density is a function of radius only, we treat the toroid as cylindrical shells each with differential volume $dV = 2\pi r h\,dr$.

$$u_B = \frac{B^2}{2\mu_0} = \frac{1}{2\mu_0}\left(\frac{\mu_0 NI}{2\pi r}\right)^2 = \boxed{\frac{\mu_0 N^2 I^2}{8\pi^2 r^2}}$$

$$U = \int u_B dV = \int_{r_1}^{r_2}\frac{\mu_0 N^2 I^2}{8\pi^2 r^2}2\pi r h\,dr = \frac{\mu_0 N^2 I^2 h}{4\pi}\int_{r_1}^{r_2}\frac{dr}{r} = \boxed{\frac{\mu_0 N^2 I^2 h}{4\pi}\ln\left(\frac{r_2}{r_1}\right)}$$

25. (a) We use Eq. 30-6 to determine the energy stored in the inductor, with the current given by Eq (30-9).

$$U = \tfrac{1}{2}LI^2 = \boxed{\frac{LV_0^2}{2R^2}\left(1 - e^{-t/\tau}\right)^2}$$

(b) Set the energy from part (a) equal to 99.9% of its maximum value and solve for the time.

$$U = 0.999\frac{V_0^2}{2R^2} = \frac{V_0^2}{2R^2}\left(1 - e^{-t/\tau}\right)^2 \rightarrow t = \tau\ln\left(1 - \sqrt{0.999}\right) \approx \boxed{7.6\tau}$$

31. (a) The AM station received by the radio is the resonant frequency, given by Eq. 30-14. We divide the resonant frequencies to create an equation relating the frequencies and capacitances. We then solve this equation for the new capacitance.

$$\frac{f_1}{f_2} = \frac{\dfrac{1}{2\pi}\sqrt{\dfrac{1}{LC_1}}}{\dfrac{1}{2\pi}\sqrt{\dfrac{1}{LC_2}}} = \sqrt{\frac{C_2}{C_1}} \rightarrow C_2 = C_1\left(\frac{f_1}{f_2}\right)^2 = (1350\,\text{pF})\left(\frac{550\,\text{kHz}}{1600\,\text{kHz}}\right)^2 = \boxed{0.16\,\text{nF}}$$

(b) The inductance is obtained from Eq. 30-14.

$$f = \frac{1}{2\pi}\sqrt{\frac{1}{LC_1}} \rightarrow L = \frac{1}{4\pi^2 f^2 C} = \frac{1}{4\pi^2\left(550\times10^3\,\text{Hz}\right)^2\left(1350\times10^{-12}\,\text{F}\right)} = \boxed{62\,\mu\text{H}}$$

37. As in the derivation of 30-16, we set the total energy equal to the sum of the magnetic and electric energies, with the charge given by Eq. 30-19. We then solve for the time that the energy is 75% of the initial energy.

$$U = U_E + U_B = \frac{Q^2}{2C} + \frac{LI^2}{2} = \frac{Q_0^2}{2C} e^{-\frac{R}{L}t} \cos^2\left(\omega' t + \phi\right) + \frac{Q_0^2}{2C} e^{-\frac{R}{L}t} \sin^2\left(\omega' t + \phi\right) = \frac{Q_0^2}{2C} e^{-\frac{R}{L}t}$$

$$0.75 \frac{Q_0^2}{2C} = \frac{Q_0^2}{2C} e^{-\frac{R}{L}t} \rightarrow t = -\frac{L}{R}\ln\left(0.75\right) = -\frac{L}{R}\ln\left(0.75\right) \approx \boxed{0.29\frac{L}{R}}$$

43. (a) At $\omega = 0$, the impedance of the capacitor is infinite. Therefore the parallel combination of the resistor R and capacitor C behaves as the resistor only, and so is R. Thus the impedance of the entire circuit is equal to the resistance of the two series resistors.

$$Z = \boxed{R + R'}$$

(b) At $\omega = \infty$, the impedance of the capacitor is zero. Therefore the parallel combination of the resistor R and capacitor C is equal to zero. Thus the impedance of the entire circuit is equal to the resistance of the series resistor only.

$$Z = \boxed{R'}$$

49. The impedance of the circuit is given by Eq. 30-28a without an inductive reactance. The reactance of the capacitor is given by Eq. 30-25b.

(a) $Z = \sqrt{R^2 + X_C^2} = \sqrt{R^2 + \frac{1}{4\pi^2 f^2 C^2}} = \sqrt{\left(75\,\Omega\right)^2 + \frac{1}{4\pi^2 \left(60\,\text{Hz}\right)^2 \left(6.8\times10^{-6}\,\text{F}\right)^2}} = 397\,\Omega$

$$\approx \boxed{400\,\Omega}\left(2\text{ sig. fig.}\right)$$

(b) $Z = \sqrt{R^2 + X_C^2} = \sqrt{R^2 + \frac{1}{4\pi^2 f^2 C^2}} = \sqrt{\left(75\,\Omega\right)^2 + \frac{1}{4\pi^2 \left(60000\,\text{Hz}\right)^2 \left(6.8\times10^{-6}\,\text{F}\right)^2}} = \boxed{75\,\Omega}$

55. (a) The rms current is the rms voltage divided by the impedance. The impedance is given by Eq. 30-28a with no capacitive reactance.

$$Z = \sqrt{R^2 + X_L^2} = \sqrt{R^2 + \left(2\pi fL\right)^2}\ .$$

$$I_{rms} = \frac{V_{rms}}{Z} = \frac{V_{rms}}{\sqrt{R^2 + 4\pi^2 f^2 L^2}} = \frac{120\,\text{V}}{\sqrt{\left(965\,\Omega\right)^2 + 4\pi^2 \left(60.0\,\text{Hz}\right)^2 \left(0.225\,\text{H}\right)^2}}$$

$$= \frac{120\,\text{V}}{968.7\,\Omega} = \boxed{0.124\,\text{A}}$$

(b) The phase angle is given by Eq. 30-29a with no capacitive reactance.

$$\phi = \tan^{-1}\frac{X_L}{R} = \tan^{-1}\frac{2\pi fL}{R} = \tan^{-1}\frac{2\pi\left(60.0\,\text{Hz}\right)\left(0.225\,\text{H}\right)}{965\,\Omega} = \boxed{5.02°}$$

The current is lagging the source voltage.

(c) The power dissipated is given by $P = I_{rms}^2 R = \left(0.124\,\text{A}\right)^2 \left(965\,\Omega\right) = \boxed{14.8\,\text{W}}$

(d) The rms voltage reading is the rms current times the resistance or reactance of the element.

$$V_{rms}_{R} = I_{rms}R = \left(0.124\,\text{A}\right)\left(965\,\Omega\right) = 119.7\,\text{V} \approx \boxed{120\,\text{V}}$$

$$V_{rms}_{L} = I_{rms}X_L = I_{rms}2\pi fL = \left(0.124\,\text{A}\right)2\pi\left(60.0\,\text{Hz}\right)\left(0.25\,\text{H}\right) = \boxed{10.5\,\text{V}}$$

Note that, because the maximum voltages occur at different times, the two readings do not add to the applied voltage of 120 V.

61. Using Eq. 30-23b we calculate the impedance of the inductor. Then we set the phase shift in Eq. 30-29a equal to 25° and solve for the resistance. We calculate the output voltage by multiplying the current through the circuit, from Eq. 30-27, by the inductive reactance (Eq. 30-23b).

$$X_L = 2\pi fL = 2\pi (175 \text{ Hz})(0.055 \text{ H}) = 60.48 \ \Omega$$

$$\tan \phi = \frac{X_L}{R} = \rightarrow R = \frac{X_L}{\tan \phi} = \frac{60.48 \ \Omega}{\tan 25°} = 129.7 \ \Omega \approx \boxed{130 \ \Omega}$$

$$\frac{V_{\text{output}}}{V_0} = \frac{V_R}{V_0} = \frac{IR}{IZ} = \frac{R}{Z} = \frac{129.70 \ \Omega}{\sqrt{(129.70 \ \Omega)^2 + (60.48 \ \Omega)^2}} = \boxed{0.91}$$

67. (a) We write the average power using Eq. 30-30, with the current in terms of the impedance (Eq. 30-27) and the power factor in terms of the resistance and impedance (Eq. 30-29b). Finally we write the impedance using Eq. 30-28b.

$$\bar{P} = I_{\text{rms}} V_{\text{rms}} \cos \phi = \frac{V_{\text{rms}}}{Z} V_{\text{rms}} \frac{R}{Z} = \frac{V_{\text{rms}}^2 R}{Z^2} = \boxed{\frac{V_0^2 R}{2\left[R^2 + (\omega L - 1/\omega C)^2\right]}}$$

(b) The power dissipation will be a maximum when the inductive reactance is equal to the capacitive reactance, which is the resonant frequency.

$$f = \boxed{\frac{1}{2\pi\sqrt{LC}}}$$

(c) We set the power dissipation equal to ½ of the maximum power dissipation and solve for the angular frequencies.

$$\bar{P} = \frac{1}{2}\bar{P}_{\text{max}} = \frac{V_0^2 R}{2\left[R^2 + (\omega L - 1/\omega C)^2\right]} = \frac{1}{2}\left(\frac{V_0^2 R}{2R^2}\right) \rightarrow (\omega L - 1/\omega C) = \pm R$$

$$\rightarrow 0 = \omega^2 LC \pm RC\omega - 1 \rightarrow \omega = \frac{\pm RC \pm \sqrt{R^2 C^2 + 4LC}}{2LC}$$

We require the angular frequencies to be positive and for a sharp peak, $R^2 C^2 \ll 4LC$. The angular width will then be the difference between the two positive frequencies.

$$\omega = \frac{2\sqrt{LC} \pm RC}{2LC} = \frac{1}{\sqrt{LC}} \pm \frac{R}{2L} \rightarrow \Delta\omega = \left(\frac{1}{\sqrt{LC}} + \frac{R}{2L}\right) - \left(\frac{1}{\sqrt{LC}} - \frac{R}{2L}\right) = \boxed{\frac{R}{L}}$$

73. When the currents have acquired their steady-state values, the capacitor will be fully charged, and so no current will flow through the capacitor. At this time, the voltage drop across the inductor will be zero, as the current flowing through the inductor is constant. Therefore, the current through R_1 is zero, and the resistors R_2 and R_3 can be treated as in series.

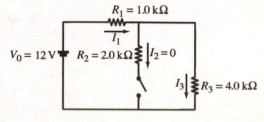

$$I_1 = I_3 = \frac{V_0}{R_1 + R_3} = \frac{12 \text{ V}}{5.0 \text{ k}\Omega} = \boxed{2.4 \text{ mA}} \ ; \ I_2 = \boxed{0}$$

79. We use Kirchhoff's loop rule to equate the input voltage to the voltage drops across the inductor and resistor. We then multiply both sides of the equation by the integrating factor $e^{\frac{Rt}{L}}$ and integrate the right-hand side of the equation using a u substitution with $u = IRe^{\frac{Rt}{L}}$ and $du = dIRe^{\frac{Rt}{L}} + Ie^{\frac{Rt}{L}}dt/L$

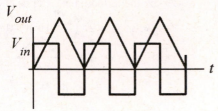

$$V_{in} = L\frac{dI}{dt} + IR \quad \rightarrow$$

$$\int V_{in}e^{\frac{Rt}{L}}dt = \int \left(L\frac{dI}{dt} + IR\right)e^{\frac{Rt}{L}}dt = \frac{L}{R}\int du = IR\frac{L}{R}e^{\frac{Rt}{L}} = V_{out}\frac{L}{R}e^{\frac{Rt}{L}}$$

For $L/R \ll t$, $e^{\frac{Rt}{L}} \approx 1$. Setting the exponential term equal to unity on both sides of the equation gives the desired results.

$$\int V_{in}dt = V_{out}\frac{L}{R}$$

85. We find the resistance using Ohm's law with the dc voltage and current. When then calculate the impedance from the ac voltage and current, and using Eq. 30-28b.

$$R = \frac{V}{I} = \frac{45\,\text{V}}{2.5\,\text{A}} = \boxed{18\,\Omega} \quad ; \quad Z = \frac{V_{rms}}{I_{rms}} = \frac{120\,\text{V}}{3.8\,\text{A}} = 31.58\,\Omega$$

$$\sqrt{R^2 + (2\pi fL)^2} \rightarrow L = \frac{\sqrt{Z^2 - R^2}}{2\pi f} = \frac{\sqrt{(31.58\,\Omega)^2 - (18\,\Omega)^2}}{2\pi(60\,\text{Hz})} = \boxed{69\,\text{mH}}$$

91. (a) The impedance of the circuit is given by Eq. 30-28b with $X_L > X_C$ and $R = 0$. We divide the magnitude of the ac voltage by the impedance to get the magnitude of the ac current in the circuit. Since $X_L > X_C$, the voltage will lead the current by $\phi = \pi/2$. No dc current will flow through the capacitor.

$$Z = \sqrt{R^2 + (\omega L - 1/\omega C)^2} = \omega L - 1/\omega C \quad I_0 = \frac{V_{20}}{Z} = \frac{V_{20}}{\omega L - 1/\omega C}$$

$$I(t) = \boxed{\frac{V_{20}}{\omega L - 1/\omega C}\sin(\omega t - \pi/2)}$$

(b) The voltage across the capacitor at any instant is equal to the charge on the capacitor divided by the capacitance. This voltage is the sum of the ac voltage and dc voltage. There is no dc voltage drop across the inductor so the dc voltage drop across the capacitor is equal to the input dc voltage.

$$V_{out,ac} = V_{out} - V_1 = \frac{Q}{C} - V_1$$

We treat the emf as a superposition of the ac and dc components. At any instant of time the sum of the voltage across the inductor and capacitor will equal the input voltage. We use Eq. 30-5 to calculate the voltage drop across the inductor. Subtracting the voltage drop across the inductor from the input voltage gives the output voltage. Finally, we subtract off the dc voltage to obtain the ac output voltage.

$$V_L = L\frac{dI}{dt} = L\frac{d}{dt}\left[\frac{V_{20}}{\omega L - 1/\omega C}\sin(\omega t - \pi/2)\right] = \frac{V_{20}L\omega}{\omega L - 1/\omega C}\cos(\omega t - \pi/2)$$

$$= \frac{V_{20}L\omega}{\omega L - 1/\omega C}\sin(\omega t)$$

$$V_{out} = V_{in} - V_L = V_1 + V_{20}\sin\omega t - \left(\frac{V_{20}L\omega}{\omega L - 1/\omega C}\sin(\omega t)\right)$$

$$= V_1 + V_{20}\left(1 - \frac{L\omega}{\omega L - 1/\omega C}\right)\sin(\omega t) = V_1 - V_{20}\left(\frac{1/\omega C}{\omega L - 1/\omega C}\right)\sin(\omega t)$$

$$V_{out,ac} = V_{out} - V_1 = -V_{20}\left(\frac{1/\omega C}{\omega L - 1/\omega C}\right)\sin(\omega t) = \boxed{\left(\frac{V_{20}}{\omega^2 LC - 1}\right)\sin(\omega t - \pi)}$$

(*c*) The attenuation of the ac voltage is greatest when the denominator is large.

$$\omega^2 LC \gg 1 \rightarrow \omega L \gg \frac{1}{\omega C} \rightarrow X_L \gg X_C$$

We divide the output ac voltage by the input ac voltage to obtain the attenuation.

$$\frac{V_{2,out}}{V_{2,in}} = \frac{\dfrac{V_{20}}{\omega^2 LC - 1}}{V_{20}} = \frac{1}{\omega^2 LC - 1} \approx \boxed{\frac{1}{\omega^2 LC}}$$

(*d*) The dc output is equal to the dc input, since there is no dc voltage drop across the inductor.

$$\boxed{V_{1,out} = V_1}$$

97. We calculate the resistance from the power dissipated and the current. Then setting the ratio of the voltage to current equal to the impedance, we solve for the inductance.

$$\overline{P} = I_{rms}^2 R \rightarrow R = \frac{\overline{P}}{I_{rms}^2} = \frac{350\,\text{W}}{(4.0\,\text{A})^2} = 21.88\,\Omega \approx \boxed{22\,\Omega}$$

$$Z = \frac{V_{rms}}{I_{rms}} = \sqrt{R^2 + (2\pi fL)^2} \rightarrow$$

$$L = \frac{\sqrt{(V_{rms}/I_{rms})^2 - R^2}}{2\pi f} = \frac{\sqrt{(120\,\text{V}/4.0\,\text{A})^2 - (21.88\,\Omega)^2}}{2\pi(60\,\text{Hz})} = \boxed{54\,\text{mH}}$$

103.(*a*) The output voltage is the voltage across the capacitor, which is the current through the circuit multiplied by the capacitive reactance. We calculate the current by dividing the input voltage by the impedance. Finally, we divide the output voltage by the input voltage to calculate the gain.

$$V_{out} = IX_C = \frac{V_{in}X_C}{\sqrt{R^2 + X_C^2}} = \frac{V_{in}}{\sqrt{(R/X_C)^2 + 1}} = \frac{V_{in}}{\sqrt{(2\pi fCR)^2 + 1}}$$

$$A = \frac{V_{out}}{V_{in}} = \boxed{\frac{1}{\sqrt{4\pi^2 f^2 C^2 R^2 + 1}}}$$

(*b*) As the frequency goes to zero, the gain becomes one. In this instance the capacitor becomes fully charged, so no current flows across the resistor. Therefore the output voltage is equal to the input voltage. As the frequency becomes very large, the capacitive reactance becomes very small, allowing a large current. In this case, most of the voltage drop is across the resistor, and the gain goes to zero.

(*c*) See the graph of the log of the gain as a function of the log of the frequency. Note that for frequencies less than about 100 Hz the gain is ~ 1. For higher frequencies the gain drops off proportionately to the frequency.

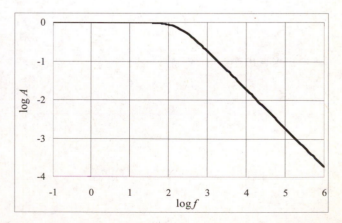

Chapter 31: Maxwell's Equations and Electromagnetic Waves

Chapter Overview and Objectives

This chapter introduces changing electric fields as a source of magnetic fields and Gauss's law for magnetic fields. Electromagnetic waves are shown to be solutions of Maxwell's equations. Some properties of electromagnetic waves are discussed.

After completing this chapter you should:

- Know that a changing electric flux produces a magnetic field.
- Know Gauss's law for magnetic fields.
- Know Maxwell's equations.
- Know that Maxwell's equations predict that electromagnetic waves exist.
- Know that the speed of light is predicted by Maxwell's equations.
- Know what the Poynting vector is, what it represents, and how to calculate it.
- Know how to calculate the intensity of electromagnetic waves given the amplitude or rms value of their electric or magnetic fields.

Summary of Equations

Magnetic field due to a changing electric flux:

$$\oint \vec{B} \cdot d\vec{\ell} = \mu_0 I_{encl} + \mu_0 \varepsilon_0 \frac{d\Phi_E}{dt}$$ (Section 31-1)

Gauss's law for magnetic fields:

$$\oint \vec{B} \cdot d\vec{A} = 0$$ (Section 31-2)

Maxwell's equations:

$$\oint \vec{E} \cdot d\vec{A} = \frac{Q}{\varepsilon_0}$$ (Section 31-3)

$$\oint \vec{B} \cdot d\vec{A} = 0$$ (Section 31-3)

$$\oint \vec{E} \cdot d\vec{\ell} = -\frac{d\Phi_B}{dt}$$ (Section 31-3)

$$\oint \vec{B} \cdot d\vec{\ell} = \mu_0 I_{encl} + \mu_0 \varepsilon_0 \frac{d\Phi_E}{dt}$$ (Section 31-3)

Electromagnetic wave equations:

$$\frac{\partial^2 E}{\partial t^2} = \frac{1}{\mu_0 \varepsilon_0} \frac{\partial^2 E}{\partial x^2}$$ (Section 31-5)

$$\frac{\partial^2 B}{\partial t^2} = \frac{1}{\mu_0 \varepsilon_0} \frac{\partial^2 B}{\partial x^2}$$ (Section 31-5)

Speed of electromagnetic waves in vacuum:

$$v = \frac{1}{\sqrt{\mu_0 \varepsilon_0}} = 3.00 \times 10^8 \text{ m/s}$$ (Section 31-5)

Speed of electromagnetic waves in a material:

$$v = \frac{1}{\sqrt{\mu \varepsilon}}$$ (Section 31-6)

Definition of the Poynting vector:

$$\vec{S} = \frac{1}{\mu_0} \vec{E} \times \vec{B}$$ (Section 31-8)

Average magnitude of Poynting vector:

$$\vec{S} = \frac{1}{2} \varepsilon_0 c E_0^2 = \frac{1}{2} \frac{c}{\mu_0} B_0^2 = \frac{E_0 B_0}{2\mu_0} = \frac{E_{rms} B_{rms}}{\mu_0}$$ (Section 31-8)

Radiation pressure of electromagnetic waves traveling perpendicular to the surface of a perfect absorber:

$$P = \frac{\overline{S}}{c}$$ (Section 31-9)

Radiation pressure of electromagnetic waves traveling perpendicular to the surface of a perfect reflector:

$$P = \frac{2\overline{S}}{c}$$ (Section 31-9)

Chapter Summary

Section 31-1. Changing Electric Fields Produce Magnetic Fields; Ampère's Law and Displacement Current

Currents are not the only source of magnetic fields. In analogy to Faraday's law that an electric field is produced by a changing magnetic flux, induced magnetic fields are produced by a changing **electric flux**. The mathematical relationship between the induced magnetic field and the changing electric flux has the same mathematical form as Faraday's law:

$$\oint \vec{B}_{\text{induced}} \cdot d\vec{\ell} = \mu_0 \varepsilon_0 \frac{d\Phi_E}{dt}$$

We can add this term to Ampere's law to write an equation for the total magnetic field

$$\oint \vec{B} \cdot d\vec{\ell} = \mu_0 I_{\text{encl}} + \mu_0 \varepsilon_0 \frac{d\Phi_E}{dt}$$

Example 31-1-A. Calculating the rate of change of electric flux. A circular path in space has a magnetic field that points tangent to the circle at all points in a counterclockwise direction, as seen in the diagram. The magnitude of the magnetic field is $B = 3.98$ µT at all points on the circle of radius $r = 5.68$ cm. A current $I = 450$ mA flows out of the page through the circle. What is the rate of change of electric flux through the circle? If the electric field is pointing out of the page through the circle, is the magnitude of the electric field increasing or decreasing with time?

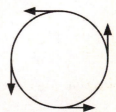

Approach: The path integral of the magnetic field depends on the enclosed current and the rate of change of the electric flux. The problem provides information about the magnetic field and the enclosed current, and we can thus determine the rate of change of the electric flux.

Solution: To calculate the change of the electric flux through the loop we use the following relation:

$$\oint \vec{B} \cdot d\vec{\ell} = \mu_0 I_{\text{encl}} + \mu_0 \varepsilon_0 \frac{d\Phi_E}{dt} \quad \Rightarrow \quad \frac{d\Phi_E}{dt} = \frac{1}{\mu_0 \varepsilon_0} \left(\oint \vec{B} \cdot d\vec{\ell} - \mu_0 I_{\text{encl}} \right)$$

We will evaluate the path integral in the counterclockwise direction. With this choice of direction, the enclosed current, which is coming out of the page, will be positive. To calculate the path integral of the magnetic field around the circle, we first consider the contribution of a small segment along the path. Along this small segment, $d\vec{\ell}$ and $\vec{B}$ are parallel and $\vec{B} \cdot d\vec{\ell} = B\, dl$.

Since the magnitude of the magnetic field is constant along the path, we can easily evaluate the path integral around the closed path:

$$\oint \vec{B} \cdot d\vec{\ell} = \oint B\, d\ell = B \oint d\ell = B(2\pi R)$$

The rate of change of the electric flux is thus equal to

$$\frac{d\Phi_E}{dt} = \frac{1}{\mu_0 \varepsilon_0}\left(2\pi RB - \mu_0 I_{encl}\right) = \frac{1}{\left(4\pi \times 10^{-7}\right)\left(8.85 \times 10^{-12}\right)}\left\{2\pi(0.0568)\left(3.98 \times 10^{-6}\right) - \left(4\pi \times 10^{-7}\right)(0.450)\right\} =$$

$$= 7.69 \times 10^{10}\ \text{V} \cdot \text{m/s}$$

Since $d\Phi_E/dt$ is positive, the electric flux is increasing in the out-of-the-page direction. That implies that the electric field is increasing in magnitude.

Section 31-2. Gauss's Law for Magnetism

There is a certain degree of symmetry between the electric field and the magnetic field, e.g., a changing magnetic flux induces an electric field and a changing electric flux induces a magnetic field. For electric fields, Gauss's law relates the electric flux through a surface to the enclosed electric charge. A similar relation exists for the magnetic flux but since no magnetic charges have ever been observed, the surface integral of the magnetic field over a closed surface is equal to zero. This conclusion is often called **Gauss's law for magnetism**:

$$\oint \vec{B} \cdot d\vec{A} = 0$$

Section 31-3. Maxwell's Equations

The equations that relate the electric and magnetic fields to their sources are called the **Maxwell's equations**:

$$\oint \vec{E} \cdot d\vec{A} = \frac{Q}{\varepsilon_0}$$

$$\oint \vec{B} \cdot d\vec{A} = 0$$

$$\oint \vec{E} \cdot d\vec{\ell} = -\frac{d\Phi_B}{dt}$$

$$\oint \vec{B} \cdot d\vec{\ell} = \mu_0 I_{encl} + \mu_0 \varepsilon_0 \frac{d\Phi_E}{dt}$$

Section 31-4. Production of Electromagnetic Waves

One of the consequences of the Maxwell's equations was the prediction that changes in the electric and magnetic fields at a particular location produce **electromagnetic (EM) waves** that travel through space with a finite velocity. The electromagnetic waves are created by accelerating charges.

If we look at great distances from the source of electromagnetic waves, large compared to the spatial extent of the source, the solutions of Maxwell's equations take a relatively simple form. In this large-distance limit, the changing electric and magnetic fields take the form of **radiation fields**. The general form of radiation fields are outward propagating waves of electric and magnetic fields. The changing electric and magnetic fields are directed perpendicular to the direction of propagation of the waves. The magnitudes of the electric and magnetic fields decrease with distance as $1/r$, where r is the distance from the source. The electric and magnetic fields have a time dependence that has the same form as the time dependence of the source.

Section 31-5. Electromagnetic Waves, and Their Speed, from Maxwell's Equations

Maxwell's equations are interdependent. A changing magnetic field induces an electric field and a changing electric field induces a magnetic field. The interdependence between the electric and magnetic fields can be removed by relating the time dependence of the electric or magnetic fields to their position dependence. This results in the following **wave equations** for the electric and magnetic fields:

$$\frac{\partial^2 E}{\partial t^2} = \frac{1}{\mu_0 \varepsilon_0} \frac{\partial^2 E}{\partial x^2}$$

$$\frac{\partial^2 B}{\partial t^2} = \frac{1}{\mu_0 \varepsilon_0} \frac{\partial^2 B}{\partial x^2}$$

Both of these equations are one-dimensional wave equations, identical in form to the wave equation from Chapter 15, Section 5. The solutions of these equations are waves that propagate with a velocity v where

$$v^2 = \frac{1}{\mu_0 \varepsilon_0} \quad \Rightarrow \quad v = \frac{1}{\sqrt{\mu_0 \varepsilon_0}} = 3.00 \times 10^8 \text{ m/s}$$

The velocity v is the speed of light.

Section 31-6. Light as an Electromagnetic Wave and the Electromagnetic Spectrum

The waves that are solutions of the Maxwell's equations have a propagation speed consistent with the measured speed of visible light. Visible light is just one example of the type of electromagnetic waves predicted by Maxwell's equations. Radio waves, microwaves, infrared light, visible light, ultraviolet light, X-rays, and gamma rays are all electromagnetic waves. The only difference between these different waves is their wavelength. They are classified differently because their different wavelengths cause them to interact with matter in different ways.

The waves predicted by Maxwell's equations can also propagate in materials. The wave equations are altered by the presence of charges and currents that are driven by the electric and magnetic fields of the electromagnetic waves, which in turn create additional electromagnetic waves. Maxwell's equations and the equations of motion of the charged particles must be solved consistently. In uniform, isotropic materials, the wave equations are similar to the wave equations in vacuum except that the velocity of propagation v is equal to

$$v = \frac{1}{\sqrt{\mu \varepsilon}}$$

where μ is the magnetic permeability of the material and ε is the electric permittivity of the material.

Section 31-7. Measuring the Speed of Light

The speed of light can be determined by measuring the time required for light to travel a specific distance. Since the speed of light is high, accurate timing techniques must be used in order to accurately determine the speed of light. The speed of light in vacuum is

$$c = 2.99792458 \times 10^8 \text{ m/s}$$

The known speed of light can be used to measure distances accurately. For example, the distance between the Earth and the Moon is monitored frequently by measuring the time required for a laser beam to travel from Earth to a mirror located on the moon and back to Earth.

Section 31-8. Energy in EM Waves; the Poynting Vector

The energy density u in regions with non-zero electric or magnetic fields has already been discussed in earlier chapters (e.g., Section 30.3).

In the case of electromagnetic waves, the magnitude of the electric and magnetic fields are related so we can write the energy density in terms of electric fields, magnetic fields, or both:

$$u = \varepsilon_0 E^2 = \frac{B^2}{\mu_0} = \sqrt{\frac{\varepsilon_0}{\mu_0}} EB$$

A vector quantity that has a magnitude equal to the energy carried by the electromagnetic wave per unit time per unit area and directed in the direction of the energy flow is the **Poynting vector**, $\vec{S}$. The Poynting vector can be written in terms of the electric and magnetic fields:

$$\vec{S} = \frac{1}{\mu_0} \vec{E} \times \vec{B}$$

The Poynting vector is time dependent at a given point in space because the electric and magnetic fields vary with time as the wave moves past the given point. A useful quantity for calculating the energy transfer by electromagnetic fields is the magnitude of the time-averaged Poynting vector. This is a useful quantity because we often look at the energy transferred over many periods of the electromagnetic wave. The magnitude of the time-averaged Poynting vector for sinusoidal electromagnetic waves is

$$\bar{S} = \frac{1}{2} \varepsilon_0 c E_0^2 = \frac{1}{2} \frac{c}{\mu_0} B_0^2 = \frac{E_0 B_0}{2\mu_0} = \frac{E_{rms} B_{rms}}{2\mu_0}$$

where E_0 and B_0 are the amplitudes of the electric and magnetic fields, respectively, and E_{rms} and B_{rms} are the rms values of the electric and magnetic fields, respectively. The final expression is valid regardless of whether the waves are sinusoidal or not. The time-averaged Poynting vector is the **intensity** of the electromagnetic wave.

Example 31-8-A. Determining the properties of an electromagnetic wave. An electromagnetic wave has a magnetic field that points to the East and an electric field that points to the North. The magnitude of the magnetic field is 3.56 μT. Determine the magnitude of the electric field of the electromagnetic wave. Determine the Poynting vector of the electromagnetic wave when the electric and magnetic fields are at their maximum. Determine the intensity of the electromagnetic wave.

Approach: The properties of the electric and magnetic fields of an electromagnetic field are coupled. If we know the properties of the magnetic field, we can determine the properties of the electric field. Once the properties of the electric and the magnetic field are known, we can determine the Poynting vector.

Solution: The amplitude E_0 of the electric field amplitude and the amplitude B_0 of the magnetic field are related in the following way:

$$B_0 = \frac{E_0}{c}$$

Since we know the amplitude of the magnetic field we can determine the amplitude of the electric field:

$$E_0 = B_0 c = \left(3.56 \times 10^{-6}\right)\left(3.00 \times 10^8\right) = 1.07 \times 10^3 \text{ V/m}$$

In order to describe the properties of the electromagnetic waves, we need to define a coordinate system. The coordinate system used to solve this problem is shown in the Figure on the right. In this coordinate system, the directions of the magnetic and electric fields are parallel to the x and the y axes, respectively:

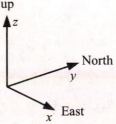

$$\vec{B}_0 = \left(3.56 \times 10^{-6}\right)\hat{i}$$

and

$$\vec{E}_0 = \left(1.07 \times 10^3\right)\hat{j}$$

The maximum Poynting vector is thus equal to

$$\vec{S}_0 = \frac{1}{\mu_0}\vec{E}_0 \times \vec{B}_0 = \frac{1}{\left(4\pi \times 10^{-7}\right)}\left\{\left(1.07 \times 10^3\right)\hat{\mathbf{j}} \times \left(3.56 \times 10^{-6}\right)\hat{\mathbf{i}}\right\} = -\left(3.03 \times 10^3\right)\hat{\mathbf{k}}$$

The average intensity of the electromagnetic radiation is the magnitude of the average Poynting vector:

$$\overline{S} = \frac{E_0 B_0}{2\mu_0} = \frac{\left(1.07 \times 10^3\right)\left(3.56 \times 10^{-6}\right)}{2\left(4\pi \times 10^{-7}\right)} = 1.51 \times 10^3 \text{ W/m}^2$$

The average intensity of the Poynting vector should be one-half of the amplitude of the Poynting vector.

Section 31-9. Radiation Pressure

Electromagnetic waves transfer momentum as well as energy. **Radiation pressure** is the force per unit area on a surface due to the momentum transferred to the surface by the electromagnetic radiation incident upon it. The radiation pressure P on a surface that completely absorbs the incident electromagnetic radiation when the electromagnetic waves are traveling perpendicular to the surface is equal to

$$P = \frac{\overline{S}}{c}$$

The radiation pressure on a surface that completely reflects the incident electromagnetic radiation when the electromagnetic waves are traveling perpendicular to the surface is equal to

$$P = \frac{2\overline{S}}{c}$$

Example 31-9-A. Providing propulsion with electromagnetic radiation. A future design of space ships or airplanes may use electromagnetic radiation to provide propulsion. What intensity of light would need to be emitted to provide a thrust of 1.00×10^5 N?

Approach: In this Section we determined the pressure that electromagnetic radiation exerts on a surface when it is absorbed by this surface. In this problem, electromagnetic radiation is emitted. Since absorption is the reverse of emission, the relation between radiation pressure and the Poynting vector can be used to determine the radiation pressure in both situations.

Solution: The required radiation pressure can be obtained from the required thrust F and the area A that emits the radiation:

$$P = \frac{F}{A}$$

Since the radiation pressure is also equal to the ratio of the average Poynting vector and the speed of light, we can use this relation to determine the average Poynting vector:

$$\overline{S} = Pc = \frac{cF}{A}$$

Assuming that the section that emits the electromagnetic radiation has an area of 1 m^2 we find that the required average Poynting vector is equal to

$$\overline{S} = \frac{cF}{A} = \frac{\left(3.00 \times 10^8\right)\left(1.00 \times 10^5\right)}{1} = 3.0 \times 10^{13} \text{ W/m}^2$$

Section 31-10. Radio and Television; Wireless Communication

Electromagnetic waves are used to carry information between two distant points. To allow multiple communications of information to take place simultaneously, users of electromagnetic waves are assigned a **carrier frequency** and a

bandwidth to use. The frequencies of the electromagnetic radiation sent/received by a user must fall between an upper and lower limit in frequency. The information is usually sent out by starting with a sinusoidal signal with the carrier frequency and then altering the signal slightly through a process called **modulation** to encode information in the electromagnetic wave.

There are many methods of modulating the carrier wave to add the information to the carrier signal. In **amplitude modulation (AM)**, the amplitude of the carrier wave is varied. AM radio stations use carrier frequencies between 530 kHz and 1700 kHz. In **frequency modulation (FM)**, the frequency of the wave is varied. FM radio stations use carrier frequencies between 88 MHz and 108 MHz. Whether amplitude, frequency, or phase modulation is used, the modulation causes the electromagnetic wave do have a range of frequencies present in it. The modulating signal is limited so the frequencies in the electromagnetic wave do not spread beyond the bandwidth allowed. Different bands of the electromagnetic spectrum are assigned to different forms of communications (e.g., cell phone service, aviation, military, law enforcement, satellite, television, etc.).

A receiver of the electromagnetic waves must perform the process of **demodulation** to obtain the information from the modulated carrier wave. The receiver must be tuned to the proper frequency to "receive" the modulated carrier wave.

Practice Quiz

1. There are no currents present, but there is a non-zero magnetic field. What else must be present?
   a) Electric charge.
   b) Electric field.
   c) Changing electric field.
   d) It is impossible to have a magnetic field without a current.

2. If the amplitude of the electric field in an electromagnetic wave is doubled, what happens to the intensity of the electromagnetic wave?
   a) Doubles.
   b) Halves.
   c) Quadruples.
   d) Doesn't change.

3. Which travels faster in a vacuum, X-rays or radio waves?
   a) X-rays.
   b) Radio waves.
   c) It depends on whether the radio waves are AM or FM.
   d) They travel at the same speed.

4. An electromagnetic wave travels toward the North. At a given instant, the electric field in the electromagnetic wave is toward the East. What is the direction of the magnetic field in the electromagnetic wave at that instant?
   a) West.
   b) South.
   c) Up.
   d) Down.

5. Transparent materials have a dielectric constant greater than one at the frequency of visible light. The magnetic susceptibility of these materials is much less than one. What can you say about the speed of light in these materials?
   a) The speed of light in the material is the same as the speed of light in vacuum.
   b) The speed of light in the material is greater than the speed of light in vacuum.
   c) The speed of light in the material is less than the speed of light in vacuum.
   d) The speed of light in the material might be less than or greater than the speed of light in vacuum.

6. If the amplitude of the electric field in an electromagnetic wave is doubled, what happens to the amplitude of the magnetic field in the electromagnetic wave?
   a) Doubles.
   b) Halves.
   c) Quadruples.
   d) Doesn't change.

7. In what direction is an electromagnetic wave traveling that has an electric field pointing East and a magnetic field pointing North?
   a) Upward.
   b) Downward.
   c) Northeast.
   d) Southwest.

8. Two surfaces of equal area are illuminated by the same source of electromagnetic radiation. Surface A absorbs 30% of the electromagnetic energy incident on it and reflects the remainder. Surface B absorbs 60% of the electromagnetic energy incident on it and reflects the remainder. Which surface has a greater radiation pressure acting on it?
   a) Surface A.
   b) Surface B.
   c) Both surfaces have the same radiation pressure acting on them.
   d) There is insufficient information to answer the question.

9. The magnetic field at all points on the circle shown in the diagram points tangent to the circle in the counterclockwise direction. No current flows through the circle, but there is an electric field pointing upward out of the page. Is the magnitude of the electric field increasing or decreasing?
   a) Increasing.
   b) Decreasing.
   c) Remaining constant.
   d) There is insufficient information to answer the question.

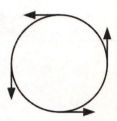

10. The magnetic field at all points on the circle shown in the diagram is equal to zero. There is an electric field pointing out of the page and it is increasing in magnitude at a uniform rate. What else must be true?
    a) There is a constant current flowing into the page through the circle.
    b) There is an increasing current flowing into the page through the circle.
    c) There is a constant current flowing out of the page through the circle.
    d) There is an increasing current flowing into the page through the circle.

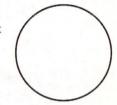

11. A circular area of radius 12 cm has a uniform electric field pointing through it, perpendicular to the area, with a magnitude of 4.0×10^3 V/m. The electric field decreases uniformly in magnitude to 1.0×10^3 V/m in 0.12 s. What is the magnetic field along the perimeter of the circular area?

12. Determine the radiation pressure on the Earth due to the Sun, assuming the radiation from the Sun is completely absorbed by the Earth. The intensity of the light reaching the Earth from the Sun is 1.5 kW/m².

13. Determine the rms electric field and rms magnetic field due to the electromagnetic radiation at a distance of 1.00 m from a 100-W light bulb. Assume all of the energy going into the bulb is radiated as electromagnetic radiation.

14. Determine the wavelength of a radio station, transmitting with a carrier frequency of 830 kHz in the AM radio band.

15. Calculate the speed of an electromagnetic wave in a dielectric material with a dielectric constant $n = 3.22$. Assume $\mu = \mu_0$.

Responses to Select End-of-Chapter Questions

1. The magnetic field will be clockwise in both cases. In the first case, the electric field is away from you and is increasing. The direction of the displacement current (proportional to $d\Phi_E / dt$) is therefore away from you and the corresponding magnetic field is clockwise. In the second case, the electric field is directed towards you and is decreasing; the displacement current is still away from you, and the magnetic field is still clockwise.

7. EM waves are self-propagating and can travel through a perfect vacuum. Sound waves are mechanical waves which require a medium, and therefore cannot travel through a perfect vacuum.

13. Yes, although the wavelengths for radio waves will be much longer than for sound waves, since the radio waves travel at the speed of light.

Solutions to Select End-of-Chapter Problems

1. The electric field between the plates is given by $E = V/d$, where d is the distance between the plates.

$$E = \frac{V}{d} \quad \rightarrow \quad \frac{dE}{dt} = \frac{1}{d}\frac{dV}{dt} = \left(\frac{1}{0.0011\,\text{m}}\right)(120\,\text{V/s}) = \boxed{1.1 \times 10^5\,\frac{\text{V}}{\text{m}\cdot\text{s}}}$$

7. (a) We follow the development and geometry given in Example 31-1, using R for the radial distance. The electric field between the plates is given by $E = V/d$, where d is the distance between the plates.

$$\oint \vec{B}\cdot d\vec{l} = \mu_0\varepsilon_0\frac{d\Phi_E}{dt} \quad \rightarrow \quad B(2\pi R_{\text{path}}) = \mu_0\varepsilon_0\frac{d(\pi R_{\text{flux}}^2 E)}{dt}$$

The subscripts are used on the radial variable because there might not be electric field flux through the entire area bounded by the amperian path. The electric field between the plates is given by $E = \dfrac{V}{d} = \dfrac{V_0\sin(2\pi ft)}{d}$, where d is the distance between the plates.

$$B(2\pi R_{\text{path}}) = \mu_0\varepsilon_0\frac{d(\pi R_{\text{flux}}^2 E)}{dt} \quad \rightarrow$$

$$B = \frac{\mu_0\varepsilon_0\pi R_{\text{flux}}^2}{2\pi R_{\text{path}}}\frac{d(E)}{dt} = \frac{\mu_0\varepsilon_0\pi R_{\text{flux}}^2}{2\pi R_{\text{path}}}\frac{2\pi fV_0}{d}\cos(2\pi ft) = \frac{\mu_0\varepsilon_0 R_{\text{flux}}^2}{R_{\text{path}}}\frac{\pi fV_0}{d}\cos(2\pi ft)$$

We see that the functional form of the magnetic field is $\boxed{B = B_0(R)\cos(2\pi ft)}$.

(b) If $R \le R_0$, then there is electric flux throughout the area bounded by the amperian loop, and so $R_{\text{path}} = R_{\text{flux}} = R$.

$$B_0(R \le R_0) = \frac{\mu_0\varepsilon_0 R_{\text{flux}}^2}{R_{\text{path}}}\frac{\pi fV_0}{d} = \mu_0\varepsilon_0\frac{\pi fV_0}{d}R = \frac{\pi(60\,\text{Hz})(150\,\text{V})}{(3.00\times10^8\,\text{m/s})^2(5.0\times10^{-3}\,\text{m})}R$$

$$= (6.283\times10^{-11}\,\text{T/m})R \approx \boxed{(6.3\times10^{-11}\,\text{T/m})R}$$

If $R > R_0$, then there is electric flux only out a radial distance of R_0. Thus $R_{\text{path}} = R$ and $R_{\text{flux}} = R_0$.

$$B_0(R > R_0) = \frac{\mu_0\varepsilon_0 R_{\text{flux}}^2}{R_{\text{path}}}\frac{\pi fV_0}{d} = \mu_0\varepsilon_0\frac{\pi fV_0 R_0^2}{d}\frac{1}{R} =$$

$$= \frac{\pi(60\,\text{Hz})(150\,\text{V})(0.030\,\text{m})^2}{(3.00\times10^8\,\text{m/s})^2(5.0\times10^{-3}\,\text{m})}\frac{1}{R} =$$

$$= (5.655\times10^{-14}\,\text{T}\cdot\text{m})\frac{1}{R} \approx \boxed{(5.7\times10^{-14}\,\text{T}\cdot\text{m})\frac{1}{R}}$$

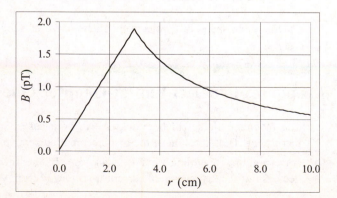

(c) See the adjacent graph. Note that the magnetic field is continuous at the transition from "inside" to "outside" the capacitor radius.

13. Use Eq. 31-14 to find the frequency of the microwave.

$$c = \lambda f \quad \rightarrow \quad f = \frac{c}{\lambda} = \frac{\left(3.00 \times 10^8 \text{ m/s}\right)}{\left(1.50 \times 10^{-2} \text{ m}\right)} = \boxed{2.00 \times 10^{10} \text{ Hz}}$$

19. (a) The radio waves travel at the speed of light, and so $\Delta d = v \Delta t$. The distance is found from the radii of the orbits. For Mars when nearest the Earth, the radii should be subtracted.

$$\Delta t = \frac{\Delta d}{c} = \frac{\left(227.9 \times 10^9 \text{ m} - 149.6 \times 10^9 \text{ m}\right)}{\left(3.000 \times 10^8 \text{ m/s}\right)} = \boxed{261 \text{ s}}$$

(b) For Mars when farthest from Earth, the radii should be added.

$$\Delta t = \frac{\Delta d}{c} = \frac{\left(227.9 \times 10^9 \text{ m} + 149.6 \times 10^9 \text{ m}\right)}{\left(3.000 \times 10^8 \text{ m/s}\right)} = \boxed{1260 \text{ s}}$$

25. The intensity is the power per unit area, and also is the time averaged value of the Poynting vector. The area is the surface area of a sphere, since the wave is spreading spherically.

$$\overline{S} = \frac{P}{A} = \frac{\left(1500 \text{ W}\right)}{\left[4\pi \left(5.0 \text{ m}\right)^2\right]} = 4.775 \text{ W/m}^2 \approx \boxed{4.8 \text{ W/m}^2}$$

$$\overline{S} = c\varepsilon_0 E_{\text{rms}}^2 \quad \rightarrow \quad E_{\text{rms}} = \sqrt{\frac{\overline{S}}{c\varepsilon_0}} = \sqrt{\frac{4.775 \text{ W/m}^2}{\left(3.00 \times 10^8 \text{ m/s}\right)\left(8.85 \times 10^{-12} \text{ C}^2/\text{N} \cdot \text{m}^2\right)}} = \boxed{42 \text{ V/m}}$$

31. In each case, the required area is the power requirement of the device divided by 10% of the intensity of the sunlight.

(a) $A = \dfrac{P}{I} = \dfrac{50 \times 10^{-3} \text{ W}}{100 \text{ W/m}^2} = 5 \times 10^{-4} \text{ m}^2 = \boxed{5 \text{ cm}^2}$

A typical calculator is about 17 cm x 8 cm, which is about 140 cm^2. So $\boxed{\text{yes}}$, the solar panel can be mounted directly on the calculator.

(b) $A = \dfrac{P}{I} = \dfrac{1500 \text{ W}}{100 \text{ W/m}^2} = 15 \text{ m}^2 \approx \boxed{20 \text{ m}^2}$ (to one sig. fig.)

A house of floor area 1000 ft^2 would have on the order of 100 m^2 of roof area. So $\boxed{\text{yes}}$, a solar panel on the roof should be able to power the hair dryer.

(c) $A = \dfrac{P}{I} = \dfrac{20 \text{ hp} \left(746 \text{ W/hp}\right)}{100 \text{ W/m}^2} = 149 \text{ m}^2 \approx \boxed{100 \text{ m}^2}$ (to one sig. fig.)

This would require a square panel of side length about 12 m. So $\boxed{\text{no}}$, this panel could not be mounted on a car and used for real-time power.

37. The intensity from a point source is inversely proportional to the distance from the source.

$$\frac{I_{\text{Earth}}}{I_{\text{Jupiter}}} = \frac{r_{\text{Sun-Jupiter}}^2}{r_{\text{Sun-Earth}}^2} = \frac{\left(7.78 \times 10^{11} \text{ m}\right)^2}{\left(1.496 \times 10^{11} \text{ m}\right)^2} = 27.0$$

So it would take an area of $\boxed{27 \text{ m}^2}$ at Jupiter to collect the same radiation as a 1.0-m^2 solar panel at the Earth.

43. The electric field is found from the desired voltage and the length of the antenna. Then use that electric field to calculate the magnitude of the Poynting vector.

$$E_{rms} = \frac{V_{rms}}{d} = \frac{1.00 \times 10^{-3}\,\text{V}}{1.60\,\text{m}} = \boxed{6.25 \times 10^{-4}\,\text{V/m}}$$

$$S = c\varepsilon_0 E_{rms}^2 = c\varepsilon_0 \frac{V_{rms}^2}{d^2} = \left(3.00 \times 10^8\,\text{m/s}\right)\left(8.85 \times 10^{-12}\,\text{C}^2/\text{N}\cdot\text{m}^2\right)\frac{\left(1.00 \times 10^{-3}\,\text{V}\right)^2}{\left(1.60\,\text{m}\right)^2}$$

$$= \boxed{1.04 \times 10^{-9}\,\text{W/m}^2}$$

49. The light has the same intensity in all directions, so use a spherical geometry centered on the source to find the value of the Poynting vector. Then use Eq. 31-19a to find the magnitude of the electric field, and Eq. 31-11 with $v = c$ to find the magnitude of the magnetic field.

$$S = \frac{P_0}{A} = \frac{P_0}{4\pi r^2} = \tfrac{1}{2}c\varepsilon_0 E_0^2 \quad \rightarrow$$

$$E_0 = \sqrt{\frac{P_0}{2\pi r^2 c\varepsilon_0}} = \sqrt{\frac{\left(75\,\text{W}\right)}{2\pi\left(2.00\,\text{m}\right)^2\left(3.00 \times 10^8\,\text{m/s}\right)\left(8.85 \times 10^{-12}\,\text{C}^2/\text{n}\cdot\text{m}^2\right)}} = 33.53\,\text{V/m}$$

$$\approx \boxed{34\,\text{V/m}}$$

$$B_0 = \frac{E_0}{c} = \frac{\left(33.53\,\text{V/m}\right)}{\left(3.00 \times 10^8\,\text{m/s}\right)} = \boxed{1.1 \times 10^{-7}\,\text{T}}$$

55. The light has the same intensity in all directions. Use a spherical geometry centered at the source with the definition of the Poynting vector.

$$S = \frac{P_0}{A} = \frac{P_0}{4\pi r^2} = \tfrac{1}{2}c\varepsilon_0 E_0^2 = \tfrac{1}{2}c\left(\frac{1}{c^2\mu_0}\right)E_0 \quad \rightarrow \quad \tfrac{1}{2}c\left(\frac{1}{c^2\mu_0}\right)E_0 = \frac{P_0}{4\pi r^2} \quad \rightarrow \quad \boxed{E_0 = \sqrt{\frac{\mu_0 c P_0}{2\pi r^2}}}$$

61. (a) We note that $-\alpha x^2 - \beta^2 t^2 + 2\alpha\beta xt = -\left(\alpha x - \beta t\right)^2$ and so $E_y = E_0 e^{-\left(\alpha x - \beta t\right)^2} = E_0 e^{-\alpha^2\left(x - \frac{\beta}{\alpha}t\right)^2}$. Since the wave is of the form $f\left(x - vt\right)$, with $v = \beta/\alpha$, the wave is moving in the $\boxed{+x\ \text{direction}}$.

(b) The speed of the wave is $v = \beta/\alpha = c$, and so $\boxed{\beta = \alpha c}$.

(c) The electric field is in the y direction, and the wave is moving in the x direction. Since $\vec{E} \times \vec{B}$ must be in the direction of motion, the magnetic field must be in the z direction. The magnitudes are related by $\left|\vec{B}\right| = \left|\vec{E}\right|/c$.

$$\boxed{B_z = \frac{E_0}{c}e^{-\left(\alpha x - \beta t\right)^2}}$$

Chapter 32: Light: Reflection and Refraction

Chapter Overview and Objectives

This chapter introduces the ray model of light. It discusses properties of the propagation of light as it interacts with boundaries between two materials. The laws of reflection and refraction are discussed, as is the phenomenon of total internal reflection. Image formation by reflecting and refracting surfaces is described.

After completing this chapter you should:
- Know what the index of refraction of a material is.
- Know what the ray model of light is and when it applies.
- Know the law of reflection.
- Know the difference between specular and diffuse reflection.
- Know what an object and an image are.
- Know what a focal point is and what the corresponding focal length is.
- Know the image formation equation for a mirror.
- Know how to calculate the focal length of a spherical mirror.
- Know how to calculate the lateral magnification of an image.
- Know what real and virtual images are.
- Know what total internal reflection and critical angles are.
- Know what dispersion is.
- Know how to determine the image position formed by a refracting surface.

Summary of Equations

Law of reflection:

$$\theta_i = \theta_r$$ (Section 32-2)

Focal length of a spherical mirror:

$$f = \frac{r}{2}$$ (Section 32-3)

Mirror equation:

$$\frac{1}{f} = \frac{1}{d_o} + \frac{1}{d_i}$$ (Section 32-3)

Lateral magnification of an image:

$$m = \frac{h_i}{h_o} = -\frac{d_i}{d_o}$$ (Section 32-3)

Definition of the index of refraction:

$$n = \frac{c}{v}$$ (Section 32-4)

Snell's law or law of refraction:

$$n_1 \sin\theta_1 = n_2 \sin\theta_2$$ (Section 32-5)

Critical angle for total internal reflection:

$$\theta_C = \arcsin\frac{n_2}{n_1}$$ (Section 32-7)

Relationship between image distance, object distance, and radius of curvature of a spherical refracting surface:

$$\frac{n_1}{d_o} + \frac{n_2}{d_i} = \frac{n_2 - n_1}{R}$$ (Section 32-8)

Chapter Summary

Section 32-1. The Ray Model of Light

The **ray model** of light makes the assumption that light follows definite paths, just as particles do. We call these paths **rays**. In uniform materials, these paths are straight lines. The ray model of light usually is applicable when all objects and apertures in the path of the light rays are large compared to the wavelength of the light. The field of **geometric optics** deals with problems concerning light under these circumstances.

Section 32-2. Reflection; Image Formation by a Plane Mirror

When light reaches a boundary between two materials, some of the energy of the light is reflected back into the material through which it reached the boundary. This is called **reflection**.

To express the relationship between the directions in which the light travels before and after the reflection from the boundary, we need a convention to specify the direction of travel relative to the boundary. All our directions are specified by measuring the direction of travel with respect to the normal of the boundary. Light rays traveling perpendicular to a surface have an angle $\theta = 0°$; light rays traveling parallel to surface have an angle $\theta = 90°$.

A light ray that travels toward the boundary surface is called the incident light ray. Its direction, relative to the normal, is called the **angle of incidence**, denoted by θ_i. A light ray that travels away from the boundary after reflection is called the reflected light ray. Its direction, relative to the normal, is called the **angle of reflection**, denoted by θ_r. The **law of reflection** states that **the angle of incidence is equal to the angle of reflection**:

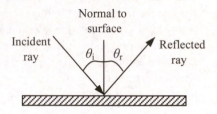

$$\theta_i = \theta_r$$

When a surface is rough, the normal to the surface changes direction by large amounts for small displacements along the surface. Parallel light rays slightly displaced from each other are reflected into many different directions. This process is called **diffuse reflection**. When the surface is very smooth, the normal to the surface points in almost the same direction over an extended area of the surface. Parallel light rays that are displaced slightly from each other are reflected into almost identical directions. This process is called **specular reflection**. When you look at a surface from which the reflection is specular, you are able to see images of the light sources that are reflecting from the surface. This is why you see yourself when you look in a mirror. When you look at a surface from which the reflection is diffuse, you do not see images of the light sources. This is why you do not see yourself when you look at a piece of white paper, even though the white paper reflects more light than a mirror does.

The distance of an object from the front of the mirror is called the **object distance** d_o. The distance of the image of this object in a flat mirror is called the **image distance** d_i. For a flat mirror, the object distance is equal to the image distance. The image of the object generated by a flat mirror is called a **virtual image** because the light rays do not travel from the location of the image, but follow a path after striking the mirror as if they came from the image behind the mirror. When a **real image** is formed, the light rays pass through the image point.

Section 32-3. Formation of Images by Spherical Mirrors

A common shape for non-flat mirrors is a spherical shape. A mirror surface that is the inside surface of a sphere is called a **concave** mirror and a mirror surface that is the outside surface of a sphere is called a **convex** mirror.

A mirror can be characterized by its **focal length**. The focal length can be determined by looking at the reflection of a beam of light rays, traveling parallel to the optical axis, that strike the mirror. The focal length f is the distance from the center of the mirror to the point at which these light rays converge. The point of convergence is called the **focal point F**.

For a spherical mirror, the focal length f is one half the radius of curvature r of the mirror:

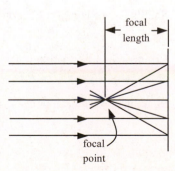

$$f = \frac{r}{2}$$

A spherical mirror will form an approximate point image of a point object. The reason the image is not a perfect point is a result of **spherical aberration**.

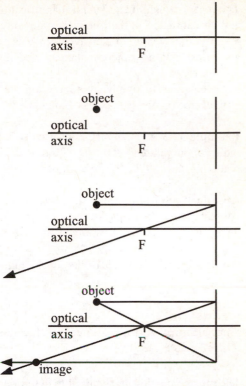

We can locate the image formed by a mirror by using a ray diagram. A ray diagram is a drawing of the mirror and the object used to determine the position of the image. Unless we are interested in spherical aberrations, we only need draw two **principal rays** to determine the location of the image. The procedure to be used is as follows:

1. Draw a line representing the optical axis of the mirror and a second perpendicular line at the position of the mirror. This second line represents the mirror.
2. Mark the position of the focal point on the optical axis.
3. Mark the position of the object in the diagram.
4. Draw a ray from the object to the mirror, parallel to the optical axis. The reflected ray leaves the mirror and passes through the focal point of the mirror.
5. Draw a second ray from the object to the mirror, passing through the focal point of the mirror. This ray leaves the mirror and travels parallel to the optical axis.
6. The location of the image is the point where the two reflected rays intersect.

If real light rays pass through the image, the image is called a **real image**. If the real light rays do not pass through their intersection point, the image is called a **virtual image**. Consider the ray diagram below, for a mirror with a negative focal length (a convex mirror). Note that the focal point F is on the back side of the mirror because of its negative focal length. Follow the instructions above to see how the ray diagram below was constructed for this mirror. For this mirror, the reflected rays appear to come from a point behind the mirror. Obviously, the rays do not travel through the image point, and the image is a virtual image.

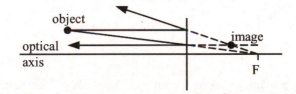

The distance of the object from the mirror, parallel to the optical axis, is called the **object distance** d_o. The distance of the image from the mirror, parallel to the optical axis, is called the **image distance** d_i. The perpendicular distance of the object from the optical axis is called the **object height** h_o and the perpendicular distance of the image from the optical axis is called the **image height** h_i. The **mirror equation** relates the object distance, the image distance, and the focal length of the mirror:

$$\frac{1}{f} = \frac{1}{d_o} + \frac{1}{d_i}$$

The **lateral magnification** m of the mirror is defined as the ratio of the image height to the object height:

$$m = \frac{h_i}{h_o}$$

The lateral magnification is related to the object distance and the image distance:

$$m = \frac{h_i}{h_o} = -\frac{d_i}{d_o}$$

The image will be upright if the lateral magnification is positive and the image will be inverted if the lateral magnification is negative.

The mirror equation and lateral magnification relationship can be used for both concave and convex mirrors and for real and virtual images if the following sign conventions are observed:

- The focal length of concave mirrors is positive and the focal length of convex mirrors is negative.
- Focal lengths, object distances, and image distances are positive in front of the mirror and negative behind the mirror.
- Object and image heights have the same sign if located on the same side of the optical axis and opposite signs if located on opposite sides of the optical axis.

Example 32-3-A: Properties of a convex lens. An object is placed a distance $d_o = 40$ cm from a convex lens with a radius of curvature $r = 50$ cm. Where is the image formed? What is the lateral magnification of the image? Is the image real or virtual? Is the image upright or inverted?

Approach: Since we are dealing with a convex lens, the focal length, which can be calculated from the known radius of curvature, will be negative. Using the mirror equation, we can determine the location of the image. The lateral magnification can be determined using the object and image distances.

Solution: The magnitude of the focal length of the mirror is equal to

$$|f| = \frac{r}{2} = 25 \text{ cm}$$

Since a convex mirror has a negative focal length we conclude that

$$f = -25 \text{ cm}$$

The mirror equation can be used to determine where the image is formed:

$$\frac{1}{f} = \frac{1}{d_o} + \frac{1}{d_i} \quad \Rightarrow \quad d_i = \frac{fd_o}{d_o - f} = \frac{(-25)(40)}{(40)-(-25)} = -15 \text{ cm}$$

Since d_i is negative, the image is located behind the mirror and the image is a virtual image. The lateral magnification of the mirror is equal to

$$m = -\frac{d_i}{d_o} = -\frac{(-15)}{(40)} = 0.38$$

Since the magnification is positive, the image height is positive and the image is upright.

Section 32-4. Index of Refraction

The speed of light in a vacuum, usually denoted as c, is 2.99792458×10^8 m/s ($\sim 3.00 \times 10^8$ m/s). The speed of light in materials is less than the speed of light in vacuum. The ratio of the speed of light in vacuum and the speed of light in a material is called the **index of refraction** of the material, n:

$$n = \frac{c}{v}$$

Because the speed of light in materials is less than the speed of light in vacuum, the refractive indices of materials are greater than one. The index of refraction of a material depends on the wavelength of the light.

Example 32-4-A. Calculating the index of refraction. An optical fiber used for communication purposes carries a light signal a distance of 1.34 km in a time of 6.52 μs. What is the index of refraction of the optical fiber?

Approach: The information provided allows us to determine the speed of light in the fiber. The ratio of the speed of light in vacuum and the speed of light in the fiber allows us to determine the index of refraction.

Solution: The speed of the light through the fiber is equal to

$$v = \frac{d}{t} = \frac{1.34 \times 10^3}{6.52 \times 10^{-6}} = 2.06 \times 10^8 \text{ m/s}$$

The index of refraction can now be calculated:

$$n = \frac{c}{v} = \frac{3.00 \times 10^8}{2.06 \times 10^8} = 1.46$$

Section 32-5. Refraction: Snell's Law

When light is transmitted across a boundary between two materials with different refractive indices, the light changes direction. This phenomenon is called **refraction**. The angle of the refracted light ray with respect to the normal is called the **angle of refraction** θ_2. The angle of the incident light ray with respect to the normal is called the **angle of incidence** θ_1. The relation between these angles is called the **law of refraction** or **Snell's law**:

$$n_1 \sin\theta_1 = n_2 \sin\theta_2$$

where n_1 and n_2 are the indices of refraction of the materials on each side of the boundary.

Example 32-5-A. Refraction of light passing through a block. A rectangular block of material has an index of refraction $n = 1.28$. A light ray, incident from air on the end of the block at an angle $\theta_i = 18°$, refracts through the end and reaches the side of the block, where it refracts again, passing back out of the block (see Figure). What is the angle θ between the direction of the ray exiting the block and its surface?

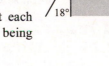

Approach: To solve this problem, we need to apply Snell's law twice (once at each surface). We need to make sure that the proper angles and indices of refraction are being used. The index of refraction of air is assumed to be 1.

Solution: The first step of the solution is to determine the direction of the light ray after it has passed through the first surface. All angles are measured with respect to the normal of the surface. The incident angle at the first surface is the complement of 18°, or 72°. To determine the angle of refraction, we use Snell's law:

$$n_1 \sin\theta_1 = n_2 \sin\theta_2 \quad\Rightarrow$$

$$\theta_2 = \arcsin\left(\frac{n_1}{n_2}\sin\theta_1\right) = \arcsin\left(\frac{(1.00)}{(1.28)}\sin 72°\right) = 48°$$

The Figure on the right can be used to show that the angle of incidence of the light ray at the second interface is $90° - 48° = 42°$. Using Snell's law, the angle of refraction of the ray leaving the block can now be determined:

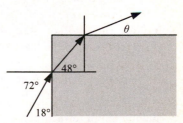

$$n_1 \sin\theta_1 = n_2 \sin\theta_2 \quad\Rightarrow$$

$$\theta_2 = \arcsin\left(\frac{n_1}{n_2}\sin\theta_1\right) = \arcsin\left(\frac{(1.28)}{(1.00)}\sin 42°\right) = 59°$$

The angle θ is the complement of the refracted angle and is thus equal to

$$\theta = 90° - 59° = 31°$$

Section 32-6. Visible Spectrum and Dispersion

Our eyes are sensitive to light with wavelengths between 400 nm and 750 nm. Light that falls between these two wavelengths is part of the **visible spectrum**. Light with wavelengths greater than 750 nm is called **infrared light** and light with wavelengths less than 400 nm is called **ultraviolet light**.

Because the index of refraction of a material depends on the wavelength of light, the direction of a refracted light ray also depends on the wavelength. This is what allows prisms to disperse light of different wavelengths in different directions. This process of separating different wavelengths into different directions is called **dispersion**.

Section 32-7. Total Internal Reflection; Fiber Optics

Consider a light ray traveling inside a material with index of refraction n_1 towards an interface with a material with an index of refraction n_2. Snell's law tells us that

$$n_1 \sin\theta_1 = n_2 \sin\theta_2$$

This relation can be rewritten as

$$\sin\theta_2 = \frac{n_1}{n_2}\sin\theta_1$$

If $n_1 > n_2$ there will be angles θ_1 for which

$$\frac{n_1}{n_2}\sin\theta_1 > 1$$

For these angles, there will be no solutions to Snell's law; the incident ray is totally reflected. The incident angle for which $\sin\theta_2 = 1$ is called the **critical angle** θ_C, the maximum incident angle for which Snell's law has a solution:

$$n_1 \sin\theta_C = n_2 \cdot 1 \quad\Rightarrow\quad \theta_C = \mathrm{asin}\frac{n_2}{n_1}$$

For incident angles greater than the critical angle, all of the light is reflected from the surface back into the material on the incident side of the boundary. This phenomenon is called **total internal reflection**. The transmission of light in optical fibers is based on this phenomenon.

Example 32-7-A. Using internal reflection to determine the index of refraction. A light ray is incident on a right-angle block in air, as shown in the diagram on the right. It internally reflects when it reaches the second surface. What is the minimum index of refraction of the block so that total internal reflection takes place at the second surface?

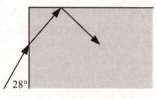

28°

Approach: The angle of incidence on the second surface can be determined using an approach similar to what has been used in Example 31-5-A. As the index of refraction of the material increases, the angle of refraction decreases. As we saw in Example 31-5-A, a decreasing angle of refraction at the first surface will result in an increasing angle of incidence on the second surface. In order to find the minimum index of refraction of the block, we need to determine what index of refraction will result in a critical angle of incidence on the second surface.

Solution: The angle of incidence on the first surface is $\theta_1 = 90° - 28° = 62°$. Using Snell's law we can determine the angle of reflection of the incident light ray:

$$n_1 \sin\theta_1 = n_2 \sin\theta_2 \quad\Rightarrow\quad \sin\theta_2 = \frac{1.00}{n_2}\sin\theta_1$$

At the minimum refractive index, this angle of refraction will result in an angle of incidence at the second surface equal to the critical angle θ_C. The critical angle θ_C satisfies the following relation:

$$\sin\theta_C = \frac{1.00}{n_2}$$

The critical angle θ_C is the complement of the refracted angle at the first surface: $\theta_C = 90° - \theta_2$. The sine of the critical angle can thus be written as

$$\sin\theta_C = \sin\left(90° - \theta_2\right) = \cos\theta_2 = \sqrt{1 - \sin^2\theta_2} = \sqrt{1 - \left(\frac{1.00}{n_2}\right)^2 \sin^2\theta_1}$$

Combining the two expressions for $\sin\theta_C$ we obtain:

$$\sin\theta_C = \frac{1}{n_2} = \sqrt{1-\left(\frac{1}{n_2}\right)^2 \sin^2\theta_1} \quad\Rightarrow\quad \left(\frac{1}{n_2}\right)^2 = 1-\left(\frac{1}{n_2}\right)^2\sin^2\theta_1 \quad\Rightarrow\quad \left(\frac{1}{n_2}\right)^2\left(1+\sin^2\theta_1\right)=1$$

The last equation can be solved for n_2:

$$\left(\frac{1}{n_2}\right)^2 = \frac{1}{\left(1+\sin^2\theta_1\right)} \quad\Rightarrow\quad n_2 = \sqrt{\left(1+\sin^2\theta_1\right)} = 1.33$$

Section 32-8. Refraction at a Spherical Surface

A spherical refractive surface can form an image. If the index of refraction of the material where the point source of light is located is n_1 and the index of refraction on the other side of the refracting surface of radius R is n_2, then the object distance is related to the image distance by

$$\frac{n_1}{d_o} + \frac{n_2}{d_i} = \frac{n_2 - n_1}{R}$$

The sign convention for this image formation system is different than that for the mirror equation:
1. If the surface is convex, its radius of curvature is positive. If the surface is concave, its radius of curvature is negative.
2. The object distance is positive if the object is located on the side from which the incident light is coming; otherwise the object distance is negative.
3. The image distance is positive if the image is located on the side from which the incident light is coming. The image distance is negative if the image is located on the opposite side from which the incident light is coming.

Example 32-8-A. Calculating the required radius of curvature of the cornea. Assume an eye has no lens and the image on the retina at the back of the eye is formed as a result of the refraction that occurs at the curved surface of the cornea. Water, which you can consider the material inside the eye to be, has an index of refraction of $n_w = 1.33$. Assume the eye has a diameter of $D = 3.2$ cm. What must be the radius of curvature of the cornea to form an image on the retina if the object is located $d_o = 50$ cm from the eye?

Approach: Since the image must be formed on the retina, the image distance is equal to the diameter of the eye. We thus know the object and image distances and the indices of refraction of the two regions. The refraction formula, discussed in this section, can be used to determine the required radius of curvature of the cornea.

Solution: Based on the information provided we know that $d_o = 50$ cm, $d_i = D = 3.2$ cm, $n_1 = n_{air} = 1.00$, and $n_2 = n_{water} = 1.33$. The image formation equation can now be used to determine the radius of curvature of the cornea:

$$\frac{n_1}{d_o} + \frac{n_2}{d_i} = \frac{n_2 - n_1}{R} \quad\Rightarrow\quad R = \frac{(n_2-n_1)d_o d_i}{n_1 d_i + n_2 d_o} = \frac{(1.33-1.00)(50)(3.2)}{(1.00)(50)+(1.33)(3.2)} = 0.97\,\text{cm}$$

Practice Quiz

1. Under what conditions is the ray model of light applicable?
    a) All dimensions in the system are small compared to the wavelength of light.
    b) All dimensions in the system are large compared to the wavelength of light.
    c) The light is of a single wavelength.
    d) There are no reflecting surfaces in the system.

2. Under what conditions does a concave mirror form a virtual image?
    a) When the object is closer to the mirror than the radius of curvature of the mirror.
    b) When the object is farther from the mirror than the radius of curvature of the mirror.
    c) When the object is closer to the mirror than the focal length of the mirror.
    d) When the object is farther from the mirror than the focal length of the mirror.

3. Which of the following best describes the reason a prism sends different wavelengths of light in a different direction from incident white light?
   a) Total internal reflection.
   b) Diffuse reflection.
   c) Critical angle.
   d) Dispersion.

4. Which of the following wavelengths of light is not visible to human eyes?
   a) 440 nm.
   b) 20 μm.
   c) 0.060 μm.
   d) 1.8×10^{-6} ft.

5. In the diagram to the right, a light ray enters a material with an index of refraction n_2 from a material with an index of refraction n_1. Which statement is necessarily true about the refractive indices?
   a) $n_2 > n_1$.
   b) $n_1 > n_2$.
   c) $n_2 < 1$ and $n_1 > 1$.
   d) $n_1 < 1$ and $n_2 > 1$.

6. Why can't you see your image in a reflection from a piece of paper?
   a) The paper does not reflect enough light.
   b) The reflection is a specular reflection.
   c) The reflection is a diffuse reflection.
   d) The dispersion that occurs separates all of the different colored images, making it impossible to see the image.

7. When the sun is close to the horizon, what does refraction of light in the atmosphere do to the position of the sun? (The index of refraction of the air is larger, the nearer to the surface of the Earth you are.)
   a) The position of the image of the Sun is closer to the horizon than the actual position of the Sun.
   b) The position of the image of the Sun is farther from the horizon than the actual position of the Sun.
   c) The position of the image of the Sun is the same as the actual position of the Sun.
   d) The answer depends on whether it is morning or evening.

8. In the ray diagram to the right, the vertical line represents a mirror. The paths of two rays from an object are shown. What type of mirror does the vertical line represent?
   a) Concave.
   b) Convex.
   c) Flat.
   d) Broken.

9. What type of image is formed with the mirror described in Question 8 above?
   a) Real and inverted.
   b) Virtual and inverted.
   c) Real and upright.
   d) Virtual and upright.

10. When you look at a pencil in a glass of water it appears bent. Which phenomenon explains why the pencil appears bent?
    a) Reflection.
    b) Refraction.
    c) Dispersion.
    d) Total internal reflection.

11. Determine the speed of light in a material that has a critical angle of 48° in air.

12. A light ray is incident on a material's surface from air at an incident angle of 30°. The direction of the light ray is 10° toward the normal to the surface after it passes through the surface. What is the index of refraction of the material?

13. A convex mirror with a radius of curvature of 50.0 cm has an object located 25.0 cm in front of it. Where is the image of this object formed? What is the lateral magnification of the image? Is the image upright or inverted? Is the image real or virtual?

14. A mirror is used to form an image of an object that is 30.0 cm from the mirror. The image is formed 40.0 cm in front of the mirror and is inverted. What is the radius of curvature of the mirror? Is it a concave mirror or a convex mirror?

15. A glass marble of diameter 0.75 in has a small seed of width 0.86 mm at its center. The glass the marble is made of has an index of refraction 1.55. If you look at the seed inside the marble, where is its image formed?

Responses to Select End-of-Chapter Questions

1. (a) The Moon would look just like it does now, since the surface is rough. Reflected sunlight is scattered by the surface of the Moon in many directions, making the surface appear white.
    (b) With a polished, mirror-like surface, the Moon would reflect an image of the Sun, the stars, and the Earth. The appearance of the Moon would be different as seen from different locations on the Earth.

7. Yes. When a concave mirror produces a real image of a real object, both d_o and d_i are positive. The magnification equation, $m = -\dfrac{d_i}{d_o}$, results in a negative magnification, which indicates that the image is inverted.

19. No. The refraction of light as it enters the pool will make the object look smaller. See Figure 32-32 and Conceptual Example 32-11.

Solutions to Select End-of-Chapter Problems

1. Because the angle of incidence must equal the angle of reflection, we see from the ray diagrams that the ray that reflects to your eye must be as far below the horizontal line to the reflection point on the mirror as the top is above the line, regardless of your position.

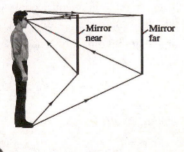

7. See the "top view" ray diagram.

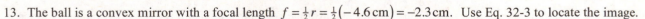

13. The ball is a convex mirror with a focal length $f = \frac{1}{2}r = \frac{1}{2}(-4.6\,\text{cm}) = -2.3\,\text{cm}$. Use Eq. 32-3 to locate the image.

$$\frac{1}{d_o} + \frac{1}{d_i} = \frac{1}{f} \quad \rightarrow \quad d_i = \frac{d_o f}{d_o - f} = \frac{(25.0\,\text{cm})(-2.3\,\text{cm})}{25.0\,\text{cm} - (-2.3\,\text{cm})} = -2.106\,\text{cm} \approx -2.1\,\text{cm}$$

The image is $\boxed{\text{2.1 cm behind the surface of the ball, virtual, and upright}}$. Note that the magnification is

$$m = -\frac{d_i}{d_o} = \frac{-(-2.106\,\text{cm})}{(25.0\,\text{cm})} = +0.084.$$

19. Take the object distance to be ∞, and use Eq. 32-3. Note that the image distance is negative since the image is behind the mirror.

$$\frac{1}{d_o}+\frac{1}{d_i}=\frac{1}{f} \;\rightarrow\; \frac{1}{\infty}+\frac{1}{d_i}=\frac{1}{f} \;\rightarrow\; f=d_i=-16.0\,\text{cm} \;\rightarrow\; r=2f=\boxed{-32.0\,\text{cm}}$$

Because the focal length is negative, the mirror is $\boxed{\text{convex.}}$

25. (a) To produce a smaller image located behind the surface of the mirror requires a $\boxed{\text{convex mirror.}}$
    (b) Find the image distance from the magnification.

$$m=\frac{h_i}{h_o}=\frac{-d_i}{d_o} \;\rightarrow\; d_i=-\frac{d_o h_i}{h_o}=-\frac{(26\,\text{cm})(3.5\,\text{cm})}{(4.5\,\text{cm})}=-20.2\,\text{cm}\approx\boxed{-20\,\text{cm}}\;(2\text{ sig. fig.})$$

As expected, $d_i<0$. The image is located $\boxed{20\text{ cm behind the surface.}}$
    (c) Find the focal length from Eq. 32.3.

$$\frac{1}{d_o}+\frac{1}{d_i}=\frac{1}{f} \;\rightarrow\; f=\frac{d_o d_i}{d_o+d_i}=\frac{(26\,\text{cm})(-20.2\,\text{cm})}{(26\,\text{cm})+(-20.2\,\text{cm})}=-90.55\,\text{cm}\approx\boxed{-91\,\text{cm}}$$

    (d) The radius of curvature is twice the focal length.

$$r=2f=2(-90.55\,\text{cm})=-181.1\,\text{cm}\approx\boxed{-180\,\text{cm}}$$

31. The lateral magnification of an image equals the height of the image divided by the height of the object. This can be written in terms of the image distance and focal length with Eqs. 32-2 and 32-3.

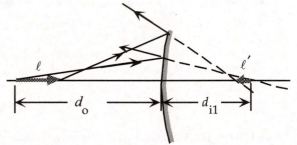

$$\frac{1}{f}=\frac{1}{d_i}+\frac{1}{d_o} \;\rightarrow\; d_i=\left(\frac{f d_o}{d_o-f}\right)$$

$$m=\frac{-d_i}{d_o}=-\frac{f}{d_o-f}$$

The longitudinal magnification will be the difference in image distances of the two ends of the object divided by the length of the image. Call the far tip of the wire object 1 with object distance d_o. The close end of the wire will be object 2 with object distance $d_o-\ell$. Using Eq. 32-2 we can find the image distances for both ends.

$$\frac{1}{f}=\frac{1}{d_o}+\frac{1}{d_{i1}} \;\rightarrow\; d_{i1}=\frac{d_o f}{d_o-f} \;;\; \frac{1}{f}=\frac{1}{d_o-\ell}+\frac{1}{d_{i2}} \;\rightarrow\; d_{i2}=\frac{(d_o-\ell)f}{d_o-\ell-f}$$

Taking the difference in image distances and dividing by the object length gives the longitudinal magnification.

$$m_\ell=\frac{d_{i1}-d_{i2}}{\ell}=\frac{1}{\ell}\left(\frac{d_o f}{d_o-f}-\frac{(d_o-\ell)f}{d_o-\ell-f}\right)=\frac{d_o f(d_o-\ell-f)-(d_o-f)(d_o-\ell)f}{\ell(d_o-f)(d_o-\ell-f)}$$

$$=\frac{-f^2}{(d_o-f)(d_o-\ell-f)}$$

Set $\ell\ll d_o$, so that the ℓ drops out of the second factor of the denominator. Then rewrite the equation in terms of the lateral magnification, using the expression derived at the beginning of the problem.

$$m_\ell=\frac{-f^2}{(d_o-f)^2}=-\left[\frac{f}{(d_o-f)}\right]^2=\boxed{-m^2}$$

The negative sign indicates that the $\boxed{\text{image is reversed front to back}}$, as shown in the diagram.

37. The length in space of a burst is the speed of light times the elapsed time.

$$d = ct = \left(3.00 \times 10^8 \text{ m/s}\right)\left(10^{-8}\text{s}\right) = \boxed{3\,\text{m}}$$

43. The beam forms the hypotenuse of two right triangles as it passes through the plastic and then the glass. The upper angle of the triangle is the angle of refraction in that medium. Note that the sum of the opposite sides is equal to the displacement D. First, we calculate the angles of refraction in each medium using Snell's Law (Eq. 32-5).

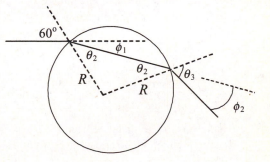

$$\sin 45 = n_1 \sin \theta_1 = n_2 \sin \theta_2$$

$$\theta_1 = \sin^{-1}\left(\frac{\sin 45}{n_1}\right) = \sin^{-1}\left(\frac{\sin 45}{1.62}\right) = 25.88°$$

$$\theta_2 = \sin^{-1}\left(\frac{\sin 45}{n_2}\right) = \sin^{-1}\left(\frac{\sin 45}{1.47}\right) = 28.75°$$

We then use the trigonometric identity for tangent to calculate the two opposite sides, and sum to get the displacement.

$$D = D_1 + D_2 = h_1 \tan \theta_1 + h_1 \tan \theta_1 = \left(2.0\,\text{cm}\right)\tan 25.88° + \left(3.0\,\text{cm}\right)\tan 28.75° = \boxed{2.6\,\text{cm}}$$

49. Since the angle of incidence at the base of the prism is $0°$, the rays are undeflected there. The angle of incidence at the upper face of the prism is $30°$. Use Snell's law to calculate the angle of refraction as the light exits the prism.

$$n_1 \sin \theta_1 = \sin \theta_r \rightarrow \theta_r = \sin^{-1}\left(1.52 \sin 30°\right) = 49.46°$$

From the diagram, note that a normal to either top surface makes a $30°$ angle from the vertical. Subtracting $30°$ from the refracted angle will give the angle of the beam with respect to the vertical. By symmetry, the angle ϕ is twice the angle of the refracted beam from the vertical.

$$\phi = 2\left(\theta_r - 30°\right) = 2\left(49.46° - 30°\right) = \boxed{38.9°}$$

55. At the first surface, the angle of incidence $\theta_1 = 60°$ from air $\left(n_1 = 1.000\right)$ and the angle of refraction θ_2 into water $\left(n_2 = n\right)$ is found using Snell's law.

$$n_1 \sin \theta_1 = n_2 \sin \theta_2 \rightarrow$$
$$\left(1.000\right)\sin 60° = \left(n\right)\sin \theta_2 \rightarrow$$
$$\theta_2 = \sin^{-1}\left(\frac{\sin 60°}{n}\right)$$

Note that at this surface the ray has been deflected from its initial direction by angle $\phi_1 = 60° - \theta_2$.

From the figure we see that the triangle that is interior to the drop is an isosceles triangle, so the angle of incidence from water $\left(n_2 = n\right)$ at the second surface is θ_2 and angle of refraction is θ_3 into air $\left(n_3 = 1.000\right)$. This relationship is identical to the relationship at the first surface, showing that the refracted angle as the light exits the drop is again $60°$.

The angle of refraction θ_3 into the air can thus be determined:

$$n_2 \sin\theta_2 = n_3 \sin\theta_3 \rightarrow (n)\sin\theta_2 = (1.000)\sin\theta_3 \rightarrow \sin\theta_3 = n\sin\theta_2 \rightarrow$$

$$\sin\theta_3 = n\left(\frac{\sin 60°}{n}\right) = \sin 60° \rightarrow \theta_3 = 60°$$

Note that at this surface the ray has been deflected from its initial direction by the angle $\phi_2 = \theta_3 - \theta_2 = 60° - \theta_2$. The total deflection of the ray is equal to the sum of the deflections at each surface.

$$\phi = \phi_1 + \phi_2 = (60° - \theta_2) + (60° - \theta_2) = 120° - 2\theta_2 = 120° - 2\sin^{-1}\left(\frac{\sin 60°}{n}\right)$$

Inserting the indices of refraction for the two colors and subtracting the angles gives the difference in total deflection.

$$\Delta\phi = \phi_{\text{violet}} - \phi_{\text{red}} = \left\{120° - 2\sin^{-1}\left[\frac{\sin 60°}{n_{\text{violet}}}\right]\right\} - \left\{120° - 2\sin^{-1}\left[\frac{\sin 60°}{n_{\text{red}}}\right]\right\}$$

$$= 2\left\{\sin^{-1}\left[\frac{\sin 60°}{n_{\text{red}}}\right] - \sin^{-1}\left[\frac{\sin 60°}{n_{\text{violet}}}\right]\right\} = 2\left\{\sin^{-1}\left[\frac{\sin 60°}{1.330}\right] - \sin^{-1}\left[\frac{\sin 60°}{1.341}\right]\right\} = \boxed{0.80°}$$

61. We find the angle of incidence from the distances.

$$\tan\theta_1 = \frac{l}{h} = \frac{(7.6\,\text{cm})}{(8.0\,\text{cm})} = 0.95 \rightarrow \theta_1 = 43.53°$$

The relationship for the maximum incident angle for refraction from liquid into air gives this.

$$n_{\text{liquid}} \sin\theta_1 = n_{\text{air}} \sin\theta_2 \rightarrow n_{\text{liquid}} \sin\theta_{1\max} = (1.00)\sin 90° \rightarrow \sin\theta_{1\max} = \frac{1}{n_{\text{liquid}}}$$

Thus we have the following.

$$\sin\theta_1 \geq \sin\theta_{1\max} = \frac{1}{n_{\text{liquid}}} \rightarrow \sin 43.53° = 0.6887 \geq \frac{1}{n_{\text{liquid}}} \rightarrow \boxed{n_{\text{liquid}} \geq 1.5}$$

67. We find the location of the image of a point on the bottom from the refraction from water to glass, using Eq. 32-8, with $R = \infty$.

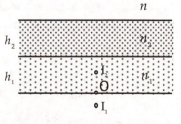

$$\frac{n_1}{d_o} + \frac{n_2}{d_i} = \frac{n_2 - n_1}{R} = 0 \rightarrow$$

$$d_i = -\frac{n_2 d_o}{n_1} = -\frac{1.58(12.0\,\text{cm})}{1.33} = -14.26\,\text{cm}$$

Using this image distance from the top surface as the object for the refraction from glass to air gives the final image location, which is the apparent depth of the water.

$$\frac{n_2}{d_{o2}} + \frac{n_3}{d_{i2}} = \frac{n_3 - n_2}{R} = 0 \rightarrow d_{i2} = -\frac{n_3 d_{o2}}{n_2} = -\frac{1.00(13.0\,\text{cm} + 14.26\,\text{cm})}{1.58} = -17.25\,\text{cm}$$

Thus the bottom appears to be $\boxed{17.3\,\text{cm below the surface of the glass}}$. In reality it is 25 cm.

73. (a) The first image seen will be due to a single reflection off the front glass. This image will be equally far behind the mirror as you are in front of the mirror.

$$D_1 = 2 \times 1.5 \text{ m} = \boxed{3.0 \text{ m}}$$

The second image seen will be the image reflected once off the front mirror and once off the back mirror. As seen in the diagram, this image will appear to be twice the distance between the mirrors.

$$D_2 = 1.5 \text{ m} + 2.2 \text{ m} + (2.2 \text{ m} - 1.5 \text{ m}) = 2 \times 2.2 \text{ m} = \boxed{4.4 \text{ m}}$$

The third image seen will be the image reflected off the front mirror, the back mirror, and off the front mirror again. As seen in the diagram this image distance will be the sum of twice your distance to the mirror and twice the distance between the mirrors.

$$D_3 = 1.5 \text{ m} + 2.2 \text{ m} + 2.2 \text{ m} + 1.5 \text{ m} = 2 \times 1.5 \text{ m} + 2 \times 2.2 \text{ m} = \boxed{7.4 \text{ m}}$$

The actual person is to the far right in the diagram.

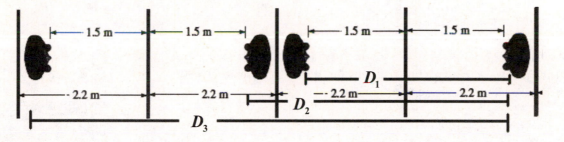

(b) We see from the diagram that the first image is facing $\boxed{\text{toward you}}$; the second image is facing $\boxed{\text{away from you}}$; and the third image is facing $\boxed{\text{toward you.}}$

79. The total deviation of the beam is the sum of the deviations at each surface. The deviation at the first surface is the refracted angle θ_2 subtracted from the incident angle θ_1. The deviation at the second surface is the incident angle θ_3 subtracted from the refracted angle θ_4. This gives the total deviation.

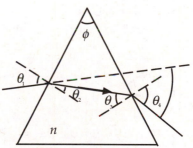

$$\delta = \delta_1 + \delta_2 = \theta_1 - \theta_2 + \theta_4 - \theta_3$$

We will express all of the angles in terms of θ_2. To minimize the deviation, we will take the derivative of the deviation with respect to θ_2, and then set that derivative equal to zero. Use Snell's law at the first surface to write the incident angle in terms of the refracted angle.

$$\sin \theta_1 = n \sin \theta_2 \;\; \rightarrow \;\; \theta_1 = \sin^{-1}\left(n \sin \theta_2\right)$$

The angle of incidence at the second surface is found using complementary angles, such that the sum of the refracted angle from the first surface and the incident angle at the second surface must equal the apex angle.

$$\phi = \theta_2 + \theta_3 \;\; \rightarrow \;\; \theta_3 = \phi - \theta_2$$

The refracted angle from the second surface is again found using Snell's law with the deviation in angle equal to the difference between the incident and refracted angles at the second surface.

$$n \sin \theta_3 = \sin \theta_4 \;\; \rightarrow \;\; \theta_4 = \sin^{-1}\left(n \sin \theta_3\right) = \sin^{-1}\left(n \sin\left(\phi - \theta_2\right)\right)$$

Inserting each of the angles into the deviation and setting the derivative equal to zero allows us to solve for the angle at which the deviation is a minimum.

$$\delta = \sin^{-1}\left(n\sin\theta_2\right) - \theta_2 + \sin^{-1}\left(n\sin(\phi-\theta_2)\right) - (\phi-\theta_2)$$

$$= \sin^{-1}\left(n\sin\theta_2\right) + \sin^{-1}\left(n\sin(\phi-\theta_2)\right) - \phi$$

$$\frac{d\delta}{d\theta_2} = \frac{n\cos\theta_2}{\sqrt{1-n^2\sin^2\theta_2}} - \frac{n\cos(\phi-\theta_2)}{\sqrt{1-n^2\sin^2(\phi-\theta_2)}} = 0 \quad\rightarrow\quad \theta_2 = \phi-\theta_2 \quad\rightarrow\quad \theta_2 = \theta_3 = \tfrac{1}{2}\phi$$

In order for $\theta_2 = \theta_3$, the ray must pass through the prism horizontally, which is perpendicular to the bisector of the apex angle ϕ. Set $\theta_2 = \tfrac{1}{2}\phi$ in the deviation equation (for the minimum deviation, δ_m) and solve for the index of refraction.

$$\delta_m = \sin^{-1}\left(n\sin\theta_2\right) + \sin^{-1}\left(n\sin(\phi-\theta_2)\right) - \phi$$

$$= \sin^{-1}\left(n\sin\tfrac{1}{2}\phi\right) + \sin^{-1}\left(n\sin\tfrac{1}{2}\phi\right) - \phi = 2\sin^{-1}\left(n\sin\tfrac{1}{2}\phi\right) - \phi$$

$$\rightarrow \boxed{n = \frac{\sin\left(\tfrac{1}{2}(\delta_m+\phi)\right)}{\sin\tfrac{1}{2}\phi}}$$

85. If total internal reflection fails at all, it fails for $\alpha \approx 90°$. Assume $\alpha = 90°$ and use Snell's law to determine the maximum β.

$$n_2\sin\beta = n_1\sin\alpha = n_1\sin 90° = n_1 \quad\rightarrow\quad \sin\beta = \frac{n_1}{n_2}$$

Snell's law can again be used to determine the angle δ for which light (if not totally internally reflected) would exit the top surface, using the relationship $\beta + \gamma = 90°$ since they form two angles of a right triangle.

$$n_1\sin\delta = n_2\sin\gamma = n_2\sin(90°-\beta) = n_2\cos\beta \quad\rightarrow\quad \sin\delta = \frac{n_2}{n_1}\cos\beta$$

Using the trigonometric relationship $\cos\beta = \sqrt{1-\sin^2\beta}$ we can solve for the exiting angle in terms of the indices of refraction.

$$\sin\delta = \frac{n_2}{n_1}\sqrt{1-\sin^2\beta} = \frac{n_2}{n_1}\sqrt{1-\left(\frac{n_1}{n_2}\right)^2}$$

Insert the values for the indices ($n_1 = 1.00$ and $n_2 = 1.51$) to determine the sine of the exit angle.

$$\sin\delta = \frac{1.51}{1.00}\sqrt{1-\left(\frac{1.00}{1.51}\right)^2} = 1.13$$

Since the sine function has a maximum value of 1, the light totally internally reflects at the glass–air interface for any incident angle of light.

If the glass is immersed in water, then $n_1 = 1.33$ and $n_2 = 1.51$.

$$\sin\delta = \frac{1.51}{1.33}\sqrt{1-\left(\frac{1.33}{1.51}\right)^2} = 0.538 \quad\rightarrow\quad \delta = \sin^{-1}0.538 = 32.5°$$

Light entering the glass from water at $90°$ can escape out the top at $32.5°$, therefore total internal reflection only occurs for incident angles $\boxed{\leq 32.5°}$.

Chapter 33: Lenses and Optical Instruments

Chapter Overview and Objectives

This chapter introduces the optics of thin lenses. Procedures on how to locate the position of images formed by thin lenses and the properties of these images are discussed. The various properties of optical instruments, including the human eye, are described.

After completing this chapter you should:
- Know what the focal point of a lens is.
- Know how the focal length of a lens is defined.
- Know how to construct ray diagrams to locate images formed by lenses.
- Know how to use the lens equation to locate images formed by lenses.
- Know how to determine the lateral magnification of an image.
- Know how to use the lensmaker's equation.
- Know how various optical instruments form images.
- Know that thin lenses are not perfect image-forming systems, but generate aberrations and distortions.

Summary of Equations

Definition of power of a lens:

$$P = \frac{1}{f}$$

(Section 33-1)

Relationship between image and object positions:

$$\frac{1}{f} = \frac{1}{d_\text{o}} + \frac{1}{d_\text{i}}$$

(Section 33-2)

Definition of lateral magnification:

$$m = \frac{h_\text{i}}{h_\text{o}} = -\frac{d_\text{i}}{d_\text{o}}$$

(Section 33-2)

Lensmaker's equation:

$$\frac{1}{f} = (n-1)\left(\frac{1}{R_1} + \frac{1}{R_2}\right)$$

(Section 33-4)

Definition of f-stop number:

$$f\text{-stop} = \frac{f}{D}$$

(Section 33-5)

Definition of angular magnification:

$$M = \frac{\theta'}{\theta}$$

(Section 33-7)

Angular magnification with image at an infinite distance:

$$M = \frac{N}{f}$$

(Section 33-7)

Angular magnification with image at the near point:

$$M = \frac{N}{f} + 1$$

(Section 33-7)

Angular magnification of telescope:

$$M \approx -\frac{f_\text{o}}{f_\text{e}}$$

(Section 33-8)

Angular magnification of a compound microscope:

$$M \approx \frac{N}{f_\text{e}} \frac{\ell - f_\text{e}}{d_\text{o}} \approx \frac{N\ell}{f_\text{e} f_\text{o}}$$

(Section 33-9)

Chapter Summary

Section 33-1. Thin Lenses; Ray Tracing

A lens has two surfaces that form images by refracting light. A lens is considered a thin lens if its thickness is much smaller than the radii of curvature of the refracting surfaces.

A lens can be characterized by its focal length. The **focal length** f is the distance from the center of the lens to the point at which light rays, traveling initially parallel to the optical axis of the lens, converge after passing through the lens. This point is called the **focal point F**. The reciprocal of the focal length is called the **power P** of the lens:

$$P = \frac{1}{f}$$

The unit of power of a lens is the diopter (D): 1 diopter = 1 m^{-1}.

If the lens has a positive focal length, it is called a **converging lens** because it causes the light rays to turn toward each other. If the lens has a negative focal length, it is called a **diverging lens** because it causes the light rays to turn away from each other.

A **ray diagram** can be used to determine the location of an image for a given object location. To draw a ray diagram we follow the following procedure:

1. Draw a line representing the optical axis of the lens and a second perpendicular line at the position of the lens to represent the lens.

2. Mark the position of the two focal points on the optical axis, each a focal length away from the lens.

3. Mark the position of the object in the diagram.

4. Draw a ray from the object to the lens, parallel to the optical axis. This ray exits the lens and passes through the focal point F' of the lens.

5. Draw a second ray, traveling from the object through the center of the lens and continuing in the same direction.

6. Draw a third ray from the object to the lens, passing through the focal point F of the lens. This ray leaves the lens and continues parallel to the optical axis.

7. The image is the point where the rays intersect.

If the real light rays pass through the image point, the image is called a **real image**. If the real light rays do not pass through the image point, the image is called a **virtual image**.

Consider the ray diagram on the right for a lens with a negative focal length. Note that the focal point F' is on the same side of the lens as the object because of the negative focal length. Follow the instructions above to see how the ray diagram on the right was constructed for a negative focal length lens.

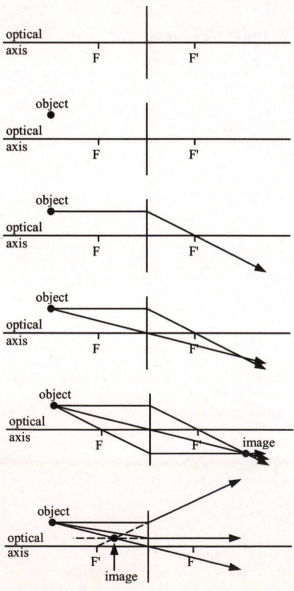

Section 33-2. The Thin Lens Equation; Magnification

There is a simple relationship between the position of the object along the optical axis, d_o, the position of the image along the optical axis, d_i, and the focal length, f, of a thin lens. This relationship is called the **thin lens equation**:

$$\frac{1}{f} = \frac{1}{d_o} + \frac{1}{d_i}$$

There is also a simple relationship between the position perpendicular to the optical axis for the object, h_o, the position perpendicular to the axis for the image, h_i, and their corresponding positions along the optical axis:

$$\frac{h_i}{h_o} = -\frac{d_i}{d_o}$$

The ratio of the positions perpendicular to the axis is called the lateral magnification m:

$$m = \frac{h_i}{h_o} = -\frac{d_i}{d_o}$$

All of these equations apply to all situations involving lenses if the following sign conventions are used:
1. The focal length of a converging lens is positive and the focal length of a diverging lens is negative.
2. The object distance is positive if the object is on the side of the lens from which the light is approaching the lens.
3. The image distance is positive if the image is on the side of the lens from which the light is leaving the lens.
4. The image height is positive if the image is upright and the image height is negative if the image is inverted.

Example 33-2-A. The image of an object created with a converging lens. An object is placed a distance of $d_o = 36.0$ cm from a lens with a focal length $f = 20.0$ cm. Where is the image formed? What is the lateral magnification of the image? Is the image upright or inverted? Is the image real or virtual?

Approach: In this problem we are provided with the focal length of the lens and the object distance. Using the lens equation, we can use this information to determine the image distance. Using the object and image distances, we can determine the lateral magnification m. If $m > 0$ the image is upright and virtual; if $m < 0$ the image is inverted and real.

Solution: The lens equation can be used to determine the image distance:

$$\frac{1}{f} = \frac{1}{d_o} + \frac{1}{d_i} \quad \Rightarrow \quad d_i = \frac{f d_o}{d_o - f} = \frac{(20.0)(36.0)}{36.0 - 20.0} = 45.0\,\text{cm}$$

The lateral magnification of the image is

$$m = -\frac{d_i}{d_o} = -\frac{45.0}{36.0} = -1.25$$

Since the lateral magnification is negative, the image is inverted and real.

Example 33-2-B. The image of an object created with a diverging lens. A lens of focal length $f = -12.5$ cm is used to form the image of an object that is $d_o = 40.0$ cm from the lens. Where is the image located? What is the lateral magnification of the image? Is the image upright or inverted? Is the image virtual or real?

Approach: In this problem we are provided with the focal length of the lens and the object distance. Using the lens equation, we can use this information to determine the image distance. Using the object and image distances, we can determine the lateral magnification m. If $m > 0$ the image is upright and virtual; if $m < 0$ the image is inverted and real.

Solution: The lens equation can be used to determine the image distance:

$$\frac{1}{f} = \frac{1}{d_o} + \frac{1}{d_i} \quad \Rightarrow \quad d_i = \frac{f d_o}{d_o - f} = \frac{(-12.5)(40.0)}{40.0 - (-12.5)} = -9.52\,\text{cm}$$

The lateral magnification of the image is

$$m = -\frac{d_i}{d_o} = -\frac{-9.52}{40.0} = +0.238$$

Since the lateral magnification is positive, the image is upright and virtual.

Section 33-3. Combinations of Lenses

If there is more than one lens in an optic system, the object for each succeeding lens is the image of the previous lens.

Example 33-3-A. Image formation by a system of two lenses. A lens with a focal length $f_1 = +10.0$ cm is placed 20.0 cm from a lens with a focal length $f_2 = -25.0$ cm. An object is placed 12.5 cm from the converging lens. Where is the final image of the two-lens system formed? What is the lateral magnification of the final image? Is the final image upright or inverted? Is the final image real or virtual?

Approach: In order to determine the location of the image of the two-lens system we have to determine the imaging properties of each of the individual lenses. For lens 1 we can determine the location of the image using the information provided. Since the relative position of the two lenses is known, we can use the image distance for lens 1 to calculate the corresponding object distance for lens 2, and thus the location of the image of lens 2. By comparing the location of the object in front of lens 1 and the image of lens 2 we can answer the questions raised. Care must be taken to ensure that the signs of the various distances are correct.

Solution: The object is placed a distance $d_o = 12.5$ cm in front of lens 1. The location of the image of this object as a result of lens 1 can be determined using the lens equation:

$$\frac{1}{f_1} = \frac{1}{d_{o,1}} + \frac{1}{d_{i,1}} \quad \Rightarrow \quad d_{i,1} = \frac{f_1 d_{o,1}}{d_{o,1} - f_1} = \frac{(10.0)(12.5)}{12.5 - 10.0} = 50.0 \, \text{cm}$$

The lateral magnification of the lens 1 is equal to

$$m_1 = -\frac{d_{i,1}}{d_{o,1}} = -\frac{50.0}{12.5} = -4.00$$

Since the magnification is negative, the image of lens 1 will be inverted, as shown in the Figure below. As can be seen in the Figure, the location of the image generated by lens 1 creates an object for lens 2 that is located 30 cm past lens 2. The corresponding object distance is thus negative: $d_{o,2} = -30$ cm. The location of the image generated by lens 2 can be found by using the lens equation:

$$\frac{1}{f_2} = \frac{1}{d_{o,2}} + \frac{1}{d_{i,2}} \quad \Rightarrow \quad d_{i,2} = \frac{f_2 d_{o,2}}{d_{o,2} - f_2} = \frac{(-25.0)(-30.0)}{-30.0 - (-25.0)} = -150 \, \text{cm}$$

Since the image distance of lens 2 is negative, the image will be located towards the left of lens 1 (see Figure). The magnification of lens 2 is equal to

$$m_2 = -\frac{d_{i,2}}{d_{o,2}} = -\frac{(-150)}{(-30)} = -5$$

Since the magnification is negative, the image of lens 2 is inverted.

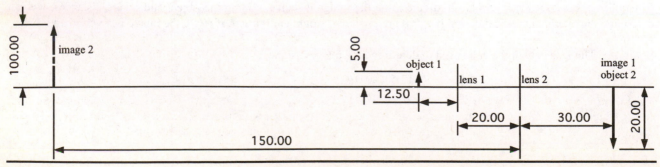

The overall magnification of the two-lens system is the product of the two individual magnifications:

$$m_{total} = m_1 m_2 = (-4.00)(-5.00) = 20.0$$

The overall magnification is positive and the final image is upright. Since the final image distance is negative, the image is virtual.

Section 33-4. Lensmaker's Equation

A thin lens is characterized by the radii of curvature of the front and the back surface. By adjusting these radii, the focal length of the lens can be adjusted. The focal length f of the thin lens in air or vacuum is given by the **lensmaker's equation**:

$$\frac{1}{f} = (n-1)\left(\frac{1}{R_1} + \frac{1}{R_2}\right)$$

where n is the index of refraction of the lens material, R_1 is the radius of curvature of the first lens surface, and R_2 is the radius of curvature of the second lens surface. A convex surface has a positive radius of curvature; a concave surface has a negative radius of curvature.

Example 33-4-A. Designing a lens with a focal length of +25.6 cm. A lens with a focal length of $f_f = +25.6$ cm is needed. An existing lens has one surface that is convex with a radius of curvature of $R_1 = 40.8$ cm and a second surface that is flat. Currently, the lens has a focal length of $f_i = 80.0$ cm. To what radius of curvature must the convex surface be ground to give the lens the necessary focal length? If instead of the convex surface being ground, the flat surface is ground, what radius of curvature should it be given to give the desired focal length?

Approach: The problem provides us with information about the properties of the existing lens. Based on the known focal length and the radii, the index of refraction of the lens material can be determined. When one or both surfaces of the lens are machined, we can adjust the focal length of the lens.

Solution: The refractive index n of the material of the lens can be determined using the lensmaker's equation and the information provided in the problem:

$$\frac{1}{f} = (n-1)\left(\frac{1}{R_1} + \frac{1}{R_2}\right) \quad \Rightarrow \quad n = \frac{1}{f}\left(\frac{1}{R_1} + \frac{1}{R_2}\right)^{-1} + 1 = \frac{1}{80.0}\left(\frac{1}{40.8} + 0\right)^{-1} + 1 = 1.51$$

Using this index of refraction we can now determine what value of R_1 will result in a focal length of $f_f = 25.6$ cm:

$$\frac{1}{f_f} = (n-1)\left(\frac{1}{R_1} + \frac{1}{R_2}\right) \quad \Rightarrow \quad R_1 = \left(\frac{1}{f_f(n-1)} - \frac{1}{R_2}\right)^{-1} = \left(\frac{1}{(25.6)(1.51-1)} - 0\right)^{-1} = 13.1\,\text{cm}$$

If we grind the flat surface while keeping the radius of curvature of surface 1 constant at $R_1 = 40.8$ cm, we can use a similar procedure to determine the required radius of curvature R_2:

$$R_2 = \left(\frac{1}{f_f(n-1)} - \frac{1}{R_1}\right)^{-1} = \left(\frac{1}{(25.6)(1.51-1)} - \frac{1}{40.8}\right)^{-1} = 19.2\,\text{cm}$$

Section 33-5. Cameras: Film and Digital

A camera is a device that has an optical system designed to form real images on either photographic film or an electronic detection system. When a camera uses photographic film, light-sensitive elements on the film change depending on the amount and color of the light that strikes it. During the development process, chemical reactions are used to create a negative of the image that is no longer light sensitive. The negative can be used to expose light-sensitive paper which can be processed to create photographic prints of the image. When a camera uses an electronic detection system, the incident light strikes the pixels of a CCD sensor, which converts the information contained in the light (intensity and color) into digital information that can be recorded. The quality of the image depends on the number of pixels and their sensitivity.

There are three main adjustments that can be made on most high-quality cameras:

1. **Shutter speed.** The shutter speed of a camera is the length of time the film or sensor is exposed to light passing through the lenses of the camera. The shutter speed is set to allow the correct amount of light to reach the film or sensor. In addition, short shutter speeds prevents moving subjects from forming blurred images.

2. **Aperture.** The aperture of the lens is the effective area of the lens that is allowed to collect light. It is specified in terms of a number called the *f***-stop number**. The *f*-stop number is defined to be

$$f\text{-stop} = \frac{f}{D}$$

where *f* is the focal length of the lens and *D* is the diameter of the opening. The aperture is adjusted along with the shutter speed to control the amount of light reaching the film or sensor. It is also adjusted to control the **depth of field** of the camera. The depth of field is the range of object distances that form a sharp image on the film or sensor. As the aperture decreases, the depth of field increases. For very small apertures, diffraction decreases the sharpness of the image.

3. **Focus.** The focusing controls of a camera allow the user to ensure that the desired objects form a sharp image. The film or sensor must be at the correct image distance for the given object distance. Good quality cameras allow a means of adjusting the optical elements to ensure that the image is placed on the film or the sensor.

Section 33-6. The Human Eye; Corrective Lenses

The human eye forms a real image on the retina. Light enters the eye through the **pupil** in the **iris**. The iris adjusts its size automatically to control the intensity of the light entering the eye. The light that passes through the pupil is focused on the **retina**. The retina detects the light that falls on it and converts it into signals that can be processed by the brain.

The eye is able to adjust the focal length of its image formation system so that objects located at different distances from the eye can form a sharp image on the retina. This is called **accommodation**. The eye's accommodation ability is measured by determining the closest object for which the eye can form a sharp image on the retina, called the **near point**, and the farthest object for which the eye can form a sharp image on the retina, called the **far point**. A near point of 25 cm and a far point that is at infinity are considered normal. By adding corrective lenses in front of the eye, corrections can be made to the near and far points of the eye.

If the far point is not an infinite distance away, the eye cannot form a sharp image of distant objects on the retina. This condition is called **nearsightedness** or **myopia**. A negative focal length lens is needed to correct for nearsightedness.

If the near point is farther from the eye than 25 cm, the eye cannot form a sharp image of objects that are closer than the far point. This condition is called **farsightedness** or **hyperopia**. Everyday close-up activities, such as reading normal print sizes, are hindered by this condition and it is desirable to correct this condition. A positive focal length lens is required to correct this condition.

Astigmatism is a condition caused by a lens or cornea that is not spherical in shape, but has different radii of curvature in different planes passing perpendicularly through the surface. This implies that light rays lying in different planes have different focal lengths and, therefore, different image positions. A sharp image cannot be formed on the retina. A corrective lens for astigmatism is aspherical.

Section 33-7. Magnifying Glass

A magnifying glass or simple magnifier is a single simple convex lens used to view objects so that an enlarged upright image of the object can be viewed by the eye. The important quantity that measures how enlarged an image looks to the eye is the angular magnification. The angular size θ of an object or image to the eye will be maximum when the object is located at the near point of the eye

(= 25 cm for normal eye)

$$\theta = \arctan\frac{h}{N} \approx \frac{h}{N}$$

where *h* is the actual height of the object or image and *N* is the distance from the eye to the object or image when it is located at the near point of the eye.

The angular magnification M is defined as the ratio of angular sizes of the object without the magnifying glass, θ, and the angular size of the object with the magnifying glass, θ':

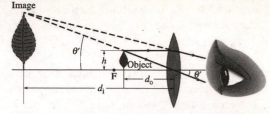

$$M = \frac{\theta'}{\theta}$$

The angular size of the object with the magnifying glass is h/d_o (see Figure). If the eye is relaxed when the magnifying glass is used, the object will be located at the focal point of the magnifying glass and $d_o = f$. The angular magnification can be rewritten as

$$M = \frac{h/f}{h/N} = \frac{N}{f}$$

If the magnifier is used so that the image is formed at the near point of the eye, then the angular magnification is found to be equal to

$$M = \frac{N}{f} + 1$$

The magnification is thus greater when the image is formed at the near point.

Example 33-7-A. Angular magnification of a magnifier. A simple magnifier is used with a power of +5.0 diopters. What is the angular magnification of this magnifier if it forms an image an infinite distance from the eye? What is the angular magnification if it forms an image at the near point of 25 cm?

Approach: The problem provides us with information about the power of the magnifier, and thus its focal length. Based on this information and the known near point of the eye, we can use the relations discussed in this Section to calculate the magnification for the two different image distances.

Solution: The focal length of the magnifier is $1/P = 0.2$ m. When the image is formed an infinite distance from the eye, the magnification is equal to

$$M = \frac{N}{f} = \frac{0.25}{0.2} = 1.25$$

The angular magnification of the magnifier for an image formed at the near point of the eye is equal to

$$M = \frac{N}{f} + 1 = \frac{0.25}{0.2} + 1 = 2.25$$

Section 33-8. Telescopes

A telescope is used to view objects that are far away. There are many different arrangements of optical elements used to make a telescope. To convey the basic principles of the design of a telescope, we consider the design of a **Keplerian** refracting telescope. This telescope contains two converging lenses, one is called the **objective lens** and the other is called the **eyepiece lens**. The objective lens forms a real image of the distant object. The eyepiece lens is used as a simple magnifier to increase the angular size of the real image formed by the objective lens. The approximate angular magnification of a telescope is

$$M \approx -\frac{f_o}{f_e}$$

where f_o is the focal length of the objective lens and f_e is the focal length of the eyepiece lens. The minus sign indicates that the image is inverted.

In order to view faint stars, the objective lens must be large. Due to the difficulty of grinding large lenses, the largest telescopes use curved mirrors instead of objective lenses. This type of telescope is called a **reflecting telescope**.

A **terrestrial telescope** is a telescope that is designed to view objects on the Earth. It relies on the same principle as the previously discussed telescopes, except that the configuration of the eyepiece is changed in order to produce an upright image.

Example 33-8-A. Designing a Keplerian telescope. A Keplerian telescope has an angular magnification of -100. The distance between the objective lens and the eyepiece lens is 1.60 m. What are the focal lengths of the objective lens and the eyepiece lens?

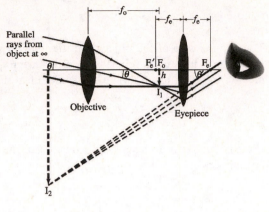

Approach: The required magnification of the telescope tells us something about the ratio of the focal lengths of the objective lens and the eyepiece lens. As can be seen from the Figure on the right, the distance between the objective lens and the eyepiece lens is equal to the sum of the focal lengths of these lenses. These two requirements provide us with two equations with two unknown, which can be solved.

Solution: The relationship between the angular magnification M and the focal lengths of the two lenses requires that

$$M \approx -\frac{f_o}{f_e}$$

The focal length of the objective lens f_o is thus equal to $-Mf_e$.
The distance L is equal to the sum of the two focal lengths:

$$L = f_o + f_e$$

We can eliminate the focal length of the objective lens from this equation and solve for the focal length of the eyepiece:

$$L = -Mf_e + f_e \quad \Rightarrow \quad f_e = \frac{L}{1-M} = \frac{1.60}{1-(-100)} = 1.58 \times 10^{-2} \text{ m}$$

The focal length of the objective lens can now be determined:

$$f_o = -Mf_e = -(-100)(1.58 \times 10^{-2}) = 1.58 \text{ m}$$

Section 33-9. Compound Microscope

A microscope is used to form enlarged images of nearby objects. A compound microscope is made from two or more simple lenses. A two-lens compound microscope has an objective lens and an eyepiece lens like the telescope. Again, the objective lens forms a real image of the object and the eyepiece lens acts as a simple magnifier of the real image. When the final image of a compound microscope is formed an infinite distance from the eye, so that the eye is relaxed when viewing the image, the angular magnification of the image is approximately equal to

$$M = \frac{N}{f_e} \frac{\ell - f_e}{d_o}$$

where N is the near point distance of the eye, ℓ is the distance between the objective lens and the eyepiece lens, f_e is the focal length of the eyepiece lens, and d_o is the distance from the objective lens to the object being viewed. In cases in which the focal lengths of the lenses is small compared to the distance between the lenses, the magnification can be approximated by

$$M \approx \frac{N\ell}{f_e f_o}$$

Example 33-9-A. Examining the approximations of the magnification M. For a compound microscope, determine the difference between the two approximate angular magnifications of the microscope. The microscope has an eyepiece lens with a focal length of 4.0 cm, an objective lens with a focal length of 4.8 mm, a separation between the two lenses of 18 cm, and the user has a near point distance of 25 cm. Assume the object is located 4.9 mm from the objective.

Approach: In this section, two approximate expressions for the magnification M were derived. In this problem, the values for these two expressions for the conditions specified are compared.

Solution: The first approximation gives us the following magnification:

$$M \approx \frac{N}{f_e} \frac{\ell - f_e}{d_o} = \frac{25}{4.0} \frac{18 - 4.0}{0.49} = 179$$

The second approximation predicts

$$M \approx \frac{N\ell}{f_e f_o} = \frac{(25)(18)}{(4.0)(0.48)} = 234$$

Comparing these two values, we note a substantial difference between their predictions. The second approximation is only valid if the focal lengths of both lenses are very small compared to the distance between the lenses. In this problem, the distance between the lenses is 18 cm and the focal length of the eyepiece is not small compared to this dimension and the second expression can thus not be used to make predictions concerning the magnification of the microscope.

Section 33-10. Aberrations of Lenses and Mirrors

There are several different identifiable reasons why lenses do not form exact point images of point objects. These are called **lens aberrations**. Some of these aberrations arise when the object location is moved off the optical axis. There are two common aberrations that occur for a point object, even if it is located on the optical axis.

- **Spherical aberrations**. These are the result of using spherically ground surfaces. Rays that are incident on the outer regions of the lens are focused on a slightly different image point than rays that pass through the center of the lens.
- **Chromatic aberrations**. The refractive index of the lens material is, in general, dependent on the wavelength of the light. This causes images of different wavelengths to be focused at slightly different positions. This aberration is easily correctable by using two lenses such that one chromatic aberration is cancelled by the chromatic aberration of the other one. Such a pair of lenses is called an **achromatic doublet**.

Two off-axis aberrations are **coma** and **astigmatism**. Coma creates image shapes that are asymmetrically shaped for small circular objects located off of the optical axis. Astigmatism causes rays in one plane to have a different image location than rays in a perpendicular plane.

Another type of aberrations are the so-called **distortions.** Distortions are variations in the magnification of the image depending on the location of the object.

Much of the aberrations and distortions can be removed by using compound lenses consisting of several simple lens elements. Spherical aberration can be removed and some other aberrations reduced by using **aspherical** lenses or mirrors.

Practice Quiz

1. A single-lens system produces a real image a distance of 30.0 cm from the lens. What does this imply about the focal length f of the lens?
   a) $0 < f \leq 30$ cm.
   b) $f < -30.0$ cm.
   c) 30.0 cm $< f$.
   d) -30.0 cm $\leq f < 0$.

2. A lens with a positive focal length forms a real image. If the object moves farther from the lens, the image
   a) moves farther from the lens and has greater magnification.
   b) moves farther from the lens and has less magnification.
   c) moves closer to the lens and has greater magnification.
   d) moves closer to the lens and has less magnification.

3. Nearsighted people can use which type of lens to correct their vision?
   a) A convex lens.
   b) A concave lens.
   c) A plano-convex lens.
   d) All of the above.

4. What type of image is formed by a simple magnifier?
    a) An inverted image.
    b) A real image.
    c) A virtual image.
    d) No image is formed.

5. Consider a lens with a negative focal length. If the object distance of this lens is positive, the image is
    a) real and upright.
    b) real and inverted.
    c) virtual and upright.
    d) virtual and inverted.

6. If the object distance is positive, where must the object be located so that a converging lens forms a virtual image?
    a) Farther from the lens than two focal lengths.
    b) Farther from the lens than one focal length.
    c) Closer to the lens than one-half focal length.
    d) Closer to the lens than one focal length.

7. A person is diagnosed with hyperopia. What does that mean?
    a) The person can't see far away objects sharply.
    b) The person can't see nearby objects sharply.
    c) The person can't sit still.
    d) The person has bad-smelling feet.

8. When you look through a thin lens at a distant object, you see an upright image that is smaller than its normal size. If you reverse the lens so that you are looking through it in the opposite direction, how will the image now appear?
    a) Inverted and larger than its normal size.
    b) Inverted and smaller than its normal size.
    c) Upright and larger than its normal size.
    d) Upright and smaller than its normal size.

9. A lens is made from a material with a refractive index of 1.500 and has a focal length f. If the lens would have been made from a material with a refractive index 2.000, what would its focal length be?
    a) $\frac{4}{3}f$
    b) $\frac{3}{4}f$
    c) $2f$
    d) $\frac{1}{2}f$

10. Which aberration does a lens have that a mirror does not?
    a) Spherical aberration.
    b) Chromatic aberration.
    c) Coma.
    d) Astigmatism.

11. A +4.5 diopter lens forms a real image a distance of +56.4 cm from the lens. Where is the object located? What is the magnification of the image?

12. You need a lens with a focal length of –36 cm. The lens blank has a refractive index of 1.59. What should the radii of curvature of the lens surfaces be if the lens is ground so both surfaces have the same radius of curvature?

13. A person is farsighted with a near point of 200 cm to be corrected so that the corrected near point is 25 cm. What should the focal length of the lens be?

14. A Keplerian telescope is to be built with an angular magnification of 80. The objective lens of the telescope has a focal length of 2.5 m. What should the focal length of the eyepiece lens be? How far apart do the eyepiece and objective lens need to be placed?

15. What is the magnification of a compound microscope made with an objective lens with a focal length of +2.00 mm and an eyepiece lens with a focal length of +4.80 cm? The length of the microscope tube is 18.6 cm and the object is located so that the final image is 25 cm from the eye.

Responses to Select End-of-Chapter Questions

1. The film must be placed behind the lens at the focal length of the lens.

7. (a) Yes. The image moves farther from the lens.
   (b) Yes. The image also gets larger.

13. The technique for determining the focal length of the diverging lens in Example 33-6 requires the combination of the two lenses together to project a real image of the sun onto a screen. The focal length of the lens combination can be measured. If the focal length of the converging lens is longer than the focal length of the diverging lens (the converging lens is weaker than the diverging lens), then the lens combination will be diverging, and will not form a real image of the sun. In this case the focal length of the combination of lenses cannot be measured, and the focal length of the diverging lens alone cannot be determined.

19. Nearsighted. Diverging lenses are used to correct nearsightedness and converging lenses are used to correct farsightedness. If the person's face appears narrower through the glasses, then the image of the face produced by the lenses is smaller than the face, virtual, and upright. Thus, the lenses must be diverging, and therefore the person is nearsighted.

25. The curved surface should face the object. If the flat surface faces the object and the rays come in parallel to the optical axis, then no bending will occur at the first surface and all the bending will occur at the second surface. Bending at the two surfaces will clearly not be equal in this case. The bending at the two surfaces may be equal if the curved surface faces the object.

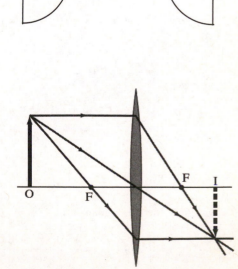

If the parallel rays from the distant object come in above or below the optical axis with the flat side towards the object, then the first bending is actually away from the axis. In this case also, bending at both surfaces can be equal if the curved side of the lens faces the object.

Solutions to Select End-of-Chapter Problems

1. (a) From the ray diagram, the object distance is about 480 cm.
   (b) We find the object distance from Eq. 33-2.

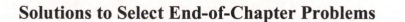

$$\frac{1}{d_o}+\frac{1}{d_i}=\frac{1}{f} \quad \rightarrow$$

$$d_o = \frac{fd_i}{d_i - f} = \frac{(215\,\text{mm})(373\,\text{mm})}{373\,\text{mm} - 215\,\text{mm}} = \boxed{508\,\text{mm}}$$

NOTE: In the first printing of the textbook, a different set of values was given: $f = 75.0\,\text{mm}$ and $d_i = 88.0\,\text{mm}$. Using that set of values gives the same object distance as above. But the ray diagram would be much more elongated, with the object distance almost 7 times the focal length.

7. (a) The image should be upright for reading. The image will be virtual, upright, and magnified .
    (b) To form a virtual, upright magnified image requires a converging lens .
    (c) We find the image distance, then the focal length, and then the power of the lens. The object distance is given.

$$m = -\frac{d_i}{d_o} \quad \rightarrow \quad d_i = -md_o$$

$$P = \frac{1}{f} = \frac{1}{d_o} + \frac{1}{d_i} = \frac{d_i + d_o}{d_o d_i} = \frac{-md_o + d_o}{d_o\left(-md_o\right)} = \frac{m-1}{md_o} = \frac{2.5-1}{(2.5)(0.090\,\text{m})} = \boxed{6.7\,\text{D}}$$

13. The sum of the object and image distances must be the distance between object and screen, which we label as d_T. We solve this relationship for the image distance, and use that expression in Eq. 33-2 in order to find the object distance.

$$d_o + d_i = d_T \quad \rightarrow \quad d_i = d_T - d_o \quad ; \quad \frac{1}{d_o} + \frac{1}{d_i} = \frac{1}{d_o} + \frac{1}{\left(d_T - d_o\right)} = \frac{1}{f} \quad \rightarrow \quad d_o^2 - d_T d_o + f d_T = 0 \quad \rightarrow$$

$$d_o = \frac{d_T \pm \sqrt{d_T^2 - 4 f d_T}}{2} = \frac{(86.0\,\text{cm}) \pm \sqrt{(86.0\,\text{cm})^2 - 4(16.0\,\text{cm})(86.0\,\text{cm})}}{2} = \boxed{21.3\,\text{cm}, 64.7\,\text{cm}}$$

Note that to have real values for d_o, we must in general have $d_T^2 - 4 f d_T > 0 \quad \rightarrow \quad d_T > 4f$.

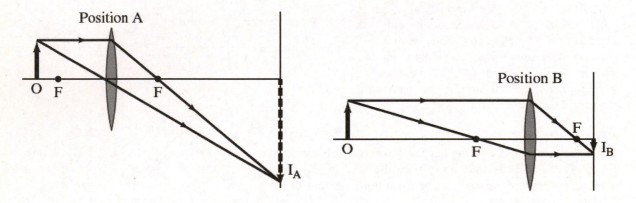

19. (a) With the definitions as given in the problem, $x = d_o - f \;\rightarrow\; d_o = x + f$ and $x' = d_i - f \;\rightarrow\; d_i = x' + f$. Use Eq. 33-2.

$$\frac{1}{d_o} + \frac{1}{d_i} = \frac{1}{x+f} + \frac{1}{x'+f} = \frac{1}{f} \quad \rightarrow \quad \frac{(x'+f)+(x+f)}{(x+f)(x'+f)} = \frac{1}{f} \quad \rightarrow$$

$$(2f + x + x')f = (x+f)(x'+f) \quad \rightarrow \quad 2f^2 + xf + x'f = x'x + xf + fx' + f^2 \quad \rightarrow \quad \boxed{f^2 = x'x}$$

(b) Use Eq. 33-2.

$$\frac{1}{d_o} + \frac{1}{d_i} = \frac{1}{f} \quad \rightarrow \quad d_i = \frac{d_o f}{d_o - f} = \frac{(48.0\,\text{cm})(38.0\,\text{cm})}{48.0\,\text{cm} - 38.0\,\text{cm}} = \boxed{182\,\text{cm}}$$

(c) Use the Newtonian form.

$$xx' = f^2 \quad \rightarrow \quad x' = \frac{f^2}{x} = \frac{(38.0\,\text{cm})^2}{(48.0\,\text{cm} - 38.0\,\text{cm})} = 144.2\,\text{cm}$$

$$d_i = x' + f = 144.2\,\text{cm} + 38.0\,\text{cm} = \boxed{182\,\text{cm}}$$

25. (a) The first lens is the converging lens. Find the image formed by the first lens.

$$\frac{1}{d_{o1}} + \frac{1}{d_{i1}} = \frac{1}{f_1} \rightarrow d_{i1} = \frac{d_{o1}f_1}{d_{o1} - f_1} = \frac{(60.0\,\text{cm})(20.0\,\text{cm})}{(60.0\,\text{cm}) - (20.0\,\text{cm})} = 30.0\,\text{cm}$$

This image is the object for the second lens. Since this image is behind the second lens, the object distance for the second lens is negative, and so $d_{o2} = 25.0\,\text{cm} - 30.0\,\text{cm} = -5.0\,\text{cm}$. Use Eq. 33-2.

$$\frac{1}{d_{o2}} + \frac{1}{d_{i2}} = \frac{1}{f_2} \rightarrow d_{i2} = \frac{d_{o2}f_2}{d_{o2} - f_2} = \frac{(-5.0\,\text{cm})(-10.0\,\text{cm})}{(-5.0\,\text{cm}) - (-10.0\,\text{cm})} = 10\,\text{cm}$$

Thus the final image is real, $\boxed{10\,\text{cm beyond the second lens}}$. The distance has two significant figures.
(b) The total magnification is the product of the magnifications for the two lenses:

$$m = m_1 m_2 = \left(-\frac{d_{i1}}{d_{o1}}\right)\left(-\frac{d_{i2}}{d_{o2}}\right) = \frac{d_{i1}d_{i2}}{d_{o1}d_{o2}} = \frac{(30.0\,\text{cm})(10.0\,\text{cm})}{(60.0\,\text{cm})(-5.0\,\text{cm})} = \boxed{-1.0\times}$$

(c) Note: Lens B should be drawn as a diverging lens.

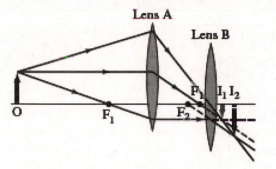

31. The plane surface has an infinite radius of curvature. Let the plane surface be surface 2, so $R_2 = \infty$. The index of refraction is found in Table 32-1.

$$\frac{1}{f} = (n-1)\left(\frac{1}{R_1} + \frac{1}{R_2}\right) = \frac{1}{f} = (n-1)\left(\frac{1}{R_1} + \frac{1}{\infty}\right) = \frac{(n-1)}{R_1} \rightarrow$$
$$R_1 = (n-1)f = (1.46 - 1)(18.7\,\text{cm}) = \boxed{8.6\,\text{cm}}$$

37. We calculate the effective f-number for the pinhole camera by dividing the focal length by the diameter of the pinhole. The focal length is equal to the image distance. Setting the exposures equal for both cameras, where the exposure is proportional to the product of the exposure time and the area of the lens opening (which is inversely proportional to the square of the f-stop number), we solve for the exposure time.

$$f\text{-stop}_2 = \frac{f}{D} = \frac{(70\,\text{mm})}{(1.0\,\text{mm})} = \frac{f}{70}.$$

$$t_1\left(f\text{-stop}_1\right)^{-2} = t_2\left(f\text{-stop}_2\right)^{-2} \rightarrow t_2 = t_1\left(\frac{f\text{-stop}_2}{f\text{-stop}_1}\right)^2 = \frac{1}{250\,\text{s}}\left(\frac{70}{11}\right)^2 = 0.16\,\text{s} \approx \boxed{\tfrac{1}{6}\,\text{s}}$$

49. Find the object distance for the contact lens to form an image at the eye's near point, using Eqs. 33-2 and 33-1.

$$\frac{1}{d_o} + \frac{1}{d_i} = \frac{1}{f} = P \rightarrow d_o = \frac{d_i}{Pd_i - 1} = \frac{-0.106\,\text{m}}{(-4.00\,\text{D})(-0.106\,\text{m}) - 1} = 0.184\,\text{m} = \boxed{18.4\,\text{cm}}$$

Likewise find the object distance for the contact lens to form an image at the eye's far point.

$$d_o = \frac{d_i}{Pd_i - 1} = \frac{-0.200\,\text{m}}{(-4.0\,\text{D})(-0.200\,\text{m}) - 1} = 1.00\,\text{m} = \boxed{100\,\text{cm}} \qquad (3\text{ sig. fig.})$$

55. (a) We find the image distance from Eq. 33-2.

$$\frac{1}{d_o} + \frac{1}{d_i} = \frac{1}{f} \rightarrow$$

$$d_i = \frac{fd_o}{d_o - f} = \frac{(6.00\,\text{cm})(5.85\,\text{cm})}{5.85\,\text{cm} - 6.00\,\text{cm}} = \boxed{-234\,\text{cm}}$$

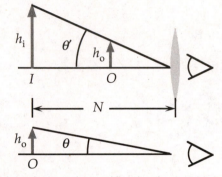

(b) The angular magnification is given by Eq. 33-6a, since the eye will have to focus over 2 m away.

$$M = \frac{N}{f} = \frac{25.0\,\text{cm}}{6.00\,\text{cm}} = \boxed{4.17\times}$$

61. We find the focal length of the eyepiece from the magnification by Eq. 33-7.

$$M = -\frac{f_o}{f_e} \rightarrow f_e = -\frac{f_o}{M} = -\frac{88\,\text{cm}}{35\times} = \boxed{2.5\,\text{cm}}$$

For both object and image far away, the separation of the lenses is the sum of the focal lengths.

$$f_o + f_e = 88\,\text{cm} + 2.5\,\text{cm} = \boxed{91\,\text{cm}}$$

67. The focal length of the mirror is found from Eq. 33-7. The radius of curvature is twice the focal length.

$$M = -\frac{f_o}{f_e} \rightarrow f_o = -Mf_e = -(120)(0.031\,\text{m}) = 3.72\,\text{m} \approx \boxed{3.7\,\text{m}} \ ; \ r = 2f = \boxed{7.4\,\text{m}}$$

73. We find the focal length of the eyepiece from the magnification of the microscope, using the approximate results of Eq. 33-10b. We already know that $f_o \ll \ell$.

$$M \approx \frac{N\ell}{f_o f_e} \rightarrow f_e = \frac{N\ell}{Mf_o} = \frac{(25\,\text{cm})(17.5\,\text{cm})}{(680)(0.40\,\text{cm})} = \boxed{1.6\,\text{cm}}$$

Note that this also satisfies the assumption that $f_e \ll \ell$.

79. For each objective lens we set the image distance equal to the sum of the focal length and 160 mm. Then, using Eq. 33-2 we write a relation for the object distance in terms of the focal length. Using this relation in Eq. 33-3 we write an equation for the magnification in terms of the objective focal length. The total magnification is the product of the magnification of the objective and focal length.

$$\frac{1}{d_o} + \frac{1}{d_i} = \frac{1}{f_o} \rightarrow \frac{1}{d_o} = \frac{1}{f_o} - \frac{1}{d_i} \rightarrow \frac{1}{d_o} = \frac{1}{f_o} - \frac{1}{f_o + 160\,\text{mm}} \rightarrow d_o = \frac{f_o(f_o + 160\,\text{mm})}{160\,\text{mm}}$$

$$m_o = \frac{d_i}{d_o} = \frac{f_o + 160\,\text{mm}}{\left[\dfrac{f_o\left(f_o + 160\,\text{mm}\right)}{160\,\text{mm}}\right]} = \frac{160\,\text{mm}}{f_o}$$

Since the objective magnification is inversely proportional to the focal length, the objective with the smallest focal length $\left(f_o = 3.9\,\text{mm}\right)$ combined with the largest eyepiece magnification $\left(M_e = 10\right)$ yields the largest overall magnification. The objective with the largest focal length $\left(f_o = 32\,\text{mm}\right)$ coupled with the smallest eyepiece magnification $\left(M_e = 5\right)$ yields the smallest overall magnification.

$$M_{\text{largest}} = \frac{160\,\text{mm}}{3.9\,\text{mm}}\left(10\times\right) = \boxed{410\times} \; ; \; M_{\text{smallest}} = \frac{160\,\text{mm}}{32\,\text{mm}}\left(5\times\right) = \boxed{25\times}$$

85. Since the object height is equal to the image height, the magnification is –1. We use Eq. 33-3 to obtain the image distance in terms of the object distance. Then we use this relationship with Eq. 33-2 to solve for the object distance.

$$m = -1 = -\frac{d_i}{d_o} \;\rightarrow\; d_i = d_o$$

$$\frac{1}{f} = \frac{1}{d_o} + \frac{1}{d_i} = \frac{1}{d_o} + \frac{1}{d_o} = \frac{2}{d_o} \;\rightarrow\; d_o = 2f = 2\left(58\,\text{mm}\right) = \boxed{116\,\text{mm}}$$

The distance between the object and the film is the sum of the object and image distances.

$$d = d_o + d_i = d_o + d_o = 2d_o = 2\left(116\,\text{mm}\right) = \boxed{232\,\text{mm}}$$

91. (a) Because the Sun is very far away, the image will be at the focal point, or $d_i = f$. We find the magnitude of the size of the image using Eq. 33-3, with the image distance equal to 28 mm.

$$\frac{h_i}{h_o} = \frac{-d_i}{d_o} \;\rightarrow\; |h_i| = \frac{h_o d_i}{d_o} = \frac{\left(1.4 \times 10^6\,\text{km}\right)\left(28\,\text{mm}\right)}{1.5 \times 10^8\,\text{km}} = \boxed{0.26\,\text{mm}}$$

(b) We repeat the same calculation with a 50 mm image distance.

$$|h_i| = \frac{\left(1.4 \times 10^6\,\text{km}\right)\left(50\,\text{mm}\right)}{1.5 \times 10^8\,\text{km}} = \boxed{0.47\,\text{mm}}$$

(c) Again, with a 135 mm image distance.

$$|h_i| = \frac{\left(1.4 \times 10^6\,\text{km}\right)\left(135\,\text{mm}\right)}{1.5 \times 10^8\,\text{km}} = \boxed{1.3\,\text{mm}}$$

(d) The equations show that image height is directly proportional to focal length. Therefore the relative magnifications will be the ratio of focal lengths.

$$\frac{28\,\text{mm}}{50\,\text{mm}} = \boxed{0.56\times} \text{ for the 28 mm lens;}$$

$$\frac{135\,\text{mm}}{50\,\text{mm}} = \boxed{2.7\times} \text{ for the 135 mm lens.}$$

97. The maximum magnification is achieved with the image at the near point, using Eq. 33-6b.

$$M_1 = 1 + \frac{N_1}{f} = 1 + \frac{\left(15.0\,\text{cm}\right)}{\left(8.5\,\text{cm}\right)} = \boxed{2.8\times}$$

For an adult we set the near point equal to 25.0 cm.

$$M_2 = 1 + \frac{N_2}{f} = 1 + \frac{(25.0\,\text{cm})}{(8.5\,\text{cm})} = \boxed{3.9\times}$$

The $\boxed{\text{person with the normal eye}}$ (adult) sees more detail.

103. The focal length of the eyepiece is found using Eq. 33-1.

$$f_e = \frac{1}{P_e} = \frac{1}{23\,\text{D}} = 4.3 \times 10^{-2}\,\text{m} = 4.3\,\text{cm}.$$

For both object and image far away, we find the focal length of the objective from the separation of the lenses.

$$l = f_o + f_e \;\rightarrow\; f_o = l - f_e = 85\,\text{cm} - 4.3\,\text{cm} = 80.7\,\text{cm}$$

The magnification of the telescope is given by Eq. 33-7.

$$M = -\frac{f_o}{f_e} = -\frac{(80.7\,\text{cm})}{(4.3\,\text{cm})} = \boxed{-19\times}$$

109.(a) We use Eqs. 33-2 and 33-3.

$$m = -\frac{d_i}{d_o} \;\rightarrow\; d_o = -\frac{d_i}{m}\;;\; \frac{1}{d_o} + \frac{1}{d_i} = \frac{1}{f} = -\frac{m}{d_i} + \frac{1}{d_i} \;\rightarrow\; m = -\frac{d_i}{f} + 1$$

This is a straight line with $\boxed{\text{slope} = -\dfrac{1}{f} \text{ and } y\text{-intercept} = 1}$.

(b) A plot of m vs. d_i is shown here.

$$f = -\frac{1}{\text{slope}} = -\frac{1}{-.0726\,\text{cm}^{-1}}$$
$$= 13.8\,\text{cm} \approx \boxed{14\,\text{cm}}$$

The y-intercept is 1.028. $\boxed{\text{Yes}}$, it is close to the expected value of 1.

(c) Use the relationship derived above.

$$m = -\frac{d_i}{f} + 1 =$$
$$= -\frac{d' + l_i}{f} + 1 =$$
$$= -\frac{d_i'}{f} + \left(1 - \frac{l_i}{f}\right)$$

$$m = -0.0726\,d_i + 1.028$$

A plot of m vs. d_i' would still have a slope of $-\dfrac{1}{f}$, so $f = \boxed{-\dfrac{1}{\text{slope}}}$ as before. The y-intercept will have changed, to

$1 - \dfrac{l_i}{f}$.

Chapter 34: The Wave Nature of Light; Interference

Chapter Overview and Objectives

This chapter introduces the wave model of light and discusses a number of applications of this model that are not easily explained by Newton's particle model of light. Several cases of two-source interference are discussed, along with the concept of constructive and destructive interference.

After completing this chapter you should:
- Know what Huygens' principle is.
- Know what constructive and destructive interference is.
- Know how to determine the position of the dark and bright fringes in a two-slit interference pattern.
- Know how to relate the intensity of a two-slit interference pattern to the position within the pattern.
- Know what thin-film interference is and how to determine the conditions of constructive and destructive interference in these films.
- Know what a Michelson interferometer is.

Summary of Equations

Constructive double-slit interference condition: $\quad d \sin\theta = m\lambda \qquad m \in \{0,1,2,...\}$ $\qquad$ (Section 34-3)

Destructive double-slit interference condition: $\quad d \sin\theta = \left(m + \tfrac{1}{2}\right)\lambda \qquad m \in \{0,1,2,...\}$ $\qquad$ (Section 34-3)

Intensity of a double-slit interference pattern: $\quad I = I_0 \cos^2\left(\dfrac{\pi d \sin\theta}{\lambda}\right) = I_0 \cos^2\left(\dfrac{\pi d}{\lambda L} y\right)$ $\qquad$ (Section 34-4)

Thin-film interference conditions when $n_1 > n_2 > n_3$ or $n_1 < n_2 < n_3$

$\qquad$ Constructive interference: $\qquad\qquad 2t = m\lambda_{n_2} \qquad m \in \{0,1,2,...\}$ $\qquad$ (Section 34-5)

$\qquad$ Destructive interference: $\qquad\qquad 2t = \left(m + \tfrac{1}{2}\right)\lambda_{n_2} \qquad m \in \{0,1,2,...\}$ $\qquad$ (Section 34-5)

Thin-film interference conditions when $n_1 > n_2$ and $n_3 > n_2$, or $n_2 > n_1$ and $n_2 > n_3$

$\qquad$ Constructive interference: $\qquad\qquad 2t = \left(m + \tfrac{1}{2}\right)\lambda_{n_2} \qquad m \in \{0,1,2,...\}$ $\qquad$ (Section 34-5)

$\qquad$ Destructive interference: $\qquad\qquad 2t = m\lambda_{n_2} \qquad m \in \{0,1,2,...\}$ $\qquad$ (Section 34-5)

Chapter Summary

Section 34-1. Waves versus Particles; Huygens' Principle and Diffraction

Huygens' principle states that each point on a wave front can be considered to be a point source of spherical waves. The continuing wave is the sum of these outgoing spherical waves in the forward direction. This principle can be used to understand the phenomenon of **diffraction**.
Diffraction is the property of wave fronts to bend in the non-forward direction when they encounter a barrier.

Section 34-2. Huygens' Principle and the Law of Refraction

Huygens' principle is consistent with the laws of reflection and refraction. This consistency relies on the fact that the frequency of light does not depend on the medium in which the light travels; the wavelength of the light on the other hand will depend on the index of refraction of the medium in which the light is traveling.

Section 34-3. Interference – Young's Double-Slit Experiment

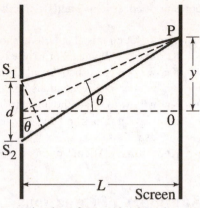

The **Young's double-slit experiment** was used to provide compelling evidence for the wave nature of light, and was used to accurately measure its wavelength. In this experiment, two slits, separated by a distance d, are illuminated by in-phase monochromatic coherent light of wavelength λ, and a pattern of bright and dark fringes is observed on a screen placed a distance L behind the slits. The bright fringes occur at those positions where there is **constructive interference** and the dark fringes occur at those positions where there is **destructive interference**. The position of the fringes can be specified in terms of the angle θ (see Figure on the right). The requirement for constructive interference is given by

$$d \sin\theta = m\lambda \qquad m \in \{0,1,2,...\}$$

The value of m is called the order of the interference fringe. The requirement for destructive interference is given by

$$d \sin\theta = \left(m + \tfrac{1}{2}\right)\lambda \qquad m \in \{0,1,2,...\}$$

These expressions can also be written in terms of the position along the screen from the center of the pattern, y. The requirement for constructive interference, in terms of y, is given by

$$d \frac{y}{L} = m\lambda \qquad m \in \{0,1,2,...\}$$

The requirement for destructive interference, in terms of y, is given by

$$d \frac{y}{L} = \left(m + \tfrac{1}{2}\right)\lambda \qquad m \in \{0,1,2,...\}$$

The requirements for constructive and destructive interference are valid when the distance between the slits and the screen is much greater than the distance between the slits and when the angle θ is small.

There is a general rule that applies to all two-source interference measurements, whether it is the two-slit interference described above or some other geometrical arrangement of two sources. Constructive interference occurs if the path length difference from the sources to the observation point is a multiple of the wavelength of the light. Destructive interference occurs if the path length difference from the sources to the observation point is an odd multiple of one-half of a wavelength of the light.

If the two slits are illuminated with light coming from the same light source, then the light emerging from the two slits is **coherent;** there is a fixed phase relationship between these two sources of light. If there is a phase relationship that fluctuates randomly, the two sources are **incoherent**. If the light coming from the two slits is incoherent, the interference pattern will change rapidly due to phase fluctuations. This usually causes the interference intensity pattern to be averaged out by the observing instrument and no interference pattern will be visible.

Example 34-3-A. Measuring the wavelength of light using interference patterns.
A two-slit interference pattern obtained with monochromatic light of wavelength λ is shown in the diagram on the right. The distance between the two slits was $d = 0.322$ mm. The pattern shown in the diagram on the right was obtained on a screen, placed a distance $L = 1.48$ m behind the slits. What is the wavelength of the light used in this measurement?

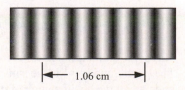

1.06 cm

Approach: The information provided in the diagram can be used to determine the distance between two interference fringes. Although we do not know the order of the interference fringes being shown, the theory discussed in this Section shows that the distance between subsequent interference fringes is constant, and the distance between fringes is thus independent of the order of the fringes. This information, combined with the known distances d and L is sufficient to determine the wavelength of the light being used.

Solution: Assume that the distance outlined in the diagram indicates the distance between a fringe of order m and a fringe of order $m + 5$. The distance between these two fringes is equal to

$$\frac{(m+5)\lambda L}{d} - \frac{m\lambda L}{d} = \frac{5\lambda L}{d} = \Delta y$$

The only unknown in this relation is the wavelength of the light, which thus can be determined:

$$\lambda = \frac{d\Delta y}{5L} = \frac{(0.0322)(1.06)}{5(148)} = 4.61 \times 10^{-5} \text{ cm} = 4.61 \times 10^{-7} \text{ m}$$

Note that in this calculation we need to make sure that the units of all distances are the same. In this example, we have converted all distances to cm in the calculation of the wavelength.

Section 34-4. Intensity in the Double-Slit Interference Pattern

Rather than only looking at the positions of the fringes in a two-slit interference pattern, we can also look at the intensity as a function of position. When the two slits are illuminated with equal intensity in phase-coherent light, the intensity as a function of position is given by

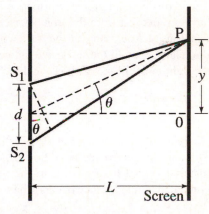

$$I = I_0 \cos^2\left(\frac{\pi d \sin\theta}{\lambda}\right) = I_0 \cos^2\left(\frac{\pi d}{\lambda L}y\right)$$

where d is the distance between the two slits, λ is the wavelength of the light, θ is the angular deflection corresponding to an on-screen displacement y, I_0 is the intensity of the light at the center of the interference pattern, and L is the distance from the slits to the plane at which the intensity is being measured. The expression for the intensity of the interference pattern is valid when the distance between the slits and the screen is much greater than the distance between the slits and when the angle θ is small.

Example 34-4-A. Predicting the intensity of an interference pattern. A two-slit interference pattern is created using two slits, separated by a distance $d = 0.556$ mm. The light used has a wavelength $\lambda = 546$ nm, and the screen on which the interference pattern is projected is located a distance $L = 98.6$ cm from the slits. The intensity at a position on the screen a distance $y_1 = 1.00$ cm from the center of the central bright fringe is $I_1 = 3.35$ W/m^2. What is the intensity a distance $y_2 = 1.25$ cm from the central bright fringe?

Approach: To solve this problem, we use the relation between intensity and displacement along the screen, as discussed in this Section. Although we do not know the intensity of the light at the center of the interference pattern, this intensity can be eliminated if we use the ratios of intensities at a distance y_1 and at a distance y_2.

Solution: The intensity at a distance y_1 is equal to

$$I_1 = I_0 \cos^2\left(\frac{\pi d}{\lambda L}y_1\right)$$

The intensity at a distance y_2 is equal to

$$I_2 = I_0 \cos^2\left(\frac{\pi d}{\lambda L}y_2\right)$$

The ratio of these two intensities is equal to

$$\frac{I_2}{I_1} = \frac{I_0 \cos^2\left(\frac{\pi d}{\lambda L} y_2\right)}{I_0 \cos^2\left(\frac{\pi d}{\lambda L} y_1\right)} = \frac{\cos^2\left(\frac{\pi d}{\lambda L} y_2\right)}{\cos^2\left(\frac{\pi d}{\lambda L} y_1\right)}$$

The only unknown in this equation is the intensity I_2, which can thus be determined:

$$I_2 = \frac{\cos^2\left(\frac{\pi d}{\lambda L} y_2\right)}{\cos^2\left(\frac{\pi d}{\lambda L} y_1\right)} I_1 = \frac{\cos^2\left(\frac{\pi\left(0.556\times10^{-3}\right)}{\left(546\times10^{-9}\right)\left(0.986\right)}0.0125\right)}{\cos^2\left(\frac{\pi\left(0.556\times10^{-3}\right)}{\left(546\times10^{-9}\right)\left(0.986\right)}0.01\right)}\left(3.35\right) = 2.72 \text{ W/m}^2$$

Section 34-5. Interference in Thin Films

The two-source interference principle applies to thin films of materials that are transparent. A thin film of material will have a reflection from the front surface and from the back surface. These two reflections will usually be approximately equal in intensity and can interfere with each other. The interference condition depends on the thickness of the film, t, the relative size of the refractive indices of the materials on each side of film and that of the film, and the wavelength of the light in the film, which is equal to

$$\lambda_{n_2} = \frac{\lambda}{n_2}$$

where λ is the wavelength of the light in vacuum.

The actual interference condition depends on the refractive indices of the incident material, n_1, the film, n_2, and the material behind the film, n_3. If the refractive indices are increasing or decreasing, $n_1 > n_2 > n_3$ or $n_1 < n_2 < n_3$, then constructive interference occurs when

$$2t = m\lambda_{n_2} \qquad m \in \{0,1,2,...\}$$

Destructive interference occurs when

$$2t = \left(m+\tfrac{1}{2}\right)\lambda_{n_2} \qquad m \in \{0,1,2,...\}$$

If the refractive indices of the materials are not in increasing or decreasing order, $n_1 > n_2$ and $n_3 > n_2$ or $n_2 > n_1$ and $n_2 > n_3$, then constructive interference occurs when

$$2t = \left(m+\tfrac{1}{2}\right)\lambda_{n_2} \qquad m \in \{0,1,2,...\}$$

Destructive interference occurs when

$$2t = m\lambda_{n_2} \qquad m \in \{0,1,2,...\}$$

Example 34-5-A. Measuring the thickness of an oil film. A thin film of oil of refractive index $n_2 = 1.28$ floats on a pool of water with refractive index $n_3 = 1.33$. Light with a wavelength $\lambda_1 = 623$ nm in air is reflected brightly by the film. What are the three smallest possible non-zero thicknesses of the oil film?

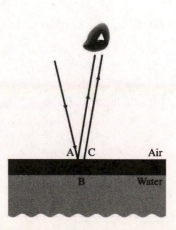

Approach: The problem states that the film reflects the light brightly. This implies that the interference is constructive. The condition for constructive interference depends on the refractive indices of the three regions. The wavelength to be used in the calculation is the wavelength of the light in the oil; this wavelength can be calculated from the wavelength of the light in air and the index of refraction of the oil.

Solution: In this problem, the refractive indices are increasing ($n_1 < n_2 < n_3$) and the condition for constructive interference is thus

$$2t = m\lambda_{n_2} \qquad m \in \{0,1,2,...\}$$

The lowest non-zero thicknesses of the oil film are obtained when $m = 1, 2,$ and 3. The wavelength of the light in the film is equal to the wavelength of light in air divided by the refractive index of the oil:

$$\lambda_{n_2} = \frac{\lambda}{n_2} = 486.7\,\text{nm}$$

The smallest three non-zero film thicknesses of the oil film are

$$t_1 = \frac{(1)(486.7)}{2} = 243\,\text{nm}$$

$$t_2 = \frac{(2)(486.7)}{2} = 487\,\text{nm}$$

$$t_3 = \frac{(3)(486.7)}{2} = 729\,\text{nm}$$

Section 34-6. Michelson Interferometer

A **Michelson interferometer** is an instrument that uses two-source interference and can be used to measure the wavelength of light, refractive indices of gases, or small changes in lengths. A **beam splitter**, M_S, is used to create two beams of light that travel along two separate paths that return to the beam splitter where they interfere, as shown in the diagram.

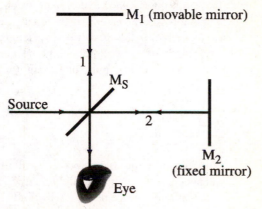

Section 34-7. Luminous Intensity

The sensitivity of our eye is not the same for all visible wavelengths. This means that two sources of light with the same intensity, but different wavelengths, do not appear to have the same brightness. It is important for lighting engineers to have a measurement that reflects the effectiveness of a light source relative to the sensitivity of the human eye. **Luminous flux**, F_ℓ, is the quantity defined to measure the brightness of a source and its unit is the **lumen (lm)**. An additional quantity that is useful is **luminous intensity**, which is the luminous flux per solid angle. The unit of luminous intensity is the **candela (cd)**. One candela is defined as one lumen per steradian. The **illuminance** of a surface, E_ℓ, is the luminous flux per unit area and is measured in units of lumens per square meter:

$$E_\ell = \frac{F_\ell}{A}$$

Illuminance depends on the distance from the source.

Practice Quiz

1. As you are viewing a two-slit interference pattern, the wavelength of the source is changed. If the interference fringes move farther apart during this change, the wavelength has
   a) increased.
   b) decreased.
   c) stayed the same.
   d) Can't tell from this information.

2. As the spacing of the slits in a two-slit interference experiment with a fixed wavelength increases, the angle at which the first-order bright fringe occurs
    a) increases.
    b) decreases.
    c) remains the same.
    d) Can't tell from this information.

3. In a two-slit interference experiment, the intensity of the light reaching the screen from either source individually would be I_0. What happens to the energy missing from the dark fringes?
    a) Energy is not conserved in wave optics.
    b) The energy is reaching the screen at the dark fringe; it just isn't seen as visible light.
    c) The intensity of the bright fringes is $4I_0$, so the missing energy is in the bright fringes.
    d) The energy never leaves the sources.

4. A wedge-shaped film of air is formed between two glass slides. What happens to the number of fringes if the index of refraction of the glass is increased?
    a) The number of fringes increases.
    b) The number of fringes decreases.
    c) The number of fringes stays the same.
    d) There is not enough information to determine the answer.

5. A two-slit interference experiment produces an interference pattern, as shown in the top figure on the right, when the slits are illuminated with light of wavelength 420 nm. Now the wavelength of the light is changed. What is the wavelength of light illuminating the slits if the interference pattern that is shown in the bottom figure on the right is produced?
    a) 280 nm.
    b) 630 nm.
    c) 840 nm.
    d) 1260 nm.

6. A monochromatic light source is illuminating one slit of a two-slit interference experiment and a second, independent, but identical, light source is illuminating the other slit. The light passing through the slits does not create an interference pattern; why not?
    a) The two sources are incoherent.
    b) The two sources do not have a definite phase relationship.
    c) The interference pattern is shifting too rapidly to see.
    d) All of the above.

7. An air gap between two glass slides gives a first-order bright reflection due to constructive interference at wavelength λ_1. If the gap between the two plates is filled with a liquid that has a refractive index greater than the refractive index of the glass, what will the wavelength of the first-order bright reflection be?
    a) $\lambda_1/2$.
    b) $2\lambda_1$.
    c) Some wavelength less than λ_1.
    d) Some wavelength greater than λ_1.

8. What is the thinnest non-zero thickness film that will show constructive interference for visible light when there is a 180°-phase change for reflections at both surfaces?
    a) One half the wavelength of the most violet visible light.
    b) One half the wavelength of the most red visible light.
    c) One fourth the wavelength of the most violet visible light.
    d) One fourth the wavelength of the most red visible light.

9. The Michelson interferometer forms a two-source interference pattern. The two sources that form this pattern are
    a) a source being studied and a source within the human eye.
    b) a single source that is split into two beams that travel along different paths.
    c) a direct source and a second source that is reflected in a mirror.
    d) a source and its image formed by a lens.

10. If the distance from a light source is r and the size of the source is small compared to r, how should the illuminance, E_ℓ, depend on r?
    a) $E_\ell \propto r$.
    b) $E_\ell \propto r^2$.
    c) $E_\ell \propto 1/r$.
    d) $E_\ell \propto 1/r^2$.

11. A two-slit interference pattern is formed on a screen a distance of 1.56 m from two slits that are separated by a distance of 0.342 mm. The 5th order bright fringe and the 12th order bright fringe are separated by a distance of 1.87 cm. What is the wavelength of the light?

12. A two-slit interference pattern is formed using a slit separation of 0.54 mm, a screen distance of 1.20 m, and a wavelength of 459 nm. How far from the center of the central bright fringe does the intensity first drop to one half the intensity at the center of the central bright spot?

13. Two plates of glass, illuminated by light with a wavelength of 560 nm, are in contact at one end and the other end is held apart by a very fine wire placed in between the plates, as shown in the diagram on the right. If the number of bright fringes per unit length is 16 fringes per cm and the glass plates are 2.54 cm long, what is the diameter of the wire?

14. A thin film of oil of refractive index 1.65 rests on a piece of plastic with a refractive index of 1.37. The thickness of the film is 1.18 μm. What are the three longest visible wavelengths of light in air that will have a bright reflection from this film due to constructive interference?

15. What is the distance between the 2nd bright fringe and the 3rd dark fringe in a two-slit interference pattern that has a slit separation of 0.78 mm, a wavelength of 433 nm, and a slit-to-screen distance of 1.44 m?

Responses to Select End-of-Chapter Questions

1. Yes, Huygens' principle applies to all waves, including sound and water waves.

7. Blue light has a shorter wavelength than red light. The angles to each of the bright fringes for the blue light would be smaller than for the corresponding orders for the red light, so the bright fringes would be closer together for the blue light.

13. Bright colored rings will occur when the path difference between the two interfering rays is $\lambda/2$, $3\lambda/2$, $5\lambda/2$, and so forth. A given ring, therefore, has a path difference that is exactly one wavelength longer than the path difference of its neighboring ring to the inside and one wavelength shorter than the path difference of its neighboring ring to the outside. Newton's rings are created by the thin film of air between a glass lens and the flat glass surface on which it is placed. Because the glass of the lens is curved, the thickness of this air film does not increase linearly. The farther a point is from the center, the less the horizontal distance that corresponds to an increase in vertical thickness of one wavelength. The horizontal distance between two neighboring rings therefore decreases with increasing distance from the center.

Solutions to Select End-of-Chapter Problems

1. Consider a wave front traveling at an angle θ_1 relative to a surface. At time $t = 0$, the wave front touches the surface at point A, as shown in the figure. After a time t, the wave front, moving at speed v, has moved forward such that the contact position has moved to point B. The distance between the two contact points is calculated using simple geometry: $AB = vt / \sin\theta_1$.
By Huygens' principle, at each point the wave front touches the surface, it creates a new wavelet. These wavelets expand out in all directions at speed v. The line passing through the surface of each of these wavelets is the reflected wave front. Using the radius of the wavelet created at $t = 0$, the center of the wavelet created at time t, and the distance between the two contact points (AB) we create a right triangle. Dividing the radius of the wavelet centered at AB (vt) by distance between the contact points gives the sine of the angle between the contact surface and the reflected wave, θ_2.

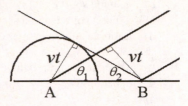

$$\sin\theta_2 = \frac{vt}{AB} = \frac{vt}{\dfrac{vt}{\sin\theta_1}} = \sin\theta_1 \quad \rightarrow \quad \boxed{\theta_2 = \theta_1}$$

Since these two angles are equal, their complementary angles (the incident and reflected angles) are also equal.

7. Using a ruler on Fig. 35-9a, the distance from the $m = 0$ fringe to the $m = 10$ fringe is found to be about 13.5 mm. For constructive interference, the path difference is a multiple of the wavelength, as given by Eq. 34-2a. The location on the screen is given by $x = \ell \tan\theta$, as seen in Fig. 34-7(c). For small angles, we have $\sin\theta \approx \tan\theta \approx x/\ell$.

$$d\sin\theta = m\lambda \quad \rightarrow \quad d\frac{x}{\ell} = m\lambda \quad \rightarrow \quad \lambda = \frac{dx}{m\ell} = \frac{dx}{m\ell} = \frac{\left(1.7\times10^{-4}\,\text{m}\right)\left(0.0135\,\text{m}\right)}{(10)(0.35\,\text{m})} = \boxed{6.6\times10^{-7}\,\text{m}}$$

13. For constructive interference, the path difference is a multiple of the wavelength, as given by Eq. 34-2a. The location on the screen is given by $x = \ell \tan\theta$, as seen in Fig. 34-7(c). For small angles, we have $\sin\theta \approx \tan\theta \approx x/\ell$. For adjacent fringes, $\Delta m = 1$.

$$d\sin\theta = m\lambda \quad \rightarrow \quad d\frac{x}{\ell} = m\lambda \quad \rightarrow \quad x = \frac{\lambda m\ell}{d} \quad \rightarrow$$

$$\Delta x = \Delta m \frac{\lambda\ell}{d} = (1)\frac{\left(544\times10^{-9}\,\text{m}\right)\left(5.0\,\text{m}\right)}{\left(1.0\times10^{-3}\,\text{m}\right)} = \boxed{2.7\times10^{-3}\,\text{m}}$$

19. The intensity of the pattern is given by Eq. 34-6. We find the angle where the intensity is half its maximum value.

$$I_\theta = I_0\cos^2\left(\frac{\pi d\sin\theta}{\lambda}\right) = \frac{1}{2}I_0 \quad \rightarrow \quad \cos^2\left(\frac{\pi d\sin\theta_{1/2}}{\lambda}\right) = \frac{1}{2} \quad \rightarrow \quad \cos\left(\frac{\pi d\sin\theta_{1/2}}{\lambda}\right) = \frac{1}{\sqrt{2}} \quad \rightarrow$$

$$\frac{\pi d\sin\theta_{1/2}}{\lambda} = \cos^{-1}\frac{1}{\sqrt{2}} = \frac{\pi}{4} \quad \rightarrow \quad \sin\theta_{1/2} = \frac{\lambda}{4d}$$

If $\lambda \ll d$, then $\sin\theta = \frac{\lambda}{4d} \ll 1$ and so $\sin\theta \approx \theta$. This is the angle from the central maximum to the location of half intensity. The angular displacement from the half-intensity position on one side of the central maximum to the half-intensity position on the other side would be twice this.

$$\Delta\theta = 2\theta_{1/2} = 2\frac{\lambda}{4d} = \boxed{\frac{\lambda}{2d}}$$

25. (a) An incident wave that reflects from the outer surface of the bubble has a phase change of $\phi_1 = \pi$. An incident wave that reflects from the inner surface of the bubble has a phase change due to the additional path length, so $\phi_2 = (2t / \lambda_{film})2\pi$.

For destructive interference with a minimum non-zero thickness of bubble, the net phase change must be π.

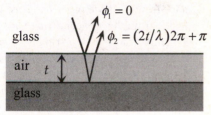

$$\phi_{net} = \phi_2 - \phi_1 = \left[\left(\frac{2t}{\lambda_{film}}\right)2\pi\right] - \pi = \pi \quad \rightarrow \quad t = \tfrac{1}{2}\lambda_{film} = \frac{\lambda}{2n} = \frac{480\,\text{nm}}{2(1.33)} = \boxed{180\,\text{nm}}$$

(b) For the next two larger thicknesses, the net phase change would be 3π and 5π.

$$\phi_{net} = \phi_2 - \phi_1 = \left[\left(\frac{2t}{\lambda_{film}}\right)2\pi\right] - \pi = 3\pi \quad \rightarrow \quad t = \lambda_{film} = \frac{\lambda}{n} = \frac{480\,\text{nm}}{(1.33)} = \boxed{361\,\text{nm}}$$

$$\phi_{net} = \phi_2 - \phi_1 = \left[\left(\frac{2t}{\lambda_{film}}\right)2\pi\right] - \pi = 5\pi \quad \rightarrow \quad t = \lambda_{film} = \frac{\lambda}{n} = \tfrac{3}{2}\frac{480\,\text{nm}}{(1.33)} = \boxed{541\,\text{nm}}$$

(c) If the thickness were much less than one wavelength, then there would be very little phase change introduced by additional path length, and so the two reflected waves would have a phase difference of about $\phi_1 = \pi$. This would produce destructive interference.

31. With respect to the incident wave, the wave that reflects from the air at the top surface of the air layer has a phase change of $\phi_1 = 0$. With respect to the incident wave, the wave that reflects from the glass at the bottom surface of the air layer has a phase change due to both the additional path length and reflection, so $\phi_2 = (2t / \lambda)2\pi + \pi$. If the interference is constructive, the net phase change must be an even non-zero integer multiple of π.

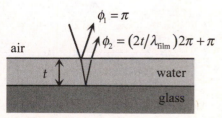

$$\phi_{net} = \phi_2 - \phi_1 = \left[\left(\frac{2t}{\lambda}\right)2\pi + \pi\right] - 0 = 2m\pi \quad \rightarrow \quad t = \tfrac{1}{2}\left(m - \tfrac{1}{2}\right)\lambda, \; m = 1, 2, \cdots.$$

The minimum thickness is with $m = 1$.

$$t_{min} = \tfrac{1}{2}(450\,\text{nm})\left(1 - \tfrac{1}{2}\right) = \boxed{113\,\text{nm}}$$

For destructive interference, the net phase change must be an odd-integer multiple of π.

$$\phi_{net} = \phi_2 - \phi_1 = \left[\left(\frac{2t}{\lambda}\right)2\pi + \pi\right] - 0 = (2m+1)\pi \quad \rightarrow \quad t = \tfrac{1}{2}m\lambda, \; = 0, 1, 2, \cdots$$

The minimum non-zero thickness is $t_{min} = \tfrac{1}{2}(450\,\text{nm})(1) = \boxed{225\,\text{nm}}$.

37. (a) Assume the indices of refraction for air, water, and glass are 1.00, 1.33, and 1.50, respectively. When illuminated from above, a ray reflected from the air-water interface undergoes a phase shift of $\phi_1 = \pi$, and a ray reflected at the water-glass interface also undergoes a phase shift of π. Thus, the two rays are unshifted in phase relative to each other due to reflection. For constructive interference, the path difference $2t$ must equal an integer number of wavelengths in water.

$$2t = m\lambda_{water} = m\frac{\lambda}{n_{water}}, m = 0, 1, 2, \cdots \quad \rightarrow \quad \boxed{\lambda = \frac{2n_{water}t}{m}}$$

(*b*) The previous relation can be solved for the *m*-value associated with the reflected color. If this *m*-value is an integer the wavelength undergoes constructive interference upon reflection.

$$\lambda = \frac{2n_{\text{water}}t}{m} \quad \rightarrow \quad m = \frac{2n_{\text{water}}t}{\lambda}$$

For a thickness $t = 200\,\mu\text{m} = 2 \times 10^5\,\text{nm}$ the *m*-values for the two wavelengths are calculated.

$$m_{700\,\text{nm}} = \frac{2n_{\text{water}}t}{\lambda} = \frac{2(1.33)(2 \times 10^5\,\text{nm})}{700\,\text{nm}} = 760$$

$$m_{400\,\text{nm}} = \frac{2n_{\text{water}}t}{\lambda} = \frac{2(1.33)(2 \times 10^5\,\text{nm})}{400\,\text{nm}} = 1330$$

Since both wavelengths yield integers for *m*, they are both reflected.

(*c*) All *m*-values between $m = 760$ and $m = 1330$ will produce reflected visible colors. There are $1330 - (760 - 1) = \boxed{571}$ such values.

(*d*) This mix of a large number of wavelengths from throughout the visible spectrum will give the thick layer a white or grey appearance.

43. There are two interference patterns formed, one by each of the two wavelengths. The fringe patterns overlap but do not interfere with each other. Accordingly, when the bright fringes of one pattern occurs at the same locations as the dark fringes of the other patterns, there will be no fringes seen, since there will be no dark bands to distinguish one fringe from the adjacent fringes.

To shift from one "no fringes" occurrence to the next, the mirror motion must produce an integer number of fringe shifts for each wavelength, and the number of shifts for the shorter wavelength must be one more than the number for the longer wavelength. From the discussion in section 34-6, we see that the path length change is twice the distance that the mirror moves. One fringe shift corresponds to a change in path length of λ, and so corresponds to a mirror motion of $\frac{1}{2}\lambda$. Let *N* be the number of fringe shifts produced by a mirror movement of Δx.

$$N_1 = 2\frac{\Delta x}{\lambda_1} \quad ; \quad N_2 = 2\frac{\Delta x}{\lambda_2} \quad ; \quad N_2 = N_1 + 1 \quad \rightarrow \quad 2\frac{\Delta x}{\lambda_2} = 2\frac{\Delta x}{\lambda_1} + 1 \quad \rightarrow$$

$$\Delta x = \frac{\lambda_1\lambda_2}{2(\lambda_1 - \lambda_2)} = \frac{(589.6\,\text{nm})(589.0\,\text{nm})}{2(0.6\,\text{nm})} = 2.89 \times 10^5\,\text{nm} \approx \boxed{2.9 \times 10^{-4}\,\text{m}}$$

49. For constructive interference, the path difference is a multiple of the wavelength, as given by Eq. 34-2a. The location on the screen is given by $x = \ell \tan\theta$, as seen in Fig. 34-7(c). For small angles, we have $\sin\theta \approx \tan\theta \approx x/\ell$. Second order means $m = 2$.

$$d\sin\theta = m\lambda \quad \rightarrow$$

$$d\frac{x}{\ell} = m\lambda \quad \rightarrow \quad x = \frac{\lambda m\ell}{d} \quad ; \quad x_1 = \frac{\lambda_1 m\ell}{d} \quad ; \quad x_2 = \frac{\lambda_2 m\ell}{d} \quad \rightarrow$$

$$\Delta x = x_1 - x_2 = \frac{\lambda_1 m\ell}{d} - \frac{\lambda_2 m\ell}{d} \quad \rightarrow$$

$$\lambda_2 = \lambda_1 - \frac{d\Delta x}{m\ell} = 690 \times 10^{-9}\,\text{m} - \frac{(6.6 \times 10^{-4}\,\text{m})(1.23 \times 10^{-3}\,\text{m})}{2(1.60\,\text{m})} = 4.36 \times 10^{-7}\,\text{m} \approx \boxed{440\,\text{nm}}$$

55. We consider a figure similar to Figure 34-12, but with the incoming rays at an angle of θ_i to the normal. Ray s_2 will travel an extra distance $\Delta\ell_1 = d\sin\theta_i$ before reaching the slits, and an extra distance $\Delta\ell_2 = d\sin\theta$ after leaving the slits. There will be a phase difference between the waves due to the path difference $\Delta\ell_1 + \Delta\ell_2$. When this total path difference is a multiple of the wavelength, constructive interference will occur.

$$\Delta\ell_1 + \Delta\ell_2 = d\sin\theta_i + d\sin\theta = m\lambda \quad \rightarrow$$

$$\sin\theta = \frac{m\lambda}{d} - \sin\theta_i, \quad m = 0, 1, 2, \cdots$$

Since the rays leave the slits at all angles in the forward direction, we could have drawn the leaving rays with a downward tilt instead of an upward tilt. This would make the ray s_2 traveling a longer distance from the slits to the screen. In this case the path difference would be $\Delta\ell_2 - \Delta\ell_1$, and would result in the following expression.

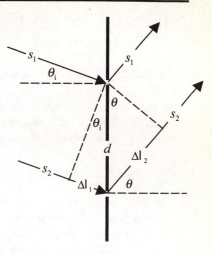

$$\Delta\ell_2 - \Delta\ell_1 = d\sin\theta - d\sin\theta_i = m\lambda \quad\rightarrow$$

$$\sin\theta = \frac{m\lambda}{d} + \sin\theta_i, \quad m = 0, 1, 2, \cdots$$

$$\Delta\ell_1 - \Delta\ell_2 = d\sin\theta_i - d\sin\theta = m\lambda \quad\rightarrow$$

$$\sin\theta = -\frac{m\lambda}{d} + \sin\theta_i, \quad m = 0, 1, 2, \cdots$$

We combine the statements as follows.

$$\boxed{\sin\theta = \frac{m\lambda}{d} \pm \sin\theta_i, \quad m = 0, 1, 2, \cdots}$$

Because of an arbitrary choice of taking $\Delta\ell_2 - \Delta\ell_1$, we could also have formulated the problem so that the result would be expressed as $\boxed{\sin\theta = \sin\theta_i \pm m\lambda/d, \quad m = 0, 1, 2, \cdots}$.

61. With respect to the incident wave, the wave that reflects from the top surface of the film has a phase change of $\phi_1 = \pi$. With respect to the incident wave, the wave that reflects from the glass ($n = 1.52$) at the bottom surface of the film has a phase change due to both the additional path length and reflection:

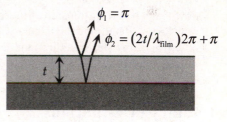

$$\phi_2 = \left(\frac{2t}{\lambda_{\text{film}}}\right)2\pi + \pi$$

For constructive interference, the net phase change must be an even non-zero integer multiple of π.

$$\phi_{\text{net}} = \phi_2 - \phi_1 = \left[\left(\frac{2t}{\lambda_{\text{film}}}\right)2\pi + \pi\right] - \pi = m2\pi \quad\rightarrow\quad t = \tfrac{1}{2}m\lambda_{\text{film}} = \tfrac{1}{2}m\frac{\lambda}{n_{\text{film}}}, \quad m = 1, 2, 3, \ldots.$$

The minimum non-zero thickness occurs for $m = 1$.

$$t_{\min} = \frac{\lambda}{2n_{\text{film}}} = \frac{643\,\text{nm}}{2(1.34)} = \boxed{240\,\text{nm}}$$

Chapter 35: Diffraction and Polarization

Chapter Overview and Objectives

This chapter introduces diffraction and polarization. The diffraction pattern of a single slit is described in detail. The relationship between diffraction and the resolution of optical instruments is discussed. Diffraction gratings and their applications to spectrometers are introduced. Polarization is described and methods of polarizing light are discussed.

After completing this chapter you should:
- Know how diffraction arises from the wave nature of light.
- Know how to determine where the single-slit diffraction intensity minima are located.
- Know how to determine the intensity of light at a given position in a single-slit diffraction pattern.
- Know how the calculation of the intensity distribution of a double-slit interference experiment is modified by the fact that a double-slit experiment must use two finite-width single slits.
- Know where the first minimum intensity in the diffraction pattern of a circular aperture is located.
- Know what it means to resolve two point sources and Rayleigh's criteria.
- Know what a diffraction grating is.
- Know how to determine the resolution and resolving power of a diffraction grating.
- Know what polarization is.
- Know that a polarizer transmits linearly polarized light.
- Know how to determine the intensity of light transmitted by a polarizer.
- Know what Brewster's angle is and how to determine it from the refractive indices of the materials on either side of a reflecting surface.

Summary of Equations

Position of minima in a single-slit diffraction pattern:

$$D \sin \theta = m\lambda \qquad m \in \{1,2,3,...\} \quad \text{(Section 35-1)}$$

Intensity distribution of a single-slit diffraction pattern:

$$I_\theta = I_0 \left[\frac{\sin\left(\frac{\pi D \sin \theta}{\lambda}\right)}{\left(\frac{\pi D \sin \theta}{\lambda}\right)} \right]^2 \qquad \text{(Section 35-2)}$$

Intensity distribution of a double-slit diffraction pattern:

$$I_\theta = I_0 \left[\frac{\sin\left(\frac{\pi D \sin \theta}{\lambda}\right)}{\left(\frac{\pi D \sin \theta}{\lambda}\right)} \right]^2 \cos^2\left(\frac{\pi d \sin \theta}{\lambda}\right)$$

$$\text{(Section 35-3)}$$

First minimum of a circular aperture diffraction pattern:

$$\theta = \frac{1.22\lambda}{D} \qquad \text{(Section 35-4)}$$

Resolving power of a microscope:

$$RP = \frac{1.22\lambda f}{D} \qquad \text{(Section 35-5)}$$

Principal maxima of diffraction gratings:

$$\Delta\theta_m = \frac{\lambda}{Nd \cos\theta_m} \qquad \text{(Section 35-7)}$$

Definition of resolving power:	$R = \dfrac{\lambda}{\Delta\lambda}$	(Section 35-9)
Resolving power of a diffraction grating:	$R = Nm$	(Section 35-9)
Bragg condition:	$m\lambda = 2d\sin\phi \quad m \in \{1,2,3,\dots\}$	(Section 35-10)
Intensity of polarized light through a polarizer:	$I = I_0 \cos^2\theta$ (Malus' law)	(Section 35-11)
Intensity of unpolarized light through a polarizer:	$I = \tfrac{1}{2}I_0$	(Section 35-11)
Brewster's angle:	$\tan\theta_p = \dfrac{n_2}{n_1}$	(Section 35-11)

Chapter Summary

Section 35-1. Diffraction by a Single Slit or Disk

Plane waves of light passing through a slit of width D produce a diffraction pattern on a screen far from the slit. The minima in the diffraction pattern occur when

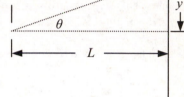

$$D\sin\theta = m\lambda \qquad m \in \{1,2,3,\dots\}$$

where λ is the wavelength of the light and θ is the angle where the minima occur (see diagram.)

Example 35-1-A. Determining the wavelength of light using single slit diffraction patterns. A single-slit diffraction pattern is formed on a screen a distance $L = 1.4$ m from the slit. The third minimum from the center peak in the diffraction pattern is located a distance $\Delta y = 1.9$ cm from the sixth minimum. If the width of the slit is $D = 0.16$ mm, what is the wavelength of the light illuminating the slit?

Approach: The third and sixth minima occur when $m = 3$ and $m = 6$. The condition for the minima can be used to determine the angles at which these minima occur. Using the information provided in the problem we can determine the positions of these minima on the screen, and thus the distance between them. Since this distance is known, we can use this information to determine the wavelength of the light.

Solution: For small angles, we can make the following approximation: $\sin\theta \approx \theta$. The requirement for the minima in the diffraction pattern can now be rewritten as:

$$D\theta = m\lambda \quad \Rightarrow \quad \theta = \frac{m\lambda}{D}$$

The positions of the minima along the screen are located at

$$y_m = \frac{m\lambda L}{D}$$

The problem states that $y_6 - y_3 = 1.9$ cm. This implies that

$$\frac{6\lambda L}{D} - \frac{3\lambda L}{D} = \Delta y$$

The previous equation can be used to calculate the wavelength of the light:

$$\lambda = \frac{D\Delta y}{3L} = \frac{\left(1.6 \times 10^{-4}\right)\left(1.9 \times 10^{-2}\right)}{3\left(1.4\right)} = 7.2 \times 10^{-7} \text{ m}$$

Section 35-2. Intensity in Single-Slit Diffraction Pattern

The intensity of the diffraction pattern as a function of θ is given by

$$I_\theta = I_0 \left[\frac{\sin\left(\dfrac{\pi D \sin\theta}{\lambda}\right)}{\left(\dfrac{\pi D \sin\theta}{\lambda}\right)} \right]^2$$

where I_0 is the intensity at $\theta = 0°$.

Example 35-2-A. Positions with specific intensities in a diffraction pattern. The intensity at the central peak of a single-slit diffraction pattern is 1.56 W/m². The wavelength of the light is 640 nm and the width of the slit is 0.24 mm. What are the first three angles away from the central maximum at which the intensity is 0.012 W/m²?

Approach: To solve the problem we use the known intensity distribution as a function of θ. This equation has one unknown, the angle θ, which can now be determined.

Solution: The expression for the intensity as a function of angle in the diffraction pattern is given by

$$I_\theta = I_0 \left[\frac{\sin\left(\dfrac{\pi a \sin\theta}{\lambda}\right)}{\left(\dfrac{\pi a \sin\theta}{\lambda}\right)} \right]^2 \quad \Rightarrow \quad \frac{\sin\left(\dfrac{\pi a \sin\theta}{\lambda}\right)}{\left(\dfrac{\pi a \sin\theta}{\lambda}\right)} = \pm\sqrt{\frac{I_\theta}{I_0}} = \pm\sqrt{\frac{0.0120}{1.56}} = \pm 0.0877$$

This equation can be rewritten as

$$\frac{\sin x}{x} = \pm 0.0877$$

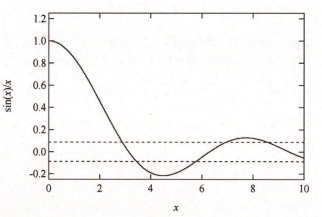

In this equation x represents $\pi a \sin\theta/\lambda$. This equation has no algebraic solution and we have to solve it numerically. An easy way to solve this equation is to make a graph of $\sin(x)/x$ as function of x and determine when the function value is ± 0.0877. The graph of $\sin(x)/x$ is shown in the Figure on the right. The two dashed horizontal lines represent the function values of ± 0.0877. As can be seen from the graph, the three lowest values of x for which $\sin(x)/x$ has a value of ± 0.0877 are 2.89, 3.45, and 5.75. Using these values of x we can now determine the corresponding angles

$$\theta = \arcsin\left(\frac{\lambda x}{\pi a}\right) = \arcsin\left(\frac{\left(640 \times 10^{-9}\right)\left(2.89\right)}{\pi\left(0.24 \times 10^{-3}\right)}\right) = 2.45 \times 10^{-3} \text{ rad}$$

$$= \arcsin\left(\frac{\left(640 \times 10^{-9}\right)\left(3.45\right)}{\pi\left(0.24 \times 10^{-3}\right)}\right) = 2.93 \times 10^{-3} \text{ rad}$$

$$= \arcsin\left(\frac{\left(640 \times 10^{-9}\right)\left(5.75\right)}{\pi\left(0.24 \times 10^{-3}\right)}\right) = 4.88 \times 10^{-3} \text{ rad}$$

Section 35-3. Diffraction in the Double-Slit Experiment

The double-slit or two-slit interference pattern of Chapter 34 is the result of light passing through two closely spaced single slits. The discussion of Chapter 34 was incomplete because it ignored the single-slit diffraction effects due to the finite width of each of the two slits.

Consider two slits, each with a width D, separated by a distance d. The top interference pattern in the figure on the right shows the double-slit interference pattern obtained in Chapter 34 for two slits, separated by a distance d. The middle diffraction pattern in the Figure shows the diffraction pattern due to a single slit of width $D = d/3$. The intensity of the double-slit interference pattern, which includes single-slit diffraction effects, is given by

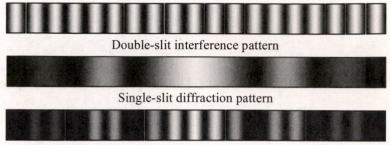

Double-slit interference pattern

Single-slit diffraction pattern

Double-slit interference pattern including single-slit diffraction

$$I_\theta = I_0 \left[\frac{\sin\left(\frac{\pi D \sin\theta}{\lambda}\right)}{\left(\frac{\pi D \sin\theta}{\lambda}\right)} \right]^2 \cos^2\left(\frac{\pi d \sin\theta}{\lambda}\right)$$

Example 35-3-A. Obtaining slit information from a diffraction pattern. A two-slit interference pattern is missing the 12th-order, 24th-order, and 36th-order constructive interference intensity peaks. What is the ratio of the separation of the slits to the width of each slit?

Approach: Some of the constructive interference peaks are missing since they coincide with the location of single-slit diffraction minima. The information provided tells us that the first single-slit diffraction minimum occurs at the location of the 12th-order two-slit interference maximum.

Solution: The location of the 12th-order double-slit interference maximum is

$$y_{12,double} = \frac{12\lambda\ell}{d}$$

The location of the 1st-order single-slit diffraction minimum is

$$y_{1,single} = \frac{\lambda\ell}{D}$$

The observation that the interference pattern is missing the 12th-order constructive interference peak requires that

$$y_{12,double} = y_{1,single} \quad \Rightarrow \quad \frac{12\lambda\ell}{d} = \frac{\lambda\ell}{D} \quad \Rightarrow \quad \frac{d}{D} = 12$$

Section 35-4. Limits of Resolution; Circular Apertures

Circular apertures also produce diffraction patterns. The diffraction patterns are circular and the locations of the minima are not uniformly spaced. The first minimum in the diffraction pattern is located at an angle θ from the center of the central bright spot where

$$\theta = \frac{1.22\lambda}{D}$$

In this equation, D is the diameter of the circular opening.

Most optical instruments have circular apertures. The quality of images formed by optical instruments is diffraction limited. The angular width of the image of a point source cannot be smaller than the diffraction limit given above. This also limits how well the optical instrument can resolve two closely spaced point sources of light. A precise limit on the resolution of the instrument depends on how precisely the intensity of the light in the diffraction pattern is measured.

Because many instruments are viewed with the eye, which may only see the intensity of the central bright spot of the diffraction pattern, an appropriate definition of the resolution is the **Rayleigh criterion**. The Rayleigh criterion states that two sources can be resolved if the diffraction pattern central bright spot of one source falls on or outside of the first diffraction pattern minimum of the other light source. This implies that the angular separation of the sources of light must be

$$\theta = \frac{1.22\lambda}{D}$$

Section 35-5. Resolution of Telescopes and Microscopes; the λ Limit

The resolution of a telescope is usually expressed in terms of the minimum angular separation of two distant sources that can be resolved by the telescope (see Section 35-4).

The **resolving power RP** of a microscope is expressed in terms of the minimum distance between two sources that are barely resolvable. If the objective lens is the smallest aperture in the microscope, the resolving power is given by

$$RP = \frac{1.22\lambda f}{D}$$

where λ is the wavelength of the light, f is the focal length of the objective lens, and D is the diameter of the aperture formed by the objective lens.

The lower limit of the focal length of the objective lens is roughly the diameter of the lens divided by 2. The corresponding resolving power is

$$RP = \frac{1.22\lambda\left(\dfrac{D}{2}\right)}{D} = 0.61\lambda$$

This results in a practical rule of thumb regarding the limit of the resolving power of optical instruments:

> **The minimum distance between two sources of light that are resolvable by an optical instrument is about one wavelength of the light used.**

Example 35-5-A. Resolving power of the eye. Assume the opening of the pupil of the eye has a diameter $D = 2.8$ mm. What is the smallest distance between two light sources, emitting light with a wavelength $\lambda = 500$ nm and located a distance $L = 1.00$ mile from the eye, that can be seen by the eye as two separate sources of light?

Approach: Rayleigh's criterion can be used to determine the minimum separation angle between the two light sources in order to be resolved by the eye. This angle, combined with the known distance between the sources and the eye, can be used to determine the minimum separation between the sources.

Solution: The minimum angular separation between the two sources is equal to

$$\theta = \frac{1.22\lambda}{D}$$

Since the wavelength of the light is much smaller than the diameter of the pupil, we can use the small angle approximation and replace θ with the ratio of the separation of the sources, d, and the distance between the sources and the eye. The Rayleigh criterion can now be rewritten as

$$\frac{d}{L} = \frac{1.22\lambda}{D} \quad \Rightarrow \quad d = \frac{1.22\lambda L}{D} = \frac{1.22\left(500 \times 10^{-9}\right)\left(1.61 \times 10^{3}\right)}{2.8 \times 10^{-3}} = 0.35\,\text{m}$$

Note that we need to convert all distances to the same units (e.g., 1 mile has been replaced with 1610 m).

Section 35-6. Resolution of the Human Eye and Useful Magnification

The limit of the resolution of the human eye is about 5×10^{-4} radians. This corresponds to a separation distance of about 0.1 mm at the near point of the human eye.

Microscopes using violet light are able to resolve objects as small as 200 nm. In order to make these objects visible to the naked eye, a magnification of a factor of 500 is required. A larger magnification will not provide the user with more details; instead, the image will be diffraction limited.

Section 35-7. Diffraction Grating

An optical device that has many equally spaced parallel slits is called a **diffraction grating**. If the diffraction grating is made of transmitting elements, it is called a **transmission grating**; if the diffraction grating is made of reflecting elements, it is called a **reflection grating**.

If the spacing between each element in the diffraction grating is d, the light that strikes the grating will interfere constructively in directions given by

$$\sin\theta = \frac{m\lambda}{d} \qquad m \in \{0,1,2,3,\dots\}$$

The intensity pattern due to a diffraction grating is different from the intensity pattern of a double-slit interference pattern; the width of the constructive interference peaks will depend on the number of elements illuminated by the incoming light and will be narrower than the width of the constructive interference peaks in a double-slit interference pattern.

Section 35-8. The Spectrometer and Spectroscopy

An instrument that is designed to determine the wavelength of light is called a **spectrometer** or **spectroscope**. The principle of operation of a spectrometer, schematically illustrated in the Figure on the right, relies in the dispersion of the light of unknown wavelength with a transmission grating. Since the constructive interference peaks are very narrow, the corresponding angles can be measured with great accuracy. Based on the measurements of the order of the interference peaks and the corresponding dispersion angle, the wavelength of the light can be calculated:

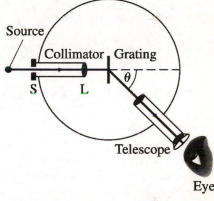

$$\lambda = \frac{d}{m}\sin\theta$$

Section 35-9. Peak Widths and Resolving Power for a Diffraction Grating

The angular width $\Delta\theta_m$ of the constructive interference peaks for a diffraction grating is given by

$$\Delta\theta_m = \frac{\lambda}{Nd\cos\theta_m},$$

where θ_m is the angular position of the mth peak, N is the number of diffraction grating elements illuminated by the light, and d is the spacing between the diffraction grating elements. The **resolving power** R of a diffraction grating is defined as the ratio of the wavelength of the light, λ, and the minimum difference in wavelengths, $\Delta\lambda$, that can be resolved by the diffraction grating:

$$R = \frac{\lambda}{\Delta\lambda}$$

The resolving power of a diffraction grating is related to the properties of the grating and the order of the constructive interference peak in the diffraction pattern:

$$R = Nm$$

Example 35-9-A. Resolving wavelengths with a diffraction grating. A transmission grating with a total width of 2.00 cm has slits separated by 180 μm. Determine the angle at which the first-order peak in the diffraction pattern occurs if the grating is illuminated by light with a wavelength of 532 nm. Determine the next largest wavelength that can be resolved from the 532 nm light by this diffraction grating.

Approach: The problem provides us with sufficient information to determine the position of the 1st-order diffraction maximum and the resolving power of the grating at the wavelength and the order of the peak under consideration.

Solution: The condition for constructive diffraction is given by

$$\sin\theta = \frac{m\lambda}{d} \quad\Rightarrow\quad \theta = \arcsin\left\{\frac{m\lambda}{d}\right\} = \arcsin\left\{\frac{1\left(532\times10^{-9}\right)}{180\times10^{-6}}\right\} = 2.96\times10^{-3}\,\text{rad}$$

The minimum difference in wavelengths that can be resolved with the grating is

$$\Delta\lambda = \frac{\lambda}{R} = \frac{\lambda}{Nm}$$

where N is the number of grating elements that are illuminated by the light. N is equal to the width of the grating divided by the distance between the elements:

$$N = \frac{w}{d} = \frac{2.00\times10^{-2}}{180\times10^{-6}} = 111$$

The minimum difference in wavelengths that can be resolved is thus equal to

$$\Delta\lambda = \frac{\lambda}{Nm} = \frac{532\times10^{-9}}{(111)(1)} = 4.79\times10^{-9}\,\text{m} \approx 5\,\text{nm}$$

Based on this value of $\Delta\lambda$, light with a wavelength of $\lambda + \Delta\lambda = 537$ nm can be resolved from light with a wavelength of $\lambda = 532$ nm.

Section 35-10. X-Rays and X-Ray Diffraction

X-rays penetrate the surface of materials and diffraction of X-rays from crystals depends not only on the spacing of atoms along the surface, but also on the spacing of the atoms below the surface. The directions of constructive interference from this three-dimensional array of atoms are given by the following condition:

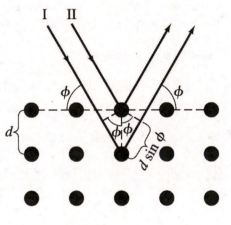

$$m\lambda = 2d\sin\phi \qquad m \in \left\{0,1,2,3,\dots\right\}$$

where d is the perpendicular distance between two planes of atoms in the array and ϕ is the angle between the incident X-ray and the surface (see Figure on the right). This condition is called the **Bragg condition** for X-ray diffraction. X-ray diffraction is commonly used to determine the arrangement of atoms in solid materials.

Section 35-11. Polarization

Because there are two independent directions that are perpendicular to the direction of propagation of an electromagnetic wave, transverse waves have two independent directions of **polarization**. If the electric field of an electromagnetic wave is oriented in a constant direction, it is said to be **plane-polarized** or **linearly polarized**. Most sources of light emit **unpolarized** light in which the electric field changes direction rapidly with time in a random fashion.

A **polarizer** can be used to create polarized light from unpolarized light. There are several different ways to construct a polarizer. A **polarizer** is a sheet of semi-transparent material that transmits light of one linear polarization and absorbs light with a perpendicular direction of polarization. The direction of the electric field of linearly polarized light that is transmitted through the polarizer is called the **axis** of the polarizer. The light transmitted by a polarizer is linearly polarized in the direction of the axis of the polarizer.

The intensity of the light transmitted by a polarizer depends on the polarization of the incident light. If the light incident on the polarizer is unpolarized, the transmitted intensity I is related to the incident intensity I_0 in the following way

$$I = \tfrac{1}{2}I_0$$

If the light incident on the polarizer is linearly polarized and the direction of polarization makes an angle θ with respect to the axis of the polarizer, the transmitted intensity is related to the incident intensity in the following way:

$$I = I_0 \cos^2 \theta$$

This relation is called **Malus' law**. By measuring the intensity of the light transmitted by the polarizer we can determine whether the light is linearly polarized or not, and determine the direction of the polarization.

Light is also polarized by reflection from surfaces, if the incident angle is not zero. The polarization is not complete unless the angle of incidence is equal to the Brewster's angle θ_p:

$$\tan\theta_p = \frac{n_2}{n_1}$$

where n_1 is the refractive index of the material on the incident side of the reflecting surface and n_2 is the refractive index on the other side of the reflecting surface.

Example 35-11-A. Transmission of polarized and unpolarized light through polarizers. Unpolarized light is incident on a polarizer. The light that passes through the polarizer is incident upon a second polarizer with its axis tilted at an angle of 50° relative to the axis of the first polarizer. What fraction of the initial intensity of the unpolarized light is transmitted through the second polarizer?

Approach: After passing through the first polarizer, the unpolarized light becomes polarized. When the now polarized light travels through the second polarizer, only its component parallel to the axis of the second polarizer will be transmitted. The problem provides sufficient information to determine the final intensity of the light.

Solution: The initial intensity of the unpolarized light is I_0. After the light passes through the first polarizer, its intensity will be reduced to I_1 where

$$I_1 = \tfrac{1}{2} I_0$$

This light will be linearly polarized in the direction of the axis of the first polarizer. The direction of polarization makes an angle of 50° with respect to the axis of the second polarizer. The intensity of the light emerging from the second polarizer is given by

$$I_2 = I_1 \cos^2\left(50^0\right) = \tfrac{1}{2} I_0 \cos^2\left(50^0\right) = 0.21 I_0$$

The light emerging from the second polarizer will be polarized parallel to the axis of the second polarizer and the intensity has been reduced by 21% of the intensity of the incident light.

Example 35-11-B. Calculating Brewster's angle using refraction. Light reaching the surface of a piece of glass from air at an incident angle of 48° is refracted at an angle of 33°. What is Brewster's angle for this surface?

Approach: The information about the refraction of the light can be used to determine the ratio of the indices of refraction of the materials. This ratio can be used to determine Brewster's angle.

Solution: The index of refraction of region 1, the air, is equal to 1.00. The index of refraction of region 2 can be determined using Snell's law since we know the angle of incidence and the angle of refraction:

$$n_1 \sin\theta_1 = n_2 \sin\theta_2 \quad\Rightarrow\quad \frac{n_2}{n_1} = \frac{\sin\theta_1}{\sin\theta_2}$$

Using this ratio of the indices of refraction, we can now determine Brewster's angle:

$$\tan\theta_p = \frac{n_2}{n_1} = \frac{\sin\theta_1}{\sin\theta_2} \quad\Rightarrow\quad \theta_p = \operatorname{atan}\left\{\frac{\sin\theta_1}{\sin\theta_2}\right\} = 54°$$

Section 35-12. Liquid Crystal Displays (LCD)

Liquid crystal displays (LCDs) rely on polarization for their operation. Each pixel of the display contains two polarizers, mounted with their axes perpendicular to each other, with a liquid crystal between them. The molecules in the liquid crystal are oriented such that they change the direction of polarization of incident light by 90°. As a result, light incident on one side of the crystal will be transmitted through the system. When an external voltage is applied across the crystal, the molecules in the crystal will realign themselves in such a way that the direction of polarization of the light is not changed, and the light is blocked by the second polarizer.

Section 35-13. Scattering of Light by the Atmosphere

Light is scattered by gas molecules due to the induced acceleration of the electrons of the molecules. These accelerating electrons radiate electromagnetic waves. The intensity of the radiated light depends on the angle between the direction of propagation and the direction of the accelerating electrons. No light is emitted in a direction parallel to the direction of the electron's acceleration. As a consequence, light scattered by the molecules in a gas will be partially polarized.

Practice Quiz

1. Near the center of a single-slit diffraction pattern, the minima may appear to be equally spaced. What happens to the spacing between the minima when the order number of the minima increases?
    a)   The spacing remains the same.
    b)   The spacing becomes smaller.
    c)   The spacing becomes larger.
    d)   The spacing can become larger or smaller, depending on the wavelength of the light.

2. The third-order single-slit diffraction minimum for light with a wavelength of 540 nm falls on the fourth-order minimum for light of unknown wavelength. What is the unknown wavelength?
    a)   720 nm.
    b)   405 nm.
    c)   640 nm.
    d)   360 nm.

3. For the double-slit interference pattern shown in the Figure below, what is the ratio of the center-to-center separation of the slits to the width of each slit?
    a)   2.
    b)   3.
    c)   4.
    d)   8.

4. You are viewing objects with your eye. Which wavelength should you use to be able to resolve the objects with the greatest amount of detail?
    a)   200 nm.
    b)   450 nm.
    c)   600 nm.
    d)   750 nm.

5. Two polarizers have their axes at right angles to one another and no light emerges from the second polarizer. Where should a third polarizer be placed and how should it be oriented in order to transmit the maximum amount of light through the three polarizers?
    a)   Place the third polarizer after the second polarizer with its axis parallel to the first polarizer's axis.
    b)   Place the third polarizer before the first polarizer with its axis parallel to the second polarizer's axis.
    c)   Place the third polarizer between the first two polarizers with its axis at 45° to the other two axes.
    d)   It is impossible to place the third polarizer so that any light is transmitted through the system.

6. A telescope with a primary mirror of diameter D is designed to be diffraction limited by a primary mirror. A suspected binary star appears as one bright spot on the photographic plates and CCD images obtained with the telescope. In order to resolve the binary star as two stars, what improvement needs to be made?
    a) A better camera must be attached to the telescope.
    b) Film with better resolution must be used.
    c) A longer exposure time must be used.
    d) A telescope with a larger diameter primary mirror must used.

7. A large concrete and steel building is directly between you and a radio tower. The radio waves cannot pass through the building, but your radio receives the station just fine. What is one possible reason your radio may still be receiving the station?
    a) The radio waves are diffracting around the building.
    b) The atoms in the building are spaced such that the Bragg scattering condition is satisfied for the wavelength of the radio station.
    c) The radio waves do not need to reach the radio receiver in order to receive the radio station.
    d) The radio waves are polarized parallel to the ground, but the building is standing vertically.

8. What models of solids are consistent with the results of X-rays scattering data off solid materials.
    a) The atoms in all solids are randomly dispersed throughout the volume of the solid.
    b) Atoms in many materials are arranged in orderly arrays.
    c) There are no such thing as atoms.
    d) Solids are only solid on the surface, but are liquid inside.

9. What do you know about the light that is transmitted by the polarizer?
    a) The transmitted light is lower in intensity than the incident light.
    b) The transmitted light is polarized in the direction of the axis of the polarizer.
    c) Both a) and b) are true.
    d) There is nothing you can conclude unless you know if the incident light was polarized or unpolarized.

10. You measure the intensity of light passing through a polarizer to be one-half the intensity of the light incident on the polarizer. What do you know?
    a) The light incident on the polarizer was unpolarized.
    b) The light incident on the polarizer was linearly polarized and its polarization direction was at an angle of 45° to the polarizer's axis.
    c) Either a) or b) is true.
    d) The polarizer is not functioning correctly.

11. Two wavelengths, λ_1 and λ_2, illuminate a single slit. The third minimum in the diffraction pattern of λ_1 is located in the same position as the sixth minimum of λ_2. If $\lambda_1 = 540$ nm, what is λ_2?

12. A single slit of width 0.180 mm is illuminated with light of wavelength 488 nm. A diffraction pattern forms on a screen that is 1.25 m away. At what distance from the central peak of the single-slit diffraction pattern is the intensity reduced to one-fourth of the intensity at the center of the central peak?

13. What is the minimum angular separation of two distant stars that can be resolved with a telescope with an effective aperture of 1.20 m when viewing the stars using light of wavelength 589 nm?

14. Unpolarized light is incident on a polarizer with its axis aligned in the vertical direction. The light then passes through a second polarizer with its axis tipped clockwise in a direction 30° to the vertical. Finally, the light passes though a third polarizer with its axis tipped clockwise in a direction of 80° to vertical. What fraction of the intensity of the initially unpolarized light is transmitted through the final polarizer?

15. Brewster's angle is determined to be 56° for the surface of a material when light is incident on the surface from air. What is the refractive index of the material?

Responses to Select End-of-Chapter Questions

1. Radio waves have a much longer wavelength than visible light and will diffract around normal-sized objects (like hills). The wavelengths of visible light are very small and will not diffract around normal-sized objects.

7. The intensity pattern is actually a function of the form $\left(\dfrac{\sin x}{x}\right)^2$ (see equations 35-7 and 35-8). The maxima of this function do not coincide exactly with the maxima of $\sin^2 x$. You can think of the intensity pattern as the combination of a $\sin^2 x$ function and a $1/x^2$ function, which forces the intensity function to zero and shifts the maxima slightly.

13. A large mirror has better resolution and gathers more light than a small mirror.

19. (*a*) Violet light will be at the top of the rainbow created by the diffraction grating. Principal maxima for a diffraction grating are at positions given by $\sin\theta = \dfrac{m\lambda}{d}$. Violet light has a shorter wavelength than red light and so will appear at a smaller angle away from the direction of the horizontal incident beam.
(*b*) Red light will appear at the top of the rainbow created by the prism. The index of refraction for violet light in a given medium is slightly greater than for red light in the same medium, and so the violet light will bend more and will appear farther from the direction of the horizontal incident beam.

Solutions to Select End-of-Chapter Problems

1. We use Eq. 35-1 to calculate the angular distance from the middle of the central peak to the first minimum. The width of the central peak is twice this angular distance.

$$\sin\theta_1 = \frac{\lambda}{D} \to \theta_1 = \sin^{-1}\left(\frac{\lambda}{D}\right) = \sin^{-1}\left(\frac{680 \times 10^{-9}\,\text{m}}{0.0365 \times 10^{-3}\,\text{m}}\right) = 1.067°$$

$$\Delta\theta = 2\theta_1 = 2(1.067°) = \boxed{2.13°}$$

7. We use the distance to the screen and half the width of the diffraction maximum to calculate the angular distance to the first minimum. Then using this angle and Eq. 35-1 we calculate the slit width. Then using the slit width and the new wavelength we calculate the angle to the first minimum and the width of the diffraction maximum.

$$\tan\theta_1 = \frac{\left(\frac{1}{2}\Delta y_1\right)}{\ell} \to \theta_1 = \tan^{-1}\frac{\left(\frac{1}{2}\Delta y_1\right)}{\ell} = \tan^{-1}\frac{\left(\frac{1}{2} \times 0.06\,\text{m}\right)}{2.20\,\text{m}} = 0.781°$$

$$\sin\theta_1 = \frac{\lambda_1}{D} \to D = \frac{\lambda_1}{\sin\theta_1} = \frac{580\,\text{nm}}{\sin 0.781°} = 42{,}537\,\text{nm}$$

$$\sin\theta_2 = \frac{\lambda_2}{D} \to \theta_2 = \sin^{-1}\left(\frac{\lambda_2}{D}\right) = \sin^{-1}\left(\frac{460\,\text{nm}}{42{,}537\,\text{nm}}\right) = 0.620°$$

$$\Delta y_2 = 2\ell\tan\theta_2 = 2(2.20\,\text{m})\tan(0.620°) = 0.0476\,\text{m} \approx \boxed{4.8\,\text{cm}}$$

13. We use Eq. 35-8 to calculate the intensity, where the angle θ is found from the displacement from the central maximum (15 cm) and the distance to the screen.

$$\tan\theta = \frac{y}{\ell} \to \theta = \tan^{-1}\left(\frac{15\,\text{cm}}{25\,\text{cm}}\right) = 31.0°$$

$$\beta = \frac{2\pi}{\lambda}D\sin\theta = \frac{2\pi}{\left(750 \times 10^{-9}\,\text{m}\right)}\left(1.0 \times 10^{-6}\,\text{m}\right)\sin 31.0° = 4.31\,\text{rad}$$

$$\frac{I}{I_0} = \left(\frac{\sin \beta/2}{\beta/2}\right)^2 = \left(\frac{\sin(4.31\,\text{rad}/2)}{4.31\,\text{rad}/2}\right)^2 = 0.1498 \approx \boxed{0.15}$$

So the light intensity at 15 cm is about 15% of the maximum intensity.

19. (a) The angle to each of the maxima of the double slit are given by Eq. 34-2a. The distance of a fringe on the screen from the center of the pattern is equal to the distance between the slit and screen multiplied by the tangent of the angle. For small angles, we can set the tangent equal to the sine of the angle. The slit spacing is found by subtracting the distance between two adjacent fringes.

$$\sin\theta_m = \frac{m\lambda}{d} \qquad y_m = \ell\tan\theta_m \approx \ell\sin\theta_m = \ell\frac{m\lambda}{d}$$

$$\Delta y = y_{m+1} - y_m = \ell\frac{(m+1)\lambda}{d} - \ell\frac{m\lambda}{d} = \frac{\ell\lambda}{d} = \frac{(1.0\,\text{m})(580\times10^{-9}\,\text{m})}{0.030\times10^{-3}\,\text{m}} = 0.019\,\text{m} = \boxed{1.9\,\text{cm}}$$

(b) We use Eq. 35-1 to determine the angle between the center and the first minimum. Then by multiplying the distance to the screen by the tangent of the angle we find the distance from the center to the first minima. The distance between the two first order diffraction minima is twice the distance from the center to one of the minima.

$$\sin\theta_1 = \frac{\lambda}{D} \rightarrow \theta_1 = \sin^{-1}\frac{\lambda}{D} = \sin^{-1}\frac{580\times10^{-9}\,\text{m}}{0.010\times10^{-3}\,\text{m}} = 3.325°$$

$$y_1 = \ell\tan\theta_1 = (1.0\,\text{m})\tan3.325° = 0.0581\,\text{m}$$

$$\Delta y = 2y_1 = 2(0.0581\,\text{m}) = 0.116\,\text{m} \approx \boxed{12\,\text{cm}}$$

25. The angular resolution is given by Eq. 35-10. The distance between the stars is the angular resolution times the distance to the stars from the Earth.

$$\theta = 1.22\frac{\lambda}{D} \; ; \; \ell = r\theta = 1.22\frac{r\lambda}{D} = 1.22\frac{(16\,\text{ly})\left(\dfrac{9.46\times10^{15}\,\text{m}}{1\,\text{ly}}\right)(550\times10^{-9}\,\text{m})}{(0.66\,\text{m})} = \boxed{1.5\times10^{11}\,\text{m}}$$

31. We use Eq. 35-13 to calculate the wavelengths from the given angles. The slit separation, d, is the inverse of the number of lines per cm, N. We assume that 12,000 is good to 3 significant figures.

$$d\sin\theta = m\lambda \rightarrow \lambda = \frac{\sin\theta}{Nm}$$

$$\lambda_1 = \frac{\sin28.8°}{12,000\,/\text{cm}} = 4.01\times10^{-5}\,\text{cm} = \boxed{401\,\text{nm}} \qquad \lambda_2 = \frac{\sin36.7°}{12,000\,/\text{cm}} = 4.98\times10^{-5}\,\text{cm} = \boxed{498\,\text{nm}}$$

$$\lambda_3 = \frac{\sin38.6°}{12,000\,/\text{cm}} = 5.201\times10^{-5}\,\text{cm} = \boxed{520\,\text{nm}} \qquad \lambda_4 = \frac{\sin47.9°}{12,000\,/\text{cm}} = 6.18\times10^{-5}\,\text{cm} = \boxed{618\,\text{nm}}$$

37. We find the second order angles for the maximum and minimum wavelengths using Eq. 35-13, where the slit separation distance is the inverse of the number of lines per centimeter. Subtracting these two angles gives the angular width.

$$d\sin\theta = m\lambda \rightarrow \theta = \sin^{-1}\left(\frac{m\lambda}{d}\right) = \sin^{-1}(m\lambda N)$$

$$\theta_1 = \sin^{-1}\left[2(4.5\times10^{-7}\,\text{m})(6.0\times10^5\,/\text{m})\right] = 32.7°$$

$$\theta_2 = \sin^{-1}\left[2(7.0\times10^{-7}\,\text{m})(6.0\times10^5\,/\text{m})\right] = 57.1°$$

$$\Delta\theta = \theta_2 - \theta_1 = 57.1° - 32.7° = \boxed{24°}$$

43. We find the angle for each "boundary" color from Eq. 35-13, and then use the fact that the displacement on the screen is given by $\tan\theta = \dfrac{y}{L}$, where y is the displacement on the screen from the central maximum, and L is the distance from the grating to the screen.

$$\sin\theta = \frac{m\lambda}{d} \quad ; \quad d = \frac{1}{610\,\text{lines/mm}}\left(\frac{1\text{m}}{10^3\text{mm}}\right) = \left(1/6.1\times10^5\right)\text{m} \quad ; \quad y = L\tan\theta = L\tan\left[\sin^{-1}\frac{m\lambda}{d}\right]$$

$$l_1 = L\tan\left[\sin^{-1}\frac{m\lambda_{\text{red}}}{d}\right] - L\tan\left[\sin^{-1}\frac{m\lambda_{\text{violet}}}{d}\right]$$

$$= \left(0.32\,\text{m}\right)\left\{\tan\left[\sin^{-1}\frac{(1)\left(700\times10^{-9}\,\text{m}\right)}{\left(1/6.1\times10^5\right)\text{m}}\right] - \tan\left[\sin^{-1}\frac{(1)\left(400\times10^{-9}\,\text{m}\right)}{\left(1/6.1\times10^5\right)\text{m}}\right]\right\}$$

$$= 0.0706\,\text{m} \approx \boxed{7\,\text{cm}}$$

$$l_2 = L\tan\left[\sin^{-1}\frac{m\lambda_{\text{red}}}{d}\right] - L\tan\left[\sin^{-1}\frac{m\lambda_{\text{violet}}}{d}\right]$$

$$= \left(0.32\,\text{m}\right)\left\{\tan\left[\sin^{-1}\frac{(2)\left(700\times10^{-9}\,\text{m}\right)}{\left(1/6.1\times10^5\right)\text{m}}\right] - \tan\left[\sin^{-1}\frac{(2)\left(400\times10^{-9}\,\text{m}\right)}{\left(1/6.1\times10^5\right)\text{m}}\right]\right\}$$

$$= 0.3464\,\text{m} \approx \boxed{35\,\text{cm}}$$

The $\boxed{\text{second order}}$ rainbow is dispersed over a larger distance.

49. We use Eq. 35-20, with $m = 1$.

$$m\lambda = 2d\sin\phi \quad \rightarrow \quad \phi = \sin^{-1}\frac{m\lambda}{2d} = \sin^{-1}\frac{(1)(0.138\,\text{nm})}{2(0.285\,\text{nm})} = \boxed{14.0°}$$

55. The light is traveling from water to diamond. We use Eq. 35-22a.

$$\tan\theta_{\text{p}} = \frac{n_{\text{diamond}}}{n_{\text{water}}} = \frac{2.42}{1.33} = 1.82 \quad \rightarrow \quad \theta_{\text{p}} = \tan^{-1}1.82 = \boxed{61.2°}$$

61. We assume vertically polarized light of intensity I_0 is incident upon the first polarizer. The angle between the polarization direction and the polarizer is θ. After the light passes that first polarizer, the angle between that light and the next polarizer will be $90° - \theta$. Apply Eq. 35-21.

$$I_1 = I_0\cos^2\theta \quad ; \quad I = I_1\cos^2\left(90° - \theta\right) = I_0\cos^2\theta\cos^2\left(90° - \theta\right) = \boxed{I_0\cos^2\theta\sin^2\theta}$$

We can also use the trigonometric identity $\sin\theta\cos\theta = \frac{1}{2}\sin2\theta$ to write the final intensity as $I = I_0\cos^2\theta\sin^2\theta = \boxed{\frac{1}{4}I_0\sin^2 2\theta}$.

$$\frac{dI}{d\theta} = \frac{d}{d\theta}\left(\tfrac{1}{4}I_0\sin^2 2\theta\right) = \tfrac{1}{4}I_0\left(2\sin2\theta\right)\left(\cos2\theta\right)2 = I_0\sin2\theta\cos2\theta = \boxed{\tfrac{1}{2}I_0\sin4\theta}$$

$$\tfrac{1}{2}I_0\sin4\theta = 0 \quad \rightarrow \quad 4\theta = 0, 180°, 360° \quad \rightarrow \quad \theta = 0, 45°, 90°$$

Substituting the three angles back into the intensity equation, we see that the angles $0°$ and $90°$ both give minimum intensity. The angle $\boxed{45°\text{ gives the maximum intensity}}$ of $\frac{1}{4}I_0$.

67. We find the angles for the first order from Eq. 35-13.

$$\theta_1 = \sin^{-1} \frac{m\lambda}{d} = \sin^{-1} \frac{(1)(4.4 \times 10^{-7} \text{ m})}{0.01 \text{ m}/7600} = 19.5°$$

$$\theta_2 = \sin^{-1} \frac{(1)(6.8 \times 10^{-7} \text{ m})}{0.01 \text{ m}/7600} = 31.1°$$

The distances from the central white line on the screen are found using the tangent of the angle and the distance to the screen.

$$y_1 = L \tan \theta_1 = (2.5 \text{ m}) \tan 19.5° = 0.89 \text{ m}$$

$$y_2 = L \tan \theta_2 = (2.5 \text{ m}) \tan 31.1° = 1.51 \text{ m}$$

Subtracting these two distances gives the linear separation of the two lines.

$$y_2 - y_1 = 1.51 \text{ m} - 0.89 \text{ m} = \boxed{0.6 \text{ m}}$$

73. The distance between lines on the diffraction grating is found by solving Eq. 35-13 for d, the grating spacing. The number of lines per meter is the reciprocal of d.

$$d = \frac{m\lambda}{\sin \theta} \quad \rightarrow \quad \frac{1}{d} = \frac{\sin \theta}{m\lambda} = \frac{\sin 21.5°}{(1)6.328 \times 10^{-7} \text{m}} = \boxed{5.79 \times 10^5 \text{ lines/m}}$$

79. For the minimum aperture the angle subtended at the lens by the smallest feature is the angular resolution, given by Eq. 35-10. We let l represent the spatial separation, and r represent the altitude of the camera above the ground.

$$\theta = \frac{1.22\lambda}{D} = \frac{\ell}{r} \quad \rightarrow \quad D = \frac{1.22\lambda r}{\ell} = \frac{1.22(580 \times 10^{-9} \text{ m})(25000 \text{ m})}{(0.05 \text{ m})} = 0.3538 \text{ m} \approx \boxed{0.4 \text{ m}}$$

85. The distance x is twice the distance to the first minima. We can write x in terms of the slit width D using Eq. 35-2, with $m = 1$. The ratio $\dfrac{\lambda}{D}$ is small, so we may approximate $\sin \theta \approx \tan \theta \approx \theta$.

$$\sin \theta = \frac{\lambda}{D} \approx \theta \quad ; \quad x = 2y = 2\ell \tan \theta = 2\ell\theta = 2\ell \frac{\lambda}{D}$$

When the plate is heated up the slit width increases due to thermal expansion. Eq. 17-1b is used to determine the new slit width, with the coefficient of thermal expansion, α, given in Table 17-1. Each slit width is used to determine a value for x. Subtracting the two values for x gives the change Δx. We use the binomial expansion to simplify the evaluation.

$$\Delta x = x - x_0 = 2\ell \left(\frac{\lambda}{D_0 (1 + \alpha\Delta T)} \right) - 2\ell \left(\frac{\lambda}{D_0} \right) = \frac{2\ell\lambda}{D_0} \left(\frac{1}{(1 + \alpha\Delta T)} - 1 \right) = \frac{2\ell\lambda}{D_0} \left((1 + \alpha\Delta T)^{-1} - 1 \right)$$

$$= \frac{2\ell\lambda}{D_0}(1 - \alpha\Delta T - 1) = -\frac{2\ell\lambda}{D_0}\alpha\Delta T = -\frac{2(2.0 \text{ m})(650 \times 10^{-9} \text{m})}{(22 \times 10^{-6} \text{m})} \left[25 \times 10^{-6} (C°)^{-1} \right](55 C°)$$

$$= \boxed{-1.7 \times 10^{-4} \text{ m}}$$

Chapter 36: Special Theory of Relativity

Chapter Overview & Objectives

This chapter introduces the special theory of relativity. It discusses the principles and postulates of special relativity and the kinematical and dynamical implications of these postulates. This chapter defines and discusses the correct relativistic expressions for familiar classical and dynamical quantities such as momentum and energy.

After completing this chapter you should:
- Know the principle of relativity.
- Know the postulates of special relativity.
- Know how to calculate time dilation and length contraction.
- Know the Lorentz transformation equations.
- Know the Lorentz velocity transformation equations.
- Know the definition of relativistic momentum.
- Know the definition of relativistic energy.
- Know the definition of relativistic kinetic energy.
- Know what rest mass is.
- Know how to calculate the Doppler shift of light.

Summary of Equations

Relativistic time dilation:

$$\Delta t = \frac{\Delta t_0}{\sqrt{1 - v^2/c^2}}$$ (Section 36-5)

Relativistic length contraction:

$$\ell = \frac{\ell_0}{\sqrt{1 - v^2/c^2}}$$ (Section 36-6)

Galilean transformation equations:

$$x = x' + v't$$ (Section 36-8)
$$y = y'$$
$$z = z'$$
$$t = t'$$

Galilean velocity transformation equations:

$$u_x = u'_x + v$$ (Section 36-8)
$$u_y = u'_y$$
$$u_z = u'_z$$

Definition of γ:

$$\gamma = \frac{1}{\sqrt{1 - v^2/c^2}}$$ (Section 36-8)

Lorentz transformation equations:

$$x = \gamma(x' + vt')$$ (Section 36-8)
$$y = y'$$
$$z = z'$$
$$t = \gamma\left(t + \frac{vx'}{c^2}\right)$$

Lorentz velocity transformation equations:	$u_x = \dfrac{u'_x + v}{1 + vu'_x/c^2}$	(Section 36-8)
	$u_y = \dfrac{u'_y \sqrt{1 - v^2/c^2}}{1 + vu'_x/c^2}$	
	$u_z = \dfrac{u'_z \sqrt{1 - v^2/c^2}}{1 + vu'_x/c^2}$	
Definition of (relativistic) linear momentum:	$\vec{\mathbf{p}} = \gamma\, m\vec{\mathbf{v}} = \dfrac{m\vec{\mathbf{v}}}{\sqrt{1 - v^2/c^2}}$	(Section 36-9)
Definition of relativistic mass:	$m_{rel} = \gamma\, m = \dfrac{m}{\sqrt{1 - v^2/c^2}}$	(Section 36-9)
Definition of (relativistic) energy:	$E = \gamma\, mc^2 = \dfrac{mc^2}{\sqrt{1 - v^2/c^2}}$	(Section 36-11)
Definition of (relativistic) kinetic energy:	$K = E - mc^2 = (\gamma - 1)mc^2$	(Section 36-11)
Relationship between energy and linear momentum:	$E^2 = p^2 c^2 + m^2 c^4$	(Section 36-11)
Relativistic Doppler shift (source approaches observer):	$f = f_0 \sqrt{\dfrac{c+v}{c-v}}$	(Section 36-12)
	$\lambda = \lambda_0 \sqrt{\dfrac{c-v}{c+v}}$	
Relativistic Doppler shift (source moves away from observer):	$f = f_0 \sqrt{\dfrac{c-v}{c+v}}$	(Section 36-12)
	$\lambda = \lambda_0 \sqrt{\dfrac{c+v}{c-v}}$	

Chapter Summary

Section 36-1. Galilean–Newtonian Relativity

The **relativity principle** states that the laws of physics are the same for all **inertial observers**. An inertial observer is an observer who observes that Newton's first law of motion is correct in his/her reference frame. In Galilean–Newtonian relativity, space and time are considered **absolute**. The length of an object and the time between two events measured by one inertial observer will be the same as the length and time measured by any other inertial observer. In Galilean–Newtonian relativity, measurements of force, mass, and acceleration made by observers in different inertial frames are the same for all observers. If Newton's second law applies in the reference frame of one inertial observer, it is valid in any other inertial reference frame. A general statement of this principle is that **all inertial reference frames are equivalent**.

A problem with Galilean–Newtonian relativity emerged when the laws of electromagnetism predicted the speed of light to be c. Since Galilean relativity implies that velocities measured by one inertial observer are different from those measured by a second inertial observer in motion relative to the first observer, there is only one "special" reference frame in which the speed of light will be c. In any other inertial reference frame, the observer will measure a different value for the speed of light and all inertial reference frames are thus not equivalent. The predictions made by the theory of electromagnetism are thus in violation of the relativity principle.

The **Michelson-Morley experiment** attempted to measure the velocity of the Earth in its motion around the Sun relative to the special reference frame. No differences in the speed of light in any direction were detected.

Section 36-2. The Michelson–Morley Experiment

The Michelson–Morley experiment used a **Michelson interferometer**, schematically shown in the Figure on the right, to measure the difference in the speed of light in different directions. The Michelson interferometer creates interference between light beams that travel along paths in two perpendicular directions. The interference condition depends on the lengths of the two paths and on the speed of light along these paths. If the interference pattern is observed for a particular orientation of the interferometer, a change in the orientation of the interferometer will change the interference conditions if the speed of light is different in different directions.

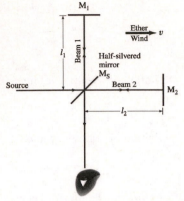

The Michelson–Morley experiment was sensitive enough to measure speed differences on the order of the speed of the Earth in its orbit about the Sun, but no changes in the interference condition were observed. The speed of light appeared to be the same in all directions.

Section 36-3. Postulates of the Special Theory of Relativity

Albert Einstein was able to reinstate the relativity principle by giving up the idea that space and time are absolutes. Instead, Einstein started with the principle that the speed of light measured by all inertial observers is the same and then derived the necessary properties of space and time from that principle. Einstein's **special theory of relativity** is based on the following two postulates:

> *First postulate (the relativity principle):*
> **The laws of physics have the same form in all inertial reference frames.**

> *Second postulate (constancy of the speed of light):*
> **Light propagates through empty space with a definite speed c, independent of the speed of the source or the observer.**

Section 36-4. Simultaneity

The special theory of relativity has certain implications for the measurements of length and time. One of those implications is that simultaneity of events at different places is dependent on the observer. Two events that occur at different positions in space may be viewed as simultaneous by one inertial observer. The two events will not necessarily be simultaneous to an inertial observer in motion relative to the first observer.

Section 36-5. Time Dilation and the Twin Paradox

The time interval between two events measured by an observer in one inertial reference frame is, in general, different from the time interval between those same two events measured by an observer in a different inertial reference frame. Consider a special inertial reference frame in which the two events occur at the same position. We call the time between these two events in this special inertial reference frame the **proper time** Δt_0. An observer in any other inertial reference frame will observe a time difference between these two events that is larger than the proper time. It is not always possible to find an inertial reference frame in which the two events occur at the same place in space.

Consider two events that occur at the same location in a certain reference frame. The time difference between these two events is the proper time Δt_0. An observer in a reference frame that moves with a velocity v with respect to the first reference frame will observe a time difference Δt between these two events where

$$\Delta t = \frac{\Delta t_0}{\sqrt{1 - v^2/c^2}}$$

If v is less than c in the above equation, the denominator is real and less than one, and Δt is greater than Δt_0. The time measured by the moving observer is greater than the proper time. This effect is called **time dilation**.

A clock always measures the proper time in its own inertial reference frame, because it is always located at the same place in its inertial reference frame. Time dilation implies that if an observer is watching a moving clock, the clock appears to be running slower. However, a person in the same inertial reference frame as the clock sees the clock running at its normal rate.

Example 36-5-A. Dilation of the US life expectancy. The approximate life expectancy in the US is 77.8 years. If observers in a spaceship, moving at a speed of 4.8×10^7 m/s relative to the Earth, were watching the Earth, what would their measurement of the average US life expectancy be?

Approach: We assume that the Earth is small enough so that the two events that define the life expectancy, birth and death, occur at the same point in space. The life expectancy quoted in the problem is the proper time. Using the time dilation formula, we can now determine the time observed by the observer moving with a velocity v with respect to the Earth.

Solution: The time dilation formula can be used to determine the time interval measured by the observers in the moving spaceship. The proper time used in the time dilation formula is 77.8 years.

$$\Delta t = \frac{\Delta t_0}{\sqrt{1 - v^2/c^2}} = \frac{77.8}{\sqrt{1 - \left(4.8 \times 10^7\right)^2 / \left(3.00 \times 10^8\right)^2}} = 78.8 \text{ yr}$$

When we do this type of calculation, we need to make sure we worry about the units. The ratio v/c is unitless and it is important that the velocity v and the speed of light c use the same units. The result of the calculation will have the same units as the units used for Δt_0.

Section 36-6. Length Contraction

The length of an object, as measured by observers in one inertial reference frame, is in general different from the length of the object measured by observers in different inertial reference frames. The length of the object in an inertial reference frame in which the object is at rest is called the **proper length** L_0. Inertial observers not at rest with respect to the object measure a length L where L is given by

$$L = L_0 \sqrt{1 - \frac{v^2}{c^2}}$$

For velocities v less than c, the square root is real and less than one. This implies that the length of an object measured by a person in motion relative to the object is shorter than the length measured by an observer at rest relative to the object. Because the length of the object measured by the moving observer is shorter than L_0, this effect is called **length contraction**.

Example 36-6-A. Using length contraction to measure velocity. You observe a spaceship moving past you at a high speed. You measure its length to be $L = 10.0$ m. You know its length in its rest frame is $L_0 = 45.0$ m. At what speed is the spaceship passing by you?

Approach: Since the spaceship is moving with respect to you, its length will be contracted. The amount of contraction depends on the velocity v. Since L and L_0 are provided, the velocity v can be determined.

Solution: The length contraction relationship relates L and L_0:

$$L = L_0 \sqrt{1 - \frac{v^2}{c^2}}$$

This is one equation with one unknown, v, which can be determined:

$$v = c\sqrt{1 - \frac{L^2}{L_0^2}} = \sqrt{1 - \frac{(10.0)^2}{(45.0)^2}}\, c = 0.975\, c$$

Section 36-7. Four-Dimensional Space–Time

When an observer in one inertial reference frame measures positions and times and then relates those measurements of position and time to an observer in another inertial reference frame, the position measurements of the second observer are related to both the position and time measurements of the first observer. Also, the time measurements of the second observer are related to both the time and position measurements of the first observer. This means that space and time are linked together inseparably. Rather than treat space and time as separate quantities, it makes sense to treat them as different parts of the same mathematical structure. The mathematical space used to describe both space and time is **four-dimensional space–time**. This structure has three spatial dimensions, as does the familiar Galilean space, and one additional dimension for time.

Section 36-8. Galilean and Lorentz Transformations

Each event we observe can be specified by its three-dimenstional position and by its time. Observers in different inertial reference frames may assign different positions and different times to the same event. There is a simple relationship, called a **transformation**, which relates the measurements of position and time made by observers in different reference frames. For the purposes of simplicity, we will choose the x-axis of both observers to be parallel to the direction of the relative velocity of the two observers. We will also choose our time reference such that at time $t = 0$ and $t' = 0$ the origins of the coordinate systems are located at the same position. We do not lose any generality by doing this, but simplify the relations that describe the transformation.

Galilean transformations relate the observations of two inertial observers in a way that preserves Galilean relativity. The measurements of position x, y, and z, and time t of one observer are related to the measurements of position x', y', and z' and time t' of a second observer via the following transformation:

$$x = x' + v't$$
$$y = y'$$
$$z = z'$$
$$t = t'$$

where v is the velocity of the second observer relative to the first observer. The Galilean transformation preserves time intervals and lengths of objects.

Galilean transformations can be used to determine how the velocity measurements made by the two inertial observers are related to each other:

$$u_x = u'_x + v$$
$$u_y = u'_y$$
$$u_z = u'_z$$

where u_x, u_y, and u_z are the x, y, and z components of the velocity measured by one inertial observer, and u'_x, u'_y, and u'_z are the x', y', and z' components of the velocity measured by a second inertial observer, moving with a velocity v with respect to the first observer. These relations are called the **Galilean velocity transformations**. Note that the Galilean transformations tell us that the speed of light observed by different inertial observers should be different.

If we start from Einstein's postulates of special relativity and try to find a linear transformation law that preserves the speed of light, we come up with a set of relations called the **Lorentz transformation equations**. These transformation equations are

$$x = \gamma\left(x' + vt'\right)$$
$$y = y'$$
$$z = z'$$
$$t = \gamma\left(t + \frac{vx'}{c^2}\right)$$

where

$$\gamma = \frac{1}{\sqrt{1 - \dfrac{v^2}{c^2}}},$$

The velocity transformation equations that result from the Lorentz transformation equations are

$$u_x = \frac{u'_x + v}{1 + v u'_x / c^2}$$

$$u_y = \frac{u'_y \sqrt{1 - v^2/c^2}}{1 + v u'_x / c^2}$$

$$u_z = \frac{u'_z \sqrt{1 - v^2/c^2}}{1 + v u'_x / c^2}$$

Example 36-8-A. Order of events. In this example we will show that the order of events in time can be different for different observers when there is no inertial reference frame in which these two events occur at the same place. Observer B moves relative to observer A with a velocity of $0.900c$ in the positive x direction. Two events occur at $(y = 0, z = 0)$ for observer A and at $(y' = 0, z' = 0)$ for observer B. Event number one occurs at a position $x = 200$ m at time $t_1 = 1.00$ μs according to observer A. Event number 2 occurs at a position $x = 600$ m at a time $t_2 = 2.00$ μs according to observer A. When and where do these two events occur according to observer B?

Approach: In this problem, the Lorentz transformation rules can be used to correlate the observations made by observer B to the observations made by observer A. Observer A observes event one to occur before event two. Note that the time difference seen by observer A is not the proper time since the two events occur at different locations in the reference frame of observer A.

Solution: The relativistic gamma factor for the transformation between the reference frames of the two observers is equal to

$$\gamma = \frac{1}{\sqrt{1 - \dfrac{v^2}{c^2}}} = \frac{1}{\sqrt{1 - \dfrac{(0.900c)^2}{c^2}}} = 2.29$$

The positions and times of the two events, as observed by observer B, can be determined from the positions and times observed by observer A, using the Lorentz transformation equations:

$$x'_1 = \gamma(x_1 - vt_1) = 2.29\left[(200) - (0.900c)(1.00 \times 10^{-6})\right] = -160\,\text{m}$$

$$t'_1 = \gamma\left(t_1 - \frac{vx_1}{c^2}\right) = 2.29\left[(1.00 \times 10^{-6}) - \frac{(0.900c)(200)}{c^2}\right] = 0.916\,\mu\text{s}$$

$$x'_2 = \gamma(x_2 - vt_2) = 2.29\left[(600) - (0.900c)(2.00 \times 10^{-6})\right] = 137\,\text{m}$$

$$t'_2 = \gamma\left(t_2 - \frac{vx_2}{c^2}\right) = 2.29\left[(2.00 \times 10^{-6}) - \frac{(0.900c)(600)}{c^2}\right] = 0.458\,\mu\text{s}$$

We see that the last event according to observer A is the first event according to observer B.

The events in this example are events that can never occur at the same position in any reference frame. Consider a reference frame that is moving with a velocity v with respect to observer A. The x positions of events 1 and 2 in this reference frame are:

$$x'_1 = \gamma(x_1 - vt_1)$$

$$x'_2 = \gamma(x_2 - vt_2)$$

If the two events occur at the same position, we can determine the required velocity v

$$\gamma(x_1 - vt_1) = \gamma(x_2 - vt_2) \quad\Rightarrow\quad x_1 - vt_1 = x_2 - vt_2 \quad\Rightarrow$$

$$v = \frac{x_2 - x_1}{t_2 - t_1} = \frac{600 - 200}{2.00 \times 10^{-6} - 1.00 \times 10^{-6}} = 4 \times 10^8\,\text{m/s} > c$$

The required velocity v is larger than the speed of light. This is not possible, and we thus conclude that we cannot find an inertial frame in which both events occur at the same position.

Example 36-8-B. Velocity addition at high speeds. You observe two spaceships passing by the Earth in the same direction. The first spaceship passes the Earth with a speed of $0.800c$. An observer on the first spaceship observes the second spaceship to be moving in the same direction the Earth is moving relative to the observer's spaceship with a speed of $0.700c$. What is the velocity of the second spaceship relative to the Earth?

Approach: The problem provides us with information about the velocity of the Earth and the second spaceship as seen by an observer on the first spaceship. According to this observer, the Earth is moving with a velocity of $+0.800c$ in the positive x direction, where the positive x direction has been chosen to be in the direction of the motion of the Earth, as seen by the observer. The second spaceship is moving with a velocity of $+0.700c$ in the (same) positive x direction. In order to answer the problem, we need to determine the velocity of the second spaceship as seen by an observer on Earth.

Solution: The velocity of the reference frame of the Earth with respect to first spaceship is $+0.800c$ in the positive x direction. The velocity of the second spaceship as seen by an observer on Earth is equal to

$$u'_x = \frac{u_x - v}{1 - u_x v / c^2} = \frac{0.700\,c - 0.800c}{1 - (0.700c)(0.800c)/c^2} = -0.227c$$

Section 36-9. Relativistic Momentum

Netwon's second law

$$\vec{F} = \frac{d\vec{p}}{dt}$$

is valid for all inertial observers if we redefine the linear **momentum** to be

$$\vec{p} = \gamma\, m\vec{v} = \frac{m\vec{v}}{\sqrt{1 - v^2/c^2}}$$

In this relation, m is the mass of the particle, as measured by an observer in the inertial frame of the particle, and $\vec{v}$ is the velocity of the particle in the reference frame of the observer studying it. The product of γ and m is called the **relativistic mass** m_{rel}:

$$m_{rel} = \gamma\, m = \frac{m}{\sqrt{1 - v^2/c^2}}$$

The quantity m is called the **rest mass** of the object, which is the mass of the object observed by an observer in an inertial frame in which the particle is at rest.

Section 36-10. The Ultimate Speed

No object can be accelerated from a speed below the speed of light to a speed greater than the speed of light. This makes the speed of light a universal speed limit.

Section 36-11. $E = mc^2$; Energy and Mass

The work-energy theorem is valid for systems that obey the Lorentz transformation equations if the **total energy** of the system is defined as

$$E = \gamma\, mc^2 = \frac{mc^2}{\sqrt{1 - v^2/c^2}}$$

For an object at rest, $\gamma = 1$, and the total energy of the system is equal to

$$E = mc^2$$

This energy is called the **rest energy**. We conclude that the object has a non-zero energy, even though it is at rest. The difference between the total energy and the rest energy is the **relativistic kinetic energy** K:

$$K = E - mc^2 = \gamma mc^2 - mc^2 = (\gamma - 1) mc^2$$

If we combine the definition of the linear momentum p and the definition of the total energy E, we can derive a relation between p and E:

$$E^2 = p^2 c^2 + m^2 c^4$$

Example 36-11-A. Relativistic energies and speeds. A particle has a kinetic energy that is three times its rest energy. Determine the energy of the particle, its linear momentum, and its speed.

Approach: The problem provides us with information about the rest energy and the kinetic energy of the particle. The total energy, which is the sum of the rest energy and the kinetic energy, can be used to determine the gamma factor and the velocity of the particle.

Solution: We know that the kinetic energy of the particle is related to its rest mass by

$$K = E - mc^2 = (\gamma - 1) mc^2$$

The problem states that

$$K = 3mc^2 = (\gamma - 1) mc^2$$

and we can conclude that $\gamma = 4$. The total energy E is thus equal to

$$E = \gamma mc^2 = 4mc^2$$

The velocity of the particle can be determined from our knowledge of γ:

$$\gamma = \frac{1}{\sqrt{1 - \frac{v^2}{c^2}}} \quad \Rightarrow \quad v = \sqrt{1 - \frac{1}{\gamma^2}}\, c = \sqrt{1 - \frac{1}{4^2}}\, c = \sqrt{\frac{15}{16}}\, c = \frac{1}{4}\sqrt{15}\, c$$

The linear momentum of the particle is

$$p = \gamma mv = 4m \left(\frac{1}{4}\sqrt{15}\, c \right) = \sqrt{15}\, mc$$

Section 36-12. Doppler Shift for Light

The Doppler effect, which was discussed in connection to sound waves in Chapter 16, also applies to light waves. The relationship between the source frequency and the observed frequency for light waves is somewhat different from the relationship discussed in Chapter 16 for sound waves. This difference is due to the requirement to use Lorentz transformations to relate the observations made by different observers. The frequency f of the light, measured by an observer, is related to the frequency f_0 of the light emitted by the source:

$$f = f_0 \sqrt{\frac{c + v}{c - v}} > f_0 \qquad \text{when the source approaches the observer at a speed } v$$

$$f = f_0 \sqrt{\frac{c - v}{c + v}} < f_0 \qquad \text{when the source moves away from the observer at a speed } v$$

These relations can be rewritten in terms of the wavelength of the light where λ is the wavelength observed by the observer and λ_0 is the wavelength of the light emitted by the source:

$$\lambda = \lambda_0 \sqrt{\frac{c-v}{c+v}} < \lambda_0 \qquad \text{when the source approaches the observer at a speed } v$$

$$\lambda = \lambda_0 \sqrt{\frac{c+v}{c-v}} > \lambda_0 \qquad \text{when the source moves away from the observer at a speed } v$$

Example 36-12-A. Doppler shift of radio waves. An FM radio station broadcasts with a carrier frequency of 101.5 MHz. An observer receives the radio station at a frequency of 101.7 MHz. What is the velocity of the observer relative to the transmitter?

Approach: The problem tells us that the frequency has increased. The source is thus approaching the observer. The change in frequency can be used to determine the source velocity v.

Solution: We can use the Doppler shift equation and determine the speed with which the observer approaches the transmitter:

$$f = f_0 \sqrt{\frac{c+v}{c-v}} \quad \Rightarrow \quad v = c \frac{f^2 - f_0^2}{f^2 + f_0^2} = (3.00 \times 10^8) \frac{(101.7)^2 - (101.5)^2}{(101.7)^2 + (101.5)^2} = 5.91 \times 10^5 \text{ m/s}$$

Since the frequencies only enter in the ratio, we need to make sure that they have the same units.

Section 36-13. The Impact of Special Relativity

The results of the theory of special relativity are consistent with all experimental evidence gathered to test this theory. It is also consistent with Galilean relativity, which the universe seems to obey in our everyday life, at relative speeds small compared to the speed of light.

Practice Quiz

1. You are in a spaceship when a second spaceship passes by at a relative speed v. You observe a meter stick oriented along the direction of the relative velocity on the second spaceship to be one-half meter in length as it passes by. What does an observer on the other spaceship observe for the length of a meter stick oriented along the direction of the relative velocity on your spaceship?
   a) One-half meter.
   b) Two meters.
   c) One and a half meters.
   d) Need to know v to answer that question.

2. Is the relativistic kinetic energy of a particle not at rest always less than, equal to, or greater than its Newtonian kinetic energy?
   a) Less than.
   b) Equal to.
   c) Greater than.
   d) Sometime less than and sometimes greater than.

3. Two events for observer A occur simultaneously and at the same position in space. Observer B is in relative motion to observer A. What statement is true about what observer B sees for the location and time of the events observed by observer A?
   a) The events are not simultaneous for observer B.
   b) The events are not located in the same position for observer B.
   c) The events are simultaneous and located in the same position for observer B.
   d) More information is needed to answer this question.

4. You are moving with a velocity $\vec{v}$ relative to another observer. What is the relativistically correct expression for the velocity of the other observer relative to you?
   a) $-\vec{v}$
   b) $-\sqrt{1-v^2/c^2}\ \vec{v}$
   c) 0
   d) $-\gamma\,\vec{v}$

5. You observe the lifetime of a particle at rest in the laboratory as τ. If the particle is moving at high speed relative to the laboratory, what will be the observed lifetime of this particle?
   a) Less than τ.
   b) Greater than τ.
   c) Equal to τ.
   d) The direction of the motion must be known in order to be able to answer the question.

6. Why are we still able to apply Newton's laws in their non-relativistic form and get results that agree with experiment even though we now know those relationships should be replaced by the relativistic expressions discussed in this chapter?
   a) The relativistic expressions are very well approximated by the non-relativistic expressions for speeds much smaller than the speed of light.
   b) The relativistic expressions only apply when speeds are greater than $0.1c$.
   c) Special relativity applies to cases where light is involved and Galilean relativity applies to mechanics.
   d) You don't really think anyone really believes this science fiction stuff, do you?

7. You are moving at a speed of $0.8c$ directly toward a star. What is the speed of the light from the star relative to you?
   a) $0.2c$
   b) $1.8c$
   c) $0.8c$
   d) $1.0c$

8. When an object, initially at rest, is acted on by a constant force, what can you say about the Newtonian acceleration of the object, as observed by an observer in an inertial reference frame?
   a) The acceleration remains constant.
   b) The acceleration increases with time.
   c) The acceleration decreases with time.
   d) The acceleration is zero.

9. Is the relativistic kinetic energy less than, equal to, or greater than the square of the relativistic linear momentum divided by twice the rest mass of the object?
   a) Less than.
   b) Equal to.
   c) Greater than.
   d) Sometimes less than and sometimes greater than.

10. Two sources of light that emit identical frequencies are both moving relative to you at speed v. One is moving directly away from you and one is moving directly toward you. Which source will exhibit the largest frequency shift?
    a) The source approaching you.
    b) The source moving away from you.
    c) Neither, the two differences will be the same magnitude.
    d) It depends on their relative distances to you.

11. An Oxygen 15 nucleus decays at rest in an average time of 122 s. How fast would this nucleus need to be moving relative to us in order to observe the average time of this nucleus to decay to be 3.00 minutes?

12. Two spaceships pass by the Earth in the same direction. To an observer on spaceship A, clocks on the Earth appear to be running at half the rate of his clock. To an observer on spaceship B, clocks on the Earth appear to be running at one third the rate of her clock. What is the speed of each spaceship relative to the Earth and what is the relative velocity of spaceship B to spaceship A?

13. A particle with rest mass m has an energy of $6mc^2$. What are the kinetic energy and momentum of this particle?

14. An observer moves with a velocity $0.782c$ in the positive x direction relative to you. You see lightning strike the location $x = 232$ m, $y = 189$ m, $z = 0$ at a time $t = 3.56$ μs. At what location and time does the other observer see the lightning strike? Assume the observers are using coordinate systems with origins that coincided at time $t = 0$.

15. The spectrum of a star that is moving along the line of sight to the star is observed. A hydrogen atom is known to emit light of a wavelength of 656 nm when it is stationary relative to the observer. The spectrum of the star shows the wavelength of this light to be 648 nm. What is the velocity of this star relative to the Earth?

Responses to Select End-of-Chapter Questions

1. No. The train is an inertial reference frame, and the laws of physics are the same in all inertial reference frames, so there is no experiment you can perform inside the train car to determine if you are moving.

7. Time *actually* passes more slowly in the moving reference frames, according to observers outside the moving frames.

13. Both the length contraction and time dilation formulas include the term $\sqrt{1 - v^2/c^2}$. If c were not the limiting speed in the universe, then it would be possible to have a situation with $v > c$. However, this would result in a negative number under the square root, which gives an imaginary number as a result, indicating that c must be the limiting speed.

19. "Energy can be neither created nor destroyed." Mass is a form of energy, and mass can be "destroyed" when it is converted to other forms of energy. The total amount of energy remains constant.

Solutions to Select End-of-Chapter Problems

1. You measure the contracted length. Find the rest length from Eq. 36-3a.

$$\ell_0 = \frac{\ell}{\sqrt{1 - v^2/c^2}} = \frac{38.2\,\text{m}}{\sqrt{1 - (0.850)^2}} = \boxed{72.5\,\text{m}}$$

7. The speed is determined from the length contraction relationship, Eq. 36-3a. Then the time is found from the speed and the contracted distance.

$$\ell = \ell_0 \sqrt{1 - v^2/c^2} \quad \rightarrow$$

$$v = c\sqrt{1 - \left(\frac{\ell}{\ell_0}\right)^2} \quad ; \quad t = \frac{\ell}{v} = \frac{\ell}{c\sqrt{1 - \left(\frac{\ell}{\ell_0}\right)^2}} = \frac{25\,\text{ly}}{c\sqrt{1 - \left(\frac{25\,\text{ly}}{65\,\text{ly}}\right)^2}} = \frac{(25\,\text{y})c}{c(0.923)} = \boxed{27\,\text{y}}$$

13. (a) In the Earth frame, the clock on the *Enterprise* will run slower. Use Eq. 36-1a.

$$\Delta t_0 = \Delta t \sqrt{1 - v^2/c^2} = (5.0 \, \text{yr}) \sqrt{1 - (0.74)^2} = \boxed{3.4 \, \text{yr}}$$

(b) Now we assume the 5.0 years is the time as measured on the *Enterprise*. Again use Eq. 36-1a.

$$\Delta t_0 = \Delta t \sqrt{1 - v^2/c^2} \rightarrow \Delta t = \frac{\Delta t_0}{\sqrt{1 - v^2/c^2}} = \frac{(5.0 \, \text{yr})}{\sqrt{1 - (0.74)^2}} = \boxed{7.4 \, \text{yr}}$$

19. We take the positive direction in the direction of the *Enterprise*. Consider the alien vessel as reference frame S, and the Earth as reference frame S′. The velocity of the Earth relative to the alien vessel is $v = -0.60\,c$. The velocity of the *Enterprise* relative to the Earth is $u'_x = 0.90\,c$. Solve for the velocity of the *Enterprise* relative to the alien vessel, u_x, using Eq. 36-7a.

$$u_x = \frac{(u'_x + v)}{\left(1 + \frac{vu'_x}{c^2}\right)} = \frac{(0.90c - 0.60c)}{[1 + (-0.60)(0.90)]} = \boxed{0.65c}$$

We could also have made the *Enterprise* as reference frame S, with $v = -0.90\,c$, and the velocity of the alien vessel relative to the Earth as $u'_x = 0.60\,c$. The same answer would result. Choosing the two spacecraft as the two reference frames would also work. Let the alien vessel be reference frame S, and the *Enterprise* be reference frame S′. Then we have the velocity of the Earth relative to the alien vessel as $u_x = -0.60\,c$, and the velocity of the Earth relative to the *Enterprise* as $u'_x = -0.90\,c$. We solve for v, the velocity of the *Enterprise* relative to the alien vessel.

$$u_x = \frac{(u'_x + v)}{\left(1 + \frac{vu'_x}{c^2}\right)} \rightarrow v = \frac{u_x - u'_x}{\left(1 - \frac{u'_x u_x}{c^2}\right)} = \frac{(-.60c) - (-0.90c)}{\left(1 - \frac{(-0.90c)(-.60c)}{c^2}\right)} = \boxed{0.65c}$$

25. (a) We take the positive direction in the direction of the first spaceship. We choose reference frame S as the Earth, and reference frame S′ as the first spaceship. So $v = 0.61c$. The speed of the second spaceship relative to the first spaceship is $u'_x = 0.87\,c$. We use Eq. 36-7a to solve for the speed of the second spaceship relative to the Earth, u.

$$u_x = \frac{(u'_x + v)}{\left(1 + \frac{vu'_x}{c^2}\right)} = \frac{(0.87c + 0.61c)}{[1 + (0.61)(0.87)]} = \boxed{0.97c}$$

(b) The only difference is now that $u'_x = -0.87\,c$.

$$u_x = \frac{(u'_x + v)}{\left(1 + \frac{vu'_x}{c^2}\right)} = \frac{(-0.87c + 0.61c)}{[1 + (0.61)(-0.87)]} = \boxed{-0.55c}$$

31. We set frame S′ as the frame moving with the observer. Frame S is the frame in which the two light bulbs are at rest. Frame S is moving with velocity v with respect to frame S′. We solve Eq. 36-6 for the time t' in terms of t, x, and v. Using the resulting equation we determine the time in frame S′ that each bulb is turned on, given that in frame S the bulbs are turned on simultaneously at $t_A = t_B = 0$. Taking the difference in these times gives the time interval as measured by the observing moving with velocity v.

$$x = \gamma(x' + vt') \rightarrow x' = \frac{x}{\gamma} - vt'$$

$$t = \gamma\left(t' + \frac{vx'}{c^2}\right) = \gamma\left[t' + \frac{v}{c^2}\left(\frac{x}{\gamma} - vt'\right)\right] = \gamma t'\left(1 - \frac{v^2}{c^2}\right) + \frac{vx}{c^2} = \frac{t'}{\gamma} + \frac{vx}{c^2} \rightarrow t' = \gamma\left(t - \frac{vx}{c^2}\right)$$

$$t'_A = \gamma\left(t_A - \frac{vx_A}{c^2}\right) = \gamma\left(0 - \frac{v \times 0}{c^2}\right) = 0 \quad ; \quad t'_B = \gamma\left(t_B - \frac{vx_B}{c^2}\right) = \gamma\left(0 - \frac{v\ell}{c^2}\right) = -\gamma\frac{v\ell}{c^2}$$

$$\Delta t' = t'_B - t'_A = \boxed{-\gamma\frac{v\ell}{c^2}}$$

According to the observer, $\boxed{\text{bulb B turned on first}}$.

37. The two momenta, as measured in the frame in which the particle was initially at rest, will be equal to each other in magnitude. The lighter particle is designated with a subscript "1", and the heavier particle with a subscript "2".

$$p_1 = p_2 \quad\rightarrow\quad \frac{m_1 v_1}{\sqrt{1 - v_1^2/c^2}} = \frac{m_2 v_2}{\sqrt{1 - v_2^2/c^2}} \quad\rightarrow$$

$$\frac{v_1^2}{\left(1 - v_1^2/c^2\right)} = \left(\frac{m_2}{m_1}\right)^2 \frac{v_2^2}{\left(1 - v_2^2/c^2\right)} = \left(\frac{6.68 \times 10^{-27}\,\text{kg}}{1.67 \times 10^{-27}\,\text{kg}}\right)^2 \left[\frac{(0.60\,c)^2}{1 - (0.60)^2}\right] = 9.0\,c^2 \quad\rightarrow$$

$$v_1 = \sqrt{0.90}\,c = \boxed{0.95\,c}$$

43. Each photon has momentum 0.50 MeV/c. Thus each photon has mass 0.50 MeV. Assuming the photons have opposite initial directions, then the total momentum is 0, and so the product mass will not be moving. Thus all of the photon energy can be converted into the mass of the particle. Accordingly, the heaviest particle would have a mass of $\boxed{1.00\,\text{MeV}/c^2}$, which is $1.78 \times 10^{-30}\,\text{kg}$.

49. We find the speed in terms of c. The kinetic energy is given by Eq. 36-10 and the momentum by Eq. 36-8.

$$v = \frac{\left(2.80 \times 10^8\,\text{m/s}\right)}{\left(3.00 \times 10^8\,\text{m/s}\right)} = 0.9333\,c$$

$$K = (\gamma - 1)mc^2 = \left(\frac{1}{\sqrt{1 - 0.9333^2}} - 1\right)(938.3\,\text{MeV}) = 1674.6\,\text{MeV} \approx \boxed{1.67\,\text{GeV}}$$

$$p = \gamma mv = \frac{1}{\sqrt{1 - 0.9333^2}}\left(938.3\,\text{MeV}/c^2\right)(0.9333\,c) = 2439\,\text{MeV}/c \approx \boxed{2.44\,\text{GeV}/c}$$

55. (a) Since the kinetic energy is half the total energy, and the total energy is the kinetic energy plus the rest energy, the kinetic energy must be equal to the rest energy. We also use Eq. 36-10.

$$K = \tfrac{1}{2}E = \tfrac{1}{2}\left(K + mc^2\right) \quad\rightarrow\quad K = mc^2$$

$$K = (\gamma - 1)mc^2 = mc^2 \quad\rightarrow\quad \gamma = 2 = \frac{1}{\sqrt{1 - v^2/c^2}} \quad\rightarrow\quad v = \sqrt{\tfrac{3}{4}}\,c = \boxed{0.866\,c}$$

(b) In this case, the kinetic energy is half the rest energy.

$$K = (\gamma - 1)mc^2 = \tfrac{1}{2}mc^2 \quad\rightarrow\quad \gamma = \tfrac{3}{2} = \frac{1}{\sqrt{1 - v^2/c^2}} \quad\rightarrow\quad v = \sqrt{\tfrac{5}{9}}\,c = \boxed{0.745\,c}$$

61. All of the energy, both rest energy and kinetic energy, becomes electromagnetic energy. We use Eq. 36-11. Both masses are the same.

$$E_{\text{total}} = E_1 + E_2 = \gamma_1 mc^2 + \gamma_2 mc^2 = (\gamma_1 + \gamma_2)mc^2 = \left(\frac{1}{\sqrt{1 - 0.43^2}} + \frac{1}{\sqrt{1 - 0.55^2}}\right)(105.7\,\text{MeV})$$

$$= 243.6\,\text{MeV} \approx \boxed{240\,\text{MeV}}$$

67. (a) We apply Eq. 36-14b to determine the received/reflected frequency f. Then we apply this same equation a second time using the frequency f as the source frequency to determine the Doppler-shifted frequency f'. We subtract the initial frequency from this Doppler-shifted frequency to obtain the beat frequency. The beat frequency will be much smaller than the emitted frequency when the speed is much smaller than the speed of light. We then set $c - v \approx c$ and solve for v.

$$f = f_0 \sqrt{\frac{c+v}{c-v}} \quad f' = f\sqrt{\frac{c+v}{c-v}} = f_0\sqrt{\frac{c+v}{c-v}}\sqrt{\frac{c+v}{c-v}} = f_0\left(\frac{c+v}{c-v}\right)$$

$$f_{\text{beat}} = f' - f_0 = f_0\left(\frac{c+v}{c-v}\right) - f_0\left(\frac{c-v}{c-v}\right) = f_0\frac{2v}{c-v} \approx f_0\frac{2v}{c} \quad \rightarrow \quad \boxed{v \approx \frac{cf_{\text{beat}}}{2f_0}}$$

$$v \approx \frac{\left(3.00\times10^8\,\text{m/s}\right)\left(6670\,\text{Hz}\right)}{2\left(36.0\times10^9\,\text{Hz}\right)} = \boxed{27.8\,\text{m/s}}$$

(b) We find the change in velocity and solve for the resulting change in beat frequency. Setting the change in the velocity equal to 1 km/h we solve for the change in beat frequency.

$$v = \frac{cf_{\text{beat}}}{2f_0} \quad \rightarrow \quad \Delta v = \frac{c\Delta f_{\text{beat}}}{2f_0} \quad \rightarrow \quad \boxed{\Delta f_{\text{beat}} = \frac{2f_0\Delta v}{c}}$$

$$\Delta f_{\text{beat}} = \frac{2\left(36.0\times10^9\,\text{Hz}\right)\left(1\,\text{km/h}\right)}{\left(3.00\times10^8\,\text{m/s}\right)}\left(\frac{1\,\text{m/s}}{3.600\,\text{km/h}}\right) = \boxed{70\,\text{Hz}}$$

73. (a) We use Eq. 36-15a. To get a longer wavelength than usual means that the object is moving away from the Earth.

$$\lambda = \lambda_0\sqrt{\frac{c+v}{c-v}} = 2.5\lambda_0 \quad \rightarrow \quad v = c\left(\frac{2.5^2-1}{2.5^2+1}\right) = \boxed{0.72c}$$

(b) We assume that the quasar is moving and the Earth is stationary. Then we use Eq. 16-9b.

$$f = \frac{f_0}{1+v/c} \quad \rightarrow \quad \frac{c}{\lambda} = \frac{c}{\lambda_0}\left(\frac{1}{1+v/c}\right) \quad \rightarrow \quad \lambda = \lambda_0\left(1+v/c\right) = 2.5\lambda_0 \quad \rightarrow \quad v = \boxed{1.5c}$$

79. The minimum energy required would be the energy to produce the pair with no kinetic energy, so the total energy is their rest energy. They both have the same mass. Use Eq. 36-12.

$$E = 2mc^2 = 2\left(0.511\,\text{MeV}\right) = \boxed{1.022\,\text{MeV}\ \left(1.64\times10^{-13}\,\text{J}\right)}$$

85. The total binding energy is the energy required to provide the increase in rest energy.

$$E = \left[\left(2m_{\text{p+e}} + 2m_{\text{n}}\right) - m_{\text{He}}\right]c^2$$

$$= \left[2\left(1.00783\,\text{u}\right) + 2\left(1.00867\,\text{u}\right) - 4.00260\,\text{u}\right]c^2\left(\frac{931.5\,\text{MeV}/c^2}{\text{u}}\right) = \boxed{28.32\,\text{MeV}}$$

91. We assume one particle is moving in the negative direction in the laboratory frame, and the other particle is moving in the positive direction. We consider the particle moving in the negative direction as reference frame S, and the laboratory as reference frame S'. The velocity of the laboratory relative to the negative-moving particle is $v = 0.85\,c$, and the velocity of the positive-moving particle relative to the laboratory frame is $u'_x = 0.85\,c$. Solve for the velocity of the positive-moving particle relative to the negative-moving particle, u_x.

$$u_x = \frac{\left(u'_x + v\right)}{\left(1 + vu'_x/c^2\right)} = \frac{\left(0.85c + 0.85c\right)}{\left[1 + \left(0.85\right)\left(0.85\right)\right]} = \boxed{0.987\,c}$$

97. We use the Lorentz transformations to derive the result.

$$x = \gamma\left(x' + vt'\right) \rightarrow \Delta x = \gamma\left(\Delta x' + v\Delta t'\right) \ ; \ t = \gamma\left(t' + \frac{vx'}{c^2}\right) \rightarrow \Delta t = \gamma\left(\Delta t' + \frac{v\Delta x'}{c^2}\right)$$

$$\left(c\Delta t\right)^2 - \left(\Delta x\right)^2 = \left[c\gamma\left(\Delta t' + \frac{v\Delta x'}{c^2}\right)\right]^2 - \left[\gamma\left(\Delta x' + v\Delta t'\right)\right]^2 = \gamma^2\left\{\left[\left(c\Delta t' + \frac{v\Delta x'}{c}\right)\right]^2 - \left(\Delta x' + v\Delta t'\right)\right\}$$

$$= \gamma^2\left\{c^2\left(\Delta t'\right)^2 + 2c\Delta t'\frac{v\Delta x'}{c} + \left(\frac{v\Delta x'}{c}\right)^2 - \left(\Delta x'\right)^2 - 2\Delta x' v\Delta t' - \left(v\Delta t'\right)^2\right\}$$

$$= \frac{1}{\left(1 - v^2/c^2\right)}\left\{\left(c^2 - v^2\right)\left(\Delta t'\right)^2 + \left[\left(\frac{v}{c}\right)^2 - 1\right]\left(\Delta x'\right)^2\right\}$$

$$= \frac{1}{1 - v^2/c^2}\left\{c^2\left(1 - v^2/c^2\right)\left(\Delta t'\right)^2 - \left(1 - v^2/c^2\right)\left(\Delta x'\right)^2\right\}$$

$$= \frac{\left(1 - v^2/c^2\right)}{\left(1 - v^2/c^2\right)}\left\{\left(c\Delta t'\right)^2 - \left(\Delta x'\right)^2\right\} = \left(c\Delta t'\right)^2 - \left(\Delta x'\right)^2$$

Chapter 37: Early Quantum Theory and Models of the Atom

Chapter Overview and Objectives

This chapter describes the phenomena, such as blackbody radiation, the photoelectric effect, the Compton effect, and atomic spectra, which could not be satisfactorily described by classical mechanics. The introduction of quantization conditions was required to understand the observations.

After completing this chapter you should:
- Know what blackbody radiation is.
- Know Wien's law.
- Know what Planck's quantization condition is.
- Know what the photoelectric effect is and its properties.
- Know what Einstein's quantization condition of the electromagnetic field is.
- Know what Compton scattering is.
- Know what pair production is.
- Know what the de Broglie wavelength of a particle is.
- Know how to calculate the energy and wavelength of photons emitted from one-electron atoms.
- Know what the Bohr model of the atom is and what Bohr's quantization condition is.
- Know how to calculate orbital energies and radii for atoms described by the Bohr model.

Summary of Equations

Wien's law:
$$\lambda_p T = 2.90 \times 10^{-3} \text{ m} \cdot \text{K}$$
(Section 37-1)

Planck's quantum hypothesis:
$$E = nhf \qquad n \in \{0,1,2,3,\ldots\}$$
(Section 37-1)

Blackbody radiation intensity per unit wavelength:
$$I(\lambda, T) = \frac{2\pi hc^2 \lambda^{-5}}{e^{hc/\lambda kT} - 1}$$
(Section 37-1)

Photon energy:
$$E = hf$$
(Section 37-2)

Maximum kinetic energy of photoelectron:
$$K_{\max} = hf - W_0$$
(Section 37-2)

Linear momentum of a photon:
$$p = \frac{E}{c} = \frac{hf}{c} = \frac{h}{\lambda}$$
(Section 37-3)

Compton shift:
$$\Delta\lambda = \lambda' - \lambda = \frac{h}{m_e c}(1 - \cos\theta)$$
(Section 37-4)

Compton wavelength:
$$\lambda_C = \frac{h}{m_e c} = 2.43 \times 10^{-12} \text{ m}$$
(Section 37-4)

de Broglie wavelength:
$$\lambda = \frac{h}{p} = \frac{h}{mv}$$
(Section 37-7)

Wavelengths in the line spectrum of hydrogen:
$$\frac{1}{\lambda} = R\left(\frac{1}{n'^2} - \frac{1}{n^2}\right) \quad n, n' \in \{1,2,3,\ldots\}$$
$$n > n'$$
(Section 37-10)

Bohr model quantization condition:
$$L = mv_n r_n = n\frac{h}{2\pi} \qquad n \in \{1,2,3,\dots\} \qquad \text{(Section 37-11)}$$

Radii of Bohr model atomic orbits:
$$r_n = \frac{n^2 h^2 \varepsilon_0}{Z\pi me^2} \approx (5.29\times10^{-11}\,\text{m})\frac{n^2}{Z} \qquad \text{(Section 37-11)}$$

Energy of Bohr model atomic orbits:
$$E_n = -\frac{me^4}{8\varepsilon_0^2 h^2}\frac{Z^2}{n^2} = (-13.6\,\text{eV})\frac{Z^2}{n^2} \qquad \text{(Section 37-11)}$$

Chapter Summary

Section 37-1. Planck's Quantum Hypothesis; Blackbody Radiation

A body at non-zero absolute temperature radiates electromagnetic waves with a wavelength distribution that is characteristic of its temperature. For an idealized object, called a **blackbody**, the intensity of the emitted electromagnetic waves per unit wavelength only depends on the absolute temperature of the body and the wavelength of the waves. This radiation is called **blackbody radiation**.

Classical electromagnetism and thermodynamics predicts that the intensity of electromagnetic radiation emitted by a blackbody at a non-zero temperature should increase as the wavelength decreases and approach infinity as the wavelength goes to zero. Experimental measurements show that the intensity per unit wavelength goes to zero as the wavelength goes to zero. To resolve this failure of electromagnetic and thermodynamic theory, Planck proposed that the vibrational energy of atoms in a blackbody is quantized to be a multiple of the Planck constant h ($h = 6.626\times10^{-34}$ J·s) and the frequency of vibration f:

$$E = nhf \qquad n \in \{0,1,2,3,\dots\}$$

The intensity of the emitted radiation per unit wavelength is given by the following expression

$$I(\lambda, T) = \frac{2\pi hc^2 \lambda^{-5}}{e^{hc/\lambda kT} - 1}$$

where λ is the wavelength of the emitted radiation, c is the speed of light, k is the Boltzmann constant, and T is the absolute temperature. This intensity distribution peaks at the wavelength λ_p, given by **Wien's law**:

$$\lambda_P T = 2.90\times10^{-3}\,\text{m}\cdot\text{K}$$

Example 37-1-A. Using color to determine temperature. Metal workers often temper or anneal metals either by heating them to a given temperature and then either quenching them in a liquid coolant or allowing them to cool slowly. They often judge the temperature by looking at the color of light emitted by the hot metal. What is the peak wavelength emitted by a blackbody at a temperature of 900°C?

Approach: The color of the radiation emitted by the body will be determined by the peak wavelength of its blackbody radiation. Using Wien's law, we can determine this wavelength. We need to make sure we express the temperature in units of Kelvin.

Solution: The absolute temperature of the surface is

$$T_K = T_C + 273.15 = 1.17\times10^3\,\text{K}$$

Using Wien's law, we can determine the peak wavelength:

$$\lambda_P = \frac{2.90\times10^{-3}\,\text{m}\cdot\text{K}}{1.17\times10^3\,\text{K}} = 2.48\times10^{-6}\,\text{m}$$

The peak wavelength is located in the infrared part of the electromagnetic spectrum. The blackbody spectrum extends far enough into the visible regime and a reddish colored glow will be seen from an object at this temperature.

Section 37-2. Photon Theory of Light and the Photoelectric Effect

Einstein proposed that, in addition to the quantization of the energy levels in atoms, the energy of the electromagnetic field is also quantized. When an atom changes from one energy level to a lower energy level, the energy difference is carried away by the emission of a **photon** of energy hf. An increase of the intensity of the electromagnetic radiation corresponds to an increase in the number of photons, without changing the energy carried away by each photon.

The quantization of electromagnetic radiation proposed by Einstein was able to explain the **photoelectric effect**. The photoelectric effect is the emission of electrons from a surface when the surface is illuminated with light. Classical electromagnetism does predict that electrons in a material can absorb the energy of the electromagnetic field and gain sufficient energy to escape from a material. It is observed that the light with wavelengths above a certain maximum wavelength is unable to release electrons from a material, independent of the intensity of the incident light. This observation is inconsistent with classical electromagnetism, which predicts that light of any wavelength should be able to eject the electrons if the light has sufficient intensity. It is also observed that the ejected electrons leave the material with a certain maximum kinetic energy. The maximum kinetic energy is a function of the wavelength of the light, independent of the intensity of the light. In classical electromagnetic theory, the electron should be able to gain more kinetic energy from a more intense electric field. The maximum kinetic energy increases with decreasing wavelength of the light.

The quantization of the light into photons results in a behavior that is consistent with the observed results. If the wavelength of light increases, the energy of the photons decreases. At one point, the energy of the photon is less than the energy required to remove an electron from the material. If the intensity of the light increases, the number of photons that reach the surface per second increases, but a given electron still only receives the energy that is carried by a single photon. The maximum kinetic energy of the emitted electrons thus does not change. The maximum kinetic energy of the electrons, K_{max}, is given by

$$K_{max} = hf - W_0 = h\frac{c}{\lambda} - W_0$$

where h is Planck's constant, f is the frequency of the light illuminating the material, and W_0 is called the **work function** of the material. The work function of the material is the minimum required energy to remove an electron from the material. The minimum frequency required to emit electrons is equal to $f_{min} = W_0/h$.

Example 37-2-A. Determining the work function of a material. Photons of wavelength 420 nm are incident on a material. The emitted electrons have a maximum kinetic energy of 2.2 eV. Determine the work function of the material.

Approach: The problem provides us with information on the frequency of the light and the maximum kinetic energy of the emitted electrons. This information is sufficient to determine the work function.

Solution: Using the expression for the maximum kinetic energy of the electron we can determine the work function of the material:

$$K_{max} = hf - W_0 = \frac{hc}{\lambda} - W_0 \quad \Rightarrow \quad W_0 = \frac{hc}{\lambda} - K_{max}$$

In order to use this relation we need to make sure that we use consistent units. We will convert the units of Planck's constant from J s to eV s and specify the wavelength in units of m.

$$E_\gamma = \frac{hc}{\lambda} = \frac{\left(\frac{6.63\times10^{-34}}{1.602\times10^{-19}}\right)(3.00\times10^8)}{420\times10^{-9}} = 3.0\,\text{eV}$$

The work function of the material is thus equal to

$$W_0 = \frac{hc}{\lambda} - K_{max} = 3.0 - 2.2 = 0.8\,\text{eV}$$

Section 37-3. Energy, Mass, and Momentum of a Photon

The energy of a single photon is hf. Each photon moves with the speed of light. Using the relativistic definition of the linear momentum of a particle, discussed in Section 36.9, we see that the linear momentum of the photon will be infinite unless the photon mass is 0. The linear momentum of the photon is $E/c = hf/c$.

Section 37-4. Compton Effect

The **Compton effect** is the scattering of photons by free or weakly bound electrons. The scattered photon is found to have a different wavelength compared to the wavelength of the incident photon, and hence, a different frequency. Classical physics is unable to describe a system in which the frequency of the outgoing electromagnetic field is different from the frequency of the incoming electromagnetic field. However, the quantization of the electromagnetic field in terms of photons can be used to explain the observed wavelength/frequency of the scattered photons. The wavelength of a Compton scattered photon, λ', is related to the incident wavelength, λ, by

$$\lambda' = \lambda + \frac{h}{m_e c}(1 - \cos\phi)$$

where λ is the wavelength of the incident photon, m_e is the rest mass of the electron, c is the speed of light, and ϕ is the angle by which the photon is scattered. The difference between the outgoing and incoming wavelengths, $\Delta\lambda$, is called the **Compton shift**:

$$\Delta\lambda = \lambda' - \lambda = \frac{h}{m_e c}(1 - \cos\theta)$$

The coefficient of the $1 - \cos\theta$ term is called the **Compton wavelength**, λ_C, of the electron:

$$\lambda_C = \frac{h}{m_e c} = 2.43 \times 10^{-12} \text{ m}$$

The maximum wavelength shift occurs when $\theta = \pi$ and is equal to

$$\Delta\lambda_{max} = \frac{2h}{m_e c}$$

Note that the maximum Compton shift depends only on the mass of the electron and not on the properties of the incident photons.

Example 37-4-A. Compton scattering of different "free" particles. Compare the wavelength of a 788 keV photon Compton scattered from a free electron and the same photon Compton scattered from a free proton when they are scattered at an angle of 90°.

Approach: The wavelength of the scattered photons will differ due to differences in the Compton wavelengths of electrons and protons.

Solution: The Compton shift for a photon scattered from an electron is

$$\Delta\lambda = \frac{h}{m_e c}(1 - \cos\theta) = \frac{6.63 \times 10^{-34}}{(9.11 \times 10^{-31})(3.00 \times 10^8)}\left(1 - \cos\frac{\pi}{2}\right) = 2.43 \times 10^{-12} \text{ m}$$

The Compton shift for a photon scattered from a proton is

$$\Delta\lambda = \frac{h}{m_p c}(1 - \cos\theta) = \frac{6.63 \times 10^{-34}}{(1.67 \times 10^{-27})(3.00 \times 10^8)}\left(1 - \cos\frac{\pi}{2}\right) = 1.32 \times 10^{-15} \text{ m}$$

The Compton shifts of photons scattered from protons is very small compared to the wavelength of the photon. Photons will thus lose more energy due to scattering of electrons compared to scattering of protons.

Section 37-5. Photon Interactions; Pair Production

The energy of the electromagnetic field, a photon, can create a pair of particles in a process called **pair production**. The easiest pair of particles to create is an electron and an anti-electron (also called a positron). The particles must be produced in pairs to conserve electric charge. To conserve energy, the minimum energy of the photon is equal to the sum of the rest masses of the particles created. For the creation of an electron-positron pair, the minimum required energy is $2m_ec^2 = 1.02$ MeV. To conserve linear momentum, another particle must be present to take some of the linear momentum of the incoming photon. The more massive this extra particle is, the less energy it takes to give it the required linear momentum.

Example 37-5-A. Pair production near a nucleus. A photon passes near a nucleus of mass 4.48×10^{-26} kg. It produces an electron-positron pair. If the electron and positron have negligible kinetic energy, what was the amount of energy given to the nucleus when the photon produced this pair?

Approach: Since the electron and position have neglible kinetic energy, their total energy is the sum of their rest energies (1.02 MeV). In order to conserve linear momentum, the linear momentum of the photon must be equal to the linear momentum of the nucleus. Conservation of energy provides us with a second equation. These two equations have two unknown and can be solved.

Solution: The linear momentum of the photon must be equal to the linear momentum of the nucleus after the pair production. This requires that

$$P_{photon} = p_{nucleus} \quad \Rightarrow \quad \frac{E_{photon}}{c} = \frac{1}{c}\sqrt{E_{nucleus}^2 - m_{nucleus}^2 c^4} \quad \Rightarrow \quad E_{photon} = \sqrt{E_{nucleus}^2 - m_{nucleus}^2 c^4}$$

Conservation of energy requires that

$$E_{photon} + m_{nucleus}c^2 = 2m_ec^2 + E_{nucleus} \quad \Rightarrow \quad E_{photon} = 2m_ec^2 + E_{nucleus} - m_{nucleus}c^2$$

The last equation can be rewritten as

$$E_{nucleus} = E_{photon} + m_{nucleus}c^2 - 2m_ec^2$$

Substituting this equation in the squared expression we obtained by applying conservation of linear momentum we obtain one equation with one unknown. The unknown, the photon energy, can now be determined:

$$E_{photon} = 2m_ec^2\left(\frac{m_{nucleus} - m_e}{m_{nucleus} - 2m_e}\right)$$

The kinetic energy of the nucleus is

$$K_{nucleus} = E_{photon} - 2m_ec^2 = 2m_ec^2\left[\left(\frac{m_{nucleus} - m_e}{m_{nucleus} - 2m_e}\right) - 1\right] = 2m_ec^2\left(\frac{m_e}{m_{nucleus} - 2m_e}\right)$$

Because of the relatively small mass of the electron compared to the nucleus, the kinetic energy of the nucleus will be small compared to the rest mass energy of the electron:

$$K_{nucleus} = 2\left(0.511 \times 10^6\right)\frac{\left(9.11 \times 10^{-31}\right)}{\left(4.48 \times 10^{-26}\right) - 2\left(9.11 \times 10^{-31}\right)} = 20.8\,\text{eV}$$

We have expressed the masses in the ratio in units of kg. The ratio is unitless and the actual units used do not matter as long as the masses are expressed in terms of the same units. The unit of kinetic energy will be the same as the units of the rest mass of the electron. In this problem we have used the eV as the unit for the rest energy. The kinetic energy of the nucleus is very small compared to the rest energy of the particle being produced.

Section 37-6. Wave-Particle Duality; The Principle of Complementarity

Light sometimes exhibits wave-like behavior and sometimes exhibits particle-like behavior. This phenomenon is called the **wave-particle duality** of light. The principle of complementarity states that experiments can be understood either in terms of the wave-nature of light or the particle-nature of light, but both descriptions cannot be used simultaneously.

Section 37-7. Wave Nature of Matter

The de Broglie wavelength of a particle is

$$\lambda = \frac{h}{p} = \frac{h}{mv}$$

Giving a wave property to matter is analogous to giving a particle property of electromagnetic waves and extends the idea of wave-particle duality to what are classically considered particles.

Example 37-7-A. Measuring the wavelength of a human body. Determine the wavelength of a $m = 100$ kg person walking with a speed of 1.00 m/s. Determine the location of the first single-slit minimum away from the center peak of the diffraction pattern when a wave of this wavelength passes through a single-slit that has a width $a = 1.00$ m (about the width of a doorway) and travels to a screen that is $L = 1.00$ km away from the door.

Approach: Using the information provided in the problem we can determine the de Broglie wavelength of the person. The diffraction theory discussed in Chapter 35 can be used to determine the location between the central peak and the first diffraction minimum.

Solution: The wavelength of the person is equal to

$$\lambda = \frac{h}{mv} = \frac{\left(6.63\times10^{-34}\right)}{(100)(1.00)} = 6.63\times10^{-36} \text{ m}$$

The angular position of the first diffraction peak minimum is equal to

$$\sin\theta = \frac{m\lambda}{a} = \frac{(1)\left(6.63\times10^{-36}\right)}{1.00} = 6.63\times10^{-36}$$

The distance y between the first minimum and the central peak divided by the distance between the slit and the screen is the tangent of the angle:

$$y = L\tan\theta$$

where $L = 1.0$ km. Since θ is small, $\tan\theta$ is approximately equal to $\sin\theta$ and y is about equal to

$$y \approx L\sin\theta = \left(1.00\times10^{3}\right)\left(6.63\times10^{-36}\right) = 6.63\times10^{-33} \text{ m}$$

This distance is much smaller than the size of a nucleus ($\sim10^{-15}$ m) and the wave nature of people will be difficult to observe.

Section 37-8. Electron Microscopes

A light microscope forms an image of light that has scattered from the object being viewed. An electron microscope relies on the same principle except that electrons are used instead of light. Since the wavelength of electrons can easily be made much smaller than the wavelength of light, an electron microscope can have a much greater resolution than an optical microscope. Electron microscopes that have better than atomic-sized resolution are available.

Section 37-9. Early Models of the Atom

Although the concept of atoms of matter goes back to ancient Greece, the first model of the atom, that is similar to the currently used model, was proposed by Ernest Rutherford. Rutherford created his nuclear model of the atom to explain the results of scattering of alpha particles from a thin gold foil. Rutherford concluded that almost all the mass of the atom is contained within the nucleus, located at the center of the atom. The radius of the atomic nucleus was about 10^{-15} m to 10^{-14} m. The radius of the atom was in the order of 10^{-10} m.

Section 37-10. Atomic Spectra: Key to the Structure of the Atom

The emission spectrum from hydrogen gas is a **line spectrum**. The light emitted from hydrogen does not have a continuum of wavelengths, but only a discrete set of wavelengths appears in the spectrum. Those wavelengths are given by the following relationship

$$\frac{1}{\lambda} = R\left(\frac{1}{n'^2} - \frac{1}{n^2}\right) \qquad n, n' \in \{1, 2, 3, \ldots\} \qquad n > n'$$

where R is called the **Rydberg constant** ($R = 1.097 \times 10^7$ m^{-1}).
The lines can be classified in groups according to the value of n'. If $n' = 1$, then the group of emission lines is called the **Lyman series**. If $n' = 2$, then the group of emission lines is called the **Balmer series**. If $n' = 3$, then the group of emission lines is called the **Paschen series**.

Example 37-10-A. Hydrogen emission lines. Determine the five longest wavelengths in the series of emission lines from hydrogen that has $n' = 4$.

Approach: The problem is solved by using the wavelength relation discussion in this Section. The largest wavelengths are associated with the smallest values of n with $n > n'$. Thus we need to determine the wavelengths associated with $n = 5$, $n = 6$, $n = 7$, $n = 8$, and $n = 9$.

Solution: We determine the wavelengths of the emission lines using the equation discussed in this Section.

For $n = 5$ $\qquad \lambda = \left[R\left(\frac{1}{n'^2} - \frac{1}{n^2}\right)\right]^{-1} = \left[\left(1.097 \times 10^7\right)\left(\frac{1}{4^2} - \frac{1}{5^2}\right)\right]^{-1} = 4.051 \times 10^{-6}$ m

For $n = 6$ $\qquad \lambda = \left[R\left(\frac{1}{n'^2} - \frac{1}{n^2}\right)\right]^{-1} = \left[\left(1.097 \times 10^7\right)\left(\frac{1}{4^2} - \frac{1}{6^2}\right)\right]^{-1} = 2.625 \times 10^{-6}$ m

For $n = 7$ $\qquad \lambda = \left[R\left(\frac{1}{n'^2} - \frac{1}{n^2}\right)\right]^{-1} = \left[\left(1.097 \times 10^7\right)\left(\frac{1}{4^2} - \frac{1}{7^2}\right)\right]^{-1} = 2.166 \times 10^{-6}$ m

For $n = 8$ $\qquad \lambda = \left[R\left(\frac{1}{n'^2} - \frac{1}{n^2}\right)\right]^{-1} = \left[\left(1.097 \times 10^7\right)\left(\frac{1}{4^2} - \frac{1}{8^2}\right)\right]^{-1} = 1.945 \times 10^{-6}$ m

For $n = 9$ $\qquad \lambda = \left[R\left(\frac{1}{n'^2} - \frac{1}{n^2}\right)\right]^{-1} = \left[\left(1.097 \times 10^7\right)\left(\frac{1}{4^2} - \frac{1}{9^2}\right)\right]^{-1} = 1.818 \times 10^{-6}$ m

All of these wavelengths are in the infrared portion of the electromagnetic spectrum.

Section 37-11. The Bohr Model

Bohr applied the **quantization condition**

$$L = mv_n r_n = n\frac{h}{2\pi} \qquad n \in \{1, 2, 3, \ldots\}$$

to the circular orbits of electrons in atoms. The value of n is called the **quantum number** of the orbit. Imposing this condition on the orbits of the electrons restricts the allowed radii of the orbits (in m):

$$r_n = \frac{n^2 h^2 \varepsilon_0}{Z\pi me^2} \approx \left(5.29 \times 10^{-11}\right)\frac{n^2}{Z}$$

The distance 5.29×10^{-11} m is called the **Bohr radius**. The speed of the electrons in these orbits (in m/s) is equal to

$$v_n = \frac{Ze^2}{2n\varepsilon_0 h} = \left(2.19 \times 10^6\right)\frac{Z}{n}$$

The energy of the electrons (in eV) that satisfies the Bohr quantization condition is equal to

$$E_n = -\frac{me^4}{8\varepsilon_0^2 h^2}\frac{Z^2}{n^2} = (-13.6)\frac{Z^2}{n^2}$$

The lowest energy state of the electron is the state with $n = 1$. This state is the **ground state** of the atom. When the electron is in a state with $n > 1$, the atom is in an **excited state**. The minimum energy required to remove the electron from the atom is the energy required to reach the $E = 0$ energy level. The required energy is called the **binding energy** or **ionization energy**.

The energy released by an electron in the atom when it makes a transition from a state with quantum number n' to a state with quantum number n is carried away by a photon. The energy of the photon (in eV) is equal to

$$\Delta E = (-13.6)\frac{1}{n'^2}Z^2 - (-13.6)\frac{1}{n^2}Z^2 = (13.6)\left(\frac{1}{n^2} - \frac{1}{n'^2}\right)Z^2$$

The wavelength of the photon can be obtained using the following relation:

$$\frac{1}{\lambda} = \frac{\Delta E}{hc} = R\left(\frac{1}{n^2} - \frac{1}{n'^2}\right)Z^2$$

The constant $R = 1.097 \times 10^7$ m^{-1} is the **Rydberg constant**. The spectrum of wavelengths that results from this relationship is in excellent agreement with experimental measurements of the emission spectrum of hydrogen atoms.

Example 37-11-A. Calculating emission lines for iron. Determine the wavelength of light emitted for a transition between the $n = 2$ and $n = 1$ orbits of a single electron in orbit around an iron nucleus.

Approach: The theory discussed in this Chapter can be applied to atoms other than hydrogen if they only have a single electron. The iron nucleus has 26 protons and we have to use $Z = 26$. Using the relations discussed in this Section, we can determine the energy levels associated with $n = 2$ and $n = 1$ and thus the energy of the photon released in this transition.

Solution: The energy of the state with $n = 2$ is equal to

$$E_2 = (-13.6)\left(\frac{1}{n^2}\right)Z^2 = (-13.6)\left(\frac{1}{2^2}\right)26^2 = -2.30 \times 10^3 \text{ eV}$$

The energy of the state with $n = 1$ is equal to

$$E_1 = (-13.6)\left(\frac{1}{n^2}\right)Z^2 = (-13.6)\left(\frac{1}{1^2}\right)26^2 = -9.19 \times 10^3 \text{ eV}$$

The energy of the photon released in the transition between $n = 2$ and $n = 1$ is equal to

$$E_{photon} = E_2 - E_1 = 6.90 \times 10^3 \text{ eV}$$

The wavelength of this photon is equal to:

$$\lambda = \frac{hc}{E} = \frac{1.240 \times 10^3}{6.90 \times 10^3} = 0.180 \text{ nm}$$

In this calculation we have used $hc = 1240 \text{ eV} \cdot \text{nm}$. This is a common choice of units since the electronic energy levels in atoms are usually expressed in units of eV and the wavelengths of the emitted photons are frequently expressed in units of nm.

Section 37-12. de Broglie's Hypothesis Applied to Atoms

de Broglie's hypothesis that the wavelength λ of a particle is given by

$$\lambda = \frac{h}{p}$$

where h is Planck's constant and p is the linear momentum of the particle, can be applied to the electron in a one-electron atom. The quantum condition of Bohr is consistent with the requirement that the electron wave correspond to a circular standing wave. This requires that a whole number of wavelengths fit in the orbit's circumference. This reinforces the notion that some type of wave theory of matter is the underlying physics of matter in atomic sized systems.

Practice Quiz

1. If the absolute temperature of a blackbody doubles, what happens to the wavelength of the peak of the intensity per unit wavelength spectrum of the blackbody radiation?
    a) It doubles in wavelength.
    b) It decreases to one-half the initial wavelength.
    c) It stays the same.
    d) It increases by a factor of e^2.

2. A blackbody glows with an emission spectrum that appears red to the human visual system. The temperature is changed so that the glow now appears yellow to the human visual system. How has the temperature changed?
    a) The temperature has increased.
    b) The temperature has decreased.
    c) The temperature has not changed.
    d) You need to know the change in area of the surface to know how the temperature changed.

3. A metal is being illuminated with light of a fixed wavelength and an intensity I. What change takes place in the electron emission if the intensity of the light is doubled to $2I$?
    a) The maximum kinetic energy of the electrons doubles.
    b) The sum of the maximum kinetic energy of the electrons and the work function of the material doubles.
    c) The number of electrons emitted per second by the surface doubles.
    d) No change occurs in the electron emission.

4. Why can't two electrons and a positron be produced by a single photon?
    a) Electric charge is a conserved quantity.
    b) Energy is a conserved quantity.
    c) Momentum is a conserved quantity.
    d) There are only two polarizations of a photon, not three.

5. When a photon Compton scatters off of a free electron, what happens to the energy lost by the photon?
    a) The energy lost by the photon is gained by the electron.
    b) The photon does not lose energy, it only changes wavelength.
    c) An additional photon is emitted with an energy equal to the energy lost by the original photon.
    d) Energy is not conserved during Compton scattering, so the energy is lost.

6. If the de Broglie wavelength of a particle is given by h/p, how does the wavelength of a particle with speed v calculated with the relativistic momentum expression compare to the wavelength calculated with the Newtonian momentum expression, mv?
    a) The wavelength calculated with the relativistic expression for the momentum is always greater than the wavelength calculated with the Newtonian expression for the wavelength.
    b) The wavelength calculated with the relativistic expression for the momentum is always shorter than the wavelength calculated with the Newtonian expression for the wavelength.
    c) The two wavelengths are identical.
    d) Sometimes one wavelength is longer, sometimes the other wavelength is longer.

7. When an electron in a hydrogen atom goes from a lower n value orbit to a higher n value orbit, what happens to the electron's kinetic energy?
   a) The kinetic energy increases.
   b) The kinetic energy decreases.
   c) The kinetic energy remains the same.
   d) The kinetic energy might increase or decrease.

8. What result of classical physics does the Bohr atom conflict with?
   a) Angular momentum is a conserved quantity.
   b) A centripetal force is necessary for an object to move on a circular path.
   c) Newton's second law.
   d) Accelerating charges radiate electromagnetic radiation.

9. The reason an electron microscope can detect much finer detail than an optical microscope is
   a) Electrons are point masses, but light is a wave.
   b) The wavelength of electrons in the microscope is smaller than the wavelength of visible light.
   c) Electron lenses are better in quality than optical lenses.
   d) The electron detector has better resolution than the human eye.

10. The $n = 1$ orbit of an electron in orbit around a hydrogen nucleus has a radius equal to one Bohr radius, r_0. What is the radius of the orbit of a single electron in the $n = 1$ orbit around a silicon nucleus?
    a) $(14)r_0$
    b) $(1/14)r_0$
    c) $(296)r_0$
    d) $(1/296)r_0$

11. What is the Celsius temperature of a blackbody if its peak intensity per unit wavelength is at a photon energy of 2.5 eV?

12. Determine the de Broglie wavelength of a 0.486 kg ball that is thrown with a speed of 40 mph.

13. What is the scattering angle of a 662 keV photon that has an energy of 584 keV after being Compton scattered from a free electron?

14. A material has a work function of 1.1 eV. What is the longest wavelength of light that will cause electrons to be emitted from the surface of this material?

15. Determine the longest three wavelengths of light emitted in the Lyman series for a single electron in orbit around a lithium nucleus.

Responses to Select End-of-Chapter Questions

1. A reddish star is the coolest, followed by a whitish-yellow star. Bluish stars have the highest temperatures. The temperature of the star is related to the frequency of the emitted light. Since red light has a lower frequency than blue light, red stars have a lower temperature than blue stars.

7. Individual photons of ultraviolet light are more energetic than photons of visible light and will deliver more energy to the skin, causing burns. UV photons also can penetrate farther into the skin, and, once at the deeper level, can deposit a large amount of energy that can cause damage to cells.

13. In the photoelectric effect the photon energy is completely absorbed by the electron. In the Compton effect, the photon is scattered from the electron and travels off at a lower energy.

19. Opposite charges attract, so the attractive Coulomb force between the positive nucleus and the negative electrons keeps the electrons from flying off into space.

25. The lines in the spectrum of hydrogen correspond to all the possible transitions that the electron can make. The Balmer lines, for example, correspond to an electron moving from all higher energy levels to the $n = 2$ level. Although an individual hydrogen atom only contains one electron, a sample of hydrogen gas contains many atoms and all the different atoms will be undergoing different transitions.

Solutions to Select End-of-Chapter Problems

In several problems, the value of hc is needed: $hc = 1240\,\text{eV}\cdot\text{nm}$.

1. We use Wien's law, Eq. 37-1.

    (a) $\lambda_P = \dfrac{(2.90\times10^{-3}\,\text{m}\cdot\text{K})}{T} = \dfrac{(2.90\times10^{-3}\,\text{m}\cdot\text{K})}{(273\,\text{K})} = 1.06\times10^{-5}\,\text{m} = \boxed{10.6\,\mu\text{m}}$

    This wavelength is in the $\boxed{\text{far infrared.}}$

    (b) $\lambda_P = \dfrac{(2.90\times10^{-3}\,\text{m}\cdot\text{K})}{T} = \dfrac{(2.90\times10^{-3}\,\text{m}\cdot\text{K})}{(3500\,\text{K})} = 8.29\times10^{-7}\,\text{m} = \boxed{829\,\text{nm}}$

    This wavelength is in the $\boxed{\text{infrared.}}$

    (c) $\lambda_P = \dfrac{(2.90\times10^{-3}\,\text{m}\cdot\text{K})}{T} = \dfrac{(2.90\times10^{-3}\,\text{m}\cdot\text{K})}{(4.2\,\text{K})} = 6.90\times10^{-4}\,\text{m} = \boxed{0.69\,\text{mm}}$

    This wavelength is in the $\boxed{\text{microwave}}$ region.

    (d) $\lambda_P = \dfrac{(2.90\times10^{-3}\,\text{m}\cdot\text{K})}{T} = \dfrac{(2.90\times10^{-3}\,\text{m}\cdot\text{K})}{(2.725\,\text{K})} = 1.06\times10^{-3}\,\text{m} = \boxed{1.06\,\text{mm}}$

    This wavelength is in the $\boxed{\text{microwave}}$ region.

7. We use Eq. 37-3 along with the fact that $f = c/\lambda$ for light. The longest wavelength will have the lowest energy.

    $E_1 = hf_1 = \dfrac{hc}{\lambda_1} = \dfrac{(6.63\times10^{-34}\,\text{J}\cdot\text{s})(3.00\times10^8\,\text{m}/\text{s})}{(410\times10^{-9}\,\text{m})} = 4.85\times10^{-19}\,\text{J}\left(\dfrac{1\,\text{eV}}{1.60\times10^{-19}\,\text{J}}\right) = 3.03\,\text{eV}$

    $E_2 = hf_2 = \dfrac{hc}{\lambda_2} = \dfrac{(6.63\times10^{-34}\,\text{J}\cdot\text{s})(3.00\times10^8\,\text{m}/\text{s})}{(750\times10^{-9}\,\text{m})} = 2.65\times10^{-19}\,\text{J}\left(\dfrac{1\,\text{eV}}{1.60\times10^{-19}\,\text{J}}\right) = 1.66\,\text{eV}$

    Thus the range of energies is $\boxed{2.7\times10^{-19}\,\text{J} < E < 4.9\times10^{-19}\,\text{J}}$ or $\boxed{1.7\,\text{eV} < E < 3.0\,\text{eV}}$.

13. The energy of the photon will equal the kinetic energy of the baseball. We use Eq. 37-3.

    $K = hf \;\rightarrow\; \tfrac{1}{2}mv^2 = h\dfrac{c}{\lambda} \;\rightarrow\; \lambda = \dfrac{2hc}{mv^2} = \dfrac{2(6.63\times10^{-34}\,\text{J}\cdot\text{s})(3.00\times10^8\,\text{m}/\text{s})}{(0.145\,\text{kg})(30.0\,\text{m/s})^2} = \boxed{3.05\times10^{-27}\,\text{m}}$

19. We use Eq. 37-4b to calculate the work function.

    $W_0 = hf - K_{\max} = \dfrac{hc}{\lambda} - K_{\max} = \dfrac{1240\,\text{eV}\cdot\text{nm}}{285\,\text{nm}} - 1.70\,\text{eV} = \boxed{2.65\,\text{eV}}$

25. (a) Since $f = c/\lambda$, the photon energy is $E = hc/\lambda$ and the largest wavelength has the smallest energy. In order to eject electrons for all possible incident visible light, the metal's work function must be less than or equal to the energy of a 750-nm photon.

The maximum value for the metal's work function W_0 is found by setting the work function equal to the energy of the 750-nm photon.

$$W_0 = \frac{hc}{\lambda} = \frac{\left(6.63 \times 10^{-34} \text{ J} \cdot \text{s}\right)\left(3.00 \times 10^8 \text{ m/s}\right)}{\left(750 \times 10^{-9} \text{ m}\right)}\left(\frac{1 \text{ eV}}{1.60 \times 10^{-19} \text{ J}}\right) = \boxed{1.66 \text{ eV}}$$

(b) If the photomultiplier is to function only for incident wavelengths less than 410-nm, then we set the work function equal to the energy of the 410-nm photon.

$$W_0 = \frac{hc}{\lambda} = \frac{\left(6.63 \times 10^{-34} \text{ J} \cdot \text{s}\right)\left(3.00 \times 10^8 \text{ m/s}\right)}{\left(410 \times 10^{-9} \text{ m}\right)}\left(\frac{1 \text{ eV}}{1.60 \times 10^{-19} \text{ J}}\right) = \boxed{3.03 \text{ eV}}$$

31. (a) In the Compton effect, the maximum change in the photon's wavelength is when scattering angle $\phi = 180°$. We use Eq. 37-6b to determine the maximum change in wavelength. Dividing the maximum change by the initial wavelength gives the maximum fractional change.

$$\Delta\lambda = \frac{h}{m_e c}(1 - \cos\theta) \quad \rightarrow$$

$$\frac{\Delta\lambda}{\lambda} = \frac{h}{m_e c \lambda}(1 - \cos\theta) = \frac{\left(6.63 \times 10^{-34} \text{ J} \cdot \text{s}\right)\left(1 - \cos 180°\right)}{\left(9.11 \times 10^{-31} \text{ kg}\right)\left(3.00 \times 10^8 \text{ m/s}\right)\left(550 \times 10^{-9} \text{ m}\right)} = \boxed{8.8 \times 10^{-6}}$$

(b) We replace the initial wavelength with $\lambda = 0.10$ nm.

$$\frac{\Delta\lambda}{\lambda} = \frac{h}{m_e c \lambda}(1 - \cos\theta) = \frac{\left(6.63 \times 10^{-34} \text{ J} \cdot \text{s}\right)\left(1 - \cos 180°\right)}{\left(9.11 \times 10^{-31} \text{ kg}\right)\left(3.00 \times 10^8 \text{ m/s}\right)\left(0.10 \times 10^{-9} \text{ m}\right)} = \boxed{0.049}$$

37. The minimum energy necessary is equal to the rest energy of the two muons.

$$E_{\min} = 2mc^2 = 2(207)(0.511 \text{ MeV}) = \boxed{212 \text{ MeV}}$$

The wavelength is given by Eq. 37-5.

$$\lambda = \frac{hc}{E} = \frac{\left(6.63 \times 10^{-34} \text{ J} \cdot \text{s}\right)\left(3.00 \times 10^8 \text{ m/s}\right)}{\left(1.60 \times 10^{-19} \text{ J/eV}\right)\left(212 \times 10^6 \text{ eV}\right)} = \boxed{5.86 \times 10^{-15} \text{ m}}$$

43. The theoretical resolution limit is the wavelength of the electron. We find the wavelength from the momentum, and find the momentum from the kinetic energy and rest energy. We use the result from Problem 94. The kinetic energy of the electron is 85 keV.

$$\lambda = \frac{hc}{\sqrt{K^2 + 2mc^2 K}} = \frac{\left(6.63 \times 10^{-34} \text{ J} \cdot \text{s}\right)\left(3.00 \times 10^8 \text{ m/s}\right)}{\left(1.60 \times 10^{-19} \text{ J/eV}\right)\sqrt{\left(85 \times 10^3 \text{ eV}\right)^2 + 2\left(0.511 \times 10^6 \text{ eV}\right)\left(85 \times 10^3 \text{ eV}\right)}}$$

$$= \boxed{4.1 \times 10^{-12} \text{ m}}$$

49. Since the particles are not relativistic, we may use $K = p^2/2m$. We then form the ratio of the wavelengths, using Eq. 37-7.

$$\lambda = \frac{h}{p} = \frac{h}{\sqrt{2mK}} \quad ; \quad \frac{\lambda_p}{\lambda_e} = \frac{\dfrac{h}{\sqrt{2m_p K}}}{\dfrac{h}{\sqrt{2m_e K}}} = \sqrt{\frac{m_e}{m_p}} < 1$$

Thus we see the proton has the shorter wavelength, since $m_e < m_p$.

55. To ionize the atom means removing the electron, or raising it to zero energy.

$$E_{ionization} = 0 - E_n = 0 - \frac{(-13.6\,\text{eV})}{n^2} = \frac{(13.6\,\text{eV})}{3^2} = \boxed{1.51\,\text{eV}}$$

61. The energy of the photon is the sum of the ionization energy of 13.6 eV and the kinetic energy of 20.0 eV. The wavelength is found from Eq. 37-3.

$$hf = \frac{hc}{\lambda} = E_{total} \rightarrow \lambda = \frac{hc}{E_{total}} = \frac{(6.63 \times 10^{-34}\,\text{J·s})(3.00 \times 10^8\,\text{m/s})}{(1.60 \times 10^{-19}\,\text{J/eV})(33.6\,\text{eV})} = 3.70 \times 10^{-8}\,\text{m} = \boxed{37.0\,\text{nm}}$$

67. The velocity is found from Eq. 37-10 evaluated for $n = 1$.

$$mvr_n = \frac{nh}{2\pi} \rightarrow$$

$$v = \frac{h}{2\pi r_1 m_e} = \frac{(6.63 \times 10^{-34}\,\text{J·s})}{2\pi(0.529 \times 10^{-10}\,\text{m})(9.11 \times 10^{-31}\,\text{kg})} = 2.190 \times 10^6\,\text{m/s} = \boxed{7.30 \times 10^{-3}\,c}$$

We see that $v \ll c$.
The relativistic factor is as follows.

$$\left[1 - \left(\frac{v}{c}\right)^2\right]^{\frac{1}{2}} \approx 1 - \frac{1}{2}\left(\frac{v}{c}\right)^2 = 1 - \frac{1}{2}\left(\frac{2.190 \times 10^6\,\text{m/s}}{3.00 \times 10^8\,\text{m/s}}\right)^2 = \boxed{1 - 2.66 \times 10^{-5}}$$

Because this is essentially 1, the use of non-relativistic formulas is $\boxed{\text{justified.}}$

73. To produce a photoelectron, the hydrogen atom must be ionized, so the minimum energy of the photon is 13.6 eV. We find the minimum frequency of the photon from Eq. 37-3.

$$E = hf \rightarrow f = \frac{E}{h} \rightarrow f_{min} = \frac{E_{min}}{h} = \frac{(13.6\,\text{eV})(1.60 \times 10^{-19}\,\text{J/eV})}{(6.63 \times 10^{-34}\,\text{J·s})} = \boxed{3.28 \times 10^{15}\,\text{Hz}}$$

79. The total energy of the two photons must equal the total energy (kinetic energy plus mass energy) of the two particles. The total momentum of the photons is 0, so the momentum of the particles must have been equal and opposite. Since both particles have the same mass and the same momentum, they each have the same kinetic energy.

$$E_{photons} = E_{particles} = 2(m_e c^2 + K) \rightarrow$$

$$K = \frac{1}{2}E_{photons} - m_e c^2 = 0.755\,\text{MeV} - 0.511\,\text{MeV} = \boxed{0.244\,\text{MeV}}$$

85. The stopping potential is the voltage that gives a potential energy change equal to the maximum kinetic energy. We use Eq. 37-4b to first find the work function, and then find the stopping potential for the higher wavelength.

$$K_{max} = eV_0 = \frac{hc}{\lambda} - W_0 \rightarrow W_0 = \frac{hc}{\lambda_0} - eV_0$$

$$eV_1 = \frac{hc}{\lambda_1} - W_0 = \frac{hc}{\lambda_1} - \left(\frac{hc}{\lambda_0} - eV_0\right) = hc\left(\frac{1}{\lambda_1} - \frac{1}{\lambda_0}\right) + eV_0$$

$$= \frac{(6.63 \times 10^{-34}\,\text{J·s})(3.00 \times 10^8\,\text{m/s})}{(1.60 \times 10^{-19}\,\text{J/eV})}\left(\frac{1}{440 \times 10^{-9}\,\text{m}} - \frac{1}{380 \times 10^{-9}\,\text{m}}\right) + 2.70\,\text{eV} = 2.25\,\text{eV}$$

The potential difference needed to cancel an electron kinetic energy of 2.25 eV is $\boxed{2.25\,\text{V.}}$

91. (a) See the adjacent figure.

(b) Absorption of a 5.1 eV photon represents a transition from the ground state to the state 5.1 eV above that, the third excited state. Possible photon emission energies are found by considering all the possible downward transitions that might occur as the electron makes its way back to the ground state.

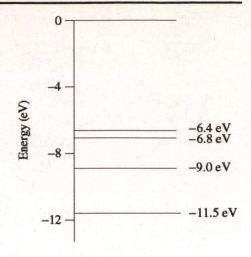

$$-6.4\,\text{eV} - (-6.8\,\text{eV}) = \boxed{0.4\,\text{eV}}$$

$$-6.4\,\text{eV} - (-9.0\,\text{eV}) = \boxed{2.6\,\text{eV}}$$

$$-6.4\,\text{eV} - (-11.5\,\text{eV}) = \boxed{5.1\,\text{eV}}$$

$$-6.8\,\text{eV} - (-9.0\,\text{eV}) = \boxed{2.2\,\text{eV}}$$

$$-6.8\,\text{eV} - (-11.5\,\text{eV}) = \boxed{4.7\,\text{eV}}$$

$$-9.0\,\text{eV} - (-11.5\,\text{eV}) = \boxed{2.5\,\text{eV}}$$

97. (a) We write the Planck time as $t_p = G^\alpha h^\beta c^\gamma$, and the units of t_p must be $[T]$.

$$t_p = G^\alpha h^\beta c^\gamma \;\rightarrow\; [T] = \left[\frac{L^3}{MT^2}\right]^\alpha \left[\frac{ML^2}{T}\right]^\beta \left[\frac{L}{T}\right]^\gamma = [L]^{3\alpha+2\beta+\gamma}[M]^{\beta-\alpha}[T]^{-2\alpha-\beta-\gamma}$$

There are no mass units in $[T]$, and so $\beta = \alpha$, and $[T] = [L]^{5\alpha+\gamma}[T]^{-3\alpha-\gamma}$. There are no length units in $[T]$, and so $\gamma = -5\alpha$ and $[T] = [T]^{-3\alpha+5\alpha} = [T]^{2\alpha}$. Thus $\alpha = \frac{1}{2} = \beta$ and $\gamma = -\frac{5}{2}$.

$$\boxed{t_p = G^{1/2}h^{1/2}c^{-5/2} = \sqrt{\frac{Gh}{c^5}}}$$

(b) $$t_p = \sqrt{\frac{Gh}{c^5}} = \sqrt{\frac{\left(6.67\times10^{-11}\,\text{N}\cdot\text{m}^2/\text{kg}^2\right)\left(6.63\times10^{-34}\,\text{J}\bullet\text{s}\right)}{\left(3.00\times10^8\,\text{m/s}\right)^5}} = \boxed{1.35\times10^{-43}\,\text{s}}$$

(c) We write the Planck length as $\lambda_p = G^\alpha h^\beta c^\gamma$, and the units of λ_p must be $[L]$.

$$\lambda_p = G^\alpha h^\beta c^\gamma \;\rightarrow\; [L] = \left[\frac{L^3}{MT^2}\right]^\alpha \left[\frac{ML^2}{T}\right]^\beta \left[\frac{L}{T}\right]^\gamma = [L]^{3\alpha+2\beta+\gamma}[M]^{\beta-\alpha}[T]^{-2\alpha-\beta-\gamma}$$

There are no mass units in $[L]$, and so $\beta = \alpha$, and $[L] = [L]^{5\alpha+\gamma}[T]^{-3\alpha-\gamma}$. There are no time units in $[L]$, and so $\gamma = -3\alpha$ and $[L] = [L]^{5\alpha-3\alpha} = [L]^{2\alpha}$. Thus $\alpha = \frac{1}{2} = \beta$ and $\gamma = -\frac{3}{2}$.

$$\boxed{t_p = G^{1/2}h^{1/2}c^{-3/2} = \sqrt{\frac{Gh}{c^3}}}$$

(d) $$\lambda_p = \sqrt{\frac{Gh}{c^3}} = \sqrt{\frac{\left(6.67\times10^{-11}\,\text{N}\cdot\text{m}^2/\text{kg}^2\right)\left(6.63\times10^{-34}\,\text{J}\bullet\text{s}\right)}{\left(3.00\times10^8\,\text{m/s}\right)^5}} = \boxed{4.05\times10^{-35}\,\text{m}}$$

103. (a) For the photoelectric effect experiment, Eq. 37-4b can be expressed as $K_{max} = hf - W_0$. The maximum kinetic energy is equal to the potential energy associated with the stopping voltage, so $K_{max} = eV_0$. We also have $f = c/\lambda$. Combine those relationships as follows.

$$K_{max} = hf - W_0 \quad \rightarrow \quad eV_0 = \frac{hc}{\lambda} - W_0 \quad \rightarrow \quad V_0 = \frac{hc}{e}\frac{1}{\lambda} - \frac{W_0}{e}$$

A plot of V_0 vs. $\dfrac{1}{\lambda}$ should yield a straight line with a slope of $\boxed{\dfrac{hc}{e}}$ and a y-intercept of $\boxed{-\dfrac{W_0}{e}}$.

(b) The graph is shown, with a linear regression fit as given by Excel. The slope is $a = \dfrac{hc}{e} = 1.24\,\text{V}\cdot\mu\text{m}$, and the y-intercept is $b = -2.31\,\text{V}$.

(c) $-\dfrac{W_0}{e} = -2.31\,\text{V} \quad \rightarrow \quad W_0 = \boxed{2.31\,\text{eV}}$

$h = \dfrac{ea}{c} =$

(d) $\quad = \dfrac{\left(1.60 \times 10^{-19}\,\text{C}\right)\left(1.24 \times 10^{-6}\,\text{V}\cdot\text{m}\right)}{3.00 \times 10^8\,\text{m/s}} =$

$\quad = \boxed{6.61 \times 10^{-34}\,\text{J}\cdot\text{s}}$

$V_0 = 1.24(1/\lambda) - 2.31$
$R^2 = 1.00$

Chapter 38: Quantum Mechanics

Chapter Overview and Objectives

In this chapter, the principles of quantum mechanics are introduced. The Heisenberg Uncertainty Principle is discussed and the Schrödinger wave equation is presented. Solutions of the Schrödinger equation for several important potentials are examined.

After completing this chapter you should:
- Know Heisenberg's Uncertainty Principle in both the position-momentum and energy-time forms.
- Know the approximate value of Planck's constant.
- Know how the wave function is related to probability density.
- Know Schrödinger's time-dependent and time-independent equations in one dimension.
- Know what the normalization condition is.
- Know the form of the solutions to Schrödinger's equation for free particles.
- Know the solutions to Schrödinger's equation for particles in an infinite square well.
- Know the continuity conditions on wave functions that are solutions to Schrödinger's equation.
- Know what quantum mechanical tunneling is and how to solve for the tunneling probability using the WKB approximation.

Summary of Equations

Heisenberg Uncertainty Principle:
$$\Delta x \Delta p \geq \frac{h}{2\pi}$$
(Section 38-3)

$$\Delta E \Delta t \geq \frac{h}{2\pi}$$
(Section 38-3)

Schrödinger's time-independent equation:
$$-\frac{\hbar^2}{2m}\frac{\partial^2 \psi(x)}{dx^2} + U(x)\psi(x) = E\psi(x)$$
(Section 38-5)

Normalization condition:
$$\int |\psi|^2 \, dV = 1$$
(Section 38-5)

Schrödinger's time-dependent equation:
$$-\frac{\hbar^2}{2m}\frac{\partial^2 \Psi(x,t)}{dx^2} + U(x)\Psi(x,t) = i\hbar \frac{\partial \Psi(x,t)}{dt}$$
(Section 38-6)

Wave number for free particle:
$$k = \sqrt{\frac{2mE}{\hbar^2}}$$
(Section 38-7)

Energy levels for particle in infinite square well:
$$E = \frac{h^2 n^2}{8m\ell^2}$$
(Section 38-8)

Wave functions for particle in infinite square well:
$$\psi(x) = \sqrt{\frac{2}{\ell}}\sin\left(\frac{n\pi x}{\ell}\right)$$
(Section 38-8)

Continuity conditions of a wave function:
$$\psi_I = \psi_{II} \text{ and } \frac{d\psi_I}{dx} = \frac{d\psi_{II}}{dx} \text{ at } x = 0$$
(Section 38-9)

$$\psi_{II} = \psi_{III} \text{ and } \frac{d\psi_{II}}{dx} = \frac{d\psi_{III}}{dx} \text{ at } x = \ell$$

Transmission probability through a barrier:

$$T \approx e^{-2GL}, \quad G = \sqrt{\frac{2m(U_0 - E)}{\hbar^2}} \qquad \text{(Section 38-10)}$$

Chapter Summary

Section 38-1. Quantum Mechanics—A New Theory

Quantum mechanics is a self-consistent theory that accommodates the apparently contradictory dual wave-like and particle-like behaviors observed in nature. Quantum mechanics is able to describe the physics of the microscopic world. When quantum mechanics is used to describe the macroscopic world, it is consistent with the laws of classical physics.

Section 38-2. The Wave Function and Its Interpretation; the Double-Slit Experiment

One approach to quantum mechanics is through a differential equation called a wave equation. The solution to the wave equation is a function of position and time and called a **wave function**, $\psi(x,t)$. The amplitude of the wave function is also called the **probability amplitude**.   The square of the magnitude, $|\psi|^2$, is called the **probability density** or **probability distribution**, and determines the probability to find a particle at a particular position x at time t.   The probability that a measurement of the position of a particle results somewhere within a volume dV around a position x at time t is equal to $|\psi(x,\ t)|^2\ dV$.

Section 38-3. The Heisenberg Uncertainty Principle

Simultaneous measurements of position and linear momentum of a particle are limited in resolution. The uncertainty in position, Δx, and the uncertainty in momentum, Δp, satisfy the inequality:

$$\Delta x \Delta p \geq \frac{h}{2\pi}$$

This inequality is called the **Heisenberg Uncertainty Principle.** The quantity h is called Planck's constant and has the value

$$h = 6.626 \times 10^{-34} \text{ J} \cdot \text{s}$$

A second form of the uncertainty principle relates the uncertainty in the energy of a particle, ΔE, to the length of time of observation of the particle, Δt:

$$\Delta E \Delta t \geq \frac{h}{2\pi}$$

Because the quantity $h/2\pi$ appears very often in quantum mechanics, it is given its own symbol, $\hbar$:

$$\hbar = \frac{h}{2\pi} = 1.055 \times 10^{-34} \text{ J} \cdot \text{s}$$

Example 38-3-A. Implications of the Heisenberg Uncertainty Principle in the macroscopic world. A ball of mass 350 g is thrown straight through a doorway with a width 0.82 m. Determine the uncertainty in the velocity of the ball parallel to the doorway after it passes through the doorway.

Approach: The Heisenberg Uncertainty Principle limits the accuracy with which we can determine the position and the linear momentum in the microscopic world and in the macroscopic world.   In this problem, the focus is on the macroscopic world. If we know that a ball is thrown through a doorway, we know its position parallel to the doorway with an uncertainty equal to the width of the doorway. The uncertainty principle tells us the corresponding uncertainty of the linear momentum of the ball.

Solution: The uncertainty in the position of the ball in a direction parallel to the doorway is 0.82 m. According to the Heisenberg Uncertainty Principle, the corresponding uncertainty in the linear momentum of the ball is

$$\Delta p \ge \frac{h}{2\pi \, \Delta x}$$

The uncertainty in the linear momentum of the ball is related to the uncertainty in its linear velocity:

$$\Delta p = m\Delta v \quad \Rightarrow \quad \Delta v = \frac{\Delta p}{m}$$

We thus conclude that

$$\Delta v = \frac{\Delta p}{m} \ge \frac{h}{2\pi m \, \Delta x} = \frac{\left(6.63 \times 10^{-34}\right)}{2\pi\left(0.350\right)\left(0.82\right)} = 3.7 \times 10^{-34} \text{ m/s}$$

The uncertainty in the velocity is much smaller than the accuracy with which we can measure the velocity of the ball, and we thus in general do not experience the constraints imposed by the Heisenberg Uncertainty Principle in the macroscopic world.

Example 38-3-B. Taking time to make observations. The mass of an object is determined in a process that observes the mass for a period of 10 μs. What is the minimum uncertainty of the mass measurement?

Approach: In the problem we observe our object for a specific time period. Based on the observations of the object during this period, we can determine its energy and thus its mass. We expect that if the period of observation increases, the uncertainty of the derived mass decreases.

Solution: The uncertainty in the time during which we observe our object is 10 μs. The Heisenberg Uncertainty Principle can be used to estimate the lower-limit of the accuracy of a measurement of the energy of the object:

$$\Delta E \ge \frac{h}{2\pi \, \Delta t}$$

Assuming the object is at rest, the uncertainty in its energy is related to the uncertainty in its mass

$$\Delta m = \frac{\Delta E}{c^2}$$

Combining these last two expressions we obtain

$$\Delta m \ge \frac{h}{2\pi c^2 \, \Delta t} = \frac{\left(6.63 \times 10^{-34}\right)}{2\pi\left(3.00 \times 10^8\right)^2 \left(10 \times 10^{-6}\right)} = 1.17 \times 10^{-46} \text{ kg}$$

This is a very small uncertainty in mass. For comparison, the mass of an electron is 9×10^{-31} kg. However, if we observe an object that has a very small lifetime, the uncertainty in its mass may become comparable to its mass.

Section 38-4. Philosophic Implications; Probability versus Determinism

In classical physics, if the positions and velocities of all particles of a system are known at some time, then, in principle, the laws of physics can be used to determine their positions and velocities at all future times. Quantum mechanics is inherently different. The solution to quantum mechanical equations provides us with a wave function that is related to a probability amplitude. As a consequence, quantum mechanics is not deterministic; that is, future events are not uniquely determined on the basis of current observation.

Section 38-5. The Schrödinger Equation in One Dimension—Time-Independent Form

The wave function is a solution to Schrödinger's wave equation. In general, a wave equation has terms with derivatives of the wave function with respect to position and terms with derivatives of the wave function with respect to time. Some wave equations can be separated into two related equations, one equation only with derivatives with respect to time and one equation only with derivatives with respect to position. In general, equations with derivatives with respect to one variable are easier to solve than equations with derivatives with respect to two different variables. We look first at such an equation, related to Schrödinger's wave equation, that only has derivatives with respect to position. This equation is called the **time-independent Schrödinger equation**:

$$-\frac{\hbar^2}{2m}\frac{\partial^2 \psi(x)}{\partial x^2} + U(x)\psi(x) = E\psi(x)$$

where $\psi(x)$ is a function related to the wave function $\Psi(x,t)$, $U(x)$ is the position-dependent potential energy of the system, and E is the total energy of the particle. The general form of Schrödinger's time-independent equation places some restrictions on the possible solutions $\psi(x)$. If the energy E is to be finite, then $\psi(x)$ must be a continuous function for all x and its derivative must be continuous, except where there is an infinite discontinuity in $U(x)$. Another constraint on $\psi(x)$ results from the interpretation of $|\psi|^2$ as the probability density. Because a measurement of the position of a particle must result in some location in space, the integral of $|\psi|^2\,dx$ must be equal to one:

$$\int |\psi|^2\,dx = 1$$

This is called the **normalization condition.**

Example 38-5-A. Normalizing the solution of the time-independent wave function. A particle has a wave function given by

$$\psi(x) = Ae^{-b|x|}$$

Determine the constant A in terms of b so that the wave function is normalized.

Approach: To ensure that the probability to find the particle somewhere is equal to 1, we must require that the integral of $|\psi|^2 dx$ must be equal to one. This integral will allow us to express A in terms of b.

Solution: The integral of $|\psi|^2 dx$ between $-\infty$ and $+\infty$ is equal to

$$\int_{-\infty}^{\infty} |\psi|^2\,dx = \int_{-\infty}^{\infty} A^2 e^{-2b|x|}\,dx = A^2\left\{\int_{-\infty}^{0} e^{2bx}\,dx + \int_{0}^{\infty} e^{-2bx}\,dx\right\} = \frac{A^2}{2b}\left\{e^{2bx}\Big|_{-\infty}^{0} - e^{-2bx}\Big|_{0}^{\infty}\right\} = \frac{A^2}{b}$$

The requirement that the integral of $|\psi|^2 dx$ is equal to one, we conclude that

$$A = \sqrt{b}$$

The wave function of the particle is thus equal to

$$\psi(x) = \sqrt{b}e^{-b|x|}$$

Section 38-6. Time-Dependent Schrödinger Equation

The complete Schrödinger equation is known as the time-dependent Schrödinger equation:

$$-\frac{\hbar^2}{2m}\frac{\partial^2 \Psi(x,t)}{\partial x^2} + U(x)\Psi(x,t) = i\hbar\frac{\partial \Psi(x,t)}{\partial t}$$

The solution of this equation is the product of a time-independent function, which is the solution of the time-independent Schrödinger equation with energy E, and a time-dependent function:

$$\Psi(x,t) = \psi_E(x)\,e^{-i\frac{E}{\hbar}t}$$

Section 38-7. Free Particles; Plane Waves and Wave Packets

For a free particle, the potential energy U is zero everywhere. Schrödinger's time-independent equation reduces to

$$-\frac{\hbar^2}{2m}\frac{\partial^2 \psi(x)}{\partial x^2} + U(x)\psi(x) = -\frac{\hbar^2}{2m}\frac{\partial^2 \psi(x)}{\partial x^2} = E\psi(x) \quad\Rightarrow\quad \frac{\partial^2 \psi(x)}{\partial x^2} = -\frac{2m}{\hbar^2}E\psi(x)$$

The most general solution of this equation is

$$\psi(x) = A \sin kx + B \cos kx$$

where the wave number k is given by

$$k = \sqrt{\frac{2mE}{\hbar^2}}$$

This solution is not localized in space; the probability density is spread out over all space. It has a definite momentum and so its uncertainty in position must be infinite to satisfy the Heisenberg Uncertainty Principle. In general, we are often interested in solutions to Schrödinger's equation that have some spatial localization. We call these types of solutions **wave packets**.

Section 38-8. Particle in an Infinitely Deep Square Well Potential (a Rigid Box)

The infinitely deep square well potential energy function is

$$U(x) = 0 \quad \text{if} \quad 0 < x < L$$
$$U(x) = \infty \quad \text{if} \quad x \leq 0 \quad \text{or} \quad x \geq L$$

The solutions to Schrödinger's time-independent equation only exist for particular values of the energy E. These values are

$$E = \frac{h^2 n^2}{8mL^2}$$

where n is an integer. The corresponding solutions to Schrödinger's time-independent equations are

$$\psi(x) = \sqrt{\frac{2}{L}} \sin\left(\frac{n\pi x}{L}\right)$$

Examples of the wave functions and the corresponding probability distributions for several different values of n are shown in the Figure on the right. Although we cannot make a definite prediction on the location of the particle, except that it will be located within the boundaries of the box, we do observe that for all specific values of n there are specific locations at which the particle will never be found (e.g., if $n = 2$, the probability to find the particle at $x = 0$, at $x = L/2$, or at $x = L$ are very small). The wave function has an amplitude of 0 for $x < 0$ or $x < L$ since the particle can never be found in these regions.

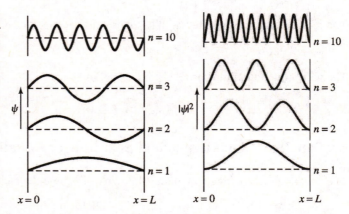

Section 38-9. Finite Potential Well

The potential energy as a function of position for a finite potential well is given by

$$U(x) = 0 \quad \text{if} \quad 0 < x < L$$
$$U(x) = U_0 \quad \text{if} \quad x \leq 0 \quad \text{or} \quad x \geq L$$

In the regions for which $x \leq 0$ or $x \geq L$, the solution to Schrödinger's equation can be written as

$$\psi(x) = Ce^{Gx} + De^{-Gx}$$

where

$$G = \sqrt{\frac{2m(U_0 - E)}{\hbar^2}}$$

Since the amplitude of the wave function should not approach infinity when x approaches either $-\infty$ or $+\infty$ we recognize that $D = 0$ in the region $x \leq 0$ and $C = 0$ in the region $x \geq L$.

In the region $0 < x < L$, we can use the solution discussed in the previous section:

$$\psi(x) = A \sin kx + B \cos kx$$

where

$$k = \sqrt{\frac{2E}{m}}$$

The parameters A, B, C, and D are not all free parameters. They are constrained as a result of the requirement that the wave function must be continuous everywhere. In addition, the derivative of the wave function must be continuous everywhere except where there is an infinite discontinuity in the potential energy. If we apply these conditions to the solutions of Schrödinger's equation in the three separate regions in the finite square well problem, we have a set of equations that can be satisfied for only particular values of E if $E < 0$ (or if $\psi = 0$ everywhere) and for all values

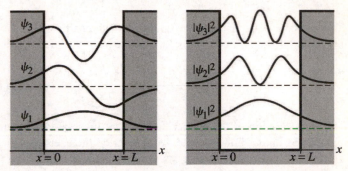

of E if $E > 0$. Examples of the wave functions and the corresponding probability distributions are shown in the Figures above. As can be seen, the probability amplitude is not zero in the region where the energy of the particle is less than the potential energy. In the classical picture, this region would not be accessible since the kinetic energy of the particle would be negative. However, according to quantum mechanics, there is a non-zero probability to observe the particle in this region.

Section 38-10. Tunneling through a Barrier

The wave function solution of Schrödinger's equation will, in general, be non-zero in all regions where the potential energy is not infinite. Of particular interest is the situation when a particle starts on one side of a potential energy barrier with less kinetic energy than the maximum potential energy of the barrier. In classical physics, the particle cannot cross the barrier. However, in quantum mechanics, this particle can appear on the opposite side of the barrier from which it started with a probability dependent on the amplitude of the wave function on the far side of the barrier. This process is called **quantum mechanical tunneling**. The probability of a particle tunneling through the barrier is approximately equal to:

$$T \approx e^{-2G\ell}$$

where

$$G = \sqrt{\frac{2m(U_0 - E)}{\hbar^2}}$$

and ℓ is the thickness of the potential energy barrier, U_0 is the height of the potential energy barrier, E is the energy of the particle, and m is the mass of the particle.

The effect of tunneling can be observed and used in many different ways. The rate with which certain atomic nuclei undergo radioactive decay depends on the tunneling probability through the **Coulomb barrier**. Based on measured rates, the shape of the barrier can be probed. The **scanning tunneling electron microscope** (STM) relies in tunneling through vacuum between a surface and a tiny probe to map the structure of a surface. The strong dependence of the transmission probability on the barrier width allows the STM to map the vertical features of the surface with a resolution of 10^{-2} to 10^{-3} nm.

Example 38-10-A. Using the tunneling probability as a probe of the height of the potential barrier. An electron with an energy $E = 32$ eV tunnels through a potential energy barrier with a thickness $\ell = 1.6$ nm. The transmission probability is measured to be $T = 0.312$. What is the height of the potential energy barrier?

Approach: Since the tunneling probability depends on both the barrier width and the barrier height, we can determine the barrier height if we know the barrier width and the tunneling probability.

Solution: The transmission probability T depends on the constant G, defined above, and the barrier thickness ℓ:

$$T \approx e^{-2G\ell}$$

Since the problem provides us with information about T and ℓ, we can determine G:

$$G = -\frac{\ln T}{2\ell} = -\frac{\ln(0.312)}{2(1.6 \times 10^{-9})} = 3.64 \times 10^9 \text{ m}^{-1} = \sqrt{\frac{2m(U_0 - E)}{\hbar^2}}$$

The constant G depends on the energy E of the particle, which is provided, and the height of the potential barrier U_0. At this point, the only unknown is U_0 which can now be determined:

$$U_0 = \frac{\hbar^2 G^2}{2m} + E = \frac{(6.63 \times 10^{-34})^2 (3.64 \times 10^9)^2}{2(9.11 \times 10^{-31})} + (32)(1.60 \times 10^{-19}) = 8.3 \times 10^{-18} \text{ J} = 52 \text{ eV}$$

In the calculation of the barrier height, care must be taken to use consistent units. It is in general easiest to convert all units to SI units. For example, we converted the energy of the particle from eV to J by using the fact that 1 eV equals 1.60×10^{-19} J. In the last step, we converted energy back from J to eV.

Practice Quiz

1. The value of the wave function at point x is equal to 2.00. What does this tell you?
   a) The probability of finding the particle at point x is 2.00.
   b) The probability of finding the particle at point x is 4.00.
   c) The probability of finding the particle between x and $x + dx$ is 4.00 dx.
   d) The value of the wave function can't be bigger than 1.00 because of normalization.

2. To be consistent with the Heisenberg Uncertainty Principle, if you improve an experiment to determine the position of a particle to a factor 10 smaller uncertainty, the smallest uncertainty in momentum you could possible measure simultaneously is
   a) a factor 10 smaller, also.
   b) a factor of 10 larger.
   c) zero.
   d) a factor 100 smaller.

3. If you take a time Δt to measure the energy of a particle, what is the minimum uncertainty in the energy measurement?
   a) $1/\Delta t$
   b) $1/\Delta t^2$
   c) h
   d) $\hbar/\Delta t$

4. A particle in an infinite square well potential has a minimum energy E. If the square well is doubled in length, what will the minimum energy be?
   a) $2E$
   b) $4E$
   c) $E/2$
   d) $E/4$

5. From what assumption does the normalization principle come from?
   a) The probability of a particle being somewhere in space must be equal to one.
   b) The uncertainty in the momentum of a particle multiplied by the uncertainty in its position divided by $\hbar$ must be equal to 1.
   c) The uncertainty in the momentum of a particle divided by the uncertainty in its position multiplied by $\hbar$ must be equal to 1.
   d) There is no assumption; one is just a convenient number to work with.

6. The time-independent Schrödinger equation has a wave function solution ψ with an energy E. What is the corresponding solution to the time-dependent Schrödinger equation?
   a) ψ
   b) $\psi\, t$
   c) $\psi\, t^2$
   d) $\psi\, e^{-iEt/\hbar}$

7. Why don't we need to be concerned with the Heisenberg Uncertainty Principle in everyday life when dealing with macroscopic objects?
   a) The Heisenberg Uncertainty Principle only applies to microscopic objects.
   b) The Heisenberg Uncertainty Principle only applies to electrons in atoms.
   c) The uncertainties required by the uncertainty principle are much smaller than the uncertainties of position and momentum that we make on macroscopic objects.
   d) The uncertainty principle only applies to measurements made by scattering light off of the object and we usually measure the position of macroscopic objects with a ruler.

8. If a particle is placed in an infinite square well of length L and a finite square well of length L, in which will it have the lower energy in the lowest energy state for that potential?
   a) The infinite square well
   b) The finite square well
   c) Both energies will be identical.
   d) Need to know the depth of the finite square well to answer the question

9. If the length of a barrier is doubled, by what factor must the difference between the energy and the barrier height change so that the tunneling probability remains the same?
   a) 4
   b) 2
   c) ½
   d) ¼

10. What is the minimum height of a potential energy barrier for which a particle with energy E will have zero probability of transmission through the barrier?
    a) $E/2$
    b) E
    c) $2E$
    d) ∞

11. What is the minimum uncertainty of the momentum of an electron in a wire that is 5 μm long?

12. You have measured the total energy of a system to within an uncertainty of 3×10^{-5} eV. What is the minimum amount of time required to make this measurement in order to obtain this uncertainty?

13. The ratio of the energies of two consecutive energy levels of a particle in an infinite square well of length 0.645 μm is exactly 1.44. The lower energy of the two is 6.24×10^{-32} J. What is the mass of the particle?

14. Determine the probability that an electron with a kinetic energy of 1.5 eV is able to tunnel through a barrier of height 3.0 eV and a thickness of 0.12 μm.

15. Determine the lowest four energy levels for an object of mass 3.8 g in a box 10 cm long. Treat the box as one-dimensional.

Responses to Select End-of-Chapter Questions

1. (a) A matter wave ψ does not need a medium as a wave on a string does. The square of the wave function for a matter wave ψ describes the probability of finding a particle within a certain spatial range, whereas the equation for a wave on a string describes the displacement of a piece of string from its equilibrium position.
    (b) An EM wave also does not need a medium. The equation for the EM wave describes the way in which the amplitudes of the electric and magnetic fields change as the wave passes a point in space. An EM wave represents a vector field and can be polarized. A matter wave is a scalar and cannot be polarized.

7. Yes. In energy form, the uncertainty principle is $\Delta E \Delta t \geq h/2\pi$. For the ground state, Δt is very large, since electrons remain in that state for a very long time, so ΔE is very small and the energy of the state can be precisely known. For excited states, which can decay to the ground state, Δt is much smaller, and ΔE is correspondingly larger. Therefore the energy of the state is less well known.

13. A particle in a box is confined to a region of space. Since the uncertainty in position is limited by the box, there must be some uncertainty in the particle's momentum, and the momentum cannot be zero. The zero point energy reflects the uncertainty in momentum.

Solutions to Select End-of-Chapter Problems

1. We find the wavelength of the neutron from Eq. 37-7. The peaks of the interference pattern are given by Eq. 34-2a and Figure 34-10. For small angles, we have $\sin\theta = \tan\theta$.

$$\lambda = \frac{h}{p} = \frac{h}{\sqrt{2m_0 K}} \quad ; \quad d\sin\theta = m\lambda, \ m = 1, 2, \dots \quad ; \quad y = \ell\tan\theta$$

$$\sin\theta = \tan\theta \quad \rightarrow \quad \frac{m\lambda}{d} = \frac{y}{\ell} \quad \rightarrow \quad y = \frac{m\lambda\ell}{d}, \ m = 1, 2, \dots \quad \rightarrow$$

$$\Delta y = \frac{\lambda\ell}{d} = \frac{h\ell}{d\sqrt{2m_0 K}} = \frac{\left(6.63\times10^{-34}\,\text{J·s}\right)\left(1.0\,\text{m}\right)}{\left(6.0\times10^{-4}\,\text{m}\right)\sqrt{2\left(1.67\times10^{-27}\,\text{kg}\right)\left(0.030\,\text{eV}\right)\left(1.60\times10^{-19}\,\text{J/eV}\right)}}$$

$$= \boxed{2.8\times10^{-7}\,\text{m}}$$

7. The uncertainty in the energy is found from the lifetime and the uncertainty principle.

$$\Delta E = \frac{\hbar}{\Delta t} = \frac{\left(1.055\times10^{-34}\,\text{J·s}\right)}{\left(12\times10^{-6}\,\text{s}\right)}\left(\frac{1\,\text{eV}}{1.60\times10^{-19}\,\text{J}}\right) = 5.49\times10^{-11}\,\text{eV}$$

$$\frac{\Delta E}{E} = \frac{5.49\times10^{-11}\,\text{eV}}{5500\,\text{eV}} = \boxed{1.0\times10^{-14}}$$

19. The general expression for the wave function of a free particle is given by Eq. 38-3a. The particles are not relativistic.

(a) $k = \dfrac{2\pi}{\lambda} = \dfrac{2\pi p}{h} = \dfrac{mv}{\hbar} = \dfrac{\left(9.11\times10^{-31}\,\text{kg}\right)\left(3.0\times10^{5}\,\text{m/s}\right)}{\left(1.055\times10^{-34}\,\text{J·s}\right)} = 2.6\times10^{9}\,\text{m}^{-1}$

$$\psi = \boxed{A\sin\left[\left(2.6\times10^{9}\,\text{m}^{-1}\right)x\right] + B\cos\left[\left(2.6\times10^{9}\,\text{m}^{-1}\right)x\right]}$$

(b) $k = \dfrac{2\pi}{\lambda} = \dfrac{2\pi p}{h} = \dfrac{mv}{\hbar} = \dfrac{\left(1.67\times10^{-27}\,\text{kg}\right)\left(3.0\times10^{5}\,\text{m/s}\right)}{\left(1.055\times10^{-34}\,\text{J}\bullet\text{s}\right)} = 4.7\times10^{12}\,\text{m}^{-1}$

$$\psi = \boxed{A\sin\left[\left(4.7\times10^{12}\,\text{m}^{-1}\right)x\right] + B\cos\left[\left(4.7\times10^{12}\,\text{m}^{-1}\right)x\right]}$$

25. We assume the particle is not relativistic. The energy levels give the kinetic energy of the particles in the box.

$$E_1 = \dfrac{h^2}{8m\ell^2} = \dfrac{p_1^2}{2m} \;\rightarrow\; p_1^2 = \dfrac{h^2}{4\ell^2} \;\rightarrow\; \left|p_1\right| = \dfrac{h}{2\ell} \;\rightarrow\; \Delta p \approx 2\left|p_1\right| = \dfrac{h}{\ell}$$

$$\Delta x \Delta p \approx \ell\dfrac{h}{\ell} = h$$

This is consistent with the uncertainty principle.

31. (a) The ground state energy is given by Eq. 38-13 with $n = 1$.

$$E_1 = \dfrac{h^2 n^2}{8m\ell^2}\Bigg|_{n=1} = \dfrac{\left(6.63\times10^{-34}\,\text{J}\bullet\text{s}\right)^2 (1)}{8\left(32\,\text{u}\right)\left(1.66\times10^{-27}\,\text{kg/u}\right)\left(4.0\times10^{-3}\,\text{m}\right)^2\left(1.60\times10^{-19}\,\text{J/eV}\right)}$$

$$= 4.041\times10^{-19}\,\text{eV} \approx \boxed{4.0\times10^{-19}\,\text{eV}}$$

(b) We equate the thermal energy expression to Eq. 38-13 in order to find the quantum number.

$$\tfrac{1}{2}kT = \dfrac{h^2 n^2}{8m\ell^2} \;\rightarrow$$

$$n = 2\sqrt{kTm}\,\dfrac{\ell}{h} = 2\sqrt{\left(1.38\times10^{-23}\,\text{J/K}\right)\left(300\,\text{K}\right)\left(32\,\text{u}\right)\left(1.66\times10^{-27}\,\text{kg/u}\right)}\,\dfrac{\left(4.0\times10^{-3}\,\text{m}\right)}{\left(6.63\times10^{-34}\,\text{J}\bullet\text{s}\right)}$$

$$= 1.789\times10^{8} \approx \boxed{2\times10^{8}}$$

(c) Use Eq. 38-13 with a large-n approximation.

$$\Delta E = E_{n+1} - E_n = \dfrac{h^2}{8m\ell^2}\left[\left(n+1\right)^2 - n^2\right] = \dfrac{h^2}{8m\ell^2}\left(2n+1\right) \approx 2n\dfrac{h^2}{8m\ell^2} = 2nE_1$$

$$= 2\left(1.789\times10^{8}\right)\left(4.041\times10^{-19}\,\text{eV}\right) = \boxed{1.4\times10^{-10}\,\text{eV}}$$

43. The transmission coefficient is given by Eqs. 38-17a and 38-17b.
    (a) The barrier height is now $1.02\left(70\,\text{eV}\right) = 71.4\,\text{eV}$.

$$2\ell\dfrac{\sqrt{2m\left(U_0 - E\right)}}{\hbar} = 2\left(0.10\times10^{-9}\,\text{m}\right)\dfrac{\sqrt{2\left(9.11\times10^{-31}\,\text{kg}\right)\left(21.4\,\text{eV}\right)\left(1.60\times10^{-19}\,\text{J/eV}\right)}}{\left(1.055\times10^{-34}\,\text{J}\bullet\text{s}\right)}$$

$$= 4.735$$

$$T = e^{-2\ell\frac{\sqrt{2m(U-E)}}{\hbar}} = e^{-4.735} = 8.782\times10^{-3}\;;\quad \dfrac{T}{T_0} = \dfrac{8.782\times10^{-3}}{0.010} = 88 \;\rightarrow\; \boxed{12\%\ \text{decrease}}$$

(b) The barrier width is now $1.02\left(0.10\,\text{nm}\right) = 0.102\,\text{nm}$.

$$2\ell\dfrac{\sqrt{2m\left(U_0 - E\right)}}{\hbar} = 2\left(0.102\times10^{-9}\,\text{m}\right)\dfrac{\sqrt{2\left(9.11\times10^{-31}\,\text{kg}\right)\left(20\,\text{eV}\right)\left(1.60\times10^{-19}\,\text{J/eV}\right)}}{\left(1.055\times10^{-34}\,\text{J}\bullet\text{s}\right)}$$

$$= 4.669$$

$$T = e^{-2\ell\frac{\sqrt{2m(U_0-E)}}{\hbar}} = e^{-4.669} = 9.382\times10^{-3}\;;\quad \dfrac{T}{T_0} = \dfrac{9.382\times10^{-3}}{0.010} = 93.8 \;\rightarrow\; \boxed{6.2\%\ \text{decrease}}$$

49. We find the wavelength of the protons from their kinetic energy, and then use the two-slit interference formulas from Chapter 34, with a small angle approximation. If the protons were accelerated by a 650-volt potential difference, then they will have 650 eV of kinetic energy.

$$\lambda = \frac{h}{p} = \frac{h}{\sqrt{2m_0 K}} \quad ; \quad d\sin\theta = m\lambda, \, m = 1, 2, \dots \quad ; \quad y = \ell\tan\theta$$

$$\sin\theta = \tan\theta \quad \rightarrow \quad \frac{m\lambda}{d} = \frac{y}{\ell} \quad \rightarrow \quad y = \frac{m\lambda\ell}{d}, \, m = 1, 2, \dots \quad \rightarrow$$

$$\Delta y = \frac{\lambda\ell}{d} = \frac{h\ell}{d\sqrt{2m_0 K}} = \frac{\left(6.63\times10^{-34}\,\text{J}\cdot\text{s}\right)\left(18\,\text{m}\right)}{\left(8.0\times10^{-4}\,\text{m}\right)\sqrt{2\left(1.67\times10^{-27}\,\text{kg}\right)\left(650\,\text{eV}\right)\left(1.60\times10^{-19}\,\text{J/eV}\right)}}$$

$$= \boxed{2.5\times10^{-8}\,\text{m}}$$

55. We model the electrons as being restricted from leaving the surface of the sodium by an energy barrier, similar to Figure 38-15a. The difference between the barrier's height and the energy of the electrons is the work function, and so $U_0 - E = W_0 = 2.28\,\text{eV}$. But quantum mechanically, some electrons will "tunnel" through that barrier without ever being given the work function energy, and thus get outside the barrier, as shown in Figure 38-15b. This is the tunneling current as indicated in Figure 38-18. The distance from the sodium surface to the tip of the microscope is the width of the barrier, ℓ. We calculate the transmission probability as a function of barrier width by Eqs. 38-17a and 38-17b. The barrier is then increased to $\ell + \Delta\ell$, which will lower the transmission probability.

$$2G\Delta\ell = 2\Delta\ell\frac{\sqrt{2m(U_0 - E)}}{\hbar} = 2\left(0.02\times10^{-9}\,\text{m}\right)\frac{\sqrt{2\left(9.11\times10^{-31}\,\text{kg}\right)\left(2.28\,\text{eV}\right)\left(1.60\times10^{-19}\,\text{J/eV}\right)}}{\left(1.055\times10^{-34}\,\text{J}\cdot\text{s}\right)}$$

$$= 0.3091$$

$$T_0 = e^{-2G\ell} \quad ; \quad T = e^{-2G(\ell+\Delta\ell)} \quad ; \quad \frac{T}{T_0} = \frac{e^{-2G(\ell+\Delta\ell)}}{e^{-2G\ell}} = e^{-2G\Delta\ell} = e^{-0.3091} = 0.734$$

The tunneling current is caused by electrons that tunnel through the barrier. Since current is directly proportional to the number of electrons making it through the barrier, any change in the transmission probability is reflected as a proportional change in current. So we see that the change in the transmission probability, which will be reflected as a change in current, is $\boxed{\text{a decrease of 27\%}}$. Note that this change is only a fraction of the size of an atom.

61. (a) See the graph.
(b) From the graph and the spreadsheet, we find these results.
$T = 10\%$ at $E/U_0 = 0.146$
$T = 20\%$ at $E/U_0 = 0.294$
$T = 50\%$ at $E/U_0 = 0.787$
$T = 80\%$ at $E/U_0 = 1.56$

Chapter 39: Quantum Mechanics of Atoms

Chapter Overview and Objectives

In this chapter the results of the application of quantum mechanics to one and many electron atoms are discussed. The solutions to Schrödinger's equation of the hydrogen atom are presented and the energy and angular momentum of the electron are studied. Examples of applied areas of atomic physics, such as lasers, are presented.

After completing this chapter you should:
- Know the allowed energies and angular momenta of an electron in a one-electron atom.
- Given a spherically symmetric wave function, be able to calculate the probability of finding the electron in a given range of radii.
- Know what the Pauli exclusion principle is and how it applies to atoms.
- Be able to determine electron configurations of atoms.
- Know how to calculate the approximate energy and wavelength of X rays.
- Know how to calculate the magnetic dipole moment of an electron.
- Know how to calculate the possible total angular momentum quantum numbers of an electron with given orbital and spin quantum numbers.
- Know what fluorescence and phosphorescence are.

Summary of Equations

Schrödinger equation for hydrogen:
$$-\frac{\hbar^2}{2m}\left(\frac{\partial^2 \psi}{\partial x^2}+\frac{\partial^2 \psi}{\partial y^2}+\frac{\partial^2 \psi}{\partial z^2}\right)-\frac{1}{4\pi\varepsilon_0}\frac{e^2}{r}\psi = E\psi \qquad \text{(Section 39-2)}$$

Electron energies in hydrogen:
$$E_n = -\frac{13.6\,eV}{n^2} \quad n\in\{1,2,3,...\} \qquad \text{(Section 39-2)}$$

Electron orbital angular momentum in hydrogen:
$$L = \hbar\sqrt{\ell(\ell+1)} \quad \ell\in\{0,1,2,...,n-1\} \qquad \text{(Section 39-2)}$$

Allowed z components of L:
$$L_z = m_\ell \hbar \quad m_l\in\{-\ell,-\ell+1,...,\ell-1,\ell\} \qquad \text{(Section 39-2)}$$

Allowed z components of S:
$$S_z = m_s \hbar \quad m_s\in\{+1/2,-1/2\} \qquad \text{(Section 39-2)}$$

Selection for single photon interaction:
$$\Delta\ell = \pm1 \qquad \text{(Section 39-2)}$$

Ground state wave function of the hydrogen atom:
$$\psi_{100} = \frac{1}{\sqrt{\pi r_0^3}}e^{-r/r_0} \qquad \text{(Section 39-3)}$$

Definition of the Bohr radius:
$$r_0 = \frac{h^2\varepsilon_0}{\pi m e^2} = 0.0529\,\text{nm} \qquad \text{(Section 39-3)}$$

Radial probability density:
$$P_r(r) = 4\pi r^2 |\psi(r)|^2 \qquad \text{(Section 39-3)}$$

Magnetic moment of electron due to orbital motion:
$$\vec{\mu} = -\frac{e}{2m}\vec{L} \qquad \text{(Section 39-7)}$$

Definition of Bohr magneton:
$$\mu_B = \frac{e\hbar}{2m} = 9.27\times10^{-27}\,\text{J/T} \qquad \text{(Section 39-7)}$$

Magnetic moment of electron due to spin:	$\mu_z = -g\mu_B m_s$	(Section 39-7)

Definition of total angular momentum of electron:	$\vec{J} = \vec{L} + \vec{S}$	(Section 39-7)

Chapter Summary

Section 39-1. Quantum-Mechanical View of Atoms

When the time-independent Schrödinger's equation is solved for a one-electron atom, the result is a three-dimensional wave function centered on the nucleus of the atom. The electron does not follow a circular path as Bohr's theory of the atom assumes. The solution to Schrödinger's equation provides us with information on the probability to find the electron within a certain spatial region. The wave function can be used to determine the most likely position of the electron.

Section 39-2. Hydrogen Atom: Schrödinger Equation and Quantum Numbers

The potential energy function of an electron in the electrostatic field of a point-like positive charge +e is given by

$$U(r) = -\frac{1}{4\pi\varepsilon_0}\frac{e^2}{r}$$

where r is the distance between the electron and the positive charge. The time-independent Schrödinger equation for this system is

$$-\frac{\hbar^2}{2m}\left(\frac{\partial^2\psi}{\partial x^2} + \frac{\partial^2\psi}{\partial y^2} + \frac{\partial^2\psi}{\partial z^2}\right) - \frac{1}{4\pi\varepsilon_0}\frac{e^2}{r}\psi = E\psi$$

The full solution of this equation lies beyond the scope of the textbook but some of the properties of the solutions are discussed.

The solutions to the time-independent Schrödinger equation for the one-electron atom have the same energy levels as the solutions to Bohr's atom. The energy is determined by an integer n which is called the **principal quantum number** of the solution and is a positive integer. The energy of the wave function with principal quantum number n, E_n, is given by

$$E_n = -\frac{13.6\,eV}{n^2} \qquad n \in \{1, 2, 3, \ldots\}$$

A second quantum number, called the **orbital quantum number** ℓ, is related to the angular momentum of the electron. The magnitude of the angular momentum of the electron due to its translational motion, called its orbital angular momentum, is given by

$$L = \hbar\sqrt{\ell(\ell+1)} \qquad \ell \in \{0, 1, 2, \ldots, n-1\}$$

Another quantum number, the **magnetic quantum number** m_ℓ, is related to the component of the orbital angular momentum in a specific direction. The usual component specified is the component along the z axis. The z component of the orbital angular momentum, L_z, is related to m_ℓ by

$$L_z = m_\ell \hbar \qquad m_\ell \in \{-\ell, -\ell+1, \ldots, \ell-1, \ell\}$$

The **spin quantum number** m_s is related to the component of the internal angular momentum of the electron, called the spin angular momentum, in a specific direction. The usual component specified is the component along the z axis. The z component of the spin angular momentum, S_z, is related to m_s by

$$S_z = m_s \hbar$$

The magnitude of the spin angular momentum for an electron is always $\hbar\sqrt{3}/2$.

Transitions between states that occur in interactions with the electromagnetic field in which a single photon is emitted or absorbed must satisfy the following **selection rule**:

$$\Delta\ell = \pm 1$$

A transition that satisfies this rule is called an **allowed transition** and a transition that does not satisfy this condition is called a **forbidden transition**. The "forbidden transitions" are not completely forbidden, but usually occur with a probability several orders of magnitude smaller than the allowed transitions.

Example 39-2-A. Transitions between excited states in the hydrogen atom. An electron in a hydrogen atom makes a transition from the $n = 3$, $\ell = 2$ state to the $n = 2$, $\ell = 3$. Is this an allowed or a forbidden transition? What is the change in the energy of the electron making this transition?

Approach: The energy of a level in the hydrogen atom is determined by the principal quantum number n. The change in the energy of the electron is thus determined by the change in the principal quantum number, and is thus independent of the orbital quantum number ℓ. The change in the orbital quantum number determines if the transition is allowed or forbidden.

Solution: The change in the orbital quantum number ℓ is equal to

$$\Delta\ell = \ell_{final} - \ell_{initial} = 3 - 2 = +1$$

Since the selection rule for an allowed transition is $\Delta\ell = \pm 1$, we conclude that the transition is an allowed transition. The change in the energy of the electron is equal to

$$\Delta E = E_{final} - E_{initial} = \left(-\frac{13.6\,\text{eV}}{n_{final}^2}\right) - \left(-\frac{13.6\,\text{eV}}{n_{initial}^2}\right) = \left(-\frac{13.6\,\text{eV}}{2^2}\right) - \left(-\frac{13.6\,\text{eV}}{3^2}\right) = -1.89\,\text{eV}$$

Section 39-3. Hydrogen Atom Wave Functions

We label the different solutions of the time-independent Schrödinger equation with subscripts n, ℓ, and m_ℓ as ψ_{nlm_ℓ}. The solution with the lowest energy is

$$\psi_{100} = \frac{1}{\sqrt{\pi r_0^3}} e^{-r/r_0}$$

where r_0 is the Bohr radius:

$$r_0 = \frac{h^2\varepsilon_0}{\pi m e^2} = 0.0529\,\text{nm}$$

The probability density is the absolute square of the wave function:

$$|\psi_{100}|^2 = \frac{1}{\pi r_0^3} e^{-2r/r_0}$$

Another function that is useful is the **radial probability distribution** P_r. The probability that a measurement of the location of the electron falls between a distance r and $r + dr$ from the nucleus is $P_r(r)\,dr$. For a wave function that only depends on the distance r and not on the direction $\hat{\mathbf{r}}$, the radial probability distribution function is given by

$$P_r(r) = 4\pi r^2 |\psi(r)|^2$$

For the ground state wave function of the hydrogen atom $P_r(r)$ is equal to

$$P_r(r) = 4\frac{r^2}{r_0^3} e^{-2r/r_0}$$

Some of the other hydrogen atom electron wave functions are

$$\psi_{200} = \frac{1}{\sqrt{32\pi r_0^3}}\left(2 - \frac{r}{r_0}\right)e^{-r/2r_0}$$

$$\psi_{210} = \frac{z}{\sqrt{32\pi\, r_0^3}}\, e^{-r/2r_0}$$

$$\psi_{211} = \frac{x+iy}{\sqrt{64\pi\, r_0^3}}\, e^{-r/2r_0}$$

$$\psi_{21-1} = \frac{x-iy}{\sqrt{64\pi\, r_0^3}}\, e^{-r/2r_0}$$

Example 39-3-A. Using the electron wave function. Determine the probability of finding an electron in the ψ_{200} wave function closer to the nucleus than one Bohr radius. Determine the probability of finding an electron in the ψ_{200} wave function farther from the nucleus than one Bohr radius, but closer to the nucleus than two Bohr radii.

Approach: The electron wave function can be used to calculate the radial probability distribution. The integral of this distribution between r_1 and r_2 is equal to the probability to find the electron between r_1 and r_2. Since we know the ψ_{200} wave function of the electron, we can calculate the radial probability distribution and determine the probability of finding the electron in various regions around the nucleus.

Solution: The radial probability distribution of the electron is given by

$$P_r(r) = 4\pi r^2 \left| \psi_{200}(r) \right|^2 =$$

$$= 4\pi r^2 \left(\frac{1}{\sqrt{32\pi\, r_0^3}} \left(2 - \frac{r}{r_0} \right) e^{-r/2r_0} \right)^2 =$$

$$= \frac{1}{8r_0^3} \left(4 - 4\frac{r}{r_0} + \frac{r^2}{r_0^2} \right) r^2 e^{-r/r_0}$$

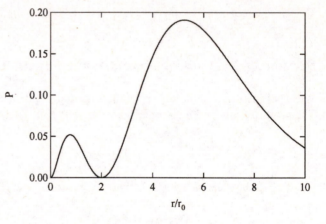

The radial probability distribution as function of r/r_0 is shown in the Figure on the right.

The probability to find the electron closer to the nucleus than the Bohr radius is equal to

$$\int_{r=0}^{r_0} P_r(r)\, dr = \frac{1}{8r_0^3} \int_{r=0}^{r_0} \left(4 - 4\frac{r}{r_0} + \frac{r^2}{r_0^2} \right) r^2 e^{-r/r_0}\, dr = -\frac{1}{8r_0^3} \left[\left(\frac{r^4}{r_0} + 4r^2 r_0 + 8r r_0^2 + 8r_0^3 \right) e^{-r/r_0} \right]_0^{r_0} =$$

$$= 1 - \tfrac{21}{8} e^{-1} \approx 0.034$$

The probability to find the electron between the Bohr radius and two Bohr radii is equal to

$$\int_{r=r_0}^{2r_0} P_r(r)\, dr = -\frac{1}{8r_0^3} \left[\left(\frac{r^4}{r_0} + 4r^2 r_0 + 8r r_0^2 + 8r_0^3 \right) e^{-r/r_0} \right]_{r_0}^{2r_0} = -\frac{1}{8r_0^3} \left[56 r_0^3 e^{-2} - 21 r_0^3 e^{-1} \right] =$$

$$= \tfrac{21}{8} e^{-1} - 7e^{-2} \approx 0.018$$

Section 39-4. Complex Atoms; the Exclusion Principle

Complex atoms are atoms with more than one electron. The **atomic number** Z of an atom is the number of protons in the nucleus of the atom, which is equal to the number of electrons in a neutral atom. When there is more than one electron near a nucleus, the solutions of the Schrödinger equation are much more difficult to determine because of the

Coulomb potential energy due to the interaction of any two electrons that must be included in the total potential energy. The solutions are still labeled with the same four quantum numbers, n, ℓ, and m_ℓ, and m_s.

The **Pauli exclusion principle** states that no two electrons can occupy the same quantum state. As a consequence, the set of quantum numbers of each electron in an atom must be different. The structure of the **periodic table** can be understood on this basis.

Section 39-5. The Periodic Table of Elements

The periodic table of the elements originally arranged the elements according to their atomic mass and their chemical properties. Since the chemical properties of an atom are determined by the electron configuration, and thus indirectly by their atomic number Z, the periodic table can also be considered an arrangement of the atoms according to their atomic number. Most periodic tables of the elements give the average atomic mass of the atoms of that element found in nature. Some periodic tables of elements also give the electron configuration of the ground state of the atom. The periodic table of the elements on the inside back cover of the text includes this information by giving the configuration of the electrons in the outermost occupied **shell**. A shell is a set of wave functions for electrons that have the same principal quantum number n. The shells have been given names. The $n = 1$ shell is called the K shell, the $n = 2$ shell is called the L shell, and the $n = 3$ shell is called the M shell.

Each shell is divided into **subshells**. A given subshell consists of all wave functions with the same principal quantum number n and the same orbital quantum number ℓ. Rather than using the ℓ values to label the subshell, a corresponding letter is used. The letter s is used for $\ell = 0$, the letter p is used for $\ell = 1$, the letter d is used for $\ell = 2$, and the letter f is used for $\ell = 3$. There are $2\ell + 1$ independent spatial wave functions for any given value of ℓ and each spatial wave function can have both a spin up and a spin down electron. Each subshell can thus have as many as $2(2\ell + 1)$ electrons in it. This means an s subshell can have two electrons, a p subshell can have six electrons, a d subshell can have ten electrons, and so on.

Example 39-5-A. Determine the ground state electron configuration for a Germanium atom.

Approach: The number of electrons in a neutral Germanium atom is 32. We start adding electrons to the subshells until the total number of electrons is equal to 32. The order in which we fill the subshells is based on the energy of the subshells.

Solution: Germanium has an atomic number of 32. The number of electrons in a neutral Germanium atom is thus equal to 32. These 32 electrons are distributed over the following shells/subshells:
- The $n = 1$ shell has two electrons.
- The $n = 2$ shell has eight electrons (two $2s$ electrons and six $2p$ electrons).
- The $n = 3$ shell has eighteen electrons (two $3s$ electrons, six $3p$ electrons, and ten $3d$ electrons).
- The $n = 4$ shell has the remaining four electrons (two $4s$ electrons and two $4p$ electrons). It should be noted that the energy level of the $4s$ subshell is below the energy level of the $3d$ subshell. Since both of them are completely filled, the filling order does not influence the final electron configuration.

We conclude that the complete electron configuration of the ground state Germanium atom is $1s^2 2s^2 2p^6 3s^2 3p^6 4s^2 3d^{10} 4p^2$.

Section 39-6. X-Ray Spectra and Atomic Number

The emission and absorption spectra of the electrons in higher quantum states in a many-electron atom are difficult to predict because the interaction energy between the electrons is on the same order of size as the transition energy and it is difficult to solve Schrödinger's equation with those interactions included. For electron transitions between $n = 2$ states and $n = 1$ states, the effect of the electron interaction is small compared to the size of the Coulomb interaction with the nucleus, especially for atoms with large atomic number. For these atoms, the transitions to have the same energy as those for a one-electron atom with an effective nuclear charge $Z_{eff} = Z - 1$.

The emission lines from inner shell electron transitions have been given names corresponding to the final electron state with a subscript referring to the initial electron state. The K_α emission line is the emission line that results from the $n = 2$ shell to the $n = 1$ (K) shell. The K_β emission line is the emission line that results from the $n = 3$ shell to the $n = 1$ (K) shell.

Example 39-6-A. X-ray emission by electrons in gold atoms. Determine the wavelength of the K_α emission from a gold atom.

Approach: The K_α X-rays are emitted when an electron transitions from the $n = 2$ shell to the $n = 1$ shell. The Bohr formula can be used to determine the energies of these states, and thus the energy released during this transition. The effective charge to be used to determine the energy levels is $79 - 1 = 78$.

Solution: The difference in energy between the two states is equal to

$$\Delta E = \left[-\frac{(13.6)(Z-1)^2}{n_{final}^2} \right] - \left[-\frac{(13.6)(Z-1)^2}{n_{initial}^2} \right] = \left[-\frac{(13.6)78^2}{1^2} \right] - \left[-\frac{(13.6)78^2}{2^2} \right] = -62{,}000 \text{ eV} = -62 \text{ keV}$$

The emitted photon thus has an energy of 62 keV and its wavelength is equal to

$$\lambda = \frac{hc}{E} = \frac{(6.63 \times 10^{-34})(3.00 \times 10^8)}{(62 \times 10^3)(1.60 \times 10^{-19})} = 2.00 \times 10^{-11} \text{ m}$$

Section 39-7. Magnetic Dipole Moments; Total Angular Momentum

The magnetic dipole moment $\vec{\mu}\,'$ of a classical electron moving in a circle at constant speed is related to its angular momentum $\vec{L}$ by

$$\vec{\mu} = ! \frac{e}{2m} \vec{L}$$

where $-e$ is the charge of the electron and m is the mass of the electron. The same result is obtained if we use the electron wave function. The potential energy U of a magnetic dipole in a magnetic field $\vec{B}$ is given by

$$U = ! \, \vec{\mu} \, \vec{B}$$

If we use the direction of the magnetic field to define the z direction, this expression reduces to

$$U = -\mu_z B = -\left(-\frac{e}{2m} L_z \right) B = \frac{e}{2m} m_\ell \hbar B$$

The **Bohr magneton** μ_B is defined as

$$\mu_B = \frac{e\hbar}{2m} = 9.27 \times 10^{-27} \text{ J/T}$$

The z component of the magnetic dipole moment, expressed in terms of the Bohr magneton, is given by

$$\mu_z = -\mu_B m_\ell$$

The potential energy of the electron can now be rewritten as

$$U = \mu_B m_\ell B$$

There is also a magnetic dipole moment related to the spin quantum number, m_s, of the electron. The z component of the spin magnetic dipole moment is given by

$$\mu_z = -g\mu_B m_s$$

where g is called the gyromagnetic ratio of the electron and has an approximate value of 2.0023 for a free electron. The total angular momentum $\vec{J}$ is the sum of the orbital angular momentum $\vec{L}$ and the spin angular momentum $\vec{S}$:

$$\vec{J} = \vec{L} + \vec{S}$$

The magnitude of the total angular momentum is quantized and is equal to

$$J = \sqrt{j(j+1)} \, \hbar$$

where $|\ell - s| \le j \le \ell + s$. The quantum number j is called **the total angular momentum quantum number**. For the electron, $s = \frac{1}{2}$ and

$$j = \ell + \tfrac{1}{2} \quad \text{or} \quad j = \ell - \tfrac{1}{2},$$

unless $\ell = 0$. If $\ell = 0$, $j = \frac{1}{2}$. The z component of the total angular momentum, J_z, is given by

$$J_z = m_j \hbar \qquad m_j \in \left\{ -j, -j+1, \ldots, j-1, j \right\}$$

The electron state can be specified using spectroscopic notation. In this notation, the principal quantum number is given first, then the capitalized letter to specify the orbital quantum number ℓ, followed by the s subscript equal to the total angular momentum quantum number j. For example, the $2P_{3/2}$ states has $n = 2$, $\ell = 1$, and $j = 3/2$.

The orbital motion of the electron creates a magnetic field in which the spin magnetic dipole moves. This adds an additional magnetic dipole interaction term that is proportional to the quantity $\bar{L} \cdot \bar{S}$ and is called the **spin-orbit interaction**.

Section 39-8. Fluorescence and Phosphorescence

When light is absorbed by an atom, often it is re-emitted at very close to the same wavelength if the electron makes the same transition between the same two corresponding states during the absorption process and the emission process. It is also possible that the re-emission process involves multiple photons if the electron makes transitions between intermediate states between the initial and final states. The photons emitted in these multiple emissions will all have lower energy and longer wavelength than the photon initially absorbed by the atom. This type of process is called **fluorescence**.

Typical lifetimes of atoms in excited states are in the order of 10^{-8} s. The emission of light by excited atoms takes place on that time scale. Certain states of atoms are **metastable states**. Atoms in these states can remain in these states much longer, even many minutes or hours. When the atoms in these states do make a transition to a lower energy level and emit light in the process, it is called **phosphorescence**.

Section 39-9. Lasers

A **laser** creates light through the process of **stimulated emission**. The word laser stands for **l**ight **a**mplification by **s**timulated **e**mission of **r**adiation. Atoms in excited states are induced to radiate when photons with the same energy as the transition pass by the excited atoms. The photon emitted by stimulated emission is in phase and moves the same direction as the stimulating photons, thus amplifying the number of photons in the light traveling in the direction of the stimulating photons.

In a gas at thermal equilibrium, there are more atoms in the lower energy states than in the higher energy states. An atom in the lower energy state can absorb the photons produced as a result of stimulated emission. This process attenuates the light, rather than amplifying it. If there are more atoms in the lower energy state than in the higher energy state, more absorption than stimulated emission occurs as photons pass through the gas. The net result is overall absorption of the photons rather than amplification. If the gas can be put into a state in which there are more atoms in the higher energy states than in the lower energy states, emission will dominate absorption. The condition in which more atoms are in the higher energy state than in the lower energy state is called a **population inversion**. The population inversion condition is obtained by different means in different types of lasers. If the energy input used to produce the population inversion is continuous, the resulting laser is continuous laser. If the energy input to cause the population inversion is pulsed, the laser is called a **pulsed laser**.

Lasers are used in many different applications. Most applications rely on the coherent narrow beam that is produced by lasers. Reflection of laser light from bar codes and the surface of CDs and DVDs is used by bar code readers and CD/DVD players, respectively, to read information. Diffraction limits the smallest features that can be read by the laser beam. The diffraction limit depends on wavelength, and CD and DVD players use different color lasers. Lasers are also used as surgical tools since they can be used to destroy tissues in a very localized area. The intense heat that can be produced across a very small area allows lasers also to be used for welding and machining metals.

Section 39-10. Holography

A **hologram** is a recording of the interference pattern between a reference source of light and the light reflected from an object. Lasers make this possible because a definite phase relationship must exist between the reference light and the reflected light. The relatively long coherence length of lasers compared to ordinary sources of light makes this possible. Holograms recorded on thin films must be viewed with a laser.

Volume or **white-light holograms** are made with a laser and recorded using a thin emulsion. They can be viewed using white light.

Practice Quiz

1. How many quantum numbers are necessary to completely specify an electron state in an atom?
    a)   1
    b)   2
    c)   3
    d)   4

2. How many possible electron states are there in the $n = 4$ shell of the atom?
    a)   4
    b)   16
    c)   32
    d)   64

3. Which of the following electron configurations is the correct ground state configuration for a fluorine atom?
    a) $1s^2 1p^6 2s^1$
    b) $1s^2 2s^2 2p^6$
    c) $1s^2 2s^2 3s^2 4s^2 5s^1$
    d) $1s^2 2s^2 2p^5$

4. How many m_l quantum numbers are the in the $l = 54$ subshell?
    a)   27
    b)   54
    c)   108
    d)   109

5. Why is $Z - 1$ used in the Bohr formula in a many electron atom when calculating the energy or wavelength of a transition from an $n = 2$ state to an $n = 1$ state rather than Z?
    a)   It is a relativistic correction because of the high kinetic energy of the $1s$ electron.
    b)   The potential energy of the interaction with the other $1s$ electron cancels out the interaction of one proton.
    c)   For larger atomic numbers, the atomic number is one greater than the number of protons in the nucleus.
    d)   The proton charge decreases when it is in a large nucleus.

6. The process when electromagnetic radiation is absorbed by an atom and then later re-emitted at longer wavelength is called
    a)   fluorescence.
    b)   spin-orbit interaction.
    c)   population inversion.
    d)   holography.

7. The condition in a laser in which more atoms are in higher energy atomic states than in lower energy atomic states is called
    a)   fluorescence.
    b)   spin-orbit interaction.
    c)   population inversion.
    d)   holography.

8. What is the approximate value of the gyromagnetic ratio of an electron?
    a) 9.11×10^{-31}
    b) 1.602×10^{-19}
    c)   1
    d)   2

9. How many different total angular momentum quantum numbers are there for a particle with an orbital angular momentum quantum number $l = 2$ and a spin quantum number $s = 3/2$?
    a) 2
    b) 3
    c) 4
    d) 6

10. Which is not a possible set of values of j and m_j resulting from an electron in an $l = 5$ state?
    a) $j = 11/2, m_j = 11/2$
    b) $j = 9/2, m_j = 11/2$
    c) $j = 9/2, m_j = 9/2$
    d) $j = 11/2, m_j = -1/2$

11. Determine the probability that an electron in the ψ_{100} wave function of the hydrogen atom is found farther from the nucleus than four Bohr radii.

12. Determine the wavelength of the K_α X-ray emission from silicon ($Z = 14$).

13. What element would have a K_α X-ray emission at a wavelength of approximately 0.12 nm?

14. Determine the most probable distance from the nucleus at which you would find the electron in the ψ_{200} wave function in a hydrogen atom.

15. Determine the j and m_j quantum numbers for all possible states of total angular momentum resulting from an $\ell = 4$ state and an $s = 1$ state.

Responses to Select End-of-Chapter Questions

1. The Bohr model placed electrons in definite circular orbits described by a single quantum number (n). The Bohr model could not explain the spectra of atoms more complex than hydrogen and could not explain fine structure in the spectra. The quantum-mechanical model uses the concept of electron "probability clouds," with the probability of finding the electron at a given position determined by the wave function. The quantum model uses four quantum numbers to describe the electron (n, l, m_l, m_s) and can explain the spectra of more complex atoms and fine structure.

7. In the time-independent Schrödinger equation, the wave function and the potential depend on the three spatial variables. The three quantum numbers result from application of boundary conditions to the wave function.

13. If there were no electron spin, then, according to the Pauli exclusion principle, s-subshells would be filled with one electron, p-subshells with three electrons, and d-subshells with five electrons. The first 20 elements of the periodic table would look like the following:

H 1 $1s^1$								
He 2 $2s^1$						Li 3 $2p^1$	Be 4 $2p^2$	B 5 $2p^3$
C 6 $3s^1$						N 7 $3p^1$	O 8 $3p^2$	F 9 $3p^3$
Ne 10 $4s^1$	Na 11 $3d^1$	Mg 12 $3d^2$	Al 13 $3d^3$	Si 14 $3d^4$	P 15 $3d^5$	S 16 $4p^1$	Cl 17 $4p^2$	Ar 18 $4p^3$
K 19 $5s^1$	Ca 20 $4d^1$							

19. In helium and other complex atoms, electrons interact with other electrons in addition to their interactions with the nucleus. The Bohr theory only works well for atoms that have a single outer electron in an s state. X-ray emissions generally involve transitions to the $1s$ or $2s$ states. In these cases the Bohr theory can be modified to correct for screening from a second electron by using the factor $Z-1$ for the nuclear charge and can yield good estimates of the transition energies. Transitions involving outer electrons in more complex atoms will be affected by additional complex screening effects and cannot be adequately described by the Bohr theory.

25. Consider a silver atom in its ground state for which the entire magnetic moment is due to the spin of only one of its electrons. In a uniform magnetic field, the dipole will experience a torque that would tend to align it with the field. In a non-uniform field, each pole of the dipole will experience a force of different magnitude. Consequently, the dipole will experience a net force that varies with the spatial orientation of the dipole. The Stern-Gerlach experiment provided the first evidence of space quantization, since it clearly indicated that there are two opposite spin orientations for the outermost electron in the silver atom.

Solutions to Select End-of-Chapter Problems

1. The value of l can range from 0 to $n-1$. Thus for $n=7$, $\boxed{l=0, 1, 2, 3, 4, 5, 6}$.

7. (a) The principal quantum number is $n=\boxed{7}$.

   (b) The energy of the state is
   $$E_7 = -\frac{(13.6\,\text{eV})}{n^2} = -\frac{(13.6\,\text{eV})}{7^2} = \boxed{-0.278\,\text{eV}}.$$

   (c) The "g" subshell has $\ell=\boxed{4}$. The magnitude of the angular momentum depends on ℓ only:
   $$L = \hbar\sqrt{\ell(\ell+1)} = \boxed{\sqrt{20}\hbar} = \sqrt{20}\left(1.055\times10^{-34}\,\text{J}\cdot\text{s}\right) = \boxed{4.72\times10^{-34}\,\text{J}\cdot\text{s}}$$

   (d) For each ℓ the value of m_l can range from $-\ell$ to $+\ell$: $\boxed{m_l = -4, -3, -2, -1, 0, 1, 2, 3, 4}$.

13. The ground state wave function is $\psi_{100} = \dfrac{1}{\sqrt{\pi r_0^3}}e^{-r/r_0}$.

   (a) $\left(\psi_{100}\right)_{r=1.5r_0} = \boxed{\dfrac{1}{\sqrt{\pi r_0^3}}e^{-1.5}}$

   (b) $\left(\left|\psi_{100}\right|^2\right)_{r=1.5r_0} = \boxed{\dfrac{1}{\pi r_0^3}e^{-3}}$

   (c) $P_r = \left(4\pi r^2\left|\psi_{100}\right|^2\right)_{r=1.5r_0} = 4\pi r_0^2\left(\dfrac{1}{\pi r_0^3}e^{-2}\right) = \boxed{\dfrac{4}{r_0}e^{-3}}$

19. We follow the directions as given in the problem. We use the first integral listed in Appendix B-5.
   $$\bar{r} = \int_0^\infty r\left|\psi_{100}\right|^2 4\pi r^2\,dr = \int_0^\infty r\frac{1}{\pi r_0^3}e^{-2\frac{r}{r_0}}4\pi r^2\,dr = 4\int_0^\infty \frac{r^3}{r_0^3}e^{-2\frac{r}{r_0}}\,dr \; ; \text{let } x = 2\frac{r}{r_0} \; \rightarrow$$
   $$\bar{r} = \tfrac{1}{4}r_0\int_0^\infty x^3 e^{-x}\,dx = \tfrac{1}{4}r_0\,(3!) = \boxed{\tfrac{3}{2}r_0}$$

25. The wave function is given in Eq. 39-5a. Note that $r = \left(x^2 + y^2 + z^2\right)^{1/2}$. We will need the derivative relationship derived in the first line below.

$$\frac{\partial r}{\partial x} = \tfrac{1}{2}\left(x^2 + y^2 + z^2\right)2x = \frac{x}{r} \quad ; \quad \psi_{100} = \frac{1}{\sqrt{\pi r_0^3}}\,e^{-\frac{r}{r_0}} \quad ; \quad \frac{\partial \psi}{\partial x} = \frac{\partial \psi}{\partial r}\frac{\partial r}{\partial x} = -\frac{1}{r_0}\frac{1}{\sqrt{\pi r_0^3}}\,e^{-\frac{r}{r_0}}\frac{x}{r}$$

$$\frac{\partial^2 \psi}{\partial x^2} = \frac{\partial}{\partial x}\left(-\frac{1}{r_0}\frac{1}{\sqrt{\pi r_0^3}}\,e^{-\frac{r}{r_0}}\frac{x}{r}\right) = \left(-\frac{1}{r_0}\frac{1}{\sqrt{\pi r_0^3}}\,e^{-\frac{r}{r_0}}\frac{1}{r}\right) + \frac{\partial}{\partial r}\left(-\frac{1}{r_0}\frac{1}{\sqrt{\pi r_0^3}}\,e^{-\frac{r}{r_0}}\frac{x}{r}\right)\frac{\partial r}{\partial x}$$

$$= \frac{-1}{rr_0}\frac{1}{\sqrt{\pi r_0^3}}\,e^{-\frac{r}{r_0}}\left[1 - x^2\left(\frac{1}{rr_0} + \frac{1}{r^2}\right)\right] = -\psi\frac{1}{rr_0}\left[1 - x^2\left(\frac{1}{rr_0} + \frac{1}{r^2}\right)\right]$$

Similarly, we would have $\dfrac{\partial r}{\partial y} = \dfrac{y}{r}$; $\dfrac{\partial^2 \psi}{\partial y^2} = -\psi\dfrac{1}{rr_0}\left[1 - y^2\left(\dfrac{1}{rr_0} + \dfrac{1}{r^2}\right)\right]$; $\dfrac{\partial r}{\partial z} = \dfrac{z}{r}$; and

$\dfrac{\partial^2 \psi}{\partial z^2} = -\psi\dfrac{1}{rr_0}\left[1 - z^2\left(\dfrac{1}{rr_0} + \dfrac{1}{r^2}\right)\right]$. Substitute into the time-independent Schrödinger equation.

$$E\psi = -\frac{\hbar^2}{2m}\left(\frac{\partial^2 \psi}{\partial x^2} + \frac{\partial^2 \psi}{\partial x^2} + \frac{\partial^2 \psi}{\partial x^2}\right) - \frac{1}{4\pi\varepsilon_0}\frac{e^2}{r}\psi$$

$$= \left(-\frac{\hbar^2}{2m}\right)\left(-\psi\frac{1}{rr_0}\right)\left\{\left[1 - x^2\left(\frac{1}{rr_0} + \frac{1}{r^2}\right)\right] + \left[1 - y^2\left(\frac{1}{rr_0} + \frac{1}{r^2}\right)\right] + \left[1 - z^2\left(\frac{1}{rr_0} + \frac{1}{r^2}\right)\right]\right\} - \frac{e^2}{4\pi\varepsilon_0 r}\psi$$

$$= \psi\left[\frac{\hbar^2}{2m}\frac{1}{rr_0}\left\{3 - \left(x^2 + y^2 + z^2\right)\left(\frac{1}{rr_0} + \frac{1}{r^2}\right)\right\} - \frac{e^2}{4\pi\varepsilon_0 r}\right]$$

$$= \psi\left[\frac{\hbar^2}{2m}\frac{1}{rr_0}\left\{3 - r^2\left(\frac{1}{rr_0} + \frac{1}{r^2}\right)\right\} - \frac{e^2}{4\pi\varepsilon_0 r}\right] = \psi\left[\frac{\hbar^2}{2m}\frac{1}{rr_0}\left(2 - \frac{r^2}{rr_0}\right) - \frac{e^2}{4\pi\varepsilon_0 r}\right]$$

$$= \psi\left[\frac{\hbar^2}{mrr_0} - \frac{\hbar^2}{2mr_0^2} - \frac{e^2}{4\pi\varepsilon_0 r}\right]$$

Since the factor in square brackets must be a constant, the terms with the r dependence must cancel.

$$\frac{\hbar^2}{mrr_0} - \frac{e^2}{4\pi\varepsilon_0 r} = 0 \quad \rightarrow \quad r_0 = \frac{4\pi\varepsilon_0\hbar^2}{me^2} = \frac{\varepsilon_0 h^2}{\pi m e^2}$$

Note from Equation 37-11 that this expression for r_0 is the same as the Bohr radius. Since those two terms cancel, we are left with the following.

$$E\psi = -\frac{\hbar^2}{2mr_0^2}\psi \quad \rightarrow \quad E = -\frac{\hbar^2}{2mr_0^2} = -\frac{\hbar^2}{2m\left(\dfrac{\varepsilon_0 h^2}{\pi m e^2}\right)^2} = \boxed{-\frac{me^4}{8h^2\varepsilon_0^2}}$$

31. Since the electron is in its lowest energy state, we must have the lowest possible value of n. Since $m_l = 2$, the smallest possible value of ℓ is $\boxed{\ell = 2}$, and the smallest possible value of n is $\boxed{n = 3}$.

37. In a filled subshell, there are an even number of electrons. All of the possible quantum number combinations for electrons in that subshell represent an electron that is present. Thus for every m_ℓ value, both values of m_s are filled, representing a spin "up" state and a spin "down" state. The total angular momentum of that pair is zero, and since all of the electrons are paired, the total angular momentum is $\boxed{\text{zero.}}$

43. The wavelength of the K_α line is calculated for molybdenum in Example 39-6. We use that same procedure. Note that the wavelength is inversely proportional to $(Z-1)^2$.

$$\frac{\lambda_{unknown}}{\lambda_{Fe}} = \frac{(Z_{Fe}-1)^2}{(Z_{unknown}-1)^2} \quad \rightarrow \quad Z_{unknown} = \left[(26-1)\sqrt{\frac{194\,pm}{229\,pm}}\right] + 1 = 24$$

The unknown material has $Z = 24$, and so is $\boxed{\text{chromium}}$.

49. (a) Refer to Figure 39-14 and the equation following it. A constant magnetic field gradient will produce a constant force on the silver atoms. Atoms with the valence electron in one of the spin states will experience an upward force, and atoms with the valence electron in the opposite spin state will experience a downward force. That constant force will produce a constant acceleration, leading to the deflection from the original direction of the atoms as they leave the oven. We assume the initial direction of the atoms is the x direction, and the magnetic field gradient is in the z direction. If undeflected, the atoms would hit the screen at $z = 0$.

$$z = \tfrac{1}{2}at^2 = \tfrac{1}{2}\frac{F}{m_{Ag}}\left(\frac{\Delta x}{v}\right)^2 = \tfrac{1}{2}\frac{\mu_z\frac{dB_z}{dz}}{m_{Ag}}\left(\frac{\Delta x}{v}\right)^2 = \tfrac{1}{2}\frac{(-g\mu_B m_s)\frac{dB_z}{dz}}{m_{Ag}}\left(\frac{\Delta x}{v}\right)^2$$

One beam is deflected up, and the other down. Their separation is the difference in the two deflections due to the two spin states.

$$\Delta z = z_{m_s=-\frac{1}{2}} - z_{m_s=\frac{1}{2}} = \tfrac{1}{2}\frac{\left[-g\mu_B\left(-\tfrac{1}{2}-\tfrac{1}{2}\right)\right]\frac{dB_z}{dz}}{m_{Ag}}\left(\frac{\Delta x}{v}\right)^2$$

$$= \tfrac{1}{2}\frac{(2.0023)(9.27\times10^{-24}\,J/T)(1800\,T/m)}{(107.87\,u)(1.66\times10^{-27}\,kg/u)}\left(\frac{0.050\,m}{780\,m/s}\right)^2 = 3.833\times10^{-4}\,m \approx \boxed{0.38\,mm}$$

(b) The separation is seen in the above equation to be proportional to the g-factor. So to find the new deflection, divide the answer to part (a) by the original g-factor.

$$\Delta z_{g=1} = \frac{3.833\times10^{-4}\,m}{2.0023} \approx \boxed{0.19\,mm}$$

55. The angular half-width of the beam can be found in Section 35-4, and is given by $\theta_{1/2} = \frac{1.22\lambda}{d}$, where d is the diameter of the diffracting circle. The angular width of the beam is twice this. The linear diameter of the beam is then the angular width times the distance from the source of the light to the observation point, $D = r\theta$. See the diagram.

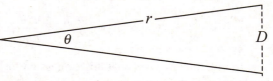

(a) $D = r\theta = r\frac{2.44\lambda}{d} = (380\times10^3\,m)\frac{2.44(694\times10^{-9}\,m)}{3.6\times10^{-3}\,m} = \boxed{180\,m}$

(b) $D = r\theta = r\frac{2.44\lambda}{d} = (384\times10^6\,m)\frac{2.44(694\times10^{-9}\,m)}{3.6\times10^{-3}\,m} = \boxed{1.8\times10^5\,m}$

61. (a) Boron has $Z = 4$, so the outermost electron has $n = 2$. We use the Bohr result with an effective Z. We might naively expect to get $Z_{eff} = 1$, indicating that the other three electrons shield the outer electron from the nucleus, or $Z_{eff} = 2$, indicating that only the inner two electrons accomplish the shielding.

$$E_2 = -\frac{(13.6\,eV)(Z_{eff})^2}{n^2} \quad \rightarrow \quad -8.26\,eV = -\frac{(13.6\,eV)(Z_{eff})^2}{2^2} \quad \rightarrow \quad Z_{eff} = \boxed{1.56}$$

This indicates that the second electron in the $n = 2$ shell does partially shield the electron that is to be removed.

(b) We find the average radius from the expression below.

$$r = \frac{n^2 r_1}{Z_{eff}} = \frac{2^2 \left(0.529 \times 10^{-10} \text{ m}\right)}{(1.56)} = \boxed{1.36 \times 10^{-10} \text{ m}}$$

67. This is very similar to Example 39-3. We find the radial probability distribution for the $n = 2$, $l = 0$ wave function, and find the position at which that distribution has a maximum. We see from Figure 39-8 that there will be two local maxima in the probability distribution function, and the global maximum is at approximately $5r_0$. The wave function is given in Eq. 39-8.

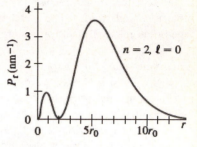

$$\psi_{200} = \frac{1}{\sqrt{32\pi r_0^3}} \left(2 - \frac{r}{r_0}\right) e^{-\frac{r}{2r_0}}$$

$$P_r = 4\pi r^2 \left|\psi_{200}\right|^2 = \frac{r^2}{8r_0^3} \left(2 - \frac{r}{r_0}\right)^2 e^{-\frac{r}{r_0}}$$

$$\frac{dP_r}{dr} = \frac{2r}{8r_0^3} \left(2 - \frac{r}{r_0}\right)^2 e^{-\frac{r}{r_0}} + \frac{r^2}{8r_0^3} 2 \left(2 - \frac{r}{r_0}\right)\left(-\frac{1}{r_0}\right) e^{-\frac{r}{r_0}} + \frac{r^2}{8r_0^3} \left(2 - \frac{r}{r_0}\right)^2 \left(-\frac{1}{r_0}\right) e^{-\frac{r}{r_0}}$$

$$= \frac{\left(2r_0 - r\right) r e^{-\frac{r}{r_0}}}{8r_0^6} \left(r^2 - 6r_0 r + 4r_0^2\right) = 0 \quad \rightarrow \quad r = 0, \; r = 2r_0, \; r^2 - 6r_0 r + 4r_0^2 = 0$$

$$r^2 - 6r_0 r + 4r_0^2 = 0 \quad \rightarrow \quad r = \frac{6r_0 \pm \sqrt{36r_0^2 - 16r_0^2}}{2} = \left(3 \pm \sqrt{5}\right) r \approx 0.764 r_0, \; 5.24 r_0$$

So there are four extrema: $r = 0, \; 0.76r_0, \; 2r_0, \; 5.2r_0$. From Figure 39-8 we see that the most probable distance is $r = \boxed{5.2r_0}$.

73. (a) The $4p \rightarrow 3p$ transition is $\boxed{\text{forbidden,}}$ because $\Delta l = 0 \neq \pm 1$.

(b) The $3p \rightarrow 1s$ transition is $\boxed{\text{allowed,}}$ because $\Delta l = -1$.

(c) The $4d \rightarrow 3d$ transition is $\boxed{\text{forbidden,}}$ because $\Delta l = 0 \neq \pm 1$.

(d) The $4d \rightarrow 3s$ transition is $\boxed{\text{forbidden,}}$ because $\Delta l = -2 \neq \pm 1$.

(e) The $4s \rightarrow 2p$ transition is $\boxed{\text{allowed,}}$ because $\Delta l = +1$.

Chapter 40: Molecules and Solids

Chapter Overview and Objectives

In this chapter topics related to the interaction between atoms are covered. Atoms can bond to each other to form molecules and solids. The mechanisms responsible for bonding and some of the properties expected of atoms bound together in molecules and solids are discussed.

After completing this chapter you should:
- Know the different types of bonds between atoms.
- Know the general form of the potential energy associated with the interaction between two atoms or molecules.
- Know what van der Waals bonding is.
- Know what contributes to the vibrational-rotational spectra of a diatomic molecule.
- Know what the free-electron theory of metals is.
- Know what the band theory of solids is.
- Know what conditions determine whether a material will be a conductor, an insulator, or a semiconductor.
- Know how semiconductor materials can be used in semiconductor components.

Summary of Equations

Simple model of interatomic potential energy:

$$U = U_{repulse} + U_{attract} = -\frac{A}{r^m} + \frac{B}{r^n}$$ (Section 40-2)

Van der Waals interaction potential:

$$U_{van\ der\ Waals} = -\frac{A}{r^6}$$ (Section 40-3)

Rotational energy levels:

$$E_{rot} = \frac{\hbar^2 \ell(\ell+1)}{2I} \qquad \ell \in \{0,1,2,...\}$$ (Section 40-4)

Definition of reduced mass:

$$\mu = \frac{m_1 m_2}{m_1 + m_2}$$ (Section 40-4)

Vibrational energy levels:

$$E_{vib} = \left(v + \tfrac{1}{2}\right)hf \qquad v \in \{0,1,2,...\}$$ (Section 40-4)

Free-electron density of states:

$$g(E) = \frac{8\sqrt{2}\pi\, m^{3/2}}{h^3} E^{1/2}$$ (Section 40-6)

Fermi-Dirac distribution function:

$$f(E) = \frac{1}{e^{(E-E_F)/kT} + 1}$$ (Section 40-6)

Chapter Summary

Section 40-1. Bonding in Molecules

When two one-electron atoms are near each other, the potential energy for either electron is altered by the presence of the other nucleus and electron. Because of the altered potential energy, the solutions to Schrödinger's equation also change. In general, the sum of the energies of the lowest energy states of the two electrons when the two atoms are near each other is lower in energy than the sum of the two energies of two isolated atoms. As a result, it requires the addition of energy to separate the atoms. The energy that must be added is called the **binding energy**.

The type of bond that is formed between the atoms is classified by how the new wave function of the electrons is distributed around the two nuclei. If the probability to find the electrons is distributed equally around the two nuclei, the bond is called a **covalent bond**. If the probability to find the electron from one atom close to the other atom is large (>50%), the bond is called an **ionic bond**. Bonds between dissimilar atoms are, in general, ionic. One consequence of the probability redistribution is the electric dipole moment of the molecule. A molecule that has an electric dipole moment is called a **polar** molecule.

Section 40-2. Potential-Energy Diagrams for Molecules

The Schrödinger equation for two or more atoms in close proximity is very difficult to solve. However, we can make some simple qualitative statements about the interactions between atoms that should apply in all cases. Although we will focus on a system with two interacting atoms, the ideas can be extended to systems with any number of atoms interacting.

For relatively large distances between two atoms, there is always an attractive force (see Section 40-3). The potential energy associated with the attractive force can be written as

$$U_{attract} = -\frac{A}{r^m}$$

where A and m are positive numbers and r is the distance between the two atoms. For the attractive Coulomb potential between two charges of opposite sign, $m = 1$.

As the atoms approach each other, it is easy to see that the electrons must remain between the two nuclei or the repulsive Coulomb force between the two nuclei will push the atoms apart. However, as the distance between the two nuclei decreases, the electrons between the two nuclei have a smaller volume to move in and still remain between the two nuclei. The uncertainty principle forces the uncertainty in momentum, and thus the kinetic energy of the electrons, to increase. This raises the total energy of the system. The conclusion is that the effective potential energy of the molecule begins to increase as the atoms approach each other. The potential energy associated with this repulsive force increases in magnitude as the distance decreases and is approximately

$$U_{repulse} = \frac{B}{r^n}$$

where B and n are positive numbers and r is the distance between the two atoms. For the repulsive Coulomb potential energy of two like charges, $n = 1$.

The total effective potential energy of the two atoms as a function of r is given by

$$U = U_{repulse} + U_{attract} = -\frac{A}{r^m} + \frac{B}{r^n}$$

To ensure that this function is repulsive for small r and attractive for large r, n must be larger than m. A function of this form is shown in the Figure on the right. The potential energy goes to zero for infinite distances. The potential energy reaches a minimum as the distance decreases; the potential energy increases as the distance decreases further. The position r_0 is the equilibrium position. At this position, the force is 0. At positions slightly smaller than r_0 the force is repulsive (directed towards larger r) while at positions slightly larger than r_0 the force is attractive (directed towards smaller r).

Remember that U is the *effective* potential energy. It includes the kinetic energy of the electrons and the Coulomb potential energies. We

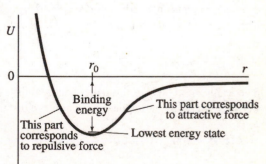

have ignored the minimum momentum of the nuclei required by the uncertainty principle at this point. Also, remember that we have based the model on the behavior at large distances and small distances. At intermediate distances, the behavior could be more complicated.

Example 40-2-A. Determining the equilibrium position. Consider the following potential energy function

$$U(r) = -\frac{A}{r^m} + \frac{B}{r^n}$$

Determine the equilibrium position of this potential and determine the potential energy at the equilibrium separation.

Approach: The equilibrium positions are the positions where the net force is 0. At these positions, $dU/dr = 0$. To determine the nature of the equilibrium position we need to determine d^2U/dr^2.

Solution: The derivative of the potential energy function U is equal to

$$\frac{dU}{dr} = m\frac{A}{r^{m+1}} - n\frac{B}{r^{n+1}}$$

The equilibrium position r_0 is the position where $dU/dr = 0$:

$$\left.\frac{dU}{dr}\right|_{r=r_0} = m\frac{A}{r_0^{m+1}} - n\frac{B}{r_0^{n+1}} = \frac{mA}{r_0^{m+1}}\left(1 - \frac{nB}{mA}\frac{1}{r_0^{n-m}}\right) = 0 \quad \Rightarrow \quad r_0 = \left(\frac{nB}{mA}\right)^{\frac{1}{(n-m)}}$$

The potential energy at this position is equal to

$$U(r_0) = -\frac{A}{\left(\dfrac{nB}{mA}\right)^{\frac{m}{n-m}}} + \frac{B}{\left(\dfrac{nB}{mA}\right)^{\frac{n}{n-m}}}$$

To determine if the equilibrium position is a position of stable or unstable equilibrium, we have to determine the value of d^2U/dr^2 at this position r_0:

$$\frac{d^2U}{dr^2} = -m(m+1)\frac{A}{r^{m+2}} + n(n+1)\frac{B}{r^{n+2}} = -m(m+1)\frac{A}{r^{m+2}}\left(1 - \frac{n(n+1)}{m(m+1)}\frac{B}{A}\frac{1}{r^{n-m}}\right)$$

$$\left.\frac{d^2U}{dr^2}\right|_{r=r_0} = -m(m+1)\frac{A}{r_0^{m+2}}\left(1 - \frac{n(n+1)}{m(m+1)}\frac{B}{A}\frac{1}{r_0^{n-m}}\right) =$$

$$= -m(m+1)\frac{A}{r_0^{m+2}}\left(1 - \frac{n(n+1)}{m(m+1)}\frac{B}{A}\frac{mA}{nB}\right) = -m(m+1)\frac{A}{r_0^{m+2}}\left(1 - \frac{(n+1)}{(m+1)}\right) > 0$$

Since $d^2U/dr^2 > 0$ the equilibrium is a stable equilibrium.

Section 40-3. Weak (van der Waals) Bonds

All atoms are attracted to each other by an induced dipole interaction called the **van der Waals force**. Fluctuations in the electron density of an atom create an electric dipole moment. These fluctuating dipole moments create induced dipole moments in nearby atoms caused by the electric field of the first atom. The force between electric dipoles is attractive. The potential energy associated with this attractive force has a $1/r^6$ dependence:

$$U_{van\ der\ Waals} = -\frac{A}{r^6}$$

The value of A depends on the polarizability of the atoms. The van der Waals force dominates the attractive force between inert gas molecules, even at close distances.

Section 40-4. Molecular Spectra

The electrons in molecules can make electronic transitions, just as the electrons in atoms do. In addition, molecules can have rotational and vibrational energy levels and the molecules can make transitions between these levels. During a transition from a higher energy state to a lower energy state, electromagnetic energy can be emitted. A transition from a lower energy to state to a higher energy state may be the result of the absorption of electromagnetic radiation.

The energy of the **rotational levels** of molecules are given by

$$E_{rot} = \frac{L^2}{2I}$$

where L is the magnitude of the angular momentum of the molecule and I is its moment of inertia about the rotation axis. Since the angular momentum is quantized, the energy levels are also quantized. If we use the angular momentum quantization condition:

$$L = \hbar \sqrt{\ell(\ell+1)}\,\hbar \qquad \ell \in \{0,1,2,\ldots\}$$

we can rewrite the energy levels in terms of the angular momentum quantum number ℓ:

$$E_{rot} = \frac{\hbar^2 \ell(\ell+\ell)}{2I}$$

Transitions between rotational energy levels that involve the emission or absorption of a single photon must satisfy the following selection rule:

$$\Delta\ell = \pm 1$$

If we calculate the transition energies allowed by this selection rule, we get the difference between two energy levels with a difference of one in their angular momentum quantum numbers. If we call the greater angular momentum quantum number ℓ, then the other angular momentum quantum number will be $\ell - 1$ and the energy difference will be equal to

$$\Delta E_{rot} = \frac{\hbar^2 \ell(\ell+1)}{2I} - \frac{\hbar^2 (\ell-1)(\ell-1+1)}{2I} = \frac{\hbar^2}{I}\ell$$

The moment of inertia of a diatomic molecule can be approximated by two point masses separated by the equilibrium distance between the atoms in the molecule. The moment of inertia about an axis through the center of mass, perpendicular to the line joining the two masses, is given by

$$I = \frac{m_1 m_2}{m_1 + m_2} r^2$$

where r is the equilibrium distance between the two atoms, and m_1 and m_2 are the masses of the two atoms. We define the reduced mass μ of the system by

$$\mu = \frac{m_1 m_2}{m_1 + m_2}$$

The moment of inertia can be written as

$$I = \mu r^2$$

The **vibrational energy levels** can be approximated by the energy levels of a simple harmonic oscillator. These energy levels are given by

$$E_{vib} = \left(v + \tfrac{1}{2}\right)hf \qquad v \in \{0,1,2,\ldots\}$$

where v is called the **vibrational quantum number**. For transitions involving the absorption or emission of a single photon, the following selection rule must be satisfied:

$$\Delta v = \pm 1$$

When a molecule changes its vibrational or rotational energies level because of absorption or emission of a single photon, both the rotational and vibrational selection rules must be satisfied. This implies that the changes in the energy of the molecule for the absorption of a photon are

$$\Delta E = \Delta E_{vib} + \Delta E_{rot} = \begin{cases} hf + (\ell+1)\dfrac{\hbar^2}{I} \\ \quad\text{or} \\ hf - \ell\dfrac{\hbar^2}{I} \end{cases}$$

where ℓ is the angular momentum quantum number of the initial state of the molecule. The changes in the energy of the molecule for the emission of a photon are

$$\Delta E = \Delta E_{vib} + \Delta E_{rot} = \begin{cases} -hf + (\ell+1)\dfrac{\hbar^2}{I} \\ \qquad \text{or} \\ -hf - \ell\dfrac{\hbar^2}{I} \end{cases}$$

where ℓ is the angular momentum quantum number of the initial state of the molecule.

Example 40-4-A. Obtaining microscopic information from macroscopic observations. A diatomic molecule is in the $\ell = 2$ angular momentum quantum state. It absorbs photons of wavelengths 2.4921 μm and 2.4936 μm. What is the vibrational frequency and the reduced mass of the diatomic molecule?

Approach: The problem provides us with information on the transition energies between two sets of states in the diatomic molecule. Since the diatomic molecule is in the $\ell = 2$ angular momentum quantum state when it absorbs the photons, the molecule must transition to states with $\ell = 3$ or $\ell = 1$. In order to satisfy the selection rule $\Delta v = \pm 1$, the molecule must absorb one quantum of vibrational energy.

Solution: The two absorption wavelengths correspond to the following transition energies

$$\Delta E_1 = \frac{hc}{\lambda_1} = 0.49757 \text{ eV} = 7.9711 \times 10^{-20} \text{ J}$$

and

$$\Delta E_2 = \frac{hc}{\lambda_2} = 0.49727 \text{ eV} = 7.9663 \times 10^{-20} \text{ J}$$

We use the two solutions for the absorbed photon energies to determine the reduced mass of the molecule:

$$\Delta E = \begin{cases} \Delta E_1 = hf + (\ell+1)\dfrac{\hbar^2}{I} = 7.9711 \times 10^{-20} \text{ J} \\ \Delta E_2 = hf - \ell\dfrac{\hbar^2}{I} = 7.9663 \times 10^{-20} \text{ J} \end{cases}$$

If we subtract the two transition energies we can determine the moment of inertia of the molecule:

$$\Delta E_1 - \Delta E_2 = h\left(f + (\ell+1)\frac{\hbar^2}{I}\right) - \left(hf - \ell\frac{\hbar^2}{I}\right) = (2\ell+1)\frac{\hbar^2}{I} \implies$$

$$I = \frac{(2l+1)\ \hbar^2}{\Delta E_1 - \Delta E_2} = \frac{5\left(1.06 \times 10^{-34}\right)^2}{7.9711 \times 10^{-20} - 7.9663 \times 10^{-20}} = 1.17 \times 10^{-45} \text{ kg} \cdot \text{m}^2$$

The vibrational frequency of the molecule can now be determined:

$$\ell\Delta E_1 + (\ell+1)\Delta E_2 = \ell hf + (\ell+1)hf = (2\ell+1)hf \implies$$

$$f = \frac{\ell\Delta E_1 + (\ell+1)\Delta E_2}{(2\ell+1)h} = \frac{2\left(7.9711 \times 10^{-20}\right) + 3\left(7.9663 \times 10^{-20}\right)}{5\left(6.63 \times 10^{-34}\right)} = 1.20 \times 10^{14} \text{ Hz}$$

Section 40-5. Bonding in Solids

A collection of positive and negative ions can bond together into a solid. The attractive potential energy can be modeled by the Coulomb potential energy. To find the total Coulomb potential energy of a collection of ions we need to find the sum

$$U = \sum \frac{kQ_i Q_j}{r_{ij}}$$

where r_{ij} is the distance between ions with charges Q_i and Q_j. The sum is over every pair of ions in the collection. If the ions are positioned in an orderly array, called a crystal, we can rewrite the sum in terms of a characteristic length of the array, r. In that case, we can write the potential energy as a function of r in the following way:

$$U = -\frac{\alpha\, ke^2}{r}$$

The constant α is called the **Madelung constant** and its value depends on the way in which the atoms are arranged in the solid.

Atoms in metals are bonded by **metallic bonds**. Metallic bonds can be considered as extreme covalent bonds. The electrons are essentially shared by all atoms in the solid.

Section 40-6. Free-Electron Theory of Metals; Fermi Energy

The potential energy of a metal can be approximated by a relatively deep square potential well in which the least tightly bound electrons move. Within the well, the potential energy is zero everywhere and the electrons are free to move around inside the well. Because the number of these electrons is very large in any macroscopic sized piece of material and the electron states are very closely spaced in energy, the energy of the states can be treated as a continuous variable, even though they are discrete. The **density of states**, $g(E)$, is the number of states per unit energy. The number of states, dN, between an energy E and an energy $E + dE$ is related to the density of states by

$$dN = g(E)\, dE$$

It is relatively easy to show that the density of states for free electrons in a square well is given by

$$g(E) = \frac{8\sqrt{2}\pi\, m^{3/2}}{h^3} E^{1/2}$$

where m is the electron mass.

At zero temperature, the electrons in a metal are in the lowest energy state. Because of the Pauli exclusion principle, the electrons fill up states to some finite energy, called the **Fermi level**, above the lowest energy state in the potential well. The energy of this highest filled state at zero temperature is called the **Fermi energy**, E_F. At temperatures above absolute zero, the probability that an electron state is occupied is given by the **Fermi-Dirac distribution function**

$$f(E) = \frac{1}{e^{(E-E_F)/kT} + 1}$$

This function replaces the Boltzmann distribution function that we used in the kinetic theory of gases. The probability distribution $f(E)$ is different from the Boltzmann distribution as a result of the Pauli exclusion principle which applies to electrons.

The **density of occupied states**, n_o, is the number of electrons in states in an energy interval between E and $E + dE$:

$$n_o(E) = g(E) f(E) = \frac{8\sqrt{2}\pi\, m^{3/2}}{h^3} \frac{E^{1/2}}{e^{(E-E_F)/kT} + 1}$$

Example 40-6-A. Calculating the properties of occupied states. Determine the probability that states at energies $E_F - 10kT$, $E_F - kT$, $E_F - 0.1kT$, E_F, $E_F + 0.1kT$, $E_F + kT$, and $E_F + 2kT$ are occupied.

Approach: The probability that a state of a given energy is occupied is given by the Fermi-Dirac distribution function. The distribution function depends on the energy difference with respect to the Fermi level and the absolute temperature.

Solution: The probability that a given state is occupied is given by the Fermi-Dirac distribution function:

$$f(E_F - 10kT) = \frac{1}{e^{(E_F - 10kT - E_F)/kT} + 1} = \frac{1}{e^{-10} + 1} = 0.99995$$

$$f(E_F - kT) = \frac{1}{e^{(E_F - kT - E_F)/kT} + 1} = \frac{1}{e^{-1} + 1} = 0.731$$

$$f\left(E_F - 0.1kT\right) = \frac{1}{e^{(E_F - 0.1kT - E_F)/kT} + 1} = \frac{1}{e^{-0.1} + 1} = 0.525$$

$$f\left(E_F\right) = \frac{1}{e^{(E_F - E_F)/kT} + 1} = \frac{1}{e^0 + 1} = 0.500$$

$$f\left(E_F + 0.1kT\right) = \frac{1}{e^{(E_F + 0.1kT - E_F)/kT} + 1} = \frac{1}{e^{0.1} + 1} = 0.475$$

$$f\left(E_F + kT\right) = \frac{1}{e^{(E_F + kT - E_F)/kT} + 1} = \frac{1}{e^1 + 1} = 0.269$$

$$f\left(E_F + 10kT\right) = \frac{1}{e^{(E_F + 10kT - E_F)/kT} + 1} = \frac{1}{e^{10} + 1} = 4.5 \times 10^{-5}$$

We see that small differences in energy around the Fermi level result in large changes of the occupation probability in the associated states (e.g., a change in energy from 1 kT below the Fermi level to 1 kT above the Fermi level changes the occupation probability by a factor of 3).

Section 40-7. Band Theory of Solids

The discrete electronic energy levels of isolated atoms are split into many closely packed energy levels when the atoms are assembled into a solid. These states are now called **energy bands**. The conductivity of a material can be understood in terms of the band theory of solids. A material that has a partially filled band in its highest occupied band is a **conductor**. The electrons in the partially filled band have plenty of empty states into which they can be excited into when an electric field is applied across the material.

The highest energy band of an **insulator** is filled. To excite an electron in the highest energy band, the applied electric field must transfer an energy equal to the energy difference between the top of the filled band, called the **valence band**, and the bottom of the next higher empty band, called the **conduction band**. If this energy exceeds the energy the electric field can transfer to the electron, no electrons will move. The difference in energy between the highest energy of the valence band and the lowest energy of the conduction band is called the **band gap** of the material.

If the band gap is not too large, a finite temperature of the material will result in a reasonable probability of electrons having an energy that will place them in the conduction band. Such materials are called **semiconductors**. The electrons that are thermally excited into the conduction band can gain momentum from an applied electric field because there are plenty of empty states available in the conduction band at nearby energies. Likewise, the empty states left behind in the valence allow the other electrons still in the valence to gain momentum from the electric field. The behavior of the motion of the charges remaining in the valence band can be understood by treating the empty states as positively charged particles called **holes**.

Section 40-8. Semiconductors and Doping

The Fermi level and the conductivity of a semiconductor can be manipulated by the addition of impurities to the semiconductor. The addition of impurities is called **doping**. Atoms with five valence electrons, called **donor impurities**, added to the semiconductor have a high probability of their fifth electron being excited to the conduction band of the semiconductor, raising its conductivity in proportion to the number of impurity atoms added, as long as the impurity concentration remains small.   The addition of donor impurities also raises the Fermi energy of the semiconductor toward the conduction band. When a semiconductor is doped in this way, it is called an **n-type semiconductor**. The *n* is used because the charge carriers are negative electrons.

Atoms with three valence electrons, called **acceptor impurities**, added to the semiconductor have a high probability of an electron from the valence band of the semiconductor to be thermally excited into their valence shell. This leaves an empty state or hole in the valence band of the semiconductor that is relatively free to move, again raising the conductivity of the semiconductor in proportion to the number of impurity atoms added, as long as the impurity concentration remains small. The addition of acceptor impurities also lowers the Fermi energy of the semiconductor toward the valence band. When a semiconductor is doped in this way, it is called a **p-type semiconductor**. The *p* is because the charge carriers are effectively the holes in the valence band and their dynamics can be understood as if they were positively charged particles.

Section 40-9. Semiconductor Diodes

A semiconductor that consists of both p and n type materials in contact forms what is called a **pn junction diode**. The boundary between the two types of materials is called a **pn junction**. Such a structure conducts when an applied potential difference across the diode is in a direction to cause conventional current to flow in the direction from p-type to the n-type material, but the structure does not conduct when the potential difference is reversed.

Semiconductor diodes have many applications in power regulation circuits. Other applications include light-emitting diodes (LEDs) and solar cells.

Section 40-10. Transistors and Integrated Circuits (Chips)

Transistors are semiconductor devices that have the ability to control either a large voltage or current with a small voltage or current. The function they are commonly used for is amplifying signals. The signal to be amplified is the small voltage or current that controls the output signal that is the large voltage or current.

Transistors can be divided into two general categories, junction transistors and field effect transistors. These two types of transistors are constructed differently, function on a different basis, and have different electrical characteristics. A bipolar junction transistor (**BJT**) is made from a semiconductor that has been doped so that an n-type semiconductor layer of material is between two p-type semiconductor layers. This is called a pnp transistor. Alternatively, the bipolar junction transistor can be made so that a p-type semiconductor layer lies between two n-type semiconductor layers. This is called an npn transistor. A terminal is connected to each of the three regions of the semiconductor. The terminals are named the emitter, base, and collector terminals. The base is always the middle layer of the structure. In a circuit, these transistors have a relatively large current that flows from the emitter terminal to the collector terminal. This current passes through the base region of the transistor. A small percentage of the carriers from the emitter recombine with the opposite type of carriers in the base as the carriers travel from the emitter to the collector. This results in a relatively small base current. If the base current is controlled, it will control the larger emitter to collector current as the percentage of the emitter to collector current that ends up as base current remains relatively constant.

Field effect transistors could also be called voltage-controlled resistors. The structure of a field effect transistor can be modeled as a slab of semiconductor of one type with a terminal connected to it at each end. These terminals are called the source and the drain. Along the length of this slab of material lies a region of the opposite type of semiconductor called the gate. The pn junction that is formed effectively reduces the cross-sectional area of the source-to-drain slab of material and increases its resistance. When the pn junction is reverse biased, the pn junction expands, further reducing the conductive cross-sectional area of the source-drain slab. This increases the resistance of the source-drain slab. Very little current flows through the reverse biased pn junction, so the gate current is much smaller than the base current of the bipolar junction transistor. In another type of field effect transistor, there is a thin insulating layer placed between the gate terminal and the pn junction, which reduces the gate current even further. The field effect transistor without the insulating layer on the gate connection is called a junction field effect transistor (**JFET**) and the field effect transistor with the insulating layer on the gate is called an insulated gate field effect transistor (**IGFET**). A common construction method for making IGFETs is to form the insulating layer by oxidizing the semiconductor. In this case the field effect transistor is called a metal oxide semiconductor field effect transistor (**MOSFET**).

Integrated circuits contain many transistors and other circuit components on a single piece of semiconductor.

Practice Quiz

1. Consider a diatomic molecule made from two atoms from the same row but two different columns from the periodic table of elements. From the combinations listed, which would be the most ionic bond (i.e., most polar molecule)?
    a) I and VII
    b) II and VI
    c) III and V
    d) IV and IV

2. The reduced mass of a two-body system with different masses is always
    a) greater than both masses.
    b) less than both masses.
    c) greater than the smaller mass and less than the larger mass.
    d) less than the smaller mass and greater than the larger mass.

3. At absolute zero temperature, what is the probability of a state with an energy greater than the Fermi energy being occupied?
    a) 1
    b) ½
    c) 0
    d) It depends on how much greater than the Fermi energy the energy of the state is.

4. The highest occupied energy band of a material has 2.98×10^{24} states for electrons. The material has 1.48×10^{24} electrons in this band. The energy gap to the next higher band is 5.8 eV. At room temperature, this material is a
    a) conductor.
    b) insulator.
    c) semiconductor.
    d) either an insulator or a semiconductor.

5. At any temperature, the probability of an electron occupying a state with an energy located near the Fermi energy is
    a) 1
    b) ½
    c) 0
    d) ¼

6. The selection laws that the change in the angular momentum quantum number and the vibrational quantum number of a molecule change by ± 1 apply to
    a) all transitions of the molecule involving rotation or vibration.
    b) only collisions of molecules with other molecules.
    c) only transitions involving the emission or absorption of a single photon.
    d) all transitions involving only rotational transitions.

7. A semiconductor that has been doped with an acceptor impurity is a(n)
    a) n-type semiconductor.
    b) p-type semiconductor.
    c) pn junction diode.
    d) intrinsic semiconductor.

8. The force between any two neutral atoms at distances greater than many atomic diameters is always
    a) repulsive.
    b) attractive.
    c) exactly zero.
    d) larger than when the atoms are closer to each other.

9. When will a material that is a semiconductor at room temperature become an insulator?
    a) When the temperature becomes low enough
    b) When the temperature becomes high enough
    c) Never
    d) When it has a few impurity atoms added to it

10. What is the largest force that holds a solid mass of ions together?
    a) Gravitational force
    b) Van der Waals force
    c) Electrostatic force
    d) Magnetic force

11. Determine the electrostatic potential energy of a pair of ions, one with charge $+e$ and one with charge $-e$, that are distance of 0.12 nm apart.

12. A metal has a Fermi energy of 3.2 eV. Determine the probability that a state with an energy of 3.4 eV is occupied at a temperature of 20° C. Determine the speed of electrons that have the Fermi energy.

13. Determine the absorption wavelengths of a diatomic molecule that has a moment of inertia of 4.6×10^{-46} kg·m^2, a vibrational frequency of 1.817×10^{14} Hz, and initially has an angular momentum quantum number $l = 3$.

14. Determine the Fermi energy of a free electron model material with an electron density of 1.2×10^{24} electrons per cubic centimeter.

15. Determine the equilibrium separation of two ions, one of charge $+e$ and one of charge $-e$, when they have a repulsive potential energy function given by

$$U_{repulsive} = \frac{1.88 \times 10^{-58} \text{ J} \cdot \text{m}^4}{r^4}$$

Responses to Select End-of-Chapter Questions

1. (a) Covalent; (b) ionic; (c) metallic.

7. The carbon atom ($Z = 6$) usually forms four bonds because carbon requires four electrons to form a closed $2p$ shell, and each hydrogen-like atom contributes one electron.

13. For an ideal pn junction diode connected in reverse bias, the holes and electrons that would normally be near the junction are pulled apart by the reverse voltage, preventing current flow across the junction. The resistance is essentially infinite. A real diode does allow a small amount of reverse current to flow if the voltage is high enough, so the resistance in this case is very high but not infinite. A pn junction diode connected in forward bias has a low resistance (the holes and electrons are close together at the junction) and current flows easily.

19. The energy comes from the power supplied by the collector/emitter voltage source. The input signal to the base just regulates how much current, and therefore power, can be drawn from the collector's voltage source.

Solutions to Select End-of-Chapter Problems

Note: A factor that appears in the analysis of electron energies is

$$\frac{e^2}{4\pi\varepsilon_0} = \left(9.00 \times 10^9 \text{ N} \cdot \text{m}^2/\text{C}^2\right)\left(1.60 \times 10^{-19} \text{C}\right)^2 = 2.30 \times 10^{-28} \text{ J} \cdot \text{m}.$$

1. We calculate the binding energy as the opposite of the electrostatic potential energy. We use Eq. 23-10 for the potential energy.

$$\text{Binding energy} = -U = -\frac{1}{4\pi\varepsilon_0}\frac{Q_1Q_2}{r} = -\frac{e^2}{4\pi\varepsilon_0}\left(\frac{1}{0.28 \times 10^{-9}\text{m}}\right) = -\frac{2.30 \times 10^{-28} \text{ J} \cdot \text{m}}{0.28 \times 10^{-9}\text{m}}$$

$$= 8.214 \times 10^{-19} \text{ J} \approx \boxed{8.2 \times 10^{-19} \text{ J}}$$

$$= 8.214 \times 10^{-19} \text{ J}\left(\frac{1\text{eV}}{1.60 \times 10^{-19}\text{ J}}\right) = 5.134 \text{ eV} \approx \boxed{5.1\text{eV}}$$

7. (a) The neutral He atom has two electrons in the ground state, $n = 1$, $\ell = 0$, $m_\ell = 0$. Thus the two electrons have opposite spins, $m_s = \pm\frac{1}{2}$. If we try to form a covalent bond, we see that an electron from one of the atoms will have the same quantum numbers as one of the electrons on the other atom. From the exclusion principle, this is not allowed, so the electrons cannot be shared.

(b) We consider the He$_2^+$ molecular ion to be formed from a neutral He atom and an He$^+$ ion. It will have three electrons. If the electron on the ion has a certain spin value, it will have the opposite spin as one of the electrons on the neutral atom. Thus those two electrons can be in the same spatial region, and so a bond can be formed.

13. The energies involved in the transitions are given in Figure 40-17. We find the rotational inertial from Eq. 40-4. The basic amount of rotational energy is $\hbar^2/I$.

$$\frac{\hbar^2}{I} = \frac{\hbar^2}{\left(\dfrac{m_1 m_2}{m_1 + m_2} r^2\right)} = \frac{\hbar^2}{\left(\frac{1}{2} m_H r^2\right)} = \frac{2\hbar^2}{m_H r^2} = \frac{2\left(1.055 \times 10^{-34} \text{ J} \cdot \text{s}\right)^2}{(1.008 \text{ u})\left(1.66 \times 10^{-27} \text{ kg/u}\right)\left(0.074 \times 10^{-9} \text{ m}\right)^2}$$

$$= 2.429 \times 10^{-21} \text{ J}$$

(a) For $l = 1$ to $l = 0$:

$$\Delta E = \frac{\hbar^2}{I} = 2.429 \times 10^{-21} \text{ J} \left(\frac{1}{1.60 \times 10^{-19} \text{ J/eV}}\right) = \boxed{1.5 \times 10^{-2} \text{ eV}}$$

$$\lambda = \frac{hc}{\Delta E} = \frac{\left(6.63 \times 10^{-34} \text{ J} \cdot \text{s}\right)\left(3.00 \times 10^8 \text{ m/s}\right)}{2.429 \times 10^{-21} \text{ J}} = \boxed{8.2 \times 10^{-5} \text{ m}}$$

(b) For $l = 2$ to $l = 1$:

$$\Delta E = 2\frac{\hbar^2}{I} = 2\left(2.429 \times 10^{-21} \text{ J}\right)\left(\frac{1}{1.60 \times 10^{-19} \text{ J/eV}}\right) = \boxed{3.0 \times 10^{-2} \text{ eV}}$$

$$\lambda = \frac{hc}{\Delta E} = \frac{\left(6.63 \times 10^{-34} \text{ J} \cdot \text{s}\right)\left(3.00 \times 10^8 \text{ m/s}\right)}{2\left(2.429 \times 10^{-21} \text{ J}\right)} = \boxed{4.1 \times 10^{-5} \text{ m}}$$

(c) For $l = 3$ to $l = 2$:

$$\Delta E = 3\frac{\hbar^2}{I} = 3\left(2.429 \times 10^{-21} \text{ J}\right)\left(\frac{1}{1.60 \times 10^{-19} \text{ J/eV}}\right) = \boxed{4.6 \times 10^{-2} \text{ eV}}$$

$$\lambda = \frac{hc}{\Delta E} = \frac{\left(6.63 \times 10^{-34} \text{ J} \cdot \text{s}\right)\left(3.00 \times 10^8 \text{ m/s}\right)}{3\left(2.429 \times 10^{-21} \text{ J}\right)} = \boxed{2.7 \times 10^{-5} \text{ m}}$$

19. Consider the system in equilibrium, to find the center of mass. See the first diagram. The dashed line represents the location of the center of mass.

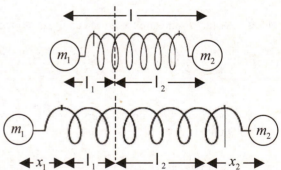

$$l = l_1 + l_2 \quad ; \quad m_1 l_1 = m_2 l_2 = m_2 \left(l - l_1\right) \quad \rightarrow$$

$$l_1 = \frac{m_2}{m_1 + m_2} l \quad ; \quad l_2 = \frac{m_1}{m_1 + m_2} l$$

Now let the spring be stretched to the left and right, but let the center of mass be unmoved.

$$x = x_1 + x_2 \quad ; \quad m_1 \left(l_1 + x_1\right) = m_2 \left(l_2 + x_2\right) \quad \rightarrow$$

$$m_1 l_1 + m_1 x_1 = m_2 l_2 + m_2 x_2 \quad \rightarrow \quad \boxed{m_1 x_1 = m_2 x_2}$$

This is the second relationship requested in the problem. Now use the differential relationships.

$$m_1 \frac{d^2 x_1}{dt^2} = -kx \quad ; \quad m_2 \frac{d^2 x_2}{dt^2} = -kx \quad \rightarrow \quad \frac{d^2 x_1}{dt^2} = -\frac{k}{m_1} x \quad ; \quad \frac{d^2 x_2}{dt^2} = -\frac{k}{m_2} x$$

$$\frac{d^2 x_1}{dt^2} + \frac{d^2 x_2}{dt^2} = -kx\left(\frac{1}{m_1} + \frac{1}{m_2}\right) \quad \rightarrow \quad \frac{d^2 \left(x_1 + x_2\right)}{dt^2} = -kx\frac{m_1 + m_2}{m_1 m_2} = -\frac{k}{\mu} x \quad \rightarrow \quad \mu\frac{d^2 x}{dt^2} = -kx$$

This last equation is the differential equation for simple harmonic motion, as in Eq. 14-3, with m replaced by μ.

The frequency is given by Eq. 14-7a, $f = \dfrac{1}{2\pi}\sqrt{\dfrac{k}{\mu}}$, which is the same as Eq. 40-5.

25. (*a*) Start with Eq. 40-9 and find the equilibrium distance, which minimizes the potential energy. Call that equilibrium distance r_0.

$$U = -\frac{\alpha}{4\pi\varepsilon_0}\frac{e^2}{r} + \frac{B}{r^m} \quad ; \quad \frac{dU}{dr}\bigg|_{r=r_0} = \left(\frac{\alpha}{4\pi\varepsilon_0}\frac{e^2}{r^2} - m\frac{B}{r^{m+1}}\right)_{r=r_0} = \frac{\alpha}{4\pi\varepsilon_0}\frac{e^2}{r_0^2} - m\frac{B}{r_0^{m+1}} = 0 \quad \rightarrow$$

$$B = \frac{\alpha}{4\pi\varepsilon_0}\frac{e^2 r_0^{m-1}}{m}$$

$$U_0 = U(r=r_0) = -\frac{\alpha}{4\pi\varepsilon_0}\frac{e^2}{r_0} + \frac{B}{r_0^m} = -\frac{\alpha}{4\pi\varepsilon_0}\frac{e^2}{r_0} + \frac{\frac{\alpha}{4\pi\varepsilon_0}\frac{e^2 r_0^{m-1}}{m}}{r_0^m} = -\frac{\alpha}{4\pi\varepsilon_0}\frac{e^2}{r_0}\left(1 - \frac{1}{m}\right)$$

(*b*) For NaI, we evaluate U_0 with $m = 10$, $\alpha = 1.75$, and $r_0 = 0.33\,\text{nm}$.

$$U_0 = -\frac{\alpha}{4\pi\varepsilon_0}\frac{e^2}{r_0}\left(1 - \frac{1}{m}\right) = -(1.75)\frac{2.30\times10^{-28}\,\text{J}\bullet\text{m}}{(0.33\times10^{-9}\,\text{m})(1.60\times10^{-19}\,\text{J/eV})}\left(1 - \tfrac{1}{10}\right) = -6.861\,\text{eV}$$

$$\approx \boxed{-6.9\,\text{eV}}$$

(*c*) For MgO, we evaluate U_0 with $m = 10$, $\alpha = 1.75$, and $r_0 = 0.21\,\text{nm}$.

$$U_0 = -\frac{\alpha}{4\pi\varepsilon_0}\frac{e^2}{r_0}\left(1 - \frac{1}{m}\right) = -(1.75)\frac{2.30\times10^{-28}\,\text{J}\bullet\text{m}}{(0.21\times10^{-9}\,\text{m})(1.60\times10^{-19}\,\text{J/eV})}\left(1 - \tfrac{1}{10}\right) = -10.78\,\text{eV}$$

$$\approx \boxed{-11\,\text{eV}}$$

(*d*) Calculate the % difference using $m = 8$ instead of $m = 10$.

$$\frac{U_{0_{m=8}} - U_{0_{m=10}}}{U_{0_{m=10}}} = \frac{-\frac{\alpha}{4\pi\varepsilon_0}\frac{e^2}{r_0}\left(1 - \frac{1}{8}\right) - -\frac{\alpha}{4\pi\varepsilon_0}\frac{e^2}{r_0}\left(1 - \frac{1}{10}\right)}{-\frac{\alpha}{4\pi\varepsilon_0}\frac{e^2}{r_0}\left(1 - \frac{1}{10}\right)} = \frac{\left(1 - \frac{1}{8}\right) - \left(1 - \frac{1}{10}\right)}{\left(1 - \frac{1}{10}\right)} = \frac{\frac{1}{10} - \frac{1}{8}}{\left(1 - \frac{1}{10}\right)} = -0.0278$$

$$\approx \boxed{-2.8\%}$$

31. The occupancy probability is given by Eq. 40-14. The Fermi level for copper is 7.0 eV.

$$f = \frac{1}{e^{(E-E_F)/kT} + 1} = \frac{1}{e^{(1.015 E_F - E_F)/kT} + 1} = \frac{1}{e^{\left[0.015(7.0\,\text{eV})(1.60\times10^{-19}\,\text{J/eV})\right]/\left[(1.38\times10^{-23}\,\text{J/K})(295\,\text{K})\right]} + 1} = 0.0159$$

$$\approx \boxed{1.6\%}$$

37. We start with Eq. 38-13 for the energy level as a function of n. If we solve for n, we have the number of levels with energies between 0 and E. Taking the differential of that expression will give the number of levels with energies between E and dE. Finally, we multiply by 2 since there can be 2 electrons (with opposite spins) in each energy level.

$$E = n^2\frac{h^2}{8ml^2} \quad \rightarrow \quad n = \sqrt{\frac{8ml^2}{h^2}}\sqrt{E} \quad \rightarrow \quad dn = \sqrt{\frac{8ml^2}{h^2}}\frac{1}{2}\frac{dE}{\sqrt{E}} \quad \rightarrow$$

$$g_l = 2\frac{dn}{dE} = 2\frac{\sqrt{\frac{8ml^2}{h^2}}\frac{1}{2}\frac{dE}{\sqrt{E}}}{dE} = \boxed{\sqrt{\frac{8ml^2}{h^2 E}}}$$

43. The photon with the longest wavelength or minimum frequency for conduction must have an energy equal to the energy gap:

$$\lambda = \frac{c}{f} = \frac{hc}{hf} = \frac{hc}{E_g} = \frac{\left(6.63 \times 10^{-34} \text{ J} \cdot \text{s}\right)\left(3.00 \times 10^8 \text{ m/s}\right)}{\left(1.60 \times 10^{-19} \text{ J/eV}\right)\left(1.14 \text{ eV}\right)} = 1.09 \times 10^{-6} \text{ m} = \boxed{1.09 \, \mu\text{m}.}$$

49. The photon will have an energy equal to the energy gap:

$$E_g = hf = \frac{hc}{\lambda} = \frac{\left(6.63 \times 10^{-34} \text{ J} \cdot \text{s}\right)\left(3.00 \times 10^8 \text{ m/s}\right)}{\left(1.60 \times 10^{-19} \text{ J/eV}\right)\left(680 \times 10^{-9} \text{ m}\right)} = \boxed{1.8 \text{ eV}}$$

55. There will be a current in the resistor while the ac voltage varies from 0.6 V to 9.0 V rms. Because the 0.6 V is small, the voltage across the resistor will be almost sinusoidal, so the rms voltage across the resistor will be close to $9.0 \text{ V} - 0.6 \text{ V} = 8.4 \text{ V}$.

(a) For a half-wave rectifier without a capacitor, the current is zero for half the time. We ignore the short time it takes for the voltage to increase from 0 to 0.6 V, and so current is flowing in the resistor for about half the time. We approximate the average current as half of the full rms current.

$$I_{\text{av}} = \frac{1}{2} \frac{V_{\text{rms}}}{R} = \frac{1}{2} \frac{8.4 \text{ V}}{0.120 \text{ k}\Omega} = \boxed{35 \text{ mA}}$$

(b) For a full-wave rectifier without a capacitor, the current is positive all the time. We ignore the short times it takes for the voltage to increase from 0 to 0.6 V, and so current is flowing in the resistor all the time. We approximate the average current as the full rms current.

$$I_{\text{av}} = \frac{V_{\text{rms}}}{R} = \frac{8.4 \text{ V}}{0.120 \text{ k}\Omega} = \boxed{70 \text{ mA}}$$

61. The arrow at the emitter terminal, E, indicates the direction of current I_E. The current into the transistor must equal the current out of the transistor.

$$\boxed{I_B + I_C = I_E}$$

67. (a) The reduced mass is defined in Eq. 40-4.

$$\mu = \frac{m_H m_{Cl}}{m_H + m_{Cl}} = \frac{\left(1.008 \text{ u}\right)\left(35.453 \text{ u}\right)}{\left(1.008 \text{ u}\right) + \left(35.453 \text{ u}\right)} = \boxed{0.9801 \text{ u}}$$

(b) We find the effective spring constant from Eq. 40-5.

$$f = \frac{1}{2\pi} \sqrt{\frac{k}{\mu}} \quad \rightarrow$$

$$k = 4\pi^2 f^2 \mu = 4\pi^2 \left(8.66 \times 10^{13} \text{ Hz}\right)^2 \left(0.9801 \text{ u}\right)\left(1.6605 \times 10^{-27} \text{ kg/u}\right) = \boxed{482 \text{ N/m}}$$

The spring constant for H_2 is estimated in Example 40-6 as 550 N/m.

$$\frac{k_{CO}}{k_{H_2}} = \frac{482 \text{ N/m}}{550 \text{ N/m}} = \boxed{0.88}$$

73. From the diagram of the cubic lattice, we see that an atom inside the cube is bonded to the six nearest neighbors. Because each bond is shared by two atoms, the number of bonds per atom is 3 (as long as the sample is large enough that most atoms are in the interior, and not on the boundary surface). We find the heat of fusion from the energy required to break the bonds:

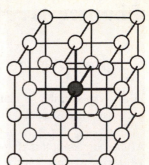

$$L_{fusion} = \left(\frac{\text{number of bonds}}{\text{atom}} \right) \left(\frac{\text{number of atoms}}{\text{mol}} \right) E_{bond}$$

$$= (3)(6.02 \times 10^{23} \text{ atoms/mol})(3.9 \times 10^{-3} \text{ eV})(1.60 \times 10^{-19} \text{ J/eV})$$

$$= 1127 \text{ J/mol} \approx \boxed{1100 \text{ J/mol}}$$

79. The photon with the maximum wavelength for absorption must have an energy equal to the energy gap.

$$E_g = hf = \frac{hc}{\lambda} = \frac{(6.63 \times 10^{-34} \text{ J·s})(3.00 \times 10^8 \text{ m/s})}{(1.60 \times 10^{-19} \text{ J/eV})(1.92 \times 10^{-3} \text{ m})} = \boxed{6.47 \times 10^{-4} \text{ eV}}$$

85. We assume the 130 V value is given to the nearest volt.
   (a) The current through the load resistor must be maintained at a constant value.

$$I_{load} = \frac{V_{output}}{R_{load}} = \frac{(130 \text{ V})}{(18.0 \text{ k}\Omega)} = 7.22 \text{ mA}$$

At the minimum supply voltage, there will be no current through the diode, so the current through R is also 7.22 mA. The supply voltage is equal to the voltage across R plus the output voltage.

$$V_R = I_{load} R = (7.22 \text{ mA})(2.80 \text{ k}\Omega) = 20.2 \text{ V} \; ; \; V_{supply \atop min} = V_R + V_{output} = \boxed{150 \text{ V}}$$

At the maximum supply voltage, the current through the diode will be 120 mA, and so the current through R is 127 mA.

$$V_R = I_{load} R = (127.22 \text{ mA})(2.80 \text{ k}\Omega) = 356 \text{ V} \; ; \; V_{supply \atop max} = V_R + V_{output} = \boxed{486 \text{ V}}$$

   (b) The supply voltage is fixed at 245 V, and the output voltage is still to be 130 V. The voltage across R is fixed at $245 \text{ V} - 130 \text{ V} = 115 \text{ V}$. We calculate the current through R.

$$I_R = \frac{V_R}{R} = \frac{(115 \text{ V})}{(2.80 \text{ k}\Omega)} = 41.1 \text{ mA}$$

If there is no current through the diode, this current will be in the load resistor.

$$R_{load} = \frac{V_{load}}{I_{load}} = \frac{130 \text{ V}}{41.1 \text{ mA}} = 3.16 \text{ k}\Omega$$

If R_{load} is less than this, there will be a greater current through R, meaning a greater voltage drop across R, and a smaller voltage across the load. Thus regulation would be lost, so 3.16 kΩ is the minimum load resistance for regulation.
If R_{load} is greater than 3.16 kΩ, the current through R_{load} will have to decrease in order for the voltage to be regulated, which means there must be current through the diode. The current through the diode is 41.1 mA when R_{load} is infinite, which is less than the diode maximum of 120 mA. Thus the range for load resistance is $\boxed{3.16 \text{ k}\Omega \leq R_{load} < \infty.}$

Chapter 41: Nuclear Physics and Radioactivity

Chapter Overview and Objectives

In this chapter, the structure of the atomic nucleus and radioactivity are described. Alpha, beta, and gamma decay of nuclei are discussed and the time dependence of the number of nuclei in a sample is derived.

After completing this chapter you should:
- Know that nuclei of atoms are made up of protons and neutrons.
- Know the definitions of atomic number, neutron number, and atomic mass number.
- Know the relationship between the atomic number, the neutron number, and the atomic mass number.
- Know the nuclear radius approximation.
- Know what a Bohr magneton is.
- Know how to calculate the binding energy of a nucleus.
- Know what alpha, beta, and gamma decay are.
- Know the definitions of the decay constant, the mean lifetime, and the half-life.
- Know how to calculate the number of nuclei and the decay rate of a sample of radioactive material, given the decay constant, mean lifetime, or half-life of the nucleus.
- Know the law of conservation of nucleon number.
- Know what radioactive dating is.
- Know how radiation is detected.

Summary of Equations

Relationship between A, N, and Z:
$$A = Z + N$$
(Section 41-1)

Approximate radius of nucleus (in m):
$$r \approx \left(1.2 \times 10^{-15}\right) A^{1/3}$$
(Section 41-1)

Definition of a nuclear magneton:
$$\mu_N = \frac{e\hbar}{2\,m_p}$$
(Section 41-1)

Binding energy of nucleus:
$$\text{BE} = \left[Zm_p + (A-Z)m_n - m_{nucleus}\right]c^2$$
(Section 41-2)

Alpha decay:
$$^A_Z N \rightarrow \,^{A-4}_{Z-2} N' + \,^4_2 He$$
(Section 41-4)

Beta decay:
$$^A_Z N \rightarrow \,^A_{Z+1} N' + \,^0_{-1} e + \bar{\nu}$$
(Section 41-5)

$$^A_Z N \rightarrow \,^A_{Z-1} N' + \,^0_{+1} e + \nu$$

Electron capture:
$$^A_Z N + \,^0_{-1} e \rightarrow \,^A_{Z-1} N' + \nu$$
(Section 41-5)

Gamma decay:
$$^A_Z N^* \rightarrow \,^A_Z N + \gamma$$
(Section 41-6)

Radioactive decay time dependence:
$$N = N_0 e^{-\lambda t}$$
(Section 41-8)

$$\frac{dN}{dt} = \left(\frac{dN}{dt}\right)_0 e^{-\lambda t} = -\lambda N_0 e^{-\lambda t}$$

Relationship between half-life, mean lifetime, and decay constant:

$$T_{1/2} = \frac{\ln 2}{\lambda} = \tau \ln 2 \qquad \text{(Section 41-8)}$$

Chapter Summary

Section 41-1. Structure and Properties of the Nucleus

The nucleus of an atom consists of **protons** and **neutrons**. The proton has a positive electric charge of 1.60×10^{-19} C and a mass of 1.67262×10^{-27} kg. The neutron has no electric charge and has a mass of 1.67493×10^{-27} kg. Both protons and neutrons are called **nucleons**.

The number of protons in a nucleus is called the **atomic number** of the nucleus and is designated by the symbol Z. The number of neutrons in a nucleus is called the **neutron number** of the nucleus and is designated by the symbol N. The total number of nucleons is called the **atomic mass number** or **nucleon number** of the nucleus and is designated by the symbol A. The relationship between these three quantities is given by

$A = Z + N$

This information is commonly expressed along with the chemical symbol of the element that the nucleus belongs to in the form

$^{A}_{Z}X$

where X is the chemical symbol of the element, Z is its atomic number, and A is its atomic mass number. The Z and the X are redundant because the atomic number determines the element of the atom and Z is often omitted. The nuclei with different values of atomic mass number but the same atomic number are called **isotopes**. Most elements exist in nature in several different isotopes.

The shape of nuclei is approximately spherical and their approximate radius, r, depends on their atomic mass number, A, and is given by

$r \approx \left(1.2 \times 10^{-15}\right) A^{1/3}$

Nuclear masses are often given in unified **atomic mass units (u)** or units of MeV/c^2. The unified atomic mass unit is defined in terms of the mass of a $^{12}_{6}C$ atom. The mass of $^{12}_{6}C$ is defined to be 12.000000 u. The relationship between u, kg, and MeV/c^2 is given by

$1\,\text{u} = 1.6605 \times 10^{-27}\,\text{kg} = 931.5\,\text{MeV}/c^2$

The proton mass is equal to

$m_p = 1.67262 \times 10^{-27}\,\text{kg} = 1.007276\,\text{u} = 938.27\,\text{MeV}/c^2$

The neutron mass is equal to

$m_n = 1.67493 \times 10^{-27}\,\text{kg} = 1.007825\,\text{u} = 939.57\,\text{MeV}/c^2$

The nucleons are, like the electron, spin ½ particles. The spin of the nucleons, along with the translational angular momentum of the nucleons within the nucleus, contributes to the **nuclear angular moment** of the nucleus.

The nucleons have magnetic moments. A convenient unit for the magnetic moment is the **nuclear magneton** μ_N which is defined as

$\mu_N = \dfrac{e\hbar}{2\,m_p}$

The magnetic moment of the proton is

$\mu_p = 2.7928\,\mu_N$

The magnetic moment of the neutron is

$$\mu_n = -1.9135 \, \mu_N$$

The minus sign in this expression indicates that the magnetic moment of the neutron is pointing in a direction opposite to the direction of the spin of the neutron.

Section 41-2. Binding Energy and Nuclear Forces

The mass of a stable nucleus is always less than the total mass of its protons and neutrons. The total binding energy of a nucleus is related to the difference between the total mass of the protons and neutrons that make up the nucleus and the mass of the nucleus:

$$\text{Binding energy} = \left[Zm_p + (A - Z)m_n - m_{nucleus} \right] c^2$$

The binding energy is the energy required to separate the nucleus into isolated protons and neutrons.
The **average binding energy per nucleon** is the binding energy divided by the number of nucleons in the nucleus. The average binding energy is greatest for the $^{56}_{26}\text{Fe}$ nucleus; $^{56}_{26}\text{Fe}$ is the most stable nucleus.
The force that is responsible for holding the nucleons inside the nucleus together is the **strong nuclear force**. The strong nuclear force is an attractive force between nucleons. The strong nuclear force is the same for neutrons and protons. It is a **short-range** force. At distances greater than about 10^{-15} m, its magnitude drops off very quickly.
There is a second type of nuclear force called the **weak nuclear force**. The weak nuclear force is involved in the decay of neutrons.

Section 41-3. Radioactivity

The decay of nuclei, along with the emission of a particle or electromagnetic radiation, is called **radioactivity**. There are three different mechanisms of nuclear decay, which will be discussed in the following sections.

Section 41-4. Alpha Decay

Some nuclei decay by emitting a ^4_2He nucleus. The ^4_2He nucleus emitted in the decay process is called an **alpha particle** and this type of decay is called **alpha decay**. The decaying nucleus, called the **parent nucleus**, has its atomic number reduced by two and its atomic mass number reduced by four. The resulting nucleus is called the **daughter nucleus**. Alpha decay can be written as

$$^A_Z\text{N} \rightarrow \, ^{A-4}_{Z-2}\text{N}' + \, ^4_2\text{He}$$

where N is the parent nucleus and N′ is the daughter nucleus. The mass of the parent nucleus will always be greater than the combined mass of the daughter nucleus and the alpha particle. The energy released during radioactive decay is called the **disintegration energy** or **Q-value**. For alpha decay, the Q value is equal to

$$Q = M_P c^2 - \left(M_D + M_\alpha \right) c^2$$

where M_P is the mass of the parent nucleus, M_D is the mass of the daughter nucleus, and M_α is the mass of the alpha particle.

Example 41-4-A. Alpha decay of ^{251}Cf. A Californium 251 nucleus alpha decays. What is the daughter nucleus? What is the disintegration energy?

Approach: A nucleus undergoing alpha decay changes its atomic number by –2 and its nucleon number by –4. The problem thus requires us to determine the atomic number of Californium and use this information to determine the atomic and atomic mass numbers of the daughter nucleus. The atomic number of the daughter nucleus will identify the element.

Solution: The alpha decay reaction of Californium 251 can be written as

$$^{251}_{98}\text{Cf} \rightarrow \, ^{247}_{96}\text{N}' + \, ^4_2\text{He}$$

where N′ is the chemical symbol of the daughter nucleus. The chemical name of an atom with atomic number 96 is Curium (Cm). The daughter nucleus is thus ^{247}Cm.

To find the disintegration energy, we determine the difference in the rest mass of the parent and the daughter and alpha particle and multiply by c^2:

$$Q = \left(M_{\text{Parent}} - M_{\text{Daughter}} - m_{\text{He}}\right)c^2$$

$$= \left(251.079580\,\text{u} - 247.070346\,\text{u} - 4.002603\,\text{u}\right)\left(\frac{931.5\,\text{MeV}/c^2}{\text{u}}\right)c^2 = 6.176\,\text{MeV}$$

The masses given in the table of masses of isotopes are the masses for the neutral atoms, including the orbital electrons. Since the number of electrons in the parent atom is equal to the sum of the number of electrons in the daughter atom and the helium atom, there is no problem using the masses of neutral atoms in the calculation of the Q value of the nuclear decay.

Section 41-5. Beta Decay

Beta decay is the emission of an electron and a **neutrino** from a nucleus. The emitted electron is called a **beta particle**, symbolized (β^-). The emitted neutrino (ν) is a particle that is difficult to detect, has a very small mass, is electrically neutral, and has an intrinsic spin of one-half. Beta decay can be written as

$$_Z^A\text{N} \to\ _{Z+1}^A\text{N}' +\ _{-1}^0 e + \bar{\nu}$$

In beta decay, the atomic mass number of the nucleus does not change; the same number of nucleons are present in the daughter nucleus as in the parent nucleus. Also notice that there is a bar placed over the neutrino because the neutrino that is emitted in beta decay is an **anti-neutrino**. Nuclei that beta decay have too many neutrons in the nucleus to be stable.

Atoms that have too few neutrons in the nucleus to be stable decay by a related decay process, called β^+ **decay**. In β^+ decay, an **anti-electron** or **positron** is emitted:

$$_Z^A\text{N} \to\ _{Z-1}^A\text{N}' +\ _{+1}^0 e + \nu$$

A nucleus that has too few neutrons to be stable can also capture an inner-shell electron in a process called **electron capture**. In electron capture, a neutrino is emitted. The general electron capture reaction can be written as

$$_Z^A\text{N} +\ _{-1}^0 e \to\ _{Z-1}^A\text{N}' + \nu$$

Section 41-6. Gamma Decay

When a nucleus undergoes alpha or beta decay, it is usually left in an excited state. Just as excited atoms often lose their excess energy by emitting electromagnetic radiation, nuclei can also lose their excess energy by emitting electromagnetic radiation. The electromagnetic radiation emitted by excited nuclei is in the range of 10 keV to 10 MeV. The process of a nucleus losing its excess energy by emitting electromagnetic radiation is called **gamma decay** and the electromagnetic radiation emitted in the process is called **gamma radiation**. The gamma decay process can be written as

$$_Z^A\text{N}^* \to\ _Z^A\text{N} + \gamma$$

where N* indicates an excited state of the nucleus N.

Section 41-7. Conservation of Nucleon Number and Other Conservation Laws

All nuclear decays conserve energy, linear momentum, angular momentum, electric charge, and nucleon number (atomic mass number). The conservation of nucleon number is called **the law of conservation of nucleon number**.

Section 41-8. Half-Life and Rate of Decay

The time at which a given nucleus decays is unpredictable. The best that can be done is to state the probability that a given nucleus will decay in a certain time interval. The decay probability of a nucleus per unit time is constant, regardless of the amount of time the nucleus has already existed. Contrast this with the process of life. A 90-year-old human has a much higher probability of expiring within the next 10 years than a 10-year-old human. On the other hand, a nucleus that has existed for 9 million years has the same probability of decaying during the next million years as it had during its first million years. The probability that a given nucleus will decay in a unit time interval is called the **decay constant** λ.

In a collection of radioactive nuclei, the time of decay of each nucleus is independent of the time of decay of the other nuclei. If the collection of nuclei is large enough, the change in the number of parent nuclei ΔN that will occur in a time interval Δt is given by

$$\Delta N = -\lambda N \Delta t$$

In the limit where Δt goes to zero ΔN is replaced by dN and Δt is replaced by dt:

$$dN = -\lambda N dt$$

ΔN and dN are negative because the number of parent nuclei decreases with increasing time. The last equation can be solved for N:

$$N(t) = N_0 e^{-\lambda t}$$

where N_0 is the number of parent nuclei at time $t = 0$. The decay rate dN/dt is equal to

$$\frac{dN}{dt} = \left(\frac{dN}{dt}\right)_{t=0} e^{-\lambda t} = -\lambda N_0 e^{-\lambda t}$$

Sometimes the exponential dependencies of the number of parent nuclei or the decay rate are written as

$$e^{-\lambda t} = e^{-t/\tau}$$

where τ is called the **mean lifetime.** It is equal to the inverse of the decay constant:

$$\tau = \frac{1}{\lambda}$$

Another frequently used parameter is the **half-life** of the nucleus $T_{1/2}$. The half-life of a nucleus is the time interval during which one-half of the parent nuclei present at the start of the time interval decay. The relationship between the half-life, the decay constant, and the mean lifetime is as follows

$$T_{1/2} = \frac{\ln 2}{\lambda} = \tau \ln 2$$

Example 41-8-A. Half life and decay constants of a nucleus obtained from the measured decay rates. The decay rate of a sample of a particular nucleus is measured to be 2.87×10^4 decays per second. 3 hours and 27 minutes later, the decay rate of the sample is measured to be 980 decays per second. What are the half-life and decay constant of the nucleus?

Approach: The decay rate of a sample depends on the initial number of parent nuclei and the decay constant. The problem provides us with information on the decay rate at two different times. This information is sufficient to determine the number of parent nuclei at time $t = 0$ and the decay constant. We are free to define our reference time and we will choose time $t = 0$ to be the time at which the decay rate is 2.87×10^4 decays per second.

Solution: The decay rate of time $t_2 = 12,420$ s (3 hours and 27 minutes) is equal to

$$(dN/dt)_{t_2} = (dN/dt)_{t=0} e^{-\lambda t_2}$$

This equation can be used to determine the decay constant of the nucleus:

$$\lambda = \frac{1}{t_2} \ln\left[\frac{(dN/dt)_{t=0}}{(dN/dt)_{t_2}}\right] = \frac{1}{12,420} \ln\left[\frac{2.87 \times 10^4}{980}\right] = 2.72 \times 10^{-4} \text{ s}^{-1}$$

The corresponding half life of the nucleus is equal to

$$T_{1/2} = \frac{\ln 2}{\lambda} = 2,549 \text{ s} = 42.5 \text{ min}$$

Section 41-9. Decay Series

When a radioactive nucleus decays, the daughter nucleus is not necessarily stable and may itself undergo radioactive decay. This may also be true of the daughter of this nucleus, and so on. The list of isotopes that the decays successively go through until a stable isotope is reached is called a **decay series**.

Section 41-10. Radioactive Dating

Ratios of the concentrations of different isotopes that are contained in a sample of material can be used to determine the age of the sample. Various isotopes can be used to carry out this **radioactive dating**.

Carbon 14 dating is used to date the remains of living creatures. During the lifetime of living creatures, the fraction of ^{14}C isotopes among the carbon nuclei is about 1.3×10^{-12}. This fraction is the same as the fraction of ^{14}C isotopes in the atmosphere. When a creature expires, it stops exchanging ^{14}C nuclei with the environment and the fraction of ^{14}C nuclei in the creature's remains begins to decrease because of the 5730-year half-life of ^{14}C. The remains of a living creature can be dated by determining the fraction of ^{14}C nuclei in the remains and calculating how many half-lives have passed since the creature expired.

Example 41-10-A. Carbon dating. Determine the age of a sample that has a fraction of 1.65×10^{-14} of its carbon nuclei that are $^{14}_{6}C$ nuclei.

Approach: To determine the age of the sample we will assume that at time $t = 0$, the ratio of ^{14}C to C is 1.3×10^{-12}. As a result of the decay of the ^{14}C, this ratio will decrease and the decrease is a measure of the age of the sample.

Solution: During the age of the sample, the number of ^{14}C atoms was reduced by a factor of $(1.65 \times 10^{-14})/(1.3 \times 10^{-12})$ = 1/78.8. The time required to achieve this reduction can be determined by solving the following equation:

$$\frac{N(t)}{N_0} = \frac{1}{78.8} = e^{-\lambda t}$$

The time t is equal to

$$t = \frac{1}{\lambda}\ln(78.8) = \frac{1}{\left(\frac{\ln 2}{T_{1/2}}\right)}\ln(78.8) = T_{1/2}\frac{\ln(78.8)}{\ln 2} = 3.6 \times 10^4 \text{ yr}$$

Section 41-11. Detection of Radiation

There are several types of detectors used for the detection of the products of radioactive decay. All of these detectors rely on the energy lost by charged particles when they travel through the detector material. The energy lost is used to excite the electrons in the detector material.

A **Geiger-Mueller tube** contains a gas and two electrodes held at different electric potentials. Radiation entering the tube has a finite probability of interacting with the gas atoms and causing them to ionize by losing an electron. The electron separated from the atom is accelerated towards the anode. If the electric potential difference between the electrodes is great enough, the electrons can gain enough kinetic energy from the electric field so that further collisions with other atoms cause those atoms to become ionized as well. An avalanche of electrons is produced quickly, and the charge pulse, collected on the anode, can be used to signal the passage of a charged particle.

A **scintillator** is an optically transparent solid or liquid material that gives off visible light when charged particles interact with the material. The number of visible photons is approximately proportional to the energy lost by the detected particles. This light can be converted with a photomultiplier tube to produce an electrical signal that is approximately proportional in amplitude to the energy lost by the charged particle. If the charged particle stops in the detector material, the amplitude of the electrical signal will be proportional to the total energy of the detected particle.

A **semiconductor detector** utilizes a reverse biased pn junction to detect charged particles. A reversed biased pn junction has only a small leakage current passing through it because the junction has no free charge carriers within it. When a charged particle enters the junction, the energy it loses is used to create free charge carriers within the pn junction. These charge carriers move and create a current that can be detected. The amplitude of the signal is approximately proportional to the energy lost by the charged particle.

Practice Quiz

Necessary information about half-lives and nuclear atomic masses can be found in Appendix F of the text.

1. Which of the following is a conserved quantity in nuclear decay processes?
   a) Number of protons.
   b) Number of neutrons.
   c) Number of neutrinos.
   d) Number of nucleons.

2. Which of the following is not a conserved quantity in nuclear decay processes?
   a) Angular momentum.
   b) Energy.
   c) Linear momentum.
   d) Number of electrons.

3. The decay rate of a sample of radioactive material has an initial activity of 1000 decays per second. Five minutes later, it has an activity of 500 decays per second. What will its activity be at the end of another five minutes of time?
   a) Zero.
   b) 125 decays per second.
   c) 250 decays per second.
   d) 500 decays per second.

4. A $^{226}_{88}$Ra nucleus alpha decays. What is the daughter nucleus?
   a) $^{226}_{89}$Ac
   b) $^{224}_{87}$Fr
   c) $^{222}_{86}$Rn
   d) $^{222}_{84}$Po

5. If you start with a sample of N nuclei of a given isotope, there are 20% remaining after 10 minutes pass by. What fraction of the nuclei are remaining after 10 minutes if you start with a sample of size $2N$?
   a) 10%
   b) 20%
   c) 40%
   d) 80%

6. A nucleus with atomic number Z and atomic mass number A has a binding energy BE. What is the binding energy per nucleon of this nucleus?
   a) BE
   b) BE/Z
   c) BE/A
   d) BE/(A + Z)

7. Almost all stable nuclei with atomic number greater than 50 have
   a) no binding energy.
   b) more protons than neutrons.
   c) more neutrons than protons.
   d) more neutrons than nucleons.

8. What is one assumption made when carbon 14 dating is used to determine when a living organism expired?
   a) The ratio of carbon 12 to carbon 14 in the environment at the time the creature lived was the same as it is today.
   b) The half-life of carbon 14 is decreasing at a uniform rate.
   c) The exchange of the carbon with the environment in the remains of the creature has continued since the time it expired.
   d) Both a) and c) above are assumed to be true.

9. If nucleus A has a probability P of decaying in time t and nucleus B has a probability 2P of decaying in time t, which statement is true about the half-lives of nucleus A and nucleus B?
   a) The half-life of nucleus A is twice the half-life of nucleus B.
   b) The half-life of nucleus B is twice the half-life of nucleus A.
   c) The half-life of nucleus A is the same as the half-life of nucleus B.
   d) The half-life of nucleus A is e^2 times greater than the half-life of nucleus B.

10. A particular nucleus has a half-life of 10 minutes. The decay rate of a given sample is 100 decays per second. What was the decay rate of the sample 20 minutes ago?
    a) 25 decays per second
    b) 200 decays per second
    c) 300 decays per second
    d) 400 decays per second

11. A sample of $^{60}_{27}$Co contains 2.78×10^{16} nuclei. How many nuclei remain after 20.0 years?

12. A promethium 145 nucleus decays to a praseodymium 141 nucleus by alpha decay:
    $$^{145}_{61}\text{Pm} \rightarrow\, ^{141}_{59}\text{Pr} \,+\, ^{4}_{2}\alpha$$
    What is the Q-value of this reaction?

13. Determine the binding energy per nucleon for $^{28}_{14}$Si and for $^{31}_{14}$Si. Which of these nuclei would you expect to be more stable? Does your answer agree with the actual stability?

14. What is the decay rate of a sample of 100.0 g of carbon from the remains from a creature that expired 16,000 years ago?

15. A strontium 90 nucleus, $^{191}_{76}$Os, decays by beta decay. What is the daughter nucleus of this decay? How much energy is released in this decay?

Responses to Select End-of-Chapter Questions

1. All isotopes of the same element have the same number of protons in their nuclei (and electrons in the atom) and will have very similar chemical properties. Isotopes of the same element have different numbers of neutrons in their nuclei and therefore different atomic masses.

7. The strong force and the electromagnetic force are two of the four fundamental forces in nature. They are both involved in holding atoms together: the strong force binds quarks into nucleons and nucleons together in the nucleus; the electromagnetic force is responsible for binding negatively charged electrons to positively charged nuclei and atoms into molecules. The strong force is the strongest of the four fundamental forces; the electromagnetic force is about 100 times weaker at distances on the order of 10^{-17} m. The strong force operates at short range and is negligible for distances greater than about the size of the nucleus. The electromagnetic force is a long-range force that decreases as the inverse square of the distance between the two interacting charged particles. The electromagnetic force operates only between charged particles. The strong force is always attractive; the electromagnetic force can be attractive or repulsive. Both these forces have mediating field particles associated with them. The gluon is the particle for the strong force and the photon is the particle for the electromagnetic force.

13. (a) Sulfur is formed: $^{32}_{15}\text{Pb} \rightarrow ^{32}_{16}\text{S} + e^- + \bar{\nu}$.

  (b) Chlorine is formed: $^{35}_{16}\text{S} \rightarrow ^{35}_{17}\text{Cl} + e^- + \bar{\nu}$.

  (c) Thallium is formed: $^{211}_{83}\text{Bi} \rightarrow ^{207}_{81}\text{Tl} + ^4_2\text{He}$.

19. No. Hydrogen has only one proton. Deuterium has one proton and one neutron. Neither has the two protons and two neutrons required to form an alpha particle.

25. In β decay, a neutrino and a β particle (electron or positron) will be emitted from the nucleus, and the number of protons in the nucleus changes. Because there are three decay products (the neutrino, the β particle, and the nucleus), the momentum of the β particle can have a range of values. In internal conversion, only an electron is emitted from the atom, and the number of protons in the nucleus stays the same. Because there are only two decay products (the electron and the nucleus), the electron will have a unique momentum and, therefore, a unique energy.

Solutions to Select End-of-Chapter Problems

1. Convert the units from MeV/c^2 to atomic mass units.

$$m = \left(139\,\text{MeV}/c^2\right)\left(\frac{1\,\text{u}}{931.49\,\text{MeV}/c^2}\right) = \boxed{0.149\,\text{u}}$$

7. (a) The mass of a nucleus with mass number A is approximately $(A\ \text{u})$ and its radius is $r = \left(1.2 \times 10^{-15}\,\text{m}\right)A^{1/3}$. Calculate the density.

$$\rho = \frac{m}{V} = \frac{A\left(1.66 \times 10^{-27}\,\text{kg/u}\right)}{\frac{4}{3}\pi r^3} = \frac{A\left(1.66 \times 10^{-27}\,\text{kg/u}\right)}{\frac{4}{3}\pi\left(1.2 \times 10^{-15}\,\text{m}\right)^3 A} = 2.293 \times 10^{17}\,\text{kg/m}^3 \approx$$

$$\approx \boxed{2.3 \times 10^{17}\,\text{kg/m}^3}$$

  We see that this is independent of A.

  (b) We find the radius from the mass and the density.

$$M = \rho\tfrac{4}{3}\pi R^3 \rightarrow R = \left(\frac{3M}{4\pi\rho}\right)^{1/3} = \left[\frac{3\left(5.98 \times 10^{24}\,\text{kg}\right)}{4\pi\left(2.293 \times 10^{17}\,\text{kg/m}^3\right)}\right]^{1/3} = 184\,\text{m} \approx \boxed{180\,\text{m}}$$

  (c) We set the density of the Earth equal to the density of the uranium nucleus. We approximate the mass of the uranium nucleus as 238 u.

$$\rho_{\text{Earth}} = \rho_U \rightarrow \frac{M_{\text{Earth}}}{\frac{4}{3}\pi R_{\text{Earth}}^3} = \frac{m_U}{\frac{4}{3}\pi r_U^3} \rightarrow$$

$$r_U = R_{\text{Earth}}\left(\frac{m_U}{M_{\text{Earth}}}\right)^{1/3} = \left(6.38 \times 10^6\,\text{m}\right)\left[\frac{238\left(1.66 \times 10^{-27}\,\text{kg}\right)}{5.98 \times 10^{24}\,\text{kg}}\right]^{1/3} = \boxed{2.58 \times 10^{-10}\,\text{m}}$$

13. From Figure 41-1, we see that the average binding energy per nucleon at $A = 63$ is about 8.7 MeV. Multiply this by the number of nucleons in the nucleus.

$$(63)(8.7\,\text{MeV}) = 548.1\,\text{MeV} \approx \boxed{550\,\text{MeV}}$$

19. (a) We find the binding energy from the masses.

$$\text{Binding Energy} = \left[2m\left({}_{2}^{4}\text{He}\right) - m\left({}_{4}^{8}\text{Be}\right)\right]c^{2}$$
$$= \left[2(4.002603\,\text{u}) - (8.005305\,\text{u})\right]c^{2}\left(931.5\,\text{MeV}/\text{u}c^{2}\right) = -0.092\,\text{MeV}$$

Because the binding energy is negative, the nucleus is unstable. It will be in a lower energy state as two alphas instead of a beryllium.

(b) We find the binding energy from the masses.

$$\text{Binding Energy} = \left[3m\left({}_{2}^{4}\text{He}\right) - m\left({}_{6}^{12}\text{C}\right)\right]c^{2}$$
$$= \left[3(4.002603\,\text{u}) - (12.000000\,\text{u})\right]c^{2}\left(931.5\,\text{MeV}/\text{u}c^{2}\right) = +7.3\,\text{MeV}$$

Because the binding energy is positive, the nucleus is $\boxed{\text{stable.}}$

25. (a) From Appendix F, ${}_{11}^{24}\text{Na}$ is a β^{-} $\boxed{\text{emitter}}$.

(b) The decay reaction is $\boxed{{}_{11}^{24}\text{Na} \rightarrow {}_{12}^{24}\text{Mg} + \beta^{-} + \overline{v}}$. We add 11 electrons to both sides in order to use atomic masses. Then the mass of the beta is accounted for in the mass of the magnesium. The maximum kinetic energy of the β^{-} corresponds to the neutrino having no kinetic energy (a limiting case). We also ignore the recoil of the magnesium.

$$K_{\beta^{-}} = \left[m\left({}_{11}^{24}\text{Na}\right) - m\left({}_{12}^{24}\text{Mg}\right)\right]c^{2}$$
$$= \left[(23.990963\,\text{u}) - (23.985042\,\text{u})\right]c^{2}\left(931.5\,\text{MeV}/\text{u}c^{2}\right) = \boxed{5.52\,\text{MeV}}$$

31. (a) We find the final nucleus by balancing the mass and charge numbers.

$$Z(X) = Z(P) - Z(e) = 15 - (-1) = 16$$
$$A(X) = A(P) - A(e) = 32 - 0 = 32$$

Thus the final nucleus is $\boxed{{}_{16}^{32}\text{S}}$.

(b) If we ignore the recoil of the sulfur and the energy of the neutrino, the maximum kinetic energy of the electron is the Q-value of the reaction. The reaction is ${}_{15}^{32}\text{P} \rightarrow {}_{16}^{32}\text{S} + \beta^{-} + \overline{v}$. We add 15 electrons to each side of the reaction, and then we may use atomic masses. The mass of the emitted beta is accounted for in the mass of the sulfur.

$$K = Q = \left[m\left({}_{15}^{32}\text{P}\right) - m\left({}_{16}^{32}\text{S}\right)\right]c^{2} \quad \rightarrow$$
$$m\left({}_{16}^{32}\text{S}\right) = m\left({}_{15}^{32}\text{P}\right) - \frac{K}{c^{2}} = \left[31.973907\,\text{u} - \frac{1.71\,\text{MeV}}{c^{2}}\left(\frac{1\,\text{u}}{931.5\,\text{MeV}/c^{2}}\right)\right] = \boxed{31.972071\,\text{u}}$$

37. Both energy and momentum are conserved. Therefore, the momenta of the product particles are equal in magnitude. We assume that the energies involved are low enough that we may use classical kinematics; in particular, $p = \sqrt{2mK}$.

$$p_{\alpha} = p_{\text{Pb}} \; ; \; K_{\text{Pb}} = \frac{p_{\text{Pb}}^{2}}{2m_{\text{Pb}}} = \frac{p_{\alpha}^{2}}{2m_{\text{Pb}}} = \frac{2m_{\alpha}K_{\alpha}}{2m_{\text{Pb}}} = \left(\frac{m_{\alpha}}{m_{\text{Pb}}}\right)K_{\alpha} = \frac{4.0026}{205.97}K_{\alpha}$$

The sum of the kinetic energies of the product particles must be equal to the Q-value for the reaction.

$$K_{\text{Pb}} + K_{\alpha} = \left[m\left({}_{84}^{210}\text{Po}\right) - m\left({}_{82}^{206}\text{Pb}\right) - m\left({}_{2}^{4}\text{He}\right)\right]c^{2} = \frac{4.0026}{205.97}K_{\alpha} + K_{\alpha} \quad \rightarrow$$

$$K_\alpha = \frac{\left[m\left(^{210}_{84}\text{Po}\right) - m\left(^{206}_{82}\text{Pb}\right) - m\left(^{4}_{2}\text{He}\right)\right]c^2}{\left(\frac{4.0026}{205.97} + 1\right)}$$

$$= \frac{\left[(209.982874\,\text{u}) - (205.974465\,\text{u}) - (4.002603\,\text{u})\right]c^2}{\left(\frac{4.0026}{205.97} + 1\right)}\left(931.5\,\text{MeV/u}c^2\right) = \boxed{5.31\,\text{MeV}}$$

43. Every half-life, the sample is multiplied by one-half.

$$\frac{N}{N_0} = \left(\tfrac{1}{2}\right)^n = \left(\tfrac{1}{2}\right)^6 = \boxed{0.015625}$$

49. We find the mass from the initial decay rate and Eq. 41-7b.

$$\left|\frac{dN}{dt}\right|_0 = \lambda N_0 = \lambda m \frac{6.02 \times 10^{23}\,\text{nuclei/mole}}{(\text{atomic weight})\,\text{g/mole}} \quad \rightarrow$$

$$m = \left|\frac{dN}{dt}\right|_0 \frac{1}{\lambda}\frac{(\text{atomic weight})}{\left(6.02 \times 10^{23}\right)} = \left|\frac{dN}{dt}\right|_0 \frac{T_{1/2}}{\ln 2}\frac{(\text{atomic weight})}{\left(6.02 \times 10^{23}\right)}$$

$$= \left(2.0 \times 10^5\,\text{s}^{-1}\right)\frac{\left(1.265 \times 10^9\,\text{yr}\right)\left(3.156 \times 10^7\,\text{s/yr}\right)}{(\ln 2)}\frac{(39.963998\,\text{g})}{\left(6.02 \times 10^{23}\right)} = \boxed{0.76\,\text{g}}$$

55. We find the mass from the activity. Note that N_A is used to represent Avogadro's number.

$$R = \lambda N = \frac{\ln 2}{T_{1/2}}\frac{mN_A}{A} \quad \rightarrow$$

$$m = \frac{RT_{1/2}A}{N_A \ln 2} = \frac{(370\,\text{decays/s})\left(4.468 \times 10^9\,\text{yr}\right)\left(3.156 \times 10^7\,\text{s/yr}\right)(238.05\,\text{g/mole})}{\left(6.02 \times 10^{23}\,\text{nuclei/mole}\right)\ln 2} = \boxed{2.98 \times 10^{-2}\,\text{g}}$$

61. The number of radioactive nuclei decreases exponentially, and every radioactive nucleus that decays becomes a daughter nucleus.

$$N = N_0 e^{-\lambda t}$$

$$N_D = N_0 - N = \boxed{N_0\left(1 - e^{-\lambda t}\right)}$$

67. Consider the reaction $\text{n} \rightarrow \text{p} + \text{e}^- + \overline{v}$. The neutron, proton, and electron are all spin $\tfrac{1}{2}$ particles. If the proton and neutron spins are aligned (both are $\tfrac{1}{2}$, for example), then the electron and neutrino spins must cancel. Since the electron is spin $\tfrac{1}{2}$, the neutrino must also be spin $\tfrac{1}{2}$ in this case.

The other possibility is if the proton and neutron spins are opposite of each other. Consider the case of the neutron having spin $\tfrac{1}{2}$ and the proton having spin $-\tfrac{1}{2}$. If the electron has spin $\tfrac{1}{2}$, then the spins of the electron and proton cancel, and the neutrino must have spin $\tfrac{1}{2}$ for angular momentum to be conserved. If the electron has spin $-\tfrac{1}{2}$, then the spin of the neutrino must be $\tfrac{3}{2}$ for angular momentum to be conserved.

A similar argument could be made for positron emission, with $\text{p} \rightarrow \text{n} + \text{e}^+ + v$.

73. We take the momentum of the nucleon to be equal to the uncertainty in the momentum of the nucleon, as given by the uncertainty principle. The uncertainty in position is estimated as the radius of the nucleus. With that momentum, we calculate the kinetic energy, using a classical formula.

$$\Delta p \Delta x \approx \hbar \;\rightarrow\; p \approx \Delta p \approx \frac{\hbar}{\Delta x} = \frac{\hbar}{r}$$

$$K = \frac{p^2}{2m} = \frac{\hbar^2}{2mr^2} = \frac{\left(\left(1.055 \times 10^{-34}\ \text{J}\bullet\text{s}\right)\right)^2}{2\left(1.67 \times 10^{-27}\ \text{kg}\right)\left[\left(56^{1/3}\right)\left(1.2 \times 10^{-15}\ \text{m}\right)\right]^2 \left(1.60 \times 10^{-13}\ \text{J/MeV}\right)}$$

$$= 0.988\ \text{MeV} \approx \boxed{1\ \text{MeV}}$$

79. The reaction is $^1_1\text{H} + ^1_0\text{n} \rightarrow {}^2_1\text{H}$. If we assume the initial kinetic energies are small, then the energy of the gamma is the Q-value of the reaction.

$$Q = \left[m\left(^1_1\text{He}\right) + m\left(^1_0\text{n}\right) - m\left(^2_1\text{He}\right) \right] c^2$$

$$= \left[\left(1.007825\ \text{u}\right) + \left(1.008665\ \text{u}\right) - \left(2.014082\ \text{u}\right) \right] c^2 \left(931.5\ \text{MeV}/\text{u}c^2\right) = \boxed{2.243\ \text{MeV}}$$

85. Since amounts are not specified, we will assume that "today" there is 0.720 g of $^{235}_{92}\text{U}$ and $100.000 - 0.720 = 99.280\ \text{g}$ of $^{238}_{92}\text{U}$. We use Eq. 41-6.

(a) Relate the amounts today to the amounts 1.0×10^9 years ago.

$$N = N_0 e^{-\lambda t} \;\rightarrow\; N_0 = N e^{\lambda t} = N e^{\frac{t}{T_{1/2}}\ln 2}$$

$$\left(N_0\right)_{235} = \left(N_{235}\right) e^{\frac{t}{T_{1/2}}\ln 2} = \left(0.720\ \text{g}\right) e^{\frac{\left(1.0\times 10^9\ \text{yr}\right)}{\left(7.04\times 10^8\ \text{yr}\right)}\ln 2} = 1.927\ \text{g}$$

$$\left(N_0\right)_{238} = \left(N_{238}\right) e^{\frac{t}{T_{1/2}}\ln 2} = \left(99.280\ \text{g}\right) e^{\frac{\left(1.0\times 10^9\right)}{\left(4.468\times 10^9\right)}\ln 2} = 115.94\ \text{g}$$

$$N_{0,238} = N_{238} e^{\frac{0.693 t}{T_{1/2}}} = \left(99.28\ \text{g}\right) e^{\frac{0.693\left(1.0\times 10^9\right)}{\left(4.468\times 10^9\right)}} = 115.937\ \text{g}.$$

The percentage of $^{235}_{92}\text{U}$ was $\dfrac{1.927}{1.927 + 115.94} \times 100\% = \boxed{1.63\%}$

(b) Relate the amounts today to the amounts 100×10^6 years from now.

$$N = N_0 e^{-\lambda t} \;\rightarrow\; \left(N_{235}\right) = \left(N_0\right)_{235} e^{-\frac{t}{T_{1/2}}\ln 2} = \left(0.720\ \text{g}\right) e^{-\frac{\left(100\times 10^6\ \text{yr}\right)}{\left(7.04\times 10^8\ \text{yr}\right)}\ln 2} = 0.6525\ \text{g}$$

$$\left(N_{238}\right) = \left(N_0\right)_{238} e^{-\frac{t}{T_{1/2}}\ln 2} = \left(99.280\ \text{g}\right) e^{-\frac{\left(100\times 10^6\ \text{yr}\right)}{\left(4.468\times 10^9\ \text{yr}\right)}\ln 2} = 97.752\ \text{g}$$

The percentage of $^{235}_{92}\text{U}$ will be $\dfrac{0.6525}{0.6525 + 97.752} \times 100\% = \boxed{0.663\%}$

Chapter 42: Nuclear Energy; Effects and Uses of Radiation

Chapter Overview and Objectives

In this chapter, the physics of nuclear energy and the uses of radiation are discussed. The two reactions that can produce nuclear energy, nuclear fission and nuclear fusion, are studied. The tools used to measure exposure to radiation are introduced, and various applications of nuclear technology are presented.

After completing this chapter you should:
* Know what nuclear fusion and nuclear fission are.
* Know how the cross section of a nuclear reaction is determined.
* Know how to calculate the Q-value of a nuclear reaction.
* Know the common and SI units for measurements of activity, absorbed dose, and effective dose.
* Know what tracers are.
* Know what the principles of tomography and nuclear magnetic resonance are.

Summary of Equations

Q-value of nuclear reaction:

$$Q = \sum M_R c^2 - \sum M_P c^2$$

(Section 42-1)

Total cross section:

$$\sigma_T = \sigma_{el} + \sigma_{inel} + \sigma_R$$

(Section 42-2)

Relationship between decay rate and decay constant:

$$\frac{dN}{dt} = -\lambda N$$

(Section 42-6)

Chapter Summary

Section 42-1. Nuclear Reactions and the Transmutation of Elements

A nuclear reaction occurs when a nucleus interacts with another nucleus or another particle or photon and the reaction products that are produced are different from those that existed before the interaction. An example of such a reaction is

$$_0^1 n + {}_{53}^{131}I \rightarrow {}_{53}^{132}I$$

In this reaction, ^{131}I captures a neutron and ^{132}I is produced.

Nuclear reactions must conserve proton number and nucleon number. This means that the sums of the lower numbers (proton number) must be the same on each side of the equation and the sums of the upper numbers (nucleon number) must be the same on both sides of the equation. Nuclear reactions also must satisfy the laws of conservation of energy and conservation of momentum. The **reaction energy** or **Q-value** of a nuclear reaction is defined as

$$Q = \sum M_R c^2 - \sum M_P c^2$$

where M_R represents the masses of the reacting particles and nuclei and M_P represents the masses of the product particles and nuclei. The Q-value is thus the difference in rest mass energy of the reactants and products. Since energy has to be conserved, the Q-value is the difference in kinetic energies of the products and the reactants

$$Q = \sum K_P - \sum K_R$$

Example 42-1-A. Calculating the Q-value. Determine the Q-value of the following reaction:

$$_{19}^{39}K + {}_1^1 p \rightarrow {}_{20}^{40}Ca$$

Approach: To determine the Q-value we need to determine the difference in the rest energy of the nuclei and particles before and after the reaction. The atomic masses of the nuclei involved can be found in Appendix F of the textbook. If we are using the atomic masses of K and Ca, we also must use the atomic mass of H.

Solution: The atomic masses are:

$^{39}_{19}K$:   38.962591 u

$^{40}_{20}Ca$:   39.963707 u

$^{1}_{1}H$:   1.007825 u

The Q-value of the reaction is thus equal to

$$Q = \sum M_R c^2 - \sum M_P c^2 = (38.962591 + 1.007825 - 39.963707)(931.5) = 6.24 \text{ MeV}$$

The conversion factor 931.5 MeV/c^2 is used to convert rest mass, expressed in units of u, to rest energy, expressed in units of MeV.

Since the Q-value is positive, energy is released when this reaction occurs.

Section 42-2. Cross Section

If projectile particles are sent at a target nucleus, some of the projectiles might not encounter the nucleus at all and others may be elastically scattered, inelastically scattered, or absorbed. The **total cross section**, σ_T, for scattering from a nucleus is the area a hard-edged disk would have that would scatter the projectiles at the same rate that the nucleus does. This area depends on the type of incoming projectile and its energy. The total cross section is equal to

$$\sigma = \frac{R}{R_0 n \ell}$$

where R is the rate of reaction, R_0 is the rate of incident particles, n is the total number of nuclei per unit volume, and ℓ is the thickness of the target.

The total cross section can be written as a sum of cross sections for each of the three processes mentioned above. The **elastic scattering cross section**, σ_{el}, is the cross section for scattering processes in which the sum of the initial kinetic energies of the projectile and the target are the same before and after the scattering process occurs. The **inelastic scattering cross section**, σ_{inel}, is the cross section for scattering processes in which the sum of the initial kinetic energies of the projectile and the target are not the same before and after the scattering process. Some of the initial kinetic energy goes into raising the projectile or the target to an excited state. A nuclear reaction may occur during a collision so that the objects that exist after the collision are not the same as the objects that existed before the collision. This results in a **reaction cross section**.

The sum of the cross sections for each process that can occur is called the **total reaction cross section**, σ_R. The total cross section is related to all of the cross sections for different types of processes by

$$\sigma_T = \sigma_{el} + \sigma_{inel} + \sigma_R$$

A unit used to measure nuclear cross sections is the barn. One barn is 10^{-28} m^2.

Example 42-2-A. Determining the total cross section by measuring the reduction in count rate. A detector detects particles directly from a source at a rate of 2.68×10^4 per second. A 2.00 cm thick aluminum block is placed between the source and the detector. The rate of detection of particles drops to 2.32×10^4 per second. What is the total reaction cross section of particles scattering from a single aluminum atom?

Approach: The problem provides us with information about R_0 and ℓ. The reaction rate R can be determined from the reduction in count rate while the number of nuclei per unit volume can be determined from the known density of aluminum and its atomic mass.

Solution: The density of aluminum is 2.70×10^3 kg/m^3 and its average atomic mass is 26.9815 u. The number density n of target nuclei is thus equal to

$$n = \frac{\text{mass density}}{\text{atomic mass}} = \frac{2.70 \times 10^3}{(26.9815)(1.6605 \times 10^{-27})} = 6.03 \times 10^{28} \text{ m}^{-3}$$

Care must be taken to use consistent units. In the calculation of n we have used kg as the unit of mass.

The rate R at which either scattering or absorption occurs is equal to the difference between the rate of particle detection with and without the scattering material in position:

$$R = R_0 - R_{\text{with scattering}} = 2.68 \times 10^4 \text{ s}^{-1} - 2.32 \times 10^4 \text{ s}^{-1} = 5.6 \times 10^3 \text{ s}^{-1}$$

The total reaction cross section can now be determined:

$$\sigma = \frac{R}{R_0 n \ell} = \frac{5.6 \times 10^3}{(2.68 \times 10^4)(6.03 \times 10^{28})(2.00 \times 10^{-2})} = 1.7 \times 10^{-28} \text{ m}^2$$

Section 42-3. Nuclear Fission; Nuclear Reactors

A few large nuclei will break apart into two approximately equal size nuclei when the nuclei are bombarded with neutrons. This process is called **nuclear fission** and the product nuclei are called **fission fragments**. One or more neutrons may also be released in the fission process. The binding energy per nucleon for fissionable nuclei is smaller than the binding energy per nuclei in the fission fragments and energy is released during this process. The fission of a $^{235}_{92}$U nucleus releases about 200 MeV of energy.

Neutrons released in the fission process can collide with other fissionable nuclei and cause further fission reactions to occur. These secondary fission reactions release additional neutrons that can induce additional fission reactions. The result is a **chain reaction**. If on average one neutron from each fission reaction induces another fission reaction, the chain reaction is a **self-sustaining chain reaction**. A **nuclear reactor** is a device used to contain and control a self-sustaining chain reaction.

In a nuclear reactor, great care is required to ensure that the reaction is self-sustaining. One must ensure the neutrons released in the fission reaction are slowed down since neutrons with large kinetic energies are not absorbed by nuclei as easily as neutrons with small kinetic energies. In order to reduce the kinetic energy of the neutrons without absorbing them, a material, called a **moderator**, is included inside the reactor. From our study of elastic collisions we know that more kinetic energy is lost in a collision with an object of similar mass than in collisions with objects with either smaller or larger masses. Moderator materials should contain nuclei with masses as similar as possible to the mass of the neutron, but not have a large cross section for absorbing the neutron. Obviously, hydrogen nuclei with only one proton meet the similar mass requirement better than any other nucleus. However, a proton has a high probability for an inelastic collision in which the neutron binds to the proton to form a deuterium nucleus. Deuterium is not as efficient at removing kinetic energy from the neutrons as hydrogen, but it does not absorb as many neutrons either and so is a better moderator. **Heavy water,** water in which the hydrogen atoms are replaced with deuterium atoms, is commonly used as a moderator in nuclear reactors.

Not all isotopes of a certain element are fissionable and the non-fissionable isotopes may absorb neutrons. In order to create a self-sustaining reaction, the abundance of the fissionable isotope may have to be increased above its naturally occurring abundance. The process of increasing the concentration of fissionable isotopes is called **enrichment**.

Neutrons can also escape from the surface of the fissionable material before colliding with a fissionable nucleus. This effect can be reduced by increasing the volume-to-surface area ratio of the fissionable material. For a given geometrical shape of the fissionable material, this is often stated in terms of the minimum mass of material for sustainable fission. This mass is called the **critical mass**.

The average number of secondary fission reactions caused by the neutrons released in a fission reaction is called the **multiplication factor** f. If f is greater than one, the rate of fission reactions increases exponentially. If f is equal to one, then fission reactions occur at a constant rate. If f is less than one, then the rate of fission reactions decreases with time. A fission reactor that is to produce a constant output power should have f equal to one. Control rods that absorb neutrons are used in nuclear reactors to adjust the rate of fission reactions to produce the desired output energy when f is equal to one. An important aspect of design of nuclear power plants is **stability**. Stability means if the rate of fission reactions increases slightly for some reason, the change in the multiplication factor caused by this increased fission rate would be negative. This would tend to decrease the fission rate back to the desired level.

Example 42-3-A. Calculating the Q-value of a fission reaction. Determine the Q-value of the fission of a $^{239}_{94}$Pu nucleus into a $^{90}_{38}$Sr nucleus, $^{138}_{56}$Ba nucleus, and 11 neutrons.

Approach: To determine the Q-value we need to determine the difference in the rest energy of the nuclei and particles before and after the reaction. The atomic masses of the nuclei involved can be found in Appendix F of the textbook. If we are using the atomic masses of Pu, Sr, and Ba we do not have to worry about the mass of the electrons since the number of electrons in Pu is the same as the total number of electrons in Sr and Ba. In addition, we need to use the mass of a neutron in our calculation.

Solution: The atomic masses and the mass of the neutrons are:

$^{239}_{94}\text{Pu}$: 239.052157 u

$^{90}_{38}\text{Sr}$: 89.907737 u

$^{138}_{56}\text{Ba}$: 137.905241 u

$^{1}_{0}\text{n}$ 1.008665 u

The Q-value of the reaction can now be determined:

$$Q = \sum M_R c^2 - \sum M_P c^2 = (239.052157 - 89.907737 - 39.963707 - 11 \times 1.008665)(931.5) = 134 \, \text{MeV}$$

The conversion factor 931.5 MeV/c^2 is used to convert rest mass, expressed in units of u, to rest energy, expressed in units of MeV.

Since the Q-value is positive, energy is released in the fission reaction. Most of this energy is released in the form of kinetic energy of the neutrons. When the neutrons are slowed down in the moderator, they transfer most of their kinetic energy into heat energy, which increases the temperature of the moderator and can be used to drive a generator to generate electricity.

Section 42-4. Fusion

The process of building nuclei of larger atomic mass from those of smaller atomic mass and individual protons and neutrons is called **nuclear fusion**. The mass of a nucleus is less than the mass of its constituent protons and neutrons, so energy is usually released in the nuclear fusion processes. However, nuclei with atomic mass numbers greater than about 60 have less binding energy per nucleon than iron 56. If these heavier nuclei are produced in a fusion process involving lighter nuclei, the reaction usually requires some additional energy besides the rest energy of the fusing nuclei.

Example 42-4-A. Calculating the Q-value of a fusion reaction. Calculate the Q-value of the following fusion reaction:

$$^{32}_{15}\text{P} + ^{32}_{15}\text{P} \rightarrow ^{64}_{30}\text{Zn}$$

Approach: To determine the Q-value we need to determine the difference in the rest energy of the nuclei and particles before and after the reaction. The atomic masses of the nuclei involved can be found in Appendix F of the textbook. If we are using the atomic masses of P and Zn we do not have to worry about the mass of the electrons since the number of electrons in two neutral P atoms is the same as the number of electrons in one neutral Zn atom.

Solution: The atomic masses are:

$^{32}_{15}\text{P}$: 31.973907 u

$^{64}_{30}\text{Zn}$: 63.929147 u

The Q-value of the reaction can now be determined:

$$Q = \sum M_R c^2 - \sum M_P c^2 = (2 \times 31.973907 - 63.929147)(931.5) = 17.4 \, \text{MeV}$$

The conversion factor 931.5 MeV/c^2 is used to convert rest mass, expressed in units of u, to rest energy, expressed in units of MeV.

Since the Q-value is positive, energy is released in the fission reaction.

Section 42-5. Passage of Radiation Through Matter; Radiation Damage

Radiation passing through matter causes damage by the energy it can give to the atoms within the matter. Radiation that has enough energy is called **ionizing radiation** because it can ionize the atoms that it interacts with. Ionization of an atom can cause chemical reactions to occur which are potentially disruptive to biological processes. Structural materials can be damaged by the disruption of their atomic structure because of collisions with high-energy particles.

Section 42-6. Measurement of Radiation—Dosimetry

The strength of a radioactive source is usually specified by reporting its **source activity**. The source activity is defined as the number of radioactive decays that occur per unit time. A commonly used unit of activity is the Curie (Ci), defined as

$$1\,\text{Ci} = 3.70 \times 10^{10} \text{ disintegration/s}$$

The SI unit for activity is the Becquerel (Bq), defined as

$$1\,\text{Bq} = 1 \text{ disintegration/s}$$

A radioactive source has a time-dependent activity. The activity of a source is $-dN/dt$ where N is the number of radioactive nuclei in the source. From Chapter 41, we know that dN/dt is proportional to the number of nuclei in the sample:

$$\frac{dN}{dt} = -\lambda N$$

where λ is the **decay constant** of the particular nuclei. As the nuclei decay, the total number of nuclei decreases and so does the total decay rate.

The damage caused by radiation depends on the amount of energy absorbed when the radiation travels through the absorbing material. There are various measures of the energy absorbed, called the **absorbed dose**, and the effectiveness of the absorbed energy to do damage. The **roentgen** (R) is the amount of radiation that deposits 8.78 mJ of energy per kilogram of air. Another measurement that is in more common use today is the **rad**. One rad is the amount of radiation that deposits 10 mJ of energy into 1 kg of any absorbing material. The SI unit of absorbed dose is the **gray** (Gy). One gray is the amount of radiation that deposits one joule of energy per kilogram of absorbing material. One gray is equal to one hundred rad.

For a given amount of kinetic energy, more massive particles travel more slowly. This results in this energy being deposited in a smaller volume of the absorbing material. The damage in this smaller volume is greater than if the same amount of energy were to be deposited spread out over a larger volume. A measure of the effectiveness in doing damage is useful, so the effective dose is defined. Each type of radiation is given a **relative biological effectiveness** (RBE) or **quality factor** (QF). The **effective dose** is defined as the quality factor multiplied by the absorbed dose. The **rad equivalent man** (**rem**) is the absorbed dose in rad multiplied by the quality factor. The SI unit for effective dose is the **sievert** (Sv). One sievert is the absorbed dose in gray multiplied by the quality factor.

Example 42-6-A. Calculating the effective dose. What is the effective dose received by an individual exposed to a gamma ray source that emits successive gamma rays with photon energies of 1.12 MeV and 0.465 MeV for 6.8 hours at a distance of 3.6 m? The activity of the source is 15 mCi. Assume that 2.8% of the 1.12 MeV gamma rays are absorbed by the body and that 4.3% of the 0.465 MeV gamma rays are absorbed by the body. Assume the effective area of the body to absorb the radiation is 0.85 m^2.

Approach: To solve this problem we need to carry out several steps. Based on the total exposure time and the source activity, we can determine the number of gamma rays that are emitted. Assuming that the radiation is emitted isotropically and assuming there is no angular correlation between the two gamma rays emitted in the decay, the total number of gamma rays that will be incident on the body can be calculated from the ratio of the effective area of the body and the surface area of a sphere of radius 3.6 m. Since the absorption rate is a function of energy, we need to calculate the effective dose for each type of gamma ray separately, and then sum the resulting values.

Solution: A source activity of 15 mCi corresponds to 5.55×10^8 decays per second. Each decay releases two photons, one of energy 1.12 MeV and one of energy 0.465 MeV. The total energy released in the form of these two gamma rays is the product of the rate of gamma-ray emission, the energy of the gamma ray, and the total exposure time, which is 6.8 hr or $6.8 \times 3600 = 2.448 \times 10^4$ s:

$$E_{1.12\,\text{MeV}} = (1.12)(5.55 \times 10^8)(2.448 \times 10^4) = 1.52 \times 10^{13} \text{ MeV}$$

$$E_{0.465\,\text{MeV}} = (0.465 \text{ MeV})(5.55 \times 10^8)(2.448 \times 10^4) = 6.32 \times 10^{12} \text{ MeV}$$

Only a fraction of the radiation will reach the body. Assuming isotropic emission of the gamma rays, this fraction is equal to the area of the body divided by the area of a sphere of radius 3.6 m:

$$\text{fraction of energy reaching body} = \frac{\text{Area of body}}{4\pi r^2} = \frac{0.85}{4\pi \, (3.6^2)} = 5.22 \times 10^{-3}$$

The energy absorbed by the body of each photon type is equal to the total energy emitted in the form of those photons, multiplied by the fraction of the 3.6-m sphere covered by the body and the absorption fraction. We thus find that

$$E_{absorbed\ 1.12\,\text{MeV}} = (1.52 \times 10^{13})(5.22 \times 10^{-3})(0.028) = 2.22 \times 10^9 \text{ MeV}$$

$$E_{absorbed\ 0.465\,\text{MeV}} = (6.32 \times 10^{12})(5.22 \times 10^{-3})(0.043) = 1.42 \times 10^9 \text{ MeV}$$

The total absorbed energy is the sum of these two energies. To determine the absorbed dose, we need to convert MeV to Joules (using the fact that $1 \text{ MeV} = 1.60 \times 10^{-13}$ J) and Joules to rad (using the fact that $1 \text{ rad} = 1.00 \times 10^{-2}$ J):

$$E_{absorbed} = (2.22 \times 10^8 + 1.42 \times 10^8)\left(\frac{1.60 \times 10^{-13}}{1}\right)\left(\frac{1}{1.00 \times 10^{-2}}\right) = 5.82 \times 10^{-3} \text{ rad}$$

The effective dose is the absorbed dose multiplied by the quality factor of the radiation. For gamma rays, the quality factor is approximately 1, and the effective dose is thus equal to 5.82×10^{-3} rem or 5.82 mrem.

Section 42-7. Radiation Therapy

The ionization capability of radiation can be used to destroy cancerous cells within the human body. This process is called **radiation therapy**. An example of the use of radiation in this manner is **proton therapy**. Radiation is also used to diagnose diseases.

Section 42-8. Tracers

Radioactive isotopes can be used to make chemical compounds that are biologically active. These compounds are introduced into biological systems and the radioactive nature of the isotopes can be used to determine what happens to the compound within the biological system. Isotopes used in this way are called **tracers**.

Section 42-9. Imaging by Tomography: CAT Scans and Emission Tomography

The familiar X-ray taken at a doctor's office is called a shadowgraph. It records the amount of X-ray energy reaching the film after passing through an object. The recorded energy density is directly related to the total absorption of X-rays along the path through the object. The recorded energy does not provide any information on the total energy absorbed at a particular point along the path.

By recording shadowgraphs through an object from many different directions, there will be enough information to reconstruct the absorption density as a function of position within that object. To facilitate this, the shadowgraph is recorded electronically rather than on film. A computer is used to calculate the adsorption density as a function of position within the object. This process is called **computed axial tomography (CAT)**.

Another important imaging technique makes use of radioactive tracers absorbed within the body. If the tracers emit radiation that can be detected outside the body, information about the location of the tracers can be determined using techniques similar to those developed for CAT scan. Two methods that make use of tracer techniques are **single photon emission tomography (SPET)** and **positron emission tomography (PET)**.

Section 42-10. Nuclear Magnetic Resonance (NMR) and Magnetic Resonance Imaging (MRI)

We know that the potential energy of a magnetic dipole with dipole moment $\vec{\mu}$ in a magnetic field $\vec{B}$ is given by

$$U = -\vec{\mu} \cdot \vec{B}$$

Nuclei have magnetic dipole moments, and the energy depends on the orientation of the dipole moment with respect to the orientation of the magnetic field. There will be discrete energy levels associated with the orientation of the nuclear magnetic dipole within the magnetic field. Transitions between energy levels can occur by absorption or emission of electromagnetic radiation.

Consider a hydrogen atom which has a single proton as its nucleus. The magnetic dipole moment of the proton has a magnitude μ_p. Like the electron, the proton is a spin ½ particle. Its angular momentum, and hence its magnetic dipole moment, can only be in one of two directions relative to the direction of the magnetic field. The difference in energy between the two orientations of the magnetic dipole moment is

$$\Delta E = U_{down} - U_{up} = \left(\mu_p B\right) - \left(-\mu_p B\right) = 2\mu_p B$$

The magnetic dipole can interact with an electromagnetic field, and photons with the energy equal to the energy differences between the two states can be absorbed by nuclei in the lowest energy state and emitted by nuclei in the higher energy state. The corresponding frequency of the photons falls in the radio frequency band of the electromagnetic spectrum.

There are several ways to utilize the interaction between the nuclear dipole moment and an electromagnetic field to make useful measurements. A simple measurement of the energy absorbed from an applied electromagnetic field at the resonant frequency will be proportional to the density of the nuclei with that resonant frequency. There is a slightly different magnetic field at the nucleus of an atom depending on to what other atoms it is chemically bound. By making a high-resolution measurement of the relative amount of emission at different frequencies from nuclei excited to their higher energy states, the relative concentrations of the different chemical bonds can be determined in a sample. If the sample is placed in a spatially varying magnetic field, then the nuclear concentration as a function of position can be determined. This process is called **magnetic resonance imaging (MRI)**.

Practice Quiz

1. Which of the following reactions is impossible?
   a) $^{132}_{54}Xe + ^{1}_{1}p \rightarrow ^{133}_{55}Cs$
   b) $^{59}_{27}Co + ^{1}_{0}n \rightarrow ^{60}_{27}Co$
   c) $^{40}_{19}K + e^{-} \rightarrow ^{40}_{20}Ca$
   d) $^{31}_{14}Si + ^{32}_{15}P \rightarrow ^{63}_{29}Cu$

2. One of the reactions below has an error in it. Which one?
   a) $^{35}_{17}Cl + ^{4}_{2}He \rightarrow ^{39}_{19}K$
   b) $^{56}_{26}Fe + ^{1}_{1}p \rightarrow ^{57}_{27}Co$
   c) $^{235}_{92}U \rightarrow ^{60}_{27}Co + ^{159}_{65}Tb + 6^{1}_{0}n$
   d) $^{235}_{92}U \rightarrow ^{102}_{42}Mo + ^{124}_{50}Sn + 9^{1}_{0}n$

3. If radiation of energy 1.64 MeV is incident on the human body, which form of radiation will cause the most damage to internal organs?
   a) Gamma radiation
   b) Beta radiation
   c) Protons
   d) Alpha radiation

4. Which of the following statements is true about nuclear fusion?
   a) The fusion of any two nuclei is always accompanied by the release of energy.
   b) The fusion of any two stable nuclei results in a stable nucleus.
   c) The fusion of any two nuclei results in a nucleus with a greater binding energy per nucleon than the two fusing nuclei.
   d) Any fusion process of stable lighter nuclei to form the $^{56}_{26}Fe$ nucleus is accompanied by the release of energy.

5. Which step is the most effective at reducing the exposure you receive from radiation?
   a) Double your distance from the source of radiation.
   b) Decrease your exposure time to one-half the time.
   c) Turn sideways to the source so less area of your body receives radiation.
   d) Close your eyes.

6. Which is not a possible pair of fission products of a plutonium nucleus?
   a) Technetium, antimony
   b) Xenon, zirconium
   c) Cesium, yttrium
   d) Promethium, bromine

7. Which pair of nuclei could fuse to form an $^{40}_{18}$Ar nucleus?
   a) $^{22}_{10}$Na and $^{18}_{8}$O
   b) $^{14}_{6}$C and $^{26}_{14}$Si
   c) $^{26}_{12}$Mg and $^{14}_{8}$O
   d) $^{39}_{19}$K and $^{1}_{1}$H

8. For a given amount of energy deposited in the body, which of the following causes the most damage?
   a) X-rays
   b) Beta particles
   c) Protons
   d) Alpha particles

9. Nucleus A has a magnetic moment twice that of nucleus B. The wavelength of radio waves that are absorbed by nucleus A to change the direction of its magnetic moment are
   a) twice the wavelength necessary to change the direction of nucleus B's magnetic moment
   b) half the wavelength necessary to change the direction of nucleus B's magnetic moment
   c) the same as the wavelength necessary to change the direction of nucleus B's magnetic moment
   d) any wavelength.

10. Estimate the total cross section for a baseball scattering off of a baseball bat.
    a) 10 b
    b) 10^{-26} b
    c) 10^{26} b
    d) 0

11. Calculate the Q-value of the following nuclear reaction:

    $$^{51}_{23}V + {}^{1}_{1}p \rightarrow {}^{52}_{24}Cr$$

12. Determine the effective dose an 84-kg person receives if their body absorbs 26 mJ of energy from beta particles.

13. How much energy is released in the following fusion reaction?

    $$^{10}_{5}B + {}^{6}_{3}Li \rightarrow {}^{16}_{8}O$$

14. A source that undergoes beta decay has an activity of 3.24 mCi. The beta particle leaves the source with an energy of 0.385 MeV. The source is a distance of 1.4 m from an individual of mass 60 kg that has an effective absorption area of 0.92 m^2. What is the effective dose received by this individual in 4.0 hours? Assume the person absorbs all of the energy in the beta particles that reach the person's body.

15. A hypothetical nucleus has spin ½ and a magnetic dipole moment of magnitude 3.421 μ_N. Determine the resonant frequency of this nucleus in a magnetic field of magnitude 1.684 T.

Responses to Select End-of-Chapter Questions

1. (a) $^{138}_{56}\text{Ba}$; (b) p or $^{1}_{1}\text{H}$; (c) γ ; (d) $^{199}_{80}\text{Hg}$

7. The energy from nuclear fission appears in the thermal (kinetic) energy of the fission fragments and the neutrons that are emitted, and in the thermal energy of nearby atoms with which they collide.

13. Ordinary water does not moderate, or slow down, neutrons as well as heavy water; more neutrons will also be lost to absorption in ordinary water. However, if the uranium in a reactor is highly enriched, there will be many fissionable nuclei available in the fuel rods. It will be likely that the few moderated neutrons will be absorbed by a fissionable nucleus, and it will be possible for a chain reaction to occur.

19. To ignite a fusion reaction, the two nuclei must have enough kinetic energy to overcome electrostatic repulsion and approach each other very closely in a collision. Electrostatic repulsion is proportional to charge and inversely proportional to the square of the distance between the centers of the charge distributions. Both deuterium and tritium have one positive charge, so the charge effect is the same for d-d and d-t ignition. Tritium has one more neutron than deuterium and thus has a larger nucleus. In the d-t ignition, the distance between the centers of the nuclei will be greater than in d-d ignition, reducing the electrostatic repulsion and requiring a lower temperature for fusion ignition.

25. Appropriate levels of radiation can kill possibly harmful bacteria and viruses on medical supplies or in food.

Solutions to Select End-of-Chapter Problems

1. By absorbing a neutron, the mass number increases by one and the atomic number is unchanged. The product nucleus is $\boxed{^{28}_{13}\text{Al}}$. Since the nucleus now has an "extra" neutron, it will decay by $\boxed{\beta^-}$, according to this reaction: $^{28}_{13}\text{Al} \rightarrow {}^{28}_{14}\text{Si} + \beta^- + \bar{\nu}_e$. Thus the product is $\boxed{^{28}_{14}\text{Si}}$.

7. (a) If the Q-value is positive, then no threshold energy is needed.

$$Q = m_p c^2 + m_{^7_3\text{Li}} c^2 - m_{^4_2\text{He}} c^2 - m_\alpha c^2$$

$$= \left[1.007825\,\text{u} + 7.016005\,\text{u} - 2\left(4.002603\,\text{u}\right)\right]\left(931.5\,\frac{\text{MeV}/c^2}{\text{u}}\right)c^2 = 17.348\,\text{MeV}$$

Since the Q-value is positive, $\boxed{\text{the reaction can occur}}$.

(b) The total kinetic energy of the products will be the Q-value plus the incoming kinetic energy.

$$K_{\text{total}} = K_{\text{reactants}} + Q = 3.5\,\text{MeV} + 17.348\,\text{MeV} = \boxed{20.8\,\text{MeV}}$$

13. (a) This is called a "pickup" reaction because the helium has "picked up" a neutron from the carbon nucleus.

(b) The alpha is ^4_2He. The reactants have 8 protons and 15 nucleons, and so have 7 neutrons. Thus the products must also have 8 protons and 7 neutrons. The alpha has 2 protons and 2 neutrons, and so X must have 6 protons and 5 neutrons. Thus X is $\boxed{^{11}_{6}\text{C}}$.

(c) The Q-value tells whether the reaction requires or releases energy.

$$Q = m_{^3_2\text{He}} c^2 + m_{^{12}_6\text{C}} c^2 - m_{^{11}_6\text{C}} c^2 - m_\alpha c^2$$

$$= \left[3.016029\,\text{u} + 12.000000\,\text{u} - 11.011434\,\text{u} - 4.002603\,\text{u}\right]\left(931.5\,\frac{\text{MeV}/c^2}{\text{u}}\right)c^2 = \boxed{1.856\,\text{MeV}}$$

Since the Q-value is positive, the reaction is $\boxed{\text{exothermic}}$.

19. From the figure we see that a collision will occur if $d \leq R_1 + R_2$. We calculate the area of the effective circle presented by R_2 to the center of R_1.

$$\sigma = \pi d^2 = \pi (R_1 + R_2)^2$$

25. The power released is the energy released per reaction times the number of reactions per second.

$$P = \frac{\text{energy}}{\text{reaction}} \times \frac{\# \text{ reactions}}{s} \rightarrow$$

$$\frac{\# \text{ reactions}}{s} = \frac{P}{\frac{\text{energy}}{\text{reaction}}} = \frac{200 \times 10^6 \, W}{\left(200 \times 10^6 \, eV/\text{reaction}\right)\left(1.60 \times 10^{-19} \, J/eV\right)} = \boxed{6 \times 10^{18} \text{ reactions/s}}$$

31. We find the number of collisions from the relationship $E_n = E_0 \left(\frac{1}{2}\right)^n$, where n is the number of collisions.

$$E_n = E_0 \left(\tfrac{1}{2}\right)^n \rightarrow n = \frac{\ln \frac{E_n}{E_0}}{\ln \frac{1}{2}} = \frac{\ln \frac{0.040 \, eV}{1.0 \times 10^6 \, eV}}{\ln \frac{1}{2}} = 24.58 \approx \boxed{25 \text{ collisions}}$$

37. Calculate the Q-value for the reaction ${}_1^2 H + {}_1^2 H \rightarrow {}_2^3 He + n$

$$Q = 2m_{{}_1^2 H} c^2 - m_{{}_2^3 He} c^2 - m_n c^2$$

$$= \left[2\left(2.014082 \, u\right) - 3.016029 \, u - 1.008665 \, u\right]\left(931.5 \, \frac{MeV/c^2}{u}\right)c^2 = \boxed{3.23 \, MeV}$$

43. Assume that the two reactions take place at equal rates, so they are both equally likely. Then from the reaction of 4 deuterons, there would be a total of 7.228 MeV of energy released, or 1.807 MeV per deuteron on the average. A total power of $\frac{1250 \, MW}{0.33} = 3788 \, MW$ must be obtained from the fusion reactions to provide the required 1250 MW output, because of the 330% efficiency. We convert the power to a number of deuterons based on the energy released per reacting deuteron, and then convert that to an amount of water using the natural abundance of deuterium.

$$3788 \, MW \rightarrow \left[\left(3788 \times 10^6 \, \frac{J}{s}\right)\left(\frac{3600 \, s}{1 \, h}\right)\left(\frac{1 \, MeV}{1.60 \times 10^{-13} \, J}\right)\left(\frac{1 \, d}{1.807 \, MeV}\right)\left(\frac{1 \, H \text{ atom}}{0.000115 \text{ d's}}\right) \times\right.$$
$$\left.\left(\frac{1 \, H_2O \text{ molecule}}{2 \, H \text{ atoms}}\right)\left(\frac{0.018 \, kg \, H_2O}{6.02 \times 10^{23} \text{ molecules}}\right)\right]$$

$$= 6131 \, kg/h \approx \boxed{6100 \, kg/h}$$

49. Use Eq. 42-11b to relate Sv to Gy. From Table 42.1, the quality factor of gamma rays is 1, and so the number of Sv is equal to the number of Gy. Thus $4.0 \, Sv = \boxed{4.0 \, Gy}$.

55. We approximate the decay rate as constant, and find the time to administer 36 Gy. If that calculated time is significantly shorter than the half-life of the isotope, then the approximation is reasonable. If 1.0 mCi delivers about 10 mGy/min, then 1.6 mCi would deliver 16 mGy/min.

$$\text{dose} = \text{rate} \times \text{time} \quad \rightarrow \quad \text{time} = \frac{\text{dose}}{\text{rate}} = \frac{36\,\text{Gy}}{16 \times 10^{-3}\,\text{Gy/min}} \left(\frac{1\,\text{day}}{1440\,\text{min}} \right) = 1.56\,\text{day} \approx \boxed{1.6\,\text{day}}$$

This is only about 11% of a half-life, so our approximation is reasonable.

61. (a) The reaction has $Z = 86$ and $A = 222$ for the parent nucleus. The alpha has $Z = 2$ and $A = 4$, so the daughter nucleus must have $Z = 84$ and $A = 218$. That makes the daughter nucleus $\boxed{{}^{218}_{84}\text{Po}}$.

(b) From Figure 41-12, polonium-218 is $\boxed{\text{radioactive}}$. It decays via both $\boxed{\text{alpha and beta decay}}$, each with a half-life of $\boxed{3.1\,\text{minutes}}$.

(c) The daughter nucleus is not a noble gas, so it is $\boxed{\text{chemically reacting}}$. It is in the same group as oxygen, so it might react with many other elements chemically.

(d) The activity is given by Eq. 41-7a, $R = \lambda N = \dfrac{\ln 2}{T_{1/2}} N$.

$$R = \frac{\ln 2}{T_{1/2}} N = \frac{\ln 2}{(3.8235\,\text{d})(86400\,\text{s/d})} \left(1.6 \times 10^{-9}\,\text{g} \right) \frac{6.02 \times 10^{23}\,\text{nuclei}}{222\,\text{g}}$$

$$= 9.104 \times 10^{6}\,\text{decays/s} \approx \boxed{9.1 \times 10^{6}\,\text{Bq}} = 0.25\,\text{mCi}$$

To find the activity after 1 month, use Eq. 41-7d.

$$R = R_0 e^{-\frac{\ln 2}{T_{1/2}} t} = \left(9.104 \times 10^{6}\,\text{decays/s} \right) e^{-\frac{\ln 2}{(3.8235\,\text{d})}(30\,\text{d})} = 3.956 \times 10^{4}\,\text{decays/s}$$

$$\approx \boxed{4.0 \times 10^{4}\,\text{Bq}} = 1.1\,\mu\text{Ci}$$

67. From Eq. 18-5, the average speed of a gas molecule (root mean square speed) is inversely proportional to the square root of the mass of the molecule, if the temperature is constant. We assume that the two gases are in the same environment and so at the same temperature. We use UF_6 molecules for the calculations.

$$\frac{v_{{}^{235}_{92}UF_6}}{v_{{}^{238}_{92}UF_6}} = \sqrt{\frac{m_{{}^{238}_{92}UF_6}}{m_{{}^{235}_{92}UF_6}}} = \sqrt{\frac{238 + 6(19)}{235 + 6(19)}} = \boxed{1.0043 : 1}$$

73. (a) The mass of fuel can be found by converting the power to energy to number of nuclei to mass.

$$\left(2400 \times 10^{6}\,\text{J/s} \right) (1\,\text{y}) \left(\frac{3.156 \times 10^{7}\,\text{s}}{1\,\text{y}} \right) \left(\frac{1\,\text{MeV}}{1.60 \times 10^{-13}\,\text{J}} \right) \left(\frac{1\,\text{fission atom}}{200\,\text{MeV}} \right) \left(\frac{0.235\,\text{kg}}{6.02 \times 10^{23}\,\text{atom}} \right)$$

$$= 9.240 \times 10^{4}\,\text{kg} \approx \boxed{920\,\text{kg}}$$

(b) The product of the first 5 factors above gives the number of U atoms that fission.

$$\text{\# Sr atoms} = 0.06 \left(2400 \times 10^{6}\,\text{J/s} \right) (1\,\text{y}) \left(\frac{3.156 \times 10^{7}\,\text{s}}{1\,\text{y}} \right) \left(\frac{1\,\text{MeV}}{1.60 \times 10^{-13}\,\text{J}} \right) \left(\frac{1\,\text{fission atom}}{200\,\text{MeV}} \right)$$

$$= 1.42 \times 10^{26}\,\text{Sr atoms}$$

The activity is given by Eq. 41-7a.

$$\left| \frac{dN}{dt} \right| = \lambda N = \frac{\ln 2}{T_{1/2}} N = \frac{\ln 2}{(29\,\text{yr})(3.156 \times 10^{7}\,\text{s/yr})} \left(1.42 \times 10^{26} \right) = 1.076 \times 10^{17}\,\text{decays/s}$$

$$= \left(1.076 \times 10^{17}\,\text{decays/s} \right) \left(\frac{1\,\text{Ci}}{3.70 \times 10^{10}\,\text{decays/s}} \right) = 2.91 \times 10^{6}\,\text{Ci} \approx \boxed{3 \times 10^{6}\,\text{Ci}}$$

79. (a) A Curie is 3.7×10^{10} decays/s.

$$\left(0.10 \times 10^{-6}\,\text{Ci}\right)\left(3.7 \times 10^{10}\,\text{decays/s}\right) = \boxed{3700\,\text{decays/s}}$$

(b) The beta particles have a quality factor of 1. We calculate the dose in gray and then convert to sieverts. The half-life is over a billion years, so we assume the activity is constant.

$$\left(3700\,\text{decays/s}\right)\left(1.4\,\text{MeV/decay}\right)\left(1.60 \times 10^{-13}\,\text{J/MeV}\right)\left(3.156 \times 10^{7}\,\text{s/y}\right)\left(\frac{1}{55\,\text{kg}}\right)$$

$$= 4.756 \times 10^{-4}\,\text{J/kg/y} = 4.756 \times 10^{-4}\,\text{Gy/y} = \boxed{4.8 \times 10^{-4}\,\text{Sv/y}}$$

This is about $\dfrac{4.756 \times 10^{-4}\,\text{Sv/y}}{3.6 \times 10^{-3}\,\text{Sv/y}} = 0.13$ or $\boxed{13\%\ \text{of the background rate}}$.

85. The whole-body dose can be converted into a number of decays, which would be the maximum number of nuclei that could be in the Tc sample. The quality factor of gammas is 1.

$$50\,\text{mrem} = 50\,\text{mrad} \;\rightarrow\; \left(50 \times 10^{-3}\,\text{rad}\right)\left(\frac{1\,\text{J/kg}}{100\,\text{rad}}\right)\left(60\,\text{kg}\right)\left(\frac{1\,\text{eV}}{1.60 \times 10^{-19}\,\text{J}}\right) = 1.875 \times 10^{17}\,\text{eV}$$

$$\left(1.875 \times 10^{17}\,\text{eV}\right)\left(\frac{1\,\text{effective}\,\gamma}{140 \times 10^{3}\,\text{eV}}\right)\left(\frac{2\,\gamma\,\text{decays}}{1\,\text{effective}\,\gamma}\right)\left(\frac{1\,\text{nucleus}}{1\,\gamma\,\text{decay}}\right) = 2.679 \times 10^{12}\,\text{nuclei}$$

This is the total number of decays that will occur. The activity for this number of nuclei can be calculated from Eq. 41-7a.

$$R = \lambda N = \frac{\ln 2}{T_{1/2}}N = \frac{\ln 2\left(2.679 10^{12}\,\text{decays}\right)}{\left(6\,\text{h}\right)\left(3600\,\text{s/h}\right)}\left(\frac{1\,\text{Ci}}{3.70 \times 10^{10}\,\text{decays/s}}\right) \approx 2.32 \times 10^{-3}\,\text{Ci} \approx \boxed{2\,\text{mCi}}$$

Chapter 43: Elementary Particles

Chapter Overview and Objectives

In this chapter, the physics of elementary particles is introduced. The tools of experimental high-energy particle physics are introduced and the current models of elementary particles are presented.

After completing this chapter you should:
- Know what particle accelerators are and why they are necessary to probe small length scales.
- Know how forces can be modeled by the exchange of particles.
- Know the quantum numbers that are conserved in elementary particle interactions.
- Know that all elementary particles can be classified as leptons, hadrons, or gauge bosons.
- Know what the quark model of hadrons is.
- Know what quantum chromodynamics and electroweak theories are.

Summary of Equations

De Broglie wavelength: $$\lambda = \frac{h}{p}$$ (Section 43-1)

Chapter Summary

Section 43-1. High-Energy Particles and Accelerators

To investigate structures on a given size scale, probes must have a wavelength of about the same size or smaller. Using the de Broglie wavelength relationship,

$$\lambda = \frac{h}{p},$$

we know that we must increase the linear momentum, and, therefore, increase the kinetic energy, of our probes in order to be able to probe smaller details.

Particle accelerators are devices that increase the kinetic energy of particles so that they can be used to probe small size scales. There are several basic types of particle accelerators that can be used to accelerate particles to very large kinetic energies. Van de Graaff generators, cyclotrons, synchrotrons, and linear accelerators are different types of particle accelerators.

Particle detectors are used to determine the nature of the particles produced in particle interactions. Particle detectors are used to provide information, such as their charge-to-mass ratios and their total energy, of the reaction products. Examples of detectors that are used or have been used are photographic emulsions, cloud chambers, bubble chambers, and wire drift chambers.

Example 43-1-A. Same energy, different wavelengths. Compare the de Broglie wavelength of electrons with 2.0 GeV of kinetic energy with the de Broglie wavelength of protons with 2.0 GeV of kinetic energy.

Approach: The de Broglie wavelength of a particle depends on its linear momentum. The linear momentum depends on both the mass and the kinetic energy of the particle. Two particles with the same kinetic energy but different masses will have different wavelengths. For each particle we need to determine its linear momentum before we can calculate its wavelength.

Solution: The kinetic energy of a particle is related to its rest mass and its linear momentum:

$$K = \sqrt{m^2 c^4 + p^2 c^2} - mc^2$$

The linear momentum of the particle can thus be determined if we know its kinetic energy:

$$p = \frac{1}{c}\sqrt{\left(K + mc^2\right)^2 - m^2c^4}$$

The de Broglie wavelength of the electron is thus equal to

$$\lambda = \frac{h}{p} = \frac{hc}{\sqrt{\left(K + mc^2\right)^2 - m^2c^4}} = \frac{\left(6.63 \times 10^{-34}\right)\left(3.00 \times 10^8\right)}{\sqrt{\left((2000 + 0.511)^2 - (0.511)^2\right)}\left(1.602 \times 10^{-13}\right)} = 6.21 \times 10^{-16} \text{ m}$$

The conversion factor of 1.602×10^{-13} is used to convert the (rest) energies from MeV to J.
The de Broglie wavelength of the proton is thus equal to

$$\lambda = \frac{h}{p} = \frac{hc}{\sqrt{\left(K + mc^2\right)^2 - m^2c^4}} = \frac{\left(6.63 \times 10^{-34}\right)\left(3.00 \times 10^8\right)}{\sqrt{\left((2000 + 938.3)^2 - (938.3)^2\right)}\left(1.602 \times 10^{-13}\right)} = 4.461 \times 10^{-16} \text{ m}$$

Even though the particles have very different masses, their momenta are not very different because they are both relativistic. The very relativistic limit of the momentum becomes $p = E/c$ regardless of the mass of the particle.

Section 43-2. Beginnings of Elementary Particle Physics—Particle Exchange

In our description of elementary particle physics, the forces between particles are considered to be the result of an exchange of particles. The result of an interaction between two particles is the exchange of momentum and energy between the two particles. The momentum and energy exchanged between the particles is carried by the particle(s) being exchanged. The electric force for example is considered to be the result of the exchange of photons between two charged particles.

Each fundamental force has a corresponding force carrier or mediator. The **photon** is the mediator of the electromagnetic force. The **pi meson** or **pion** is the mediator of the strong nuclear force. The $\mathbf{W^+}$, $\mathbf{W^-}$, and $\mathbf{Z}$ **bosons** are the mediators of the weak nuclear force. The **graviton** is the mediator of the gravitational force. All of the force mediators are whole integer spin particles. They can carry intrinsic or spin angular momentum as well as momentum and energy.

Section 43-3. Particles and Antiparticles

Every particle has a corresponding **antiparticle**. The antiparticle has the opposite electric charge as the particle, but is identical in mass and intrinsic spin. Some uncharged particles, such as the photon, are their own antiparticles. Other uncharged particles, such as the neutron, have a distinct antiparticle.

Section 43-4. Particle Interactions and Conservation Laws

We already know some of the conservation laws of nature (e.g., conservation of charge, conservation of energy, conservation of translational momentum, and conservation of angular momentum). There are additional conservation laws that only become evident when investigating interactions between fundamental particles. Each particle has a **baryon number** B. Nucleons have a baryon number $B = +1$ and the antiparticles of the nucleons have a baryon number $B = -1$. Electrons and neutrinos have baryon number $B = 0$. The baryon number appears to be conserved during particle interactions.

Three other conserved quantities are the three **lepton numbers**, L_e, L_μ, L_τ. The electron lepton number L_e is 1 for the electron and the electron neutrino, –1 for their antiparticles, and 0 for all other particles. The muon lepton number L_μ is 1 for the muon and the muon neutrino, –1 for their antiparticles, and 0 for all other particles. The tau lepton number L_τ is 1 for the tau and the tau neutrino, –1 for their antiparticles, and 0 for all other particles. The muon and tau leptons are particles that are similar to the electron, but more massive.

Example 43-4-A. Conservation of baryon number. Which of the following forbidden particle reactions are not allowed because they violate conservation of baryon number?

1. $\tau^- \rightarrow \pi^- + \overline{\nu}_\tau$ 3. $n \rightarrow p + \pi^-$

2. $\tau^- \rightarrow e^- + \overline{\nu}_\tau$ 4. $\Lambda^0 \rightarrow \overline{p} + p + \pi^0$

Approach: Since we know the baryon number of all the particles involved in these reactions we can determine the initial and final baryon numbers. If the initial and final baryon numbers are the same, baryon number is conserved and the reaction does not violate the baryon number conservation law.

Solution: For each of the four reactions studied in this problem we can quickly determine the initial and the final baryon number and thus determine if baryon number is conserved:

1. The baryon number of the τ^- is 0, the baryon number of the π^- is 0, and the baryon number of the tau anti-neutrino is also 0. Baryon number is thus conserved.
2. The baryon number of the τ^- is 0, the baryon number of the e^- is 0, and the baryon number of the tau anti-neutrino is also 0. Baryon number is thus conserved.
3. The baryon number of the neutron is +1, the baryon number of the proton is +1, and the baryon number of the π^- is 0. Baryon number is thus conserved.
4. The baryon number of the Λ^0 is +1, the baryon number of the antiproton is −1, the baryon number of the proton is +1, and the baryon number of the π^0 is 0. Baryon number is thus not conserved.

Example 43-4-B. Conservation of lepton number. Which of the forbidden particle reactions in Example 43-4-A are not allowed because they violate conservation of a lepton number?

Approach: Since we know the lepton number of all the particles involved in these reactions we can determine the initial and final lepton numbers. If the initial and final lepton numbers are the same, lepton number is conserved and the reaction does not violate the lepton conservation law. Since there are three different lepton numbers, we need to determine the electron lepton number, the muon lepton number, and the tau lepton number.

Solution: For each of the four reactions studied in this problem we can quickly determine the initial and the final lepton number and thus determine if lepton number is conserved:

1. The tau lepton number of the τ^- is +1, the tau lepton number of the π^- is 0, and the tau lepton number of the tau neutrino is −1. Tau lepton number is thus not conserved.
2. The tau lepton number of the τ^- is +1, the tau lepton number of the e^- is 0, and the tau lepton number of the tau anti-neutrino is +1. Tau lepton number is thus conserved. The electron lepton number of the τ^- is 0, the electron lepton number of the e^- is +1, and the electron lepton number of the tau anti-neutrino is 0. Electron lepton number is thus not conserved.
3. All of the particles have a lepton number equal to 0 and all the lepton numbers are thus conserved.
4. All of the particles have a lepton number equal to 0 and all the lepton numbers are thus conserved.

Section 43-5. Neutrinos — Recent Results

The fusion reactions that produce energy in the Sun produce also electron neutrinos. A measurement of the energy output of the Sun can thus be used to predict the number of solar neutrinos that reach the Earth. The number of solar neutrinos detected was significantly less than the predicted number of solar neutrinos. This observation became known as the **solar neutrino problem**. The problem can be resolved if the electron neutrinos change flavor on the way from the Sun to the Earth; detectors that are only sensitive to electron neutrinos would thus record a deficit of electron neutrinos. This phenomenon is called **neutrino flavor oscillations**. The confirmation of flavor oscillations resulted in the conclusion that the electron, muon, and tau lepton numbers are not perfectly conserved, but their sum, $L_e + L_\mu + L_\tau$, is always conserved.

One important consequence of the observation of neutrino flavor oscillations is that the neutrinos cannot be massless, as was originally assumed. At least one of the neutrinos must have a mass of at least 0.04 eV/c^2.

Section 43-6. Particle Classification

The particles can be divided into categories. First of all, particles can be divided into two groups according to whether they are integer spin particles or odd half-integer spin particles. The particles with integer spins are called **bosons** and the particles with odd half-integer spins are called **fermions**. The bosons obey **Bose-Einstein statistics** and the fermions obey **Fermi-Dirac statistics**. The **gauge bosons** are force mediators. The **leptons**, which are fermions, are the electron, muon, and tau leptons and their corresponding neutrinos and antiparticles. Leptons do not interact through the strong nuclear force. **Hadrons** are the particles that interact through the strong nuclear force. The hadrons can be subdivided into **mesons** and **baryons**. Mesons have baryon number $B = 0$ and baryons have a non-zero baryon number.

Section 43-7. Particle Stability and Resonances

Most particles are unstable and they decay into two or more lighter mass particles. The lifetime of a particle is characteristic of the fundamental force that is responsible for the decay. Decays due to weak interactions have typical lifetimes greater than 10^{-13} s. Decays due to strong interactions have typical lifetimes in the range of 10^{-19} to 10^{-16} s. Some particles have such a short lifetime that they are called **resonances** rather than particles. The **width** of the resonance is directly related to the lifetime of the resonance:

$$\Delta E \approx \frac{\hbar}{\Delta t}$$

Section 43-8. Strange Particles? Charm? Towards a New Model

Conservation of baryon number and conservation of the lepton numbers allow us to see why some otherwise possible particle interactions do not occur in nature. The set of conservation laws we have looked at so far is incomplete. Some interactions that are allowed by these conservation laws do not occur in nature.

Introducing a new property called **strangeness**, along with **conservation of strangeness**, appears to be consistent with the observed behavior. Particles that have non-zero strangeness are called **strange particles**. Strangeness is conserved during strong interactions, but is not conserved during weak interactions.

Another property that is conserved is **charm**, and the corresponding conservation law is **conservation of charm**. Charm is conserved during strong interactions, but not during weak interactions.

Section 43-9. Quarks

Leptons show no internal structure down to the smallest size scales probed so far. However, it has long been apparent that hadrons have some internal structure and are not elementary particles, but are built from some "simpler" particles. In addition, the large number of different baryons appears to make nature unusually complex if each of the hadrons is an elementary particle. The **quark** model of hadrons constructs all the hadrons from six different quarks and their antiparticles. The six types or **flavors** of quarks are **up** (u), **down** (d), **strange** (s), **charmed** (c), **bottom** (b), and **top** (t). All of the quarks are spin ½ particles. Three new quantum numbers (charm, bottomness, and topness) are needed to completely specify a quark. The quantum numbers of the quarks are

Quark	Spin	Charge	Baryon Number	Strangeness	Charm	Bottomness	Topness
u	½	$+\frac{2}{3}e$	⅓	0	0	0	0
d	½	$-\frac{1}{3}e$	⅓	0	0	0	0
s	½	$-\frac{1}{3}e$	⅓	−1	0	0	0
c	½	$+\frac{2}{3}e$	⅓	0	+1	0	0
b	½	$-\frac{1}{3}e$	⅓	0	0	−1	0
t	½	$+\frac{2}{3}e$	⅓	0	0	0	+1

Each quark has a corresponding antiquark. The antiquark has the negative value of each quantum number of the corresponding quark.

In the quark model, mesons are constructed from a quark-antiquark pair and baryons are constructed from three quarks or antiquarks. Arbitrary pairs or triplets are not allowed. The quantum numbers of the pairs or triplets are the sums of the quantum numbers of the quarks. The charge of the combination of quarks must be an integer multiple of an electron charge. The baryon number of the combination of quarks must be an integer.

Section 43-10. The "Standard Model": Quantum Chromodynamics (QCD) and Electroweak Theory

The quark model of matter discussed in Section 43-9 is a model of the structure of baryons, but a theory of the interaction between the quarks is also needed to complete the theory. The model of the quark interaction is somewhat more complex than the electromagnetic force. The standard model of quark interactions is called **quantum chromodynamics**. In this model, each quark type comes in one of three different possible **color charges**. The gauge boson that mediates the **color force** between two quarks is called a **gluon**.

Electroweak theory unifies the electromagnetic force and the weak force into two different manifestations of the same fundamental interaction.

Quantum chromodynamics combined with the electroweak theory is frequently called the **standard model** of particle physics.

Section 43-11. Grand Unified Theories

Theories that attempt to treat the electromagnetic force, the weak force, and the strong force as different manifestations of the same fundamental interaction are called **grand unified theories**. These theories predict that the unification of these three forces only occurs at distances less than about 10^{-31} m. At larger distances, the symmetry is broken and instead of one unifying force we observe three distinctly different forces.

One consequence of the grand unified theories is the prediction that the proton is not stable, but may decay into a neutral pion and a positron. The grand unified theories may also be able to account for the observed **matter-antimatter problem**. At the very early stage of the development of the Universe, baryon number may not have been a conserved quantity, and this imbalance may be able to account for the dominance of matter over antimatter we observe.

Section 43-12. Strings and Supersymmetry

String theories attempt to unify all four forces in nature (note that the grant unified theories discussed in Section 43-11 do not include gravity.) In string theory, particles are envisioned not to be point-like but as very short one-dimensional strings (about 10^{-35} m long.)

Superstring theory applies the principle of **supersymmetry** to string theory. Supersymmetry predicts that interactions exist that changes fermions to bosons and vice versa. Each known fermion has a supersymmetric boson partner. For example, for each quark there is a **squark**, for each lepton there is a **slepton**.

Practice Quiz

1. Which of the following is not a conserved quantity?
    a) Baryon number.
    b) Lepton number.
    c) Hadron number.
    d) Energy.

2. Which of the conservation laws does the reaction $p + n \rightarrow p + p + p + \pi^- + \pi^-$ violate?
    a) Conservation of baryon number
    b) Conservation of hadron number
    c) Conservation of a lepton number
    d) Conservation of charge

3. Which of the following is a meson with charge $+e$, strangeness $+1$, bottomness 0, and topness 0?
    a) $u d \bar{s}$
    b) $u \bar{s}$
    c) $\bar{s} c$
    d) $u s \bar{u}$

4. Which particle is a force carrier or mediator of the weak force?
    a) Photon.
    b) Hadron.
    c) W^+ boson.
    d) Neutrino.

5. Which of the following statements is true?
    a) All particles that have integer spins are gauge bosons.
    b) All gauge particles have integer spin.
    c) All particles have integer spin.
    d) All gauge particles are massless.

6. How can the quark model allow the uncharged neutron to have a magnetic moment?
    a) The quark model cannot account for the non-zero magnetic moment of the neutron.
    b) The intrinsic magnetic moments of the quarks that a neutron is composed of do not add up to zero.
    c) The orbital motion of the charged quarks within the nucleus result in a magnetic moment of the nucleus.
    d) Both reasons in b) and c) above can contribute to the magnetic moment of the neutron.

7. Which of the following particles is not a hadron?
    a) Muon.
    b) Pion.
    c) Neutron.
    d) Proton.

8. Which type of force does not affect a μ^- particle?
    a) Electromagnetic force.
    b) Strong force.
    c) Weak force.
    d) Gravitational force.

9. Which quantity is conserved during strong interactions, but not in weak interactions?
    a) Baryon number.
    b) Charge.
    c) Lepton number.
    d) Strangeness.

10. What property must a particle have to be its own anti-particle?
    a) No spin.
    b) No mass.
    c) No charge.
    d) No magnetic moment.

11. What is the difference between the speed of light and the speed of a 2.0 GeV electron?

12. A beam of electrons collides with a beam of protons with kinetic energy 12 GeV traveling in the opposite direction as the electrons. If the total momentum of the electrons and protons is zero, what is the kinetic energy of the electrons?

13. A cyclotron is constructed with a radius of 5.82 cm. The magnetic field is 2.2 T. What must be the frequency of the accelerating voltage so that the cyclotron can accelerate protons? What is the kinetic energy of the protons when they reach the outside radius of the cyclotron?

14. If two protons collide with equal kinetic energies traveling in opposite directions, what is the minimum kinetic energy of each proton necessary for the reaction

$$p + p \rightarrow p + p + \pi^0 + \pi^0$$

to occur?

15. A Σ^- is at rest in the laboratory and decays into a neutron and a π^-. What are the momentum and energy of the two decay products?

Responses to Select End-of-Chapter Questions

1. $p + n \rightarrow p + p + \pi^-$.

7. An electron takes part in electromagnetic, weak, and gravitational interactions. A neutrino takes part in weak and gravitational interactions. A proton takes part in all four interactions: strong, electromagnetic, weak, and gravitational.

13. Based on the lifetimes listed, all of the particles in Table 43-4, except the J/ψ and the Y, decay via the weak interaction.

19. The reaction is not possible, because it does not conserve lepton number. L = 1 on the left-hand side of the reaction equation, and L = −1 on the right-hand side of the reaction equation.

Solutions to Select End-of-Chapter Problems

1. The total energy is given by Eq. 36-11a.

$$E = m_0 c^2 + K = 0.938\,\text{GeV} + 4.65\text{GeV} = \boxed{5.59\,\text{GeV}}$$

7. From Eq. 41-1, the diameter of a nucleon is about $d_{\text{nucleon}} = 2.4 \times 10^{-15}\,\text{m}$. The 25-MeV alpha particles and protons are not relativistic, so their momentum is given by $p = mv = \sqrt{2mK}$. The wavelength is given by Eq. 43-1, $\lambda = \dfrac{h}{p} = \dfrac{h}{\sqrt{2mK}}$.

$$\lambda_\alpha = \frac{h}{\sqrt{2\,m_\alpha \text{KE}}} = \frac{6.63 \times 10^{-34}\,\text{J}\cdot\text{s}}{\sqrt{2(4)\left(1.66 \times 10^{-27}\,\text{kg}\right)\left(25 \times 10^{6}\,\text{eV}\right)\left(1.6 \times 10^{-19}\,\text{J/eV}\right)}} = 2.88 \times 10^{-15}\,\text{m}$$

$$\lambda_p = \frac{h}{\sqrt{2\,m_p \text{KE}}} = \frac{6.63 \times 10^{-34}\,\text{J}\cdot\text{s}}{\sqrt{2\left(1.67 \times 10^{-27}\,\text{kg}\right)\left(25 \times 10^{6}\,\text{eV}\right)\left(1.6 \times 10^{-19}\,\text{J/eV}\right)}} = 5.75 \times 10^{-15}$$

We see that $\boxed{\lambda_\alpha \approx d_{\text{nucleon}}}$ and $\boxed{\lambda_p \approx 2d_{\text{nucleon}}}$. Thus the $\boxed{\text{alpha particle will be better}}$ for picking out details in the nucleus.

13. Start with an expression from Section 42-1, relating the momentum and radius of curvature for a particle in a magnetic field, with q replaced by e.

$$v = \frac{eBr}{m} \quad \rightarrow \quad mv = eBr \quad \rightarrow \quad p = eBr$$

In the relativistic limit, $p = E/c$ and so $\dfrac{E}{c} = eBr$. To put the energy in electron volts, divide the energy by the charge of the object.

$$\frac{E}{c} = eBr \quad \rightarrow \quad \boxed{\frac{E}{e} = Brc}$$

19. Because the two protons are heading towards each other with the same speed, the total momentum of the system is 0. The minimum kinetic energy for the collision would result in all three particles at rest, and so the minimum kinetic energy of the collision must be equal to the mass energy of the π^0. Each proton will have half of that kinetic energy. From Table 43-2, the mass of the π^0 is $135.0\,\text{MeV}/c^2$.

$$2K_{\text{proton}} = m_{\pi^0} c^2 = 135.0\,\text{MeV} \quad \rightarrow \quad K_{\text{proton}} = \boxed{67.5\,\text{MeV}}$$

25. (*a*) We work in the rest frame of the isolated electron, so that it is initially at rest. Energy conservation gives the following.

$$m_e c^2 = K_e + m_e c^2 + E_\gamma \quad \rightarrow \quad K_e = -E_\gamma \quad \rightarrow \quad K_e = E_\gamma = 0$$

Since the photon has no energy, it does not exist, and so has not been emitted.
(*b*) For the photon exchange in Figure 43-8, the photon exists for such a short time that the uncertainty principle allows energy to not be conserved during the exchange.

31. The two neutrinos must move together, in the opposite direction of the electron, in order for the electron to have the maximum kinetic energy, and thus the total momentum of the neutrinos will be equal in magnitude to the momentum of the electron. Since a neutrino is (essentially) massless, we have $E_\nu = p_\nu c$. We assume that the muon is at rest when it decays. Use conservation of energy and momentum, along with their relativistic relationship.

$$p_{e^-} = p_{\bar{\nu}_e} + p_{\nu_\mu}$$

$$m_{\mu^-} c^2 = E_{e^-} + E_{\bar{\nu}_e} + E_{\nu_\mu} = E_{e^-} + p_{\bar{\nu}_e} c + p_{\nu_\mu} c = E_{e^-} + \left(p_{\bar{\nu}_e} + p_{\nu_\mu} \right) c = E_{e^-} + p_{e^-} c \quad \rightarrow$$

$$m_{\mu^-} c^2 - E_{e^-} = p_{e^-} c \quad \rightarrow \quad \left(m_{\mu^-} c^2 - E_{e^-} \right)^2 = \left(p_{e^-} c \right)^2 = E_{e^-}^2 - m_{e^-}^2 c^4 \quad \rightarrow$$

$$m_{\mu^-}^2 c^4 - 2 m_{\mu^-} c^2 E_{e^-} + E_{e^-}^2 = E_{e^-}^2 - m_{e^-}^2 c^4 \quad \rightarrow \quad E_{e^-} = \frac{m_{\mu^-}^2 c^4 + m_{e^-}^2 c^4}{2 m_{\mu^-} c^2} = K_{e^-} + m_{e^-} c^2 \quad \rightarrow$$

$$K_{e^-} = \frac{m_{\mu^-}^2 c^4 + m_{e^-}^2 c^4}{2 m_{\mu^-} c^2} - m_{e^-} c^2 = \frac{(105.7\,\text{MeV})^2 + (0.511\,\text{MeV})^2}{2(105.7\,\text{MeV})} - (0.511\,\text{MeV}) = \boxed{52.3\,\text{MeV}}$$

37. We find the energy width from the lifetime in Table 42-2 and the uncertainty principle.

(*a*) $\Delta t = 10^{-18}\,\text{s} \quad \Delta E \approx \dfrac{h}{2\pi \Delta t} = \dfrac{6.63 \times 10^{-34}\,\text{J}\cdot\text{s}}{2\pi \left(10^{-18}\,\text{s}\right)\left(1.60 \times 10^{-19}\,\text{J/eV}\right)} = 659\,\text{eV} \approx \boxed{700\,\text{eV}}$

(*b*) $\Delta t = 10^{-23}\,\text{s} \quad \Delta E \approx \dfrac{h}{2\pi \Delta t} = \dfrac{6.63 \times 10^{-34}\,\text{J}\cdot\text{s}}{2\pi \left(10^{-23}\,\text{s}\right)\left(1.60 \times 10^{-19}\,\text{J/eV}\right)} = 6.59 \times 10^7\,\text{eV} \approx \boxed{70\,\text{MeV}}$

43. To form the D_s^+ meson, we must have a total charge of +1, a baryon number of 0, a strangeness of +1, and a charm of +1. We assume that there is no topness or bottomness. To get the charm, we must have a "c" quark, with a charge of $+\frac{2}{3} e$. To have a total charge of +1, there must be another quark with a charge of $+\frac{1}{3} e$. To have a baryon number of 0, that second quark must be an antiquark. To have a strangeness of +1, the other quark must be an anti-strange. Thus $\boxed{D_s^+ = c\,\bar{s}}$.

49. These protons will be moving at essentially the speed of light for the entire time of acceleration. The number of revolutions is the total gain in energy divided by the energy gain per revolution. Then the distance is the number of revolutions times the circumference of the ring, and the time is the distance of travel divided by the speed of the protons.

$$N = \frac{\Delta E}{\Delta E / \text{rev}} = \frac{\left(1.0 \times 10^{12}\,\text{eV} - 150 \times 10^9\,\text{eV}\right)}{2.5 \times 10^6\,\text{eV/rev}} = 3.4 \times 10^5\,\text{rev}$$

$$d = N(2\pi R) = \left(3.4 \times 10^5\right) 2\pi \left(1.0 \times 10^3\,\text{m}\right) = 2.136 \times 10^9\,\text{m} \approx \boxed{2.1 \times 10^9\,\text{m}}$$

$$t = \frac{d}{c} = \frac{2.136 \times 10^9\,\text{m}}{3.00 \times 10^8\,\text{m/s}} = \boxed{7.1\,\text{s}}$$

55. The fundamental fermions are the quarks and electrons. In a water molecule there are 2 hydrogen atoms consisting of one electron and one proton each, and 1 oxygen atom, consisting of 8 electrons, 8 protons, and 8 neutrons. Thus there are 18 nucleons, consisting of 3 quarks each, and 10 electrons. The total number of fermions is thus $18 \times 3 + 10 = \boxed{64\,\text{fermions}}$.

61. The total energy is the sum of the kinetic energy and the mass energy. The wavelength is found from the relativistic momentum.

$$E = K + mc^2 = 15 \times 10^9 \, \text{eV} + 938 \times 10^6 \, \text{eV} = 1.594 \times 10^{10} \, \text{eV} \approx \boxed{16 \, \text{GeV}}$$

$$\lambda = \frac{h}{p} = \frac{h}{\dfrac{\sqrt{E^2 - \left(mc^2\right)^2}}{c}} = \frac{hc}{\sqrt{E^2 - \left(mc^2\right)^2}}$$

$$= \frac{\left(6.63 \times 10^{-34} \, \text{J} \cdot \text{s}\right)\left(3.00 \times 10^8 \, \text{m/s}\right)}{\sqrt{\left(1.594 \times 10^{10} \, \text{eV}\right)^2 - \left(938 \times 10^6 \, \text{eV}\right)^2}} \frac{1}{\left(1.60 \times 10^{-19} \, \text{J/eV}\right)} = \boxed{7.8 \times 10^{-17} \, \text{m}}$$

67. The value of R_0 is not known until we draw the graph. We note the following:

$$R = R_0 e^{-t/\tau} \quad \rightarrow \quad R/R_0 = e^{-t/\tau} \quad \rightarrow \quad \ln R - \ln R_0 = -t/\tau \quad \rightarrow \quad \ln R = -t/\tau + \ln R_0$$

A graph of $\ln R$ vs. t should give a straight line with a slope of $-1/\tau$ and a y-intercept of $\ln R_0$. The determination of the mean life does not depend on R_0, and so to find the mean life, we may simply plot $\ln R$ vs. t. That graph is shown, along with the slope and y-intercept.

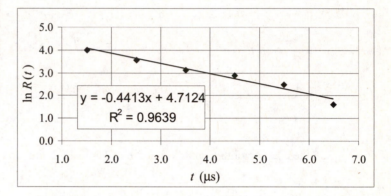

$$\tau = \frac{1}{0.4413} = 2.266 \, \mu\text{s} \approx \boxed{2.3 \, \mu\text{s}}$$

$$\% \, \text{diff} = \frac{2.266 - 2.197}{2.197} \times 100 = \boxed{3.1\%}$$

Chapter 44: Astrophysics and Cosmology

Chapter Overview and Objectives

In this chapter, the applications of the laws of physics to the universe are introduced. The ideas behind the theory of general relativity are discussed. Theories describing the formation and evolution of stars are presented. The standard model of the creation and evolution of the universe is introduced.

After completing this chapter you should:
- Be familiar with the units of distance used in astronomy.
- Know that stars go through different evolutionary processes, depending on the mass of the star.
- Know what the principle of equivalence is.
- Know that the general theory of relativity explains gravitational forces using curved space-time.
- Know that the universe is expanding and how Hubble's law describes that expansion.
- Know what the cosmic microwave background radiation is.
- Know what the big bang model of the universe is.
- Know what the critical density of the universe is and how it affects the future of the universe.

Summary of Equations

Relationship between light-year and meter:
$$1\,\text{ly} = \left(2.998 \times 10^8 \text{ m/s}\right)\left(3.156 \times 10^7 \text{ s}\right)$$
$$= 9.46 \times 10^{15} \text{ m}$$
(Section 44-1)

Apparent brightness in terms of intrinsic luminosity: $b = \dfrac{L}{4\pi d^2}$ (Section 44-2)

Relationship between parsec, light-year, and meter: $1\,\text{pc} = 3.26\,\text{ly} = 3.08 \times 10^{16} \text{ m}$ (Section 44-3)

Schwarzschild radius: $R = \dfrac{2GM}{c^2}$ (Section 44-4)

Hubble's law: $v = Hd$ (Section 44-5)

Critical density of the universe: $\rho_C \approx 10^{-26} \text{ kg/m}^3$ (Section 44-9)

Chapter Summary

Section 44-1. Stars and Galaxies

The unit that is commonly used to measure astronomical distances is the light-year. One **light-year (ly)** is the distance light travels in one year:

$$1\,\text{ly} = \left(2.998 \times 10^8 \text{ m/s}\right)\left(3.156 \times 10^7 \text{ s}\right) = 9.46 \times 10^{15} \text{ m}$$

Visible stars appear to be organized in large groups called **galaxies**. Galaxies appear to be organized in larger groups called **galactic clusters**. Galactic clusters appear to be organized in larger groups called **superclusters**.

Section 44-2. Stellar Evolution: Nucleosynthesis, and the Birth and Death of Stars

The brightness of an observed source of light depends on its radiated power output and its distance from the observer. The **intrinsic luminosity** L of a radiating source is the total power it radiates. The **apparent brightness** b is defined as the intensity of the radiated energy at the position of the observer. Assuming the energy is radiated isotropically by the source and none of the radiation is absorbed by intervening material, the apparent brightness is related to the intrinsic luminosity in the following way

$$b = \frac{L}{4\pi d^2}$$

The luminosity of a star is related to its mass: more massive stars have a greater luminosity. There is a strong correlation between the luminosity of a star and its surface temperature. This correlation is most frequently shown using a **Hertzsprung-Russell diagram**. Most stars are part of a band in this diagram called the **main sequence**.

A current model of star creation is that stars condense out of a cloud of hydrogen gas. A region with a higher than average density of hydrogen gas attracts nearby hydrogen gas, increasing its density further. As the density increases, the temperature of the gas rises. If enough gas collects, the temperature rises enough to initiate the fusion of hydrogen. The energy generated by hydrogen fusion increases the temperature and pressure in the core of the star until it balances the gravitational pressure associated with the weight of the hydrogen.

Most stars continue in this relatively stable dynamic equilibrium for most of their life. The lifetime of a star depends on its mass. Eventually, most of the hydrogen in the core is fused into helium and there is insufficient hydrogen remaining to keep the temperature and pressure high enough to prevent gravitational collapse. The collapse continues until the temperature rises enough to start the fusion of helium nuclei. The temperature must rise before helium fusion can start since the repulsive force of two helium nuclei is greater than the repulsive force of two hydrogen nuclei.

If the mass of the star is large enough, the cycles of fusion reactions continue until the product of the fusion reactions is ^{56}Fe. Since ^{56}Fe is the most tightly bound nucleus, fusion reactions of ^{56}Fe require energy instead of generating it. As a result, the ^{56}Fe core begins its final gravitational collapse. The result of this collapse depends on the mass of the star. Low-mass stars end up as **white dwarfs**. Stars with a mass larger than about 1.4 solar masses collapse into **neutron stars**. The gravitational interaction in a neutron star is strong enough to contain its matter in a very small volume (a neutron star with the mass of the Sun would have a diameter of about 20 km.) During the collapse, a tremendous amount of gravitational potential energy is liberated in an explosion called a **supernova**. Neutron stars usually have a relatively rapid rate of rotation when formed because the moment of inertia of the stars is greatly decreased during the collapse. The electromagnetic radiation of the neutron star combined with its rapid rate of rotation results in a **pulsar**. If the mass of the neutron star is great enough, the gravitational interaction will cause the star to continue to collapse endlessly. A **black hole** is formed from which not even light can escape.

Section 44-3. Distance Measurements

The distance to an object can be determined from the difference in the viewing direction from two spatially separated viewing positions. The **parallax angle** ϕ is defined in the diagram on the right. It is one-half the angular difference in the viewing directions of the object. For small angles, the distance to the object is inversely proportional to the parallax angle and the distance between the two direction measurements.

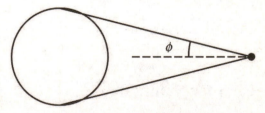

It is common to specify the distance to a star in **parsecs** (pc). One parsec is the distance to an object that has a parallax angle of one second of arc when the two viewing locations are separated by the mean diameter of the Earth's orbit around the Sun. One parsec is equal to 3.26 light-years or 3.08 × 10^{16} m:

$$1\,\text{pc} = 3.26\,\text{ly} = 3.08 \times 10^{16}\,\text{m}$$

Example 44-3-A. Measuring distances with your eyes. The human visual system uses parallax to help determine the distance to an object by the difference in direction of the light source to the two eyes. The angular resolution of the human eye is about 5 × 10^{-4} rad. Use this to determine the maximum possible distance of objects for which the visual system can use parallax to assist in determining distances. The separation of the two eyes is about 8 cm.

Approach: In this problem, the two angular position measurements are made with the two eyes. The smallest angle each eye can measure is defined by the angular resolution of the measurements made by each eye. Since this angle is 5×10^{-4} rad we can use the known separation of the two eyes to determine what distance corresponds to an angle of 5×10^{-4} rad.

Solution: Our eyes cannot detect angular differences smaller than 5×10^{-4} rad. We want to know the distance of an object such that the parallax angle of this object is equal to the angular resolution of the eye. The angular resolution of the eye is small and we can use the small angle approximation ($\tan\phi \approx \phi$):

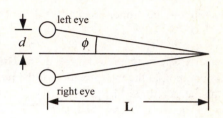

$$L = \frac{d}{\tan\phi} \approx \frac{d}{\phi} = \frac{4 \times 10^{-2}}{5 \times 10^{-4}} = 80 \text{ m}$$

The actual limiting distance for useful parallax is probably somewhat less than this, but you can easily verify that parallax is detectable to a distance of several tens of meters.

Example 44-3-B. Intrinsic luminosity. A star has an apparent brightness of 1.75×10^{-7} W/m^2 and a parallax angle of 38 seconds of arc. What is the intrinsic luminosity of the star, assuming it radiates isotropically and there is no scattering or absorbing material between the object and its observer?

Approach: The parallax angle can be used to determine the distance of the object from the Earth, as discussed in this Section. The apparent brightness provides us with information on the amount of power per unit area. Assuming that the object radiates isotropically, the intrinsic luminosity can be determined by multiplying the apparent brightness by the area of a sphere with a radius equal to the distance between the Earth and the star.

Solution: The distance that corresponds to 1 second of arc is 3.08×10^{16} m. The distance to the star is 38 times larger and is thus equal to

$$d = 38\left(3.08 \times 10^{16}\right) = 1.17 \times 10^{18} \text{ m}$$

The relationship between intrinsic luminosity and apparent brightness can be used to determine the intrinsic luminosity:

$$b = \frac{L}{4\pi d^2} \quad \Rightarrow \quad L = 4\pi d^2 b = 4\pi\left(1.17 \times 10^{18}\right)^2\left(1.75 \times 10^{-7}\right) = 3.01 \times 10^{30} \text{ W}$$

Section 44-4. General Relativity: Gravity and the Curvature of Space

The **principle of equivalence** states that it is impossible to determine from local measurements whether an observer is accelerating or moving at constant velocity in a uniform gravitational field. Based on this principle, Albert Einstein developed the **theory of general relativity**. The theory of general relativity explains gravitational interactions by assuming that the presence of mass changes the geometry of space-time. The familiar geometry that is studied in high school is Euclidean geometry. The rules of Euclidean geometry describe what is called flat space. In the general theory of relativity, the presence of mass causes space-time to be curved. One of the results of the curvature of space-time is that the shortest distance between two points is not necessarily a Euclidean straight line, but can be a curve. The general name for the shortest distance path between two points is **geodesic**. In the ray model of light, light follows geodesics. If the curvature of space-time is large enough, light may be unable to escape from a given region of space. We can think of this as happening when the escape velocity of a given mass becomes greater than the speed of light. Using Newton's law of gravitation, this happens at the surface of the star when its radius is equal to

$$R = \frac{2GM}{c^2}$$

This radius is called the **Schwarzschild radius** of mass M. The theory of general relativity predicts that if a star of mass M collapses to within its Schwarzschild radius, it continues to collapse to a single point, called a **singularity**.

Section 44-5. The Expanding Universe: Redshift and Hubble's Law

If spectra of light from distant galaxies are observed and compared to the known emission spectra of atoms, one finds the spectral lines shifted toward longer wavelengths. This **redshift** is caused by the Doppler effect. The velocity of the galaxy relative to the Earth can be calculated from the observed redshift. The velocity v of galaxies, relative to the Earth, is found to be proportional to the distance d between the galaxy and the Earth. This relation is known as Hubble's law:

$$v = Hd$$

where the constant H is known as the **Hubble parameter** and its value is approximately 22 km/s/10^6 ly.

In addition to the interpretation of the redshift in terms of the Doppler effect, there are two other effects that contribute to the measured redshift:

- The **cosmological redshift**. This redshift is a result of the expansion of space itself and this effect is responsible for the larger redshifts observed for very distant galaxies.
- The **gravitational redshift**. This redshift is a result of the increase in the gravitational potential energy of a photon leaving a massive star.

If all galaxies are moving away from one another in the manner described by Hubble's law, we can infer that there may have been a time at which they were all in the same location. If we assume that the expansion rate has remained constant, we can solve for the time at which all of the galaxy positions were the same. This time works out to be 14×10^9 yr. We might take this as an estimate of the age of the universe.

Section 44-6. The Big Bang and the Cosmic Microwave Background

No matter which direction in the sky one observes, there is a blackbody spectrum of radiation reaching the Earth. The temperature of a black body with the observed radiation spectrum is about 3 K. The peak of this radiation is in the microwave portion of the electromagnetic spectrum, and the radiation is called the **cosmic microwave background radiation**. The cosmic microwave background is consistent with an expanding universe and the big bang model of its evolution.

In 1992 the COBE experiment detected tiny inhomogeneities in the cosmic microwave background. These inhomogeneities, or **anisotropies**, in the microwave background may have provided the seeds for the formation of stars and galaxies. The measured cosmic microwave background provides strong support for the **Big Bang model**, and provides us with information about the conditions of the early universe (about 380,000 years after the Big Bang).

Section 44-7. The Standard Cosmological Model: Early History of the Universe

In the **Standard Cosmological Model**, the universe expands outward from either a zero or very small radius. As the universe expands outward, the temperature drops with time. The types of matter and interactions that dominate the universe change as the universe expands and cools. The evolution of the early universe can be divided into different eras, based on the type of matter and interactions that dominate the universe at those times and temperatures. The following eras are often used to describe the evolution of the universe:

- The Planck era. This era covers the period between the birth of the universe and 10^{-43} s. During this era there is a single unified force. At the end of the Planck era, the gravitational force "condenses out" as a separate force and symmetry is broken.
- The grand unified era. This era covers the period between 10^{-43} s and 10^{-35} s. During this era there is no distinction between quarks and leptons. At the end of the grand unified era, the strong force "condenses out" and quarks and leptons become distinct.
- The hadron era. This era covers the period between 10^{-35} s and 10^{-4} s. The quarks, which were initially free, start to condense and form hadrons.
- The lepton era. This era covers the period between 10^{-4} s and 10 s. During this era, the temperature of the universe is sufficiently low so that only electron-positron pairs are created with a significant probability. At the end of this era, the temperature reaches a level where this production mechanism is not possible anymore, and almost all electrons and positrons disappear as a result of annihilation reactions.
- The radiation era. This era covers the period between 10 s and 380,000 yr. During this period, the major constituents of the universe are photons and neutrinos. Much more energy is contained in the form of radiation than in the form of matter. The matter in the universe is dominated by hydrogen and helium nuclei and by electrons. At the end of this era, the electrons and nuclei are combining to form atoms, and the universe becomes matter dominated.

- The matter dominated era. This era covers the period from 380,000 yr until today. During this era, the atoms combine to form stars, stars combine to form galaxies, etc.

Section 44-8. Inflation: Explaining Flatness, Uniformity, and Structure

The early universe went through a period of exponential inflation. Inflation is an important ingredient of most cosmological models and can explain several remarkable features of the universe:
- Flatness. Inflation can explain why the universe is flat; why it has zero curvature.
- Cosmic microwave background uniformity. Inflation can explain the remarkable uniformity observed in the cosmic microwave background.
- Galaxy seeds, fluctuations, magnetic monopoles. Inflation can explain how the current structures of the universe developed. Inflation can also explain why magnetic monopoles have not been observed.

Section 44-9. Dark Matter and Dark Energy

Depending on the average density of the universe, the universe may go on expanding forever or it may reach a maximum size and then start contracting. If the density is greater than the **critical density** ρ_C then the universe will stop expanding at some point in the future and start contracting. If the density of the universe is less than the critical density, then the universe will go on expanding forever. If the critical density is equal to the critical velocity, the universe will go on expanding forever, but the rate of expansion will go to zero as the time increases to infinity. The critical density estimated using the current observed rate of expansion of the universe is

$$\rho_C \approx 10^{-26} \, \text{kg/m}^3$$

The amount of directly detectable matter in the universe appears to have a density significantly lower than the critical density. There is an unknown amount of matter that does not emit enough radiation to be detectable on the Earth. This is often called **dark matter**. There is some evidence that there is a significant amount of dark matter because the rotation rate of galaxies appears greater than the rate predicted if the visible mass of the galaxies is the entire mass of the galaxies.

Measurements of the rate of expansion of the universe in 1998 showed that the rate was increasing instead of slowing down. The acceleration in the expansion of the universe is due to **dark energy**.

Section 44-10. Large-Scale Structures of the Universe

The measured inhomogeneities observed in the cosmic microwave background are consistent with the large-scale distributions of the galaxies observed in the universe. The simulations that are used to predict the evolution of the universe work best if they incorporate both dark energy and cold dark matter.

Section 44-11. Finally …

The questions on which cosmology focuses are profound ones that have fascinated and will continue to fascinate human intellect. If the initial conditions would have been slightly different, we may not have been here. Is there another fundamental law that determined the initial conditions?

Practice Quiz

1. A star has a mass of 1.5 solar masses. Which object will the star eventually end up as?
   a) A white dwarf.
   b) A neutron star.
   c) A black hole.
   d) A green giant.

2. There are stars that can be observed that are close enough to determine their distance using parallax angle. If we calculate their intrinsic luminosity from their apparent brightness and their distance, their luminosity falls far below the main sequence for their surface temperature. What could cause the intrinsic luminosity of the star to be so low?
    a) There may be dust between the star and the Earth that absorbs or scatters much of the light from the star.
    b) The star may be moving toward the Earth at a relatively high speed.
    c) The star might be a red giant.
    d) The star's gravity might be so great that it is preventing the radiation from escaping efficiently.

3. What should happen to the temperature of the cosmic background radiation as the universe continues expand?
    a) The temperature of the cosmic background radiation should drop as the universe expands.
    b) The temperature of the cosmic background radiation should rise as the universe expands.
    c) The temperature of the cosmic background radiation should remain the same as the universe expands.
    d) The change in temperature as the universe expands depends on whether the universe has greater or less than the critical density.

4. If the universe is closed and begins to contract into the big crunch, which era would happen before the other?
    a) The radiation era would occur before the lepton era.
    b) The hadron era would occur before the radiation era.
    c) The lepton era would occur before the matter-dominated era.
    d) The hadron era would occur before the lepton era.

5. What is the approximate age of the universe based on Hubble's law?
    a) 1.4 million years.
    b) 14 million years.
    c) 1.4 billion years.
    d) 14 billion years.

6. Which of the following is a result of the universe having greater than the critical density?
    a) The big crunch.
    b) A supernova.
    c) Nuclear fission of hydrogen nuclei.
    d) Stellar ignition.

7. Light cannot escape from a
    a) white dwarf.
    b) black dwarf.
    c) red giant.
    d) black hole.

8. The reason nothing can be determined about the universe before a time of about 10^{-43} seconds after the big bang is due to the fact that
    a) no one was alive at that time.
    b) the universe did not exist prior to that time.
    c) a theory of quantum gravity is needed to understand processes that occurred before that time.
    d) no particles have a lifetime that short.

9. In the standard cosmological model of the universe, the reason the hadron era preceded the lepton era is
    a) Hadrons decay in a shorter time than leptons.
    b) Hadrons were created before leptons.
    c) All leptons were created from the decay of hadrons.
    d) Hadron masses are greater than the electron mass.

10. Which of the following is evidence that there is a non-negligible amount of dark matter?
    a) The luminosity of distance galaxies is lower than expected because dark matter absorbs much of the energy before it reaches the Earth.
    b) The rotation rate of galaxies is greater than expected for the amount of visible matter observed in the galaxies.
    c) The sky is darker than expected without the presence of dark matter.
    d) Many galaxies have been observed that do not emit any electromagnetic radiation.

11. Determine the distance to an object that has an intrinsic luminosity of 2.4×10^{20} W and an apparent brightness of 1.32×10^{-5} W/m^2.

12. Calculate the Schwarzschild radius of the Moon.

13. What is the approximate redshift of the 122 nm emission line from hydrogen from a galaxy that is 1.8×10^9 ly away from the Earth?

14. The product $\rho_c c^2$ is an energy density. It is the critical energy density. How many photons per cubic meter with the peak wavelength of the cosmic background radiation would be necessary to give the critical energy density? What would the approximate intensity of this electromagnetic radiation be?

15. Determine the parallax angle of an object located a distance of 34.6 light-years from the Earth.

Responses to Select End-of-Chapter Questions

1. The Milky Way appears "murky" or "milky" to the naked eye, and so before telescopes were used it was thought to be cloud-like. When viewed with a telescope, much of the "murkiness" is resolved into stars and star clusters, so we no longer consider the Milky Way to be milky.

7. Yes. Hotter stars are found on the main sequence above and to the left of cooler stars. If H–R diagrams of clusters of stars are compared, it is found that older clusters are missing the upper left portions of their main sequences. All the stars in a given cluster are formed at about the same time, and the absence of the hotter main sequence stars in a cluster indicates that they have shorter lives and have already used up their core hydrogen and become red giants. In fact, the "turn-off" point, or point at which the upper end of the main sequence stops, can be used to determine the ages of clusters.

13. They would appear to be receding. In an expanding universe, the distances between galaxies are increasing, and so the view from any galaxy is that all other galaxies are moving away.

19. The early universe was too hot for atoms to exist. The average kinetic energies of particles were high and frequent collisions prevented electrons from remaining with nuclei.

Solutions to Select End-of-Chapter Problems

1. Convert the angle to seconds of arc, reciprocate to find the distance in parsecs, and then convert to light years.

$$\phi = \left(2.9 \times 10^{-4}\right)^\circ \left(\frac{3600''}{1^\circ}\right) = 1.044''$$

$$d(\text{pc}) = \frac{1}{\phi''} = \frac{1}{1.044''} = 0.958\,\text{pc} \left(\frac{3.26\,\text{ly}}{1\,\text{pc}}\right) = \boxed{3.1\,\text{ly}}$$

7.   The apparent brightness of an object is inversely proportional to the square of the observer's distance from the object, given by Eq. 44-1. To find the relative brightness at one location as compared to another, take a ratio of the apparent brightness at each location.

$$\frac{b_{Jupiter}}{b_{Earth}} = \frac{\dfrac{L}{4\pi d_{Jupiter}^2}}{\dfrac{L}{4\pi d_{Earth}^2}} = \frac{d_{Earth}^2}{d_{Jupiter}^2} = \left(\frac{d_{Earth}}{d_{Jupiter}}\right)^2 = \left(\frac{1}{5.2}\right)^2 = \boxed{0.037}$$

13. The density is the mass divided by the volume.

$$\rho = \frac{M}{V} = \frac{M_{Sun}}{\frac{4}{3}\pi R_{Earth}^3} = \frac{1.99 \times 10^{30}\,\text{kg}}{\frac{4}{3}\pi\left(6.38 \times 10^6\,\text{m}\right)^3} = \boxed{1.83 \times 10^9\,\text{kg/m}^3}$$

Since the volumes are the same, the ratio of the densities is the same as the ratio of the masses.

$$\frac{\rho}{\rho_{Earth}} = \frac{M}{M_{Earth}} = \frac{1.99 \times 10^{30}\,\text{kg}}{5.98 \times 10^{24}\,\text{kg}} = \boxed{3.33 \times 10^5\ \text{times larger}}$$

19. The limiting value for the angles in a triangle on a sphere is $\boxed{540°}$. Imagine drawing an equilateral triangle near the north pole, enclosing the north pole. If that triangle were small, the surface would be approximately flat, and each angle in the triangle would be $60°$. Then imagine "stretching" each side of that triangle down towards the equator, while keeping sure that the north pole stayed inside the triangle. The angle at each vertex of the triangle would expand, with a limiting value of $180°$. The three $180°$ angles in the triangle would sum to $540°$.

25. We find the velocity from Hubble's law, Eq. 44-4, and the observed wavelength from the Doppler shift, Eq. 44-3.

(a) $$\frac{v}{c} = \frac{Hd}{c} = \frac{(22000\,\text{m/s/Mly})(7.0\,\text{Mly})}{3.00 \times 10^8\,\text{m/s}} = 5.133 \times 10^{-4}$$

$$\lambda = \lambda_0 \sqrt{\frac{1+v/c}{1-v/c}} = (656\,\text{nm})\sqrt{\frac{1 + 5.133 \times 10^{-4}}{1 - 5.133 \times 10^{-4}}} = 656.34\,\text{nm} \approx \boxed{656\,\text{nm}}$$

(b) $$\frac{v}{c} = \frac{Hd}{c} = \frac{(22000\,\text{m/s/Mly})(70\,\text{Mly})}{3.00 \times 10^8\,\text{m/s}} = 5.133 \times 10^{-3}$$

$$\lambda = \lambda_0 \sqrt{\frac{1+v/c}{1-v/c}} = (656\,\text{nm})\sqrt{\frac{1 + 5.133 \times 10^{-3}}{1 - 5.133 \times 10^{-3}}} = 659.38\,\text{nm} \approx \boxed{659\,\text{nm}}$$

31. Wien's law is given in Eq. 37-1.

$$\lambda_P T = 2.90 \times 10^{-3}\,\text{m·K} \;\rightarrow\; \lambda_P = \frac{2.90 \times 10^{-3}\,\text{m·K}}{T} = \frac{2.90 \times 10^{-3}\,\text{m·K}}{2.7\,\text{K}} = \boxed{1.1 \times 10^{-3}\,\text{m}}$$

37. (a) From page 1201, a white dwarf with a mass equal to that of the Sun has a radius about the size of the Earth's radius, $\boxed{6380\,\text{km}}$. From page 1202, a neutron star with a mass equal to 1.5 solar masses has a radius of about $\boxed{20\,\text{km}}$. For the black hole, we use the Schwarzschild radius formula.

$$R = \frac{2GM}{c^2} = \frac{2\left(6.67 \times 10^{-11}\,\text{N·m}^2/\text{kg}^2\right)3\left(1.99 \times 10^{30}\,\text{kg}\right)}{\left(3.00 \times 10^8\,\text{m/s}\right)^2} = 8849\,\text{m} \approx \boxed{8.85\,\text{km}}$$

(b) The ratio is $6380 : 20 : 8.85 = 721 : 2.26 : 1 \approx \boxed{700 : 2 : 1}$.

43. Use Newton's law of universal gravitation.

$$F = G\frac{m_1 m_2}{r^2} = \left(6.67 \times 10^{-11}\ \text{N} \cdot \text{m}^2/\text{kg}^2\right)\frac{\left(3 \times 10^{41}\ \text{kg}\right)^2}{\left[\left(2 \times 10^6\ \text{ly}\right)\left(9.46 \times 10^{15}\ \text{m/ly}\right)\right]^2} = 1.68 \times 10^{28}\ \text{N}$$

$$\approx \boxed{2 \times 10^{28}\ \text{N}}$$

49. (*a*) To find the energy released in the reaction, we calculate the *Q*-value for this reaction. From Eq. 42-2a, the *Q*-value is the mass energy of the reactants minus the mass energy of the products. The masses are found in Appendix F.

$$Q = 2m_C c^2 - m_{Mg}c^2 = \left[2\left(12.000000\ \text{u}\right) - 23.985042\ \text{u}\right]c^2\left(931.5\ \text{Mev}/c^2\right) = \boxed{13.93\ \text{MeV}}$$

(*b*) The total kinetic energy should be equal to the electrical potential energy of the two nuclei when they are just touching. The distance between the two nuclei will be twice the nuclear radius, from Eq. 41-1. Each nucleus will have half the total kinetic energy.

$$r = \left(1.2 \times 10^{-15}\ \text{m}\right)\left(A\right)^{1/3} = \left(1.2 \times 10^{-15}\ \text{m}\right)\left(12\right)^{1/3} \qquad U = \frac{1}{4\pi\varepsilon_0}\frac{q_{\text{nucleus}}^2}{2r}$$

$$K = \tfrac{1}{2}U = \tfrac{1}{2}\frac{1}{4\pi\varepsilon_0}\frac{q_{\text{nucleus}}^2}{2r}$$

$$= \tfrac{1}{2}\left(8.988 \times 10^9\ \text{N} \cdot \text{n}^2/\text{C}^2\right)\frac{\left(6\right)^2\left(1.60 \times 10^{-19}\ \text{C}\right)^2}{2\left(1.2 \times 10^{-15}\ \text{m}\right)\left(12\right)^{1/3}}\left(\frac{1\,\text{MeV}}{1.60 \times 10^{-13}\,\text{J}}\right) = 4.711\ \text{MeV}$$

$$\approx \boxed{4.7\ \text{MeV}}$$

(*c*) We approximate the temperature–kinetic energy relationship by $kT = K$ as given on page 1217.

$$kT = K \;\rightarrow\; T = \frac{K}{k} = \frac{\left(4.711\ \text{MeV}\right)\left(1.60 \times 10^{-13}\ \text{J/MeV}\right)}{1.38 \times 10^{-23}\ \text{J/K}} = \boxed{5.5 \times 10^{10}\ \text{K}}$$

55. Because Venus has a more negative apparent magnitude, $\boxed{\text{Venus is brighter}}$. We write the logarithmic relationship as follows, letting *m* represent the magnitude and *b* the brightness.

$$m = k\log b\ ;\ m_2 - m_1 = k\left(\log b_2 - \log b_1\right) = k\log\left(b_2/b_1\right) \;\rightarrow$$

$$k = \frac{m_2 - m_1}{\log\left(b_2/b_1\right)} = \frac{+5}{\log\left(0.01\right)} = -2.5$$

$$m_2 - m_1 = k\log\left(b_2/b_1\right) \;\rightarrow\; \frac{b_2}{b_1} = 10^{\frac{m_2 - m_1}{k}} \;\rightarrow\; \frac{b_{\text{Venus}}}{b_{\text{Sirius}}} = 10^{\frac{m_{\text{Venus}} - m_{\text{Sirius}}}{-2.5}} = 10^{\frac{-4.4 + 1.4}{-2.5}} = \boxed{16}$$

Quiz Answers

Chapter 21
1. a
2. d
3. d
4. a
5. d
6. d
7. a
8. c
9. d
10. d
11. 0.21 N west
12. 0.26 N
   133° CCW from $+x$-axis
13. $(-1.11\,\hat{\mathbf{i}} + 0.790\,\hat{\mathbf{j}} + 5.14\,\hat{\mathbf{k}}) \times 10^4$ N·m
14. 1.29×10^3 m
15. $(-0.31\,\hat{\mathbf{i}} + 0.37\,\hat{\mathbf{j}} + 0.46\,\hat{\mathbf{k}})$ N·m

Chapter 22
1. d
2. b
3. b
4. b
5. a
6. c
7. a
8. e
9. f
10. d
11. 1.99×10^{-7} N·m²/C
12. -4.17×10^{-15} C
13. $E_{r<R} =$
$$\frac{\rho_0}{\varepsilon_0}\left(\frac{r}{3} - \frac{r^2}{2R} + \frac{r^3}{5R^2}\right)$$
$$E_{r>R} = \frac{\rho_0 R^3}{30\varepsilon_0 r^2}$$
14. $2\varepsilon_0 E_0 x$
15. $4E_0 L^2/\pi^2$

Chapter 23
1. d
2. c
3. a
4. b
5. d
6. b
7. c
8. b
9. c
10. b
11. 0.775 J
12. 0.768 J
13. 480 eV
   1.30×10^7 m/s
14. At l: -25%
   At $10l$: -0.25%
15. $E_x = -V_0\left(\dfrac{y}{z^2} - \dfrac{y}{x^2}\right)$

$E_y = -V_0\left(\dfrac{x}{z^2} + \dfrac{1}{x}\right)$

$E_z = V_0\left(\dfrac{2xy}{z^3}\right)$

Chapter 24
1. c
2. b
3. a
4. b
5. a
6. c
7. b
8. a
9. d
10. a
11. 3.8×10^{-9} F
12. 22.8 µF
13. $Q_{12} = 144$ µC
   $E_{12} = 864$ µJ
   $Q_{27} = 130$ µC
   $E_{27} = 313$ µJ
   $Q_{18} = 130$ µC
   $E_{18} = 469$ µJ
14. $\frac{1}{2}(K_1 + K_2)C$
15. $\left[\dfrac{K_1 + K_2}{2K_1 K_2} - 1\right] \frac{1}{2}CV^2$

Chapter 25
1. d
2. b
3. c
4. c
5. c
6. c
7. b
8. b
9. c
10. b
11. 190 mA, 2.28 W
12. $\approx \$12$
13. $V_0/\sqrt{3}$
14. $\approx 3\ \Omega$
15. 5.2×10^{-5} m/s

Chapter 26
1. b
2. c
3. c
4. b
5. b
6. a
7. b
8. b
9. c
10. a
11. $8.9\ \Omega$
12. $65.1\ \Omega$
13. $I_{R1} = 0.19$ A
   $I_{R2} = 0.12$ A
   $I_{R3} = 0.07$ A
14. 71 ms
15. $2.50 \times 10^{-2}\ \Omega$

Chapter 27
1. b
2. d
3. d
4. d
5. c
6. d
7. d
8. c
9. c
10. b
11. 7.87 N/m $\hat{\mathbf{k}}$
12. $(-5.95\,\hat{\mathbf{i}} + 4.67\,\hat{\mathbf{j}} - 3.58\,\hat{\mathbf{k}}) \times 10^{13}$ m/s²
13. $(E_0/B_0)\,\hat{\mathbf{i}}$
14. 5.48×10^{-6} m
15. 1.32×10^{23} m⁻³

Chapter 28
1. c
2. d
3. d

4. d
5. b
6. d
7. c
8.
9. a
10. c
11. 6.0×10^{-6} T
12. $\mu_0 I^2/4\pi$
    toward wire
13. 8.9×10^3
14. $(6.0 \times 10^{-7}$ T·m$)/r$
    $(4.0 \times 10^{-7}$ T·m$)/r$
15. $4\mu_0 I/\pi a\sqrt{2}$

Chapter 29

1. d
2. a
3. b
4. a
5. c
6. a
7. b
8. b
9. a
10. a
11. 1.83 mA
12. -1.0 T·m^2
13. 29.7 Hz
14. 120
15. ½, $I_{primary} = 3.0$ A
    $I_{secondary} = 6.0$ A

Chapter 30

1. a
2. c
3. b
4. b
5. b
6. c
7. b
8. a
9. a
10. d
11. 57.4 s
12. 18.0 pF
13. 1.01×10^{-10} F
    2.18×10^{-3} H
14. 7.17×10^3 Hz
    2.06 A
15. 1.08×10^4 W

Chapter 31

1. c
2. c
3. d
4. d
5. c
6. a
7. a
8. a
9. a
10. a
11. 5.6×10^{-15} T
12. 5.0×10^{-6} N/m^2
13. 77.4 V/m
    2.58×10^{-7} T
14. 361 m
15. 1.67×10^8 m/s

Chapter 32

1. c
2. c
3. d
4. c
5. b
6. c
7. b
8. a
9. a
10. b
11. 2.23×10^8 m/s
12. 2.88
13. 12.5 cm behind mirror, 0.50,
    upright, virtual
14. 34.3 cm concave
15. 0.18 in inside

Chapter 33

1. a
2. d
3. b
4. c
5. c
6. b
7. b
8. d
9. d
10. b
11. 36.7 cm, -1.54
12. -42.5 cm
13. $+28.6$ cm
14. 3.1 cm, 2.53 m
15. ≈ 480

Chapter 34

1. a
2. a
3. c
4. c
5. b
6. d
7. d
8. a
9. b
10. d
11. 410 nm
12. 0.255 mm
13. 11.4 μm
14. 514 nm, 1.54 μm
    2.57 μm
15. 0.40 mm

Chapter 35

1. c
2. b
3. c
4. b
5. c
6. d
7. a
8. b
9. b
10. c
11. 405 nm
12. 1.96 mm
13. 2.17×10^{-4} rad
14. 0.155
15. 1.48

Chapter 36

1. a
2. c
3. c
4. a
5. b
6. a
7. a
8. c
9. c
10. a
11. 0.735c
12. $v_A = 0.866c$
    $v_B = 0.943c$
    $v_{AB} = 0.420c$
13. $K = 5mc^2$
    $p = 5.91mc$
14. $x' = -967$ m, $y' = 189$ m
    $z' = 0$, $t' = 4.74$ μs
15. 0.0122c, toward Earth

Chapter 37

1. b
2. a
3. c
4. a
5. a
6. b
7. b
8. d
9. b
10. b
11. 5.8×10^3 K
12. 7.6×10^{-35} m
13. $26.2°$
14. 1.1 μm
15. 13.5 nm, 11.4 nm
    1.08 nm

Chapter 38

1. c
2. b
3. d
4. d
5. a
6. d
7. c
8. b
9. d
10. d
11. 2.1×10^{-27} kg·m/s
12. 2.18×10^{-13} s
13. 5.31×10^{-24} kg
14. 0.22
15. 1.45×10^{-64} J
    5.80×10^{-64} J
    1.31×10^{-63} J
    2.32×10^{-63} J

Chapter 39

1. d
2. c
3. d
4. d
5. b
6. a
7. c
8. d
9. c
10. b
11. $41e^{-8} \approx 0.014$
12. 0.54 nm
13. $Z = 27$, Cobalt
14. $(3 + \sqrt{5})r_0$

15. $j = 3$,
    $m_j = \{-3, -2, \ldots, 3\}$;
    $j = 4$,
    $m_j = \{-4, -3, \ldots, 4\}$;
    $j = 5$,
    $m_j = \{-5, -4, \ldots, 5\}$;

Chapter 40

1. a
2. b
3. c
4. a
5. b
6. c
7. b
8. b
9. a
10. c
11. 1.9×10^{-18} J
12. 3.6×10^{-4}
    1.06×10^6 m/s
13. 1.652 μm
    1.650 μm
14. 6.60×10^{-18} J
15. 1.48×10^{-10} m

Chapter 41

1. d
2. d
3. c
4. c
5. b
6. c
7. c
8. a
9. a
10. d
11. 2.0×10^{15}
12. 2.32 MeV
13. $^{28}_{14}$Si : 8.45 MeV
    $^{31}_{14}$Si : 8.46 MeV
    Expect $^{31}_{14}$Si to be more stable, but it isn't.
14. 29.9 decays/s
15. 0.313 MeV

Chapter 42

1. c
2. c
3. a
4. d

5. a
6. d
7. a
8. d
9. b
10. c
11. 10.5 MeV
12. 3.1×10^{-2} rem
13. 30.9 MeV
14. 6.6×10^{-3} rem
15. 87.83 MHz

Chapter 43

1. c
2. a
3. b
4. c
5. b
6. d
7. a
8. b
9. d
10. c
11. 9.8 m/s
12. 12.9 GeV
13. 3.37×10^7 Hz
    1.26×10^{-13} J
14. 270 MeV
15. $K_\pi = 98.6$ MeV
    $p_\pi = 193$ MeV/c
    $K_n = 19.6$ MeV
    $P_n = 193$ MeV/c

Chapter 44

1. b
2. a
3. a
4. a
5. d
6. a
7. d
8. c
9. d
10. b
11. 1.20×10^{12} m
12. 1.08×10^{-4} m
13. $\Delta\lambda = 15.6$ nm
14. 4.81×1012
    10^{-18} W/m^2
15. 4.58×10^{-7} rad